Teacher's Guide • Testing Program

- Performance Assessment
  - Interview/Task Tests
  - Portfolio
  - Problem-Solving Assessment
  - Checklists
- Standardized Format Tests
- Free Response Format Tests
- Management Forms

Harcourt Brace Jovanovich, Inc.
Orlando Austin San Diego Chicago Dallas New York

Printed in the United States of America

ISBN 0-15-301021-5

1 2 3 4 5 6 7 8 9 10 082 95 94 94 92 91

Contents

I
Performance Assessment

It has been the custom in the past to use narrow approaches to testing, such as tests composed of multiple-choice questions to be answered by students in a fixed time period. Many have been uneasy about the process and have recognized that this testing format, while efficient, has little to do with what the students have learned and the ways in which they have been taught.

Multiple means of assessment for all aspects of mathematical knowledge and its connections are provided in the Performance Assessment methods described in this section. Through the evaluation of students' portfolios, problem solving processes, and interview/task items, a broader vision of what the student knows will emerge. A Class Record Form for Performance Assessment is provided to record the results of the interview/task items. These methods, together with the formal assessment, provide an accurate and thorough indication of the mathematics that students know.

The Mathematics Portfolio

What Is a Mathematics Portfolio?

The portfolio is a collection of each student's work gathered over an extended period of time. A portfolio illustrates the growth, talents, achievements, and reflections of the mathematics learner and provides a means for the teacher to assess the student's performance.

An effective portfolio will

- include items collected over the entire school year.
- show the "big picture"—providing a broad understanding of a student's mathematics language and feelings through words, diagrams, checklists, and so on.
- give students opportunities to collaborate with their peers.
- give students a chance to experience success, to develop pride in their work, and to develop positive attitudes toward mathematics.

Building a Portfolio

There are many opportunities to collect student's work throughout the year as you use *Mathematics Plus*. A list of suggested portfolio items for each chapter is given on pages 4 to 6. These suggestions are also listed on page F of each Chapter Overview in the Teacher's Edition. Students may also appreciate a chance to select some of their work to be included in the portfolio. Attitudes and performance can be recorded on checklists provided at the end of this section of the *Teacher's Guide * Testing Program*.

To begin:

- Provide a file folder for each student with the student's name clearly marked on the tab or folder.
- Explain to students that throughout the year they will save some of their work in the folder. Sometimes it will be their individual work; sometimes it will be group reports and projects, or completed checklists.
- Assign a fun activity to the entire class that can be placed into the portfolio by all students. EXAMPLE: Ask students to draw a map of the school playground in which they use at least three different shapes.
- Comment positively on the maps and reinforce the process. Have students place their maps into their portfolios.

Evaluating a Portfolio

The ultimate purpose of assessment is to enable students to evaluate themselves. Portfolios have the potential to create authentic portraits of what students learn and offer an alternative means for documenting growth, change, and risk-taking in mathematics learning. Evaluating their own growth in mathematics will be a new experience for most students. The following points made with regular portfolio evaluation will encourage growth in self-evaluation.

- Discuss the contents of the portfolio with each student as you examine it at regular intervals during the school year.
- Examine each portfolio on the basis of the growth the student has made rather than in comparison with other portfolios.
- Ask the student questions as you examine the portfolio.
- Point out the strengths and weaknesses in the work.
- Encourage and reward students by emphasizing the growth you see, the original thinking, and the completion of tasks.
- Reinforce and adjust instruction of the broad goals you want to accomplish as you evaluate the portfolios.

What Should the Teacher Look For?

Growth in mathematics is shown by:

- the non-standard responses that students make.
- the ways students solve a problem or attempt to solve a problem. Note the strategies they used and the reasons they succeeded or became confused.
- the ways students communicate their understanding of math problems. Note whether they use words, pictures, or abstract algorithms.
- unique solutions or ways of thinking.

Placing comments such as the following on student's work can have instructional benefits.

"Interesting approach to solving this problem."
"Is this the only solution to the problem?"
"Where would you take it from here?"
"Can you think of a related problem?"
"What about . . . ?"

What Should the Student Look For?

Have students look through their samples and describe the kinds of learning experiences they had. Point out to students the following statements—which are likely to represent some of the ways they will feel.

I enjoyed the problems.
I learned something new.
This repeated things I already knew.
Now I'm getting the hang of it.
I was challenged.
I discovered something new myself.

Sharing the Portfolios

- Examine the portfolio with parents or guardians to share concrete examples of the work the student is doing. Emphasize the growth you see as well as the expectations you have.
- Examine portfolios with your students to emphasize their experiences of success and to develop pride in their work and positive attitudes toward mathematics
- Examine portfolios with your supervisor to share the growth your students have made and to show the ways you have developed the curriculum objectives.

The Benefits of Mathematics Portfolios

Portfolios can be the basis for informed change in mathematics classrooms because they:

- send positive messages to students about successful processes rather than end results.
- give you better insights as to how students understand and work problems.
- focus on monitoring the development of reasoning skills.
- help students become responsible for their own learning.
- promote teacher-student dialogue.
- focus on the student rather than on the assignment.
- focus on the development of conceptual understandings rather than applications of skills and procedures.

Build a Portfolio

CHAPTER 1

Lesson 1.3 Find and Analyze Data—PE page 7
(Use small and large numbers, Exercises 7–10.)

Lesson 1.5 Make Up a Problem—PE page 11
(Write a problem.)

Lesson 1.7 Think Along—TR 1
(Complete Problem-Solving Think Along for Exercise 5 on PE page 15.)

Lesson 1.11 Talk About It—PE page 24
(Record answers to questions.)

Lesson 1.12 Writer's Corner—PE page 27
(Describe a small business.)

CHAPTER 2

Lesson 2.2 Organize Data—PE page 43
(Gather and record data.)

Lesson 2.3 Visual Thinking—PE page 45
(Make a line plot.)

Lesson 2.6 Think Along—TR 1
(Complete Problem-Solving Think Along for Exercise 8 on PE page 53.)

Lesson 2.8 Find and Organize Data—PE page 57
(Record and display data, Exercises 11–12.)

Lesson 2.12 Writer's Corner—PE page 65
(Write a story.)

CHAPTER 3

Lesson 3.2 History Connection—PE page 79
(Find prime numbers, Exercise 36.)

Lesson 3.3 Logical Reasoning—PE page 81
(Answer questions.)

Lesson 3.6 Think Along—TR 1
(Complete Problem-Solving Think Along for Exercise 5 on PE page 87.)

Lesson 3.10 Writer's Corner—PE page 97
(Write problems.)

CHAPTER 4

Lesson 4.2 Wrap Up—PE page 109
(Record answer to question.)

Lesson 4.5 Writer's Corner—PE page 115
(Write problems.)

Lesson 4.10 Checking Understanding—PE page 127
(Draw number lines to model division, Exercises 3 and 4.)

Lesson 4.13 Think Along—TR 1
(Complete Problem-Solving Think Along for Exercise 3 on PE page 133.)

CHAPTER 5

Lesson 5.1 Talk About It—PE page 142
(Record answers to questions.)

Lesson 5.5 Checking Understanding—PE page 151
(Explain models, Exercises 1–3.)

Lesson 5.10 Think Along—TR 1
(Complete Problem-Solving Think Along for Exercise 6 on PE page 163.)

Lesson 5.13 Writer's Corner—PE page 169
(Make up a mnemonic device.)

CHAPTER 6

Lesson 6.4 Career Connection—PE page 187
(Describe and draw designs.)

Lesson 6.5 Check for Understanding—PE page 189
(Construct angles, Exercises 1–3.)

Lesson 6.7 Writer's Corner—PE page 195
(Write a problem.)

Lesson 6.9 Logical Reasoning—PE page 201
(Draw triangles.)

Lesson 6.12 Think Along—TR 1
(Complete Problem-Solving Think Along fo Exercise 8 on PE page 207.)

CHAPTER 7

Lesson 7.5 Think Along—TR 1
(Complete Problem-Solving Think Along for Exercise 8 on PE page 225.)

Lesson 7.8 Number Sense • Mental Math—PE page 233
(Create similar figures.)

Lesson 7.10 Writer's Corner—PE page 237
(Write a paragraph and problems.)

Lesson 7.11 Making the Connection—PE page 239
(Draw rectangles, Exercise 4.)

CHAPTER 8

Lesson 8.6 Think Along—TR 1
(Complete Problem-Solving Think Along for Exercise 7 on PE page 259.)

Lesson 8.7 Check for Understanding—PE page 261
(Copy and complete table, Exercise 1.)

Lesson 8.12 Talk About It—PE page 272
(Record answer to question.)

Lesson 8.14 Writer's Corner—PE page 277
(Write problems.)

CHAPTER 9

Lesson 9.2 Checking Understanding—PE page 289
(Model finding sums, Exercises 1 and 2.)

Lesson 9.7 Think Along—TR 1
(Complete Problem-Solving Think Along for Exercise 4 on PE page 299.)

Lesson 9.10 Work Together—PE page 306
(Record answers to questions.)

Lesson 9.13 Checking Understanding—PE page 313
(Draw a graph and write an equation, Exercises 5–6.)

Lesson 9.14 Writer's Corner—PE page 315
(Write a paragraph.)

CHAPTER 10

Lesson 10.4 Writer's Corner—PE page 331
(Write a problem.)

Spotlight on Problem Solving—PE page 337
(Survey the class, Exercise 3.)

CHAPTER 10 *(continued)*

Lesson 10.10 Talk About It—PE page 346
(Record answers to questions.)

Lesson 10.11 History Connection—PE page 349
(Write a paragraph.)

Lesson 10.12 Think Along—TR 1
(Complete Problem-Solving Think Along for Exercise 5 on PE page 351.)

CHAPTER 11

Lesson 11.1 Visual Thinking—PE page 361
(Make a table, Exercises 10 and 11.)

Lesson 11.4 Work Together—PE page 367
(Use a spinner.)

Lesson 11.6 Writer's Corner—PE page 371
(Write a problem.)

Lesson 11.11 Think Along—TR 1
(Complete Problem-Solving Think Along for Exercise 10 on PE page 383.)

CHAPTER 12

Lesson 12.4 Visual Thinking—PE page 399
(Find perimeter.)

Lesson 12.6 Think Along—TR 1
(Complete Problem-Solving Think Along for Exercise 7 on PE page 403.)

Lesson 12.10 Writer's Corner—PE page 413
(Write a problem.)

Lesson 12.11 Make Up a Problem—PE page 415
(Use line and turn symmetry.)

Lesson 12.12 Checking Understanding—PE page 419
(Plot transformations, Exercises 7–8.)

CHAPTER 13

Lesson 13.2 Visual Thinking—PE page 435
(Draw three-dimensional figures.)

Lesson 13.3 Talk About It—PE page 436
(Record answer to question.)

Lesson 13.5 Writer's Corner—PE page 441
(Write problems.)

Lesson 13.9 Think Along—TR 1
(Complete Problem-Solving Think Along for Exercise 7 on PE page 451.)

CHAPTER 14

Lesson 14.1 Checking Understanding—PE page 461
(Draw a diagram for Exercise 2.)

Lesson 14.3 Work Together—PE page 465
(Draw tessellations.)

Lesson 14.4 Writer's Corner—PE page 467
(Write descriptions.)

Lesson 14.5 Checking Understanding—PE page 471
(Graph points, Exercises 14–17.)

Lesson 14.11 Think Along—TR 1
(Complete Problem-Solving Think Along for Exercise 7 on PE page 485.)

Assessing Problem Solving

Assessing a student's ability to solve problems involves more than checking the student's answer. It involves looking at how students process information and how they work at solving problems. The heuristic used in *Mathematics Plus*—Understand, Plan, Solve, Look Back—guides the student's thinking process and provides a structure within which the student can work toward a solution. Evaluating the student's progress through the parts of the heuristic can help you assess the student's areas of strength and weakness in solving problems.

The following instruments may be used to assess students' problem-solving abilities.

Problem Solving Think Along

Oral Response Form (See page 8)

This form can be given to the student to be used as a self-questioning instrument or as a guide for working through a problem in a cooperative learning group. This form may also be used to evaluate the oral responses of an individual student or the oral presentation of a cooperative learning group. Students need the opportunity to verbalize their thinking process as they work through a problem. This form will help them organize an oral presentation of a solution process. You may wish to use this response form with the problem-solving items in the Interview/Task Test for each chapter.

Problem Solving Think Along

Scoring Guide for Oral Responses (See page 10)

This analytic Scoring Guide, which has a criterion score for the Understand and Plan sections and one for the Solve and Look Back sections, may be used to evaluate the oral presentation of an individual or a cooperative learning group. You can evaluate and score each section as you are listening to the student's presentation.

Problem Solving Think Along

Written Response Form (See page 9)

This form, which may be used by an individual student or a cooperative learning group, provides a recording sheet for students to record their responses as they work through each section of the heuristic. This Written Response Form is also used to record students' responses to the problem-solving test items in the Interview/Task Test for each chapter.

Problem Solving Think Along

Scoring Guide for Written Responses (See page 11)

This analytic Scoring Guide, which gives a criterion score for the Understand and Plan sections and one for the Solve and Look Back sections, will help you pinpoint the parts of the problem-solving process in which your students need more instruction.

Name ______________________________

Problem Solving Think Along

Solving problems is a thinking process. Asking yourself questions as you work through the steps in solving a problem can help guide your thinking. These questions will help you understand the problem, plan how to solve it, solve it, and then look back and check your solution. These questions will also help you think about other ways to solve the problem.

Understand

1. What is the problem about?
2. What is the question?
3. What information is given in the problem?

Plan

4. What problem-solving strategies might I try to help me solve the problem?
5. About what do I think my answer will be?

Solve

6. How can I solve the problem?
7. How can I state my answer in a complete sentence?

Look Back

8. How do I know whether my answer is reasonable?
9. How else might I have solved this problem?

Name ______________________________

THINK ALONG

Problem Solving

Understand

1. Retell the problem in your own words. ______________________________

2. Restate the question as a fill-in-the-blank sentence. ______________________________

3. List the information given. ______________________________

Plan

4. List one or more problem-solving strategies that you can use. ______________________________

5. Predict what your answer will be. ______________________________

Solve

6. Show how you solved the problem. ______________________________

7. Write your answer in a complete sentence. ______________________________

Look Back

8. Tell how you know your answer is reasonable. ______________________________

9. Describe another way you could have solved the problem. ______________________________

Name ______________________________

Problem Solving Think Along

Scoring Guide • Oral Responses

Understand

Score

_____ **1.** Restate the problem in his or her own words.
0 points - No restatement given.
1 point - Incomplete problem restatement.
2 points - Complete problem restatement given.

_____ **2.** Identify the question.
0 points - No restatement of the question given.
1 point - Incomplete or incorrect restatement of the question given.
2 points - Correct restatement of the question given.

_____ **3.** State list of information needed to solve the problem.
0 points - No list given.
1 point - Incomplete list given.
2 points - Complete list given.

Criterion Score 4/6 *Pupil Score* _____

Plan

_____ **1.** State one or more strategies that might be helpful in solving the problem.
0 points - No strategies given.
1 point - One or more strategies given but are poor choices.
2 points - One or more useful strategies given.

_____ **2.** State reasonable estimated answer.
0 points - No estimated answer given.
1 point - Unreasonable estimate given.
2 points - Reasonable estimate given.

Criterion Score 3/4 *Pupil Score* _____

Solve

_____ **1.** Describe a solution method that correctly represents the information in the problem.
0 points - No solution method given.
1 point - Incorrect solution method given.
2 points - Correct solution method given.

_____ **2.** State correct answer in complete sentence.
0 points - No sentence given.
1 point - Sentence given in which either the numerical answer is incorrect or the sentence does not answer the question.
2 points - Complete sentence given with the correct numerical answer completely answering the question.

Criterion Score 3/4 *Pupil Score* _____

Look Back

_____ **1.** State sentence explaining why the answer is reasonable.
0 points - No sentence given.
1 point - Sentence given with incomplete or incorrect reason.
2 points - Complete and correct explanation given.

_____ **2.** Describe another strategy that could have been used to solve the problem.
0 points - No other strategy described.
1 point - Another strategy described, but strategy is a poor choice.
2 points - Another useful strategy described.

Criterion Score 3/4 *Pupil Score* _____

TOTAL 13/18 *Pupil Score* _____

Problem Solving Think Along

Scoring Guide • Written Responses

Understand

Indicator 1: Student restates the problem in his or her own words.	***Scoring:*** 0 points – No restatement written. 1 point – Incomplete problem restatement written. 2 points – Complete problem restatement written.
Indicator 2: Student restates the question as a fill-in-the-blank statement.	***Scoring:*** 0 points – No restatement written. 1 point – Incorrect or incomplete restatement written. 2 points – Correct restatement of the question.
Indicator 3: Student writes a complete list of the information needed to solve the problem.	***Scoring:*** 0 points – No list made. 1 point – Incomplete list made. 2 points – Complete list made.

Criterion Score 4/6

Plan

Indicator 1: Student lists one or more problem-solving strategies that might be helpful in solving the problem.	***Scoring:*** 0 points – No strategies listed. 1 point – One or more strategies listed, but strategies are poor choices. 2 points – One or more useful strategies listed.
Indicator 2: Student gives a reasonable estimated answer.	***Scoring:*** 0 points – No estimated answer given. 1 point – Incorrect number sentence written. 2 points – Reasonable estimated given.

Criterion Score 3/4

Solve

Indicator 1: Student shows a solution method that correctly represents the information in the problem.	***Scoring:*** 0 points – No sententce written. 1 point – Incorrect number sentence written. 2 points – Correct number sentence written.
Indicator 2: Student writes a complete sentence giving the correct answer.	***Scoring:*** 0 points – No sentence written. 1 point – Sentence has an incorrect numerical answer or does not answer the question. 2 points – Sentence has correct answer and completely answers the question.

Criterion Score 3/4

Look Back

Indicator 1: Student writes a sentence explaining why the answer is reasonable.	***Scoring:*** 0 points – No sentence written. 1 point – Sentence gives an incomplete or incorrect reason. 2 points – A complete and correct explanation written.
Indicator 2: Student describes another strategy that could have been used to solve the problem.	***Scoring:*** 0 points – No other strategy described. 1 point – Another strategy described, but strategy is a poor choice. 2 points – Another useful strategy described.

Criterion Score 3/4

TOTAL 13/18

Evaluating Interview/Task Items

The interview/task test items are designed to provide an optional instrument to evaluate each student's level of accomplishment for each tested objective in the *Mathematics Plus* program. These items provide opportunities for students to verbalize or write about their thinking or to use manipulatives or other pictorial representations to represent their thinking. They test students at the concrete and pictorial levels, where appropriate, so that you can assess each student's progress toward functioning at the abstract level. The items will enable you to analyze the student's thought processes as they work on different types of problems and will enable you to plan instruction that will meet your students' needs.

You may wish to use these test items as you work through the content in the chapter to determine whether students are ready to move on or whether they need additional teaching or reinforcement activities. You may also wish to use these test items with students who did not successfully pass the standardized format or free response test for the chapter to determine what types of reteaching activities are appropriate. These test items may also be used with students who have difficulty reading written material or who have learning disabilities.

The test items are designed to focus on evaluating how students think about mathematics and how they work at solving problems rather than on whether they can get the correct answer. The evaluation criteria given for each test item will help you pinpoint the errors in the students' thinking processes as they work through the problem.

A checklist of possible responses is provided to record each student's thinking processes. The Class Record Form can be used to show satisfactory completion of interview/task test items.

Evaluation of Interview/Task Test

DATE ________

STUDENT'S NAME ________________________________ CLASS ________

TEST ITEM	EVALUATE WHETHER STUDENT
1-A **Explain the method you use as you estimate 2.89 + 4.23 + 6.95.**	_____ chooses an appropriate method of estimating. _____ applies the method correctly. _____ uses mental math to estimate. _____ identifies the estimated sum as about 14 (found by using front-end digits with adjustment or rounding).
1-A **Explain the method you use as you estimate 356 × 452.**	_____ chooses an appropriate method of estimating. _____ applies the method correctly. _____ uses mental math to estimate. _____ identifies the estimated product as a number between 120,000 (found by using front-end digits) and 200,000 (found by rounding both factors to the greatest place value.)
1-B **Explain each step as you find the product of 7.42 × 3.5.**	_____ regroups only when necessary. _____ knows basic facts. _____ aligns partial products correctly. _____ places decimal point in product correctly. _____ identifies the product as 25.97.
1-B **Explain each step as you find the quotient of 21.402 ÷ 8.7.**	_____ multiplies to make the divisor a whole number. _____ places the decimal point in the quotient. _____ identifies the quotient as 2.46.
1-C **Explain how you know that the value of 2^4 is greater than the base number, 2, while the value of $(0.2)^4$ is less than the base number, (0.2).**	_____ explains that as you multiply whole numbers, the value increases. _____ explains that as you multiply decimal numbers, the value decreases. _____ identifies 2^4 as 16 and $(0.2)^4$ as 0.0016.

TEST ITEM		EVALUATE WHETHER STUDENT
1-D	**Explain how you would write 7,230,000 in scientific notation.**	_____ writes the number as a product of two factors. _____ writes, as the first factor, a number between 1 and 10. _____ writes, as the second factor, a power of 10. _____ identifies the answer as 7.23×10^6.
1-E	**Look at Exercise 2 on page 15 in your textbook. Explain how you would use the strategy *draw a diagram* to solve this problem.**	_____ answers the questions as given on the Problem-Solving Think Along. (See pages 8–9 of this book.)
1-E	**Look at Exercise 2 on page 27 in your textbook. Explain how you would use the strategy *guess and check* to solve this problem.**	_____ answers the questions as given on the Problem-Solving Think Along. (See pages 8–9 of this book.)

Evaluation of Interview/Task Test

DATE ____________

STUDENT'S NAME ______________________________ CLASS ____________

TEST ITEM	EVALUATE WHETHER STUDENT
2-A Identify the type of graph you would use to display in centimeters the heights of the students in your class.	_____ chooses an appropriate type of graph (bar graph, histogram, or stem-and-leaf plot). _____ explains how choice was made.
2-B Explain how you would survey an unbiased sample to determine the favorite school subject of the seventh graders in your school.	_____ describes a random sample. _____ restricts sample to seventh graders in his or her school. _____ prepares a frequency distribution table to gather data.
2-C Look at page 64 in your textbook. Explain how you would find the range, mean, median, and mode of the math scores for Ms. Johnson's class.	_____ identifies the range of a group of numbers as the difference between the greatest and least numbers. _____ identifies the mean as the sum of the group of numbers divided by the number of addends in the group. _____ identifies the median as the middle number in a group of numbers. _____ identifies the mode of a group of numbers as the number that occurs most often. computes correctly: _____ range is 26. _____ mean is 87.6. _____ median is 90. _____ mode is 96.
2-D Look at Exercise 3 on page 49 in your textbook. Explain how you would use the graph to help you solve this problem.	_____ answers the questions as given on the Problem-Solving Think Along. (See pages 8–9 of this book.)
2-D Look at Exercise 3 on page 65 in your textbook. Explain how you would decide the type of graph that would best display this set of data.	_____ answers the questions as given on the Problem-Solving Think Along. (See pages 8–9 of this book.)

Evaluation of Interview/Task Test

DATE ____________

STUDENT'S NAME ______________________________ CLASS ____________

TEST ITEM	EVALUATE WHETHER STUDENT
3-A Explain your thinking as you determine the prime factorization of 40.	_____ states that all factors in prime factorization must be prime numbers. _____ uses a factor tree or divides until the quotient is 1 to find the prime factors: $2 \times 2 \times 2 \times 5$, or $2^3 \times 5$.
3-B Explain your thinking as you find the greatest common factor of 24 and 32.	_____ states that the greatest common factor must be a factor of both numbers. _____ lists the factors of 24 and 32 or multiplies the common factors in the prime factorization of 24 and 32. _____ identifies the GCF as 8.
3-C Explain your thinking as you order these fractions from least to greatest: $\frac{1}{2}, \frac{3}{8}, \frac{3}{4}$.	_____ writes equivalent fractions with like denominators. _____ compares numerators of the fractions with like denominators. _____ identifies the order as: $\frac{3}{8}, \frac{1}{2}, \frac{3}{4}$.
3-D Explain your thinking as you use number lines to help you find a fraction between $\frac{2}{8}$ and $\frac{3}{8}$.	_____ states that between any two fractions there are always an infinite number of fractions. _____ draws number lines to illustrate that there are fractions between $\frac{2}{8}$ and $\frac{3}{8}$. _____ identifies a fraction between $\frac{2}{8}$ and $\frac{3}{8}$. (A possible answer is $\frac{5}{16}$.)
3-E Look at Exercise 2 on page 87 in your textbook. Explain how you would use the strategy *find a pattern* to solve this problem.	_____ answers the questions as given on the Problem-Solving Think Along. (See pages 8–9 of this book.)
3-E Look at Exercise 2 on page 97 in your textbook. Explain how you would use the strategy *draw a diagram* to solve this problem.	_____ answers the questions as given on the Problem-Solving Think Along. (See pages 8–9 of this book.)

Evaluation of Interview/Task Test

DATE ________

STUDENT'S NAME ______________________________ CLASS ________

TEST ITEM	EVALUATE WHETHER STUDENT
4-A **Explain the method you use as you estimate $3\frac{7}{8} \times 5\frac{1}{3}$.**	____ chooses an appropriate method of estimating. ____ applies the method correctly. ____ uses mental math to estimate. ____ identifies the estimated product as about 20 (found by rounding).
4-B **Explain the method you use as you estimate $22\frac{1}{2} \div 7\frac{5}{6}$.**	____ chooses an appropriate method of estimating. ____ applies the method correctly. ____ uses mental math to estimate. ____ identifies the estimated quotient as about 3 (found by using compatible numbers).
4-B **Explain each step as you find the difference of $8\frac{1}{2} - 2\frac{3}{4}$.**	____ writes equivalent fractions with like denominators for $\frac{1}{2}$ and $\frac{3}{4}$. ____ renames $8\frac{2}{4}$ as $7\frac{6}{4}$ before subtracting. ____ identifies the difference as $5\frac{3}{4}$.
4-C **Explain each step as you find the quotient of $3\frac{1}{8} \div 1\frac{2}{3}$.**	____ writes the mixed numbers as fractions. ____ uses the reciprocal to write a multiplication problem. ____ simplifies before multiplying. ____ identifies the quotient as $1\frac{7}{8}$.
4-D **Look at Exercise 2 on page 115 in your textbook. Explain how you would use the strategy *solve a simpler problem* to solve this problem.**	____ answers the questions as given on the Problem-Solving Think Along. (See pages 8–9 of this book.)
4-D **Look at Exercise 1 on page 133 in your textbook. Explain how you would solve this problem.**	____ answers the questions as given on the Problem-Solving Think Along. (See pages 8–9 of this book.)

Evaluation of Interview/Task Test

DATE ______

STUDENT'S NAME ______ CLASS ______

TEST ITEM	EVALUATE WHETHER STUDENT
5-A **Explain your thinking as you name two different word expressions for the algebraic expression 5*n*.**	____ explains that 5*n* is an expression showing multiplication. ____ states two correct word expressions. (Possible expressions are 5 times a number, *n*; the product of 5 and a number, *n*; and a number, *n*, multiplied by 5.)
5-B **Explain your thinking as you use the order of operations to find the value of this expression: $10 - (3 + 6) \div 3^2$.**	follows the order of operations: ____ 1. does the operation inside the parentheses. ____ 2. works with exponents. ____ 3. multiplies and divides from left to right. ____ 4. adds and subtracts from left to right. ____ identifies the value of the expression as 9.
5-C **Explain each step as you solve the inequality $a + 4 < 10$.**	____ subtracts 4 from each side of the inequality. ____ places the $<$ sign correctly in the solution. ____ identifies the solution as $a < 6$.
5-D **Explain each step as you solve the equation $\frac{b}{3} = 12$. Check your solution.**	____ multiplies each side of the equation by 3. ____ identifies the solution as $b = 36$. ____ checks the solution by replacing *b* with 36.
5-E **Look at Exercise 2 on page 163 in your textbook. Explain how you would use the strategy *use a formula* to solve this problem.**	____ answers the questions as given on the Problem-Solving Think Along. (See pages 8–9 of this book.)
5-E **Look at Exercise 2 on page 169 in your textbook. Explain how you would use the strategy *work backward* to solve this problem.**	____ answers the questions as given on the Problem-Solving Think Along. (See pages 8–9 of this book.)

Evaluation of Interview/Task Test

DATE ______

STUDENT'S NAME ______________________ CLASS ______

TEST ITEM	EVALUATE WHETHER STUDENT
6-A **Name objects in the classroom that suggest parallel lines, congruent angles, a regular polygon, and a parallelogram.**	selects an appropriate object for _____ parallel lines. _____ congruent angles. _____ a regular polygon. _____ a parallelogram. states characteristics to justify why object was chosen for _____ parallel lines. _____ congruent angles. _____ a regular polygon. _____ a parallelogram.
6-A **Explain each step as you construct a triangle with sides that measure 3 cm, 4 cm, and 5 cm.**	_____ uses a compass and a straightedge to construct the triangle. _____ draws a ray and uses a compass to measure one side of the triangle. _____ draws arcs correctly for the other two sides of the triangle. _____ connects the correct points to form a triangle with sides measuring 3 cm, 4 cm, and 5 cm.
6-B **Explain how you would use a Venn diagram to show the relationships among polygons, regular polygons, and quadrilaterals.**	_____ explains that some quadrilaterals are regular polygons. _____ explains that all quadrilaterals are polygons. _____ explains that some regular polygons are quadrilaterals. _____ explains that all regular polygons are polygons. _____ draws Venn diagram correctly.
6-C **Look at Exercise 1 on page 195 in your textbook. Explain how you would use the strategy *find a pattern* to solve this problem.**	_____ answers the questions as given on the Problem-Solving Think Along. (See pages 8–9 of this book.)

TEST ITEM	EVALUATE WHETHER STUDENT
6-C **Look at Exercise 1 on page 207 in your textbook. Explain how you would use the strategy *use a formula* to solve this problem.**	_____ answers the questions as given on the Problem-Solving Think Along. (See pages 8–9 of this book.)

Evaluation of Interview/Task Test

DATE ______

STUDENT'S NAME ______________________ CLASS ______

TEST ITEM	EVALUATE WHETHER STUDENT
7-A **Describe a method you could use to show that $\frac{2}{3} = \frac{6}{9}$ is a proportion.**	____ chooses an appropriate method (equivalent fractions or cross products). ____ describes how method is used.
7-B **Explain as you use a proportion to find the unit price of apples if 4 apples sell for $0.92.**	____ uses a proportion. ____ solves the proportion by using equivalent fractions or cross products. ____ identifies the unit price as $0.23.
7-C **Explain how you could use proportions to help you make a scale drawing of the classroom.**	____ measures dimensions of the classroom. ____ chooses an appropriate scale to relate the dimensions of the classroom to the dimensions of the drawing. ____ sets up proportions correctly.
7-D **Look at the triangles on page 230 in your textbook. Explain how you use proportions as you find the length of $\overline{DF}$ if you know that the length of $\overline{AC}$ is 5 inches.**	____ sets up proportion correctly. ____ solves proportion. ____ identifies the length of $\overline{DF}$ as $8\frac{1}{3}$ inches.
7-E **Look at Exercise 2 on page 225 in your textbook. Explain how you would use the strategy *use a formula* to solve this problem.**	____ answers the questions as given on the Problem-Solving Think Along. (See pages 8–9 of this book.)
7-E **Look at Exercise 2 on page 237 in your textbook. Explain how you would use the map on page 236 to help you solve this problem.**	____ answers the questions as given on the Problem-Solving Think Along. (See pages 8–9 of this book.)

Evaluation of Interview/Task Test

DATE ____________

STUDENT'S NAME ______________________________ CLASS ____________

TEST ITEM	EVALUATE WHETHER STUDENT
8-A **Model 25% on a decimal square. Explain how the decimal square can help you name a decimal and a fraction equivalent to 25%.**	____ shades 25 of 100 squares on the decimal square. ____ states that, as a decimal, 0.25 of the decimal square is shaded. ____ states that, as a fraction, $\frac{25}{100}$ or $\frac{1}{4}$ of the decimal square is shaded.
8-B **Explain your thinking as you compute to find what percent 15 is of 25.**	____ uses an equation or a proportion to solve. ____ sets up equation or proportion correctly. ____ states that 15 is 60% of 25.
8-B **Explain the method you use as you estimate 15% of 62.**	____ chooses an appropriate method of estimation. ____ applies the method correctly. ____ uses mental math to estimate. ____ identifies 15% of 62 as about 9.
8-C **Look at the circle graph on page 280 in your textbook. Explain how you can use the graph as you find how much Allen will spend on entertainment if he earns $30 one week.**	____ reads the data in the graph correctly. ____ computes 10% of 30. ____ states that Allen will spend $3 on entertainment.
8-D **Explain each step as you find the total cost of a jacket that has a regular price of $50 and a sales tax of 6%.**	____ multiplies $50 $\times$ 0.06 to find the sales tax. ____ adds the sales tax to the regular price to find the total cost. ____ identifies the total cost as $53.
8-E **Look at Exercise 9 on page 259 in your textbook. Explain how you would use the graph to solve this problem.**	____ answers the questions as given on the Problem-Solving Think Along. (See pages 8–9 of this book.)
8-E **Look at Exercise 2 on page 277 in your textbook. Explain how you would use the strategy *guess and check* to solve this problem.**	____ answers the questions as given on the Problem-Solving Think Along. (See pages 8–9 of this book.)

Evaluation of Interview/Task Test

DATE ________

STUDENT'S NAME ______________________________ CLASS ________

TEST ITEM	EVALUATE WHETHER STUDENT
9-A Explain your thinking as you draw a number line to show integers for these situations: a loss of 4 yards; 2 degrees Celsius above zero.	_____ draws a number line to show negative and positive integers.
	_____ shows "a loss of 4 yards" at $^{-}4$ on the number line.
	_____ shows "2 degrees Celsius above zero" at 2 on the number line.
9-B Explain your thinking as you use blue and red counters or a number line to find the difference of $^{-}5 - 2$.	_____ explains how a model (counters or a number line) illustrates the problem.
	_____ identifies the difference as $^{-}7$.
9-B Explain each step as you find the product of $^{-}4 \times 2$.	_____ explains that the product of a positive integer and a negative integer is negative.
	_____ identifies the product as $^{-}8$.
9-C Explain each step as you solve $x + {}^{-}2 = {}^{-}5$.	_____ adds the opposite of $^{-}2$ (2) to each side of the equation.
	_____ identifies the solution as $x = {}^{-}3$.
9-D Explain your thinking as you locate the point represented by the ordered pair $(4, {}^{-}3)$ on a coordinate plane.	_____ explains that the first number tells how far you move horizontally.
	_____ explains that the second number tells how far you move vertically.
	_____ explains that for a positive number you move to the right or up.
	_____ explains that for a negative number you move left or down.
	_____ identifies the location as 4 units to the right and 3 units down from the origin.
9-E Look at Exercise 2 on page 299 in your textbook. Explain how you would use the strategy *work backward* to solve this problem.	_____ answers the questions as given on the Problem-Solving Think Along. (See pages 8–9 of this book.)
9-E Look at Exercise 2 on page 315 in your textbook. Explain how you would use the strategy *write an equation* to solve this problem.	_____ answers the questions as given on the Problem-Solving Think Along. (See pages 8–9 of this book.)

Evaluation of Interview/Task Test

DATE ________

STUDENT'S NAME ______________________________ CLASS ________

TEST ITEM	EVALUATE WHETHER STUDENT
10-A **Explain your thinking as you give an example of each type of number: terminating decimal, repeating decimal, rational number, irrational number, and real number.**	chooses an appropriate number for _____ a terminating decimal. _____ a repeating decimal. _____ a rational number. _____ an irrational number. _____ a real number. explains why example is appropriate for _____ a terminating decimal. _____ a repeating decimal. _____ a rational number. _____ an irrational number. _____ a real number.
10-B **Explain your thinking as you order these numbers from least to greatest: 5, $\frac{-1}{2}$, $^{-}2$, $\frac{1}{2}$, 3.5.**	_____ explains that as you move to the right on the number line, the value of the integers becomes greater. _____ identifies the order as: $^{-}2$, $\frac{-1}{2}$, $\frac{1}{2}$, 3.5, 5.
10-C **Explain your thinking as you write 0.000052 in scientific notation.**	_____ writes the number as a product of two factors. _____ writes, as the first factor, a number between 1 and 10. _____ writes, as the second factor, a power of 10. _____ explains that the exponent for the power of 10 is negative because the decimal point is moved to the right. _____ identifies the answer 5.2×10^{-5}.
10-D **Explain your thinking as you name the two square roots of 25.**	_____ knows that every positive number has two square roots, one positive and one negative. _____ identifies the square roots as 5 and $^{-}5$.

Performance Assessment (See Teacher's Edition, p. 322F)

TEST ITEM		EVALUATE WHETHER STUDENT
10-E	**Look at Exercise 2 on page 331 in your textbook. Explain how you would use the strategy *act it out* to solve this problem.**	______ answers the questions as given on the Problem-Solving Think Along. (See pages 8–9 of this book.)
10-E	**Look at Exercise 2 on page 351 in your textbook. Explain how you would use the strategy *use a table* to solve this problem.**	______ answers the questions as given on the Problem-Solving Think Along. (See pages 8–9 of this book.)

Evaluation of Interview/Task Test

DATE ______

STUDENT'S NAME ______________________ CLASS ______

TEST ITEM	EVALUATE WHETHER STUDENT
11-A **Explain how you could use a tree diagram to show the number of outcomes that are possible if you toss a number cube with the numbers 1, 2, 3, 4, 5, 6 and then toss a coin.**	______ explains how to construct a tree diagram correctly.

Number Cube	Coin	Outcomes
1	H	1-H
	T	1-T
2	H	2-H
	T	2-T
3	H	3-H
	T	3-T
4	H	4-H
	T	4-T
5	H	5-H
	T	5-T
6	H	6-H
	T	6-T

______ explains how to interpret a tree diagram to show there are 12 possible outcomes.

TEST ITEM	EVALUATE WHETHER STUDENT
11-B **Use four cards labeled A, B, C, and D to model** **1. the different arrangements of four cards taken two at a time if the order does not matter.** **2. the different arrangements of three cards (A, B, C) if the order is important.**	______ shows six different arrangements to describe a combination of two cards from a set of four when the order does not matter: **(A,B), (A,C), (A,D), (B,C), (B,D), (C,D).** ______ shows six different arrangements of three cards to describe a permutation if the order is important: **ABC, ACB, BAC, BCA, CAB, CBA.**
11-C **Toss a coin ten times. Record the results of your tosses. Use the results of your tosses to name the experimental probability of tossing "Heads." Explain how your experimental probability compares to the mathematical probability.**	______ identifies the experimental probability as the ratio of the number of times the coin landed on Heads to the total number of trials $\left(\frac{H}{10}\right)$. ______ identifies the mathematical probability as the ratio of the number of favorable outcomes to the number of possible outcomes $\left(\frac{1}{2}\right)$.

TEST ITEM	EVALUATE WHETHER STUDENT
11-D **Place 3 red counters and 3 yellow counters in a bag. Explain how you could use the counters in the bag to compare two independent events and two dependent events.**	_____ explains that for independent events the results of the first event do not affect the results of the second event. _____ identifies independent events (such as drawing a counter as the first event and replacing it before drawing another counter as the second event). _____ explains that for dependent events the results of the first event do affect the results of the second event. _____ identifies dependent events (such as drawing a counter as the first event and without replacing it drawing another counter as the second event).
11-E **Look at Exercise 8 on page 371 in your textbook. Explain how you would use the strategy *make an organized list* to solve this problem.**	_____ answers the questions as given on the Problem-Solving Think Along. (See pages 8–9 of this book.)
11-E **Look at Exercise 5 on page 383 in your textbook. Explain how you would use the strategy *conduct a simulation* to solve this problem.**	_____ answers the questions as given on the Problem-Solving Think Along. (See pages 8–9 of this book.)

Evaluation of Interview/Task Test

DATE ______

STUDENT'S NAME ______ CLASS ______

TEST ITEM	EVALUATE WHETHER STUDENT
12-A **Explain how you would decide which metric unit and which customary unit to use to measure the length of the Florida coastline, the width of a stamp, and the height of the school building.**	______ chooses appropriate units for the length of the Florida coastline: km, mi. ______ chooses appropriate units for the width of a stamp: mm, in. ______ chooses appropriate units for the height of the school building: m, ft.
12-B **The length of a building is measured as 11 yards and as 34 feet. Explain how you know which measurement is more precise.**	______ explains that the smaller the unit of measure, the greater the precision. ______ states that a foot is a smaller unit of measure than a yard. ______ identifies 34 feet as more precise.
12-C **Explain your thinking as you find the perimeter of your desktop.**	______ states that perimeter is distance around. ______ names an appropriate tool for measuring. ______ states the correct perimeter.
12-D **Look at the trapezoids on page 408 in your textbook. Use the trapezoids to explain how the area of a trapezoid is related to the area of a parallelogram.**	______ explains that two congruent trapezoids can be placed to form a parallelogram. ______ explains that the area of one trapezoid is $\frac{1}{2}$ the area of the parallelogram formed by the two trapezoids, or $\frac{1}{2}(b_1 + b_2) \cdot h$.
12-E **Use graph paper and a small scalene triangle cut from paper to explain how a figure and its image are related in a translation, a reflection, and a rotation.**	______ explains that in all three transformations, the size and shape of the figure and its image remain the same. explains that in all three transformations the position of the figure and its image change: ______ in a translation, the figure is moved along a straight line. ______ in a reflection, the figure is flipped over a line of symmetry. ______ in a rotation, the figure is turned about a central point, called the turn center.

TEST ITEM		EVALUATE WHETHER STUDENT
12-F	**Look at Exercise 1 on page 403 in your textbook. Explain how you would use the strategy *use a formula* to solve this problem.**	_____ answers the questions as given on the Problem-Solving Think Along. (See pages 8–9 of this book.)
12-F	**Look at Exercise 2 on page 413 in your textbook. Explain how you would use the strategy *make a model* to solve this problem.**	_____ answers the questions as given on the Problem-Solving Think Along. (See pages 8–9 of this book.)

Evaluation of Interview/Task Test

DATE ________

STUDENT'S NAME ______________________ CLASS ________

TEST ITEM	EVALUATE WHETHER STUDENT
13-A **Name objects in the classroom or in your home that remind you of a rectangular prism, a pyramid, a cone, and a cylinder. Explain why each reminds you of the geometric figure.**	selects an appropriate object for _____ a rectangular prism. _____ a pyramid. _____ a cone. _____ a cylinder. states characteristics to justify why object was chosen for _____ a rectangular prism. _____ a pyramid. _____ a cone. _____ a cylinder.
13-B **Look at the figure in Exercise 18 on page 442 in your textbook. Explain your thinking as you find the surface area of this figure.**	_____ explains that the surface area is the sum of the areas of the faces of the solid figure. _____ identifies the surface area as 261 cm^2.
13-C **Look at the figure in Exercise 1 on page 445 in your textbook. Explain how the volume of this prism is related to the volume of a pyramid that has the same base and height as the prism.**	_____ explains that the volume of a pyramid is $\frac{1}{3}$ the volume of a prism that has the same base and height. _____ identifies the volume of the prism as 18 m^3 and the volume of a pyramid with the same base and height as 6 m^3.
13-D **Look at Exercise 2 on page 441 in your textbook. Explain how you would use the strategy *use estimation* to solve this problem.**	_____ answers the questions as given on the Problem-Solving Think Along. (See pages 8–9 of this book.)
13-D **Look at Exercise 1 on page 451 in your textbook. Explain how you would use the strategy *make a model* to solve this problem.**	_____ answers the questions as given on the Problem-Solving Think Along. (See pages 8–9 of this book.)

Evaluation of Interview/Task Test

DATE ______

STUDENT'S NAME ______________________ CLASS ______

TEST ITEM	EVALUATE WHETHER STUDENT
14-A **Draw a figure you could use to tessellate a plane. Explain how you know your figure will tessellate a plane.**	____ explains that a tessellation is an arrangement of spaces that completely covers the plane, with no gaps and no overlaps. ____ draws a shape that will tessellate a plane.
14-B **Draw rectangles on graph paper to help you explain how doubling both the length and the width of a rectangle will affect the area of the rectangle.**	____ draws appropriate rectangles to show the relationship. ____ explains that if both the length and the width of a rectangle are doubled, the area will increase 4 times.
14-C **Explain how the Pythagorean Property can be used to find the hypotenuse of a right triangle if you know that the two legs are 6 cm and 8 cm.**	____ explains that the square of the length of the hypotenuse of a right triangle is equal to the sum of the squares of the lengths of the two legs ($c^2 = a^2 + b^2$). ____ identifies the length of the hypotenuse as 10 cm.
14-D **On graph paper, draw a graph that represents a function and a graph that does not represent a function. Explain your choices.**	____ explains that a graph represents a function if each element in the domain is matched to only one value of the range. ____ explains that a graph does NOT represent a function if each element in the domain is matched to more than one value of the range. ____ draws appropriate graphs.
14-E **Look at Exercise 1 on page 467 in your textbook. Explain how you would use the strategy *solve a simpler problem* to solve this problem.**	____ answers the questions as given on the Problem-Solving Think Along. (See pages 8–9 of this book.)
14-E **Look at Exercise 3 on page 485 in your textbook. Explain how you would choose a strategy to solve this problem.**	____ answers the questions as given on the Problem-Solving Think Along. (See pages 8–9 of this book.)

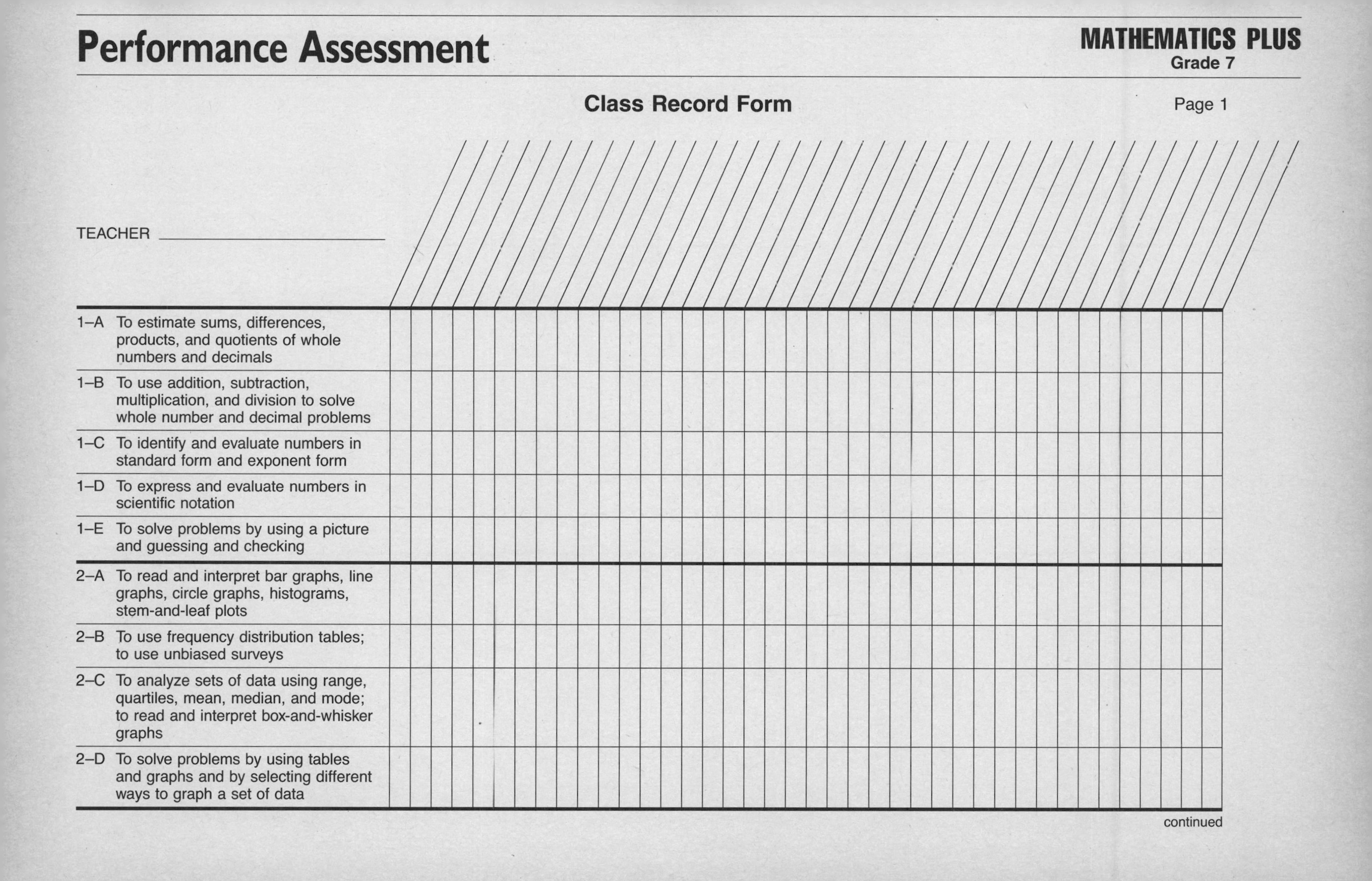

Performance Assessment

MATHEMATICS PLUS
Grade 7

Class Record Form

TEACHER ______________________

1–A	To estimate sums, differences, products, and quotients of whole numbers and decimals																																								
1–B	To use addition, subtraction, multiplication, and division to solve whole number and decimal problems																																								
1–C	To identify and evaluate numbers in standard form and exponent form																																								
1–D	To express and evaluate numbers in scientific notation																																								
1–E	To solve problems by using a picture and guessing and checking																																								
2–A	To read and interpret bar graphs, line graphs, circle graphs, histograms, stem-and-leaf plots																																								
2–B	To use frequency distribution tables; to use unbiased surveys																																								
2–C	To analyze sets of data using range, quartiles, mean, median, and mode; to read and interpret box-and-whisker graphs																																								
2–D	To solve problems by using tables and graphs and by selecting different ways to graph a set of data																																								

continued

Performance Assessment

Class Record Form

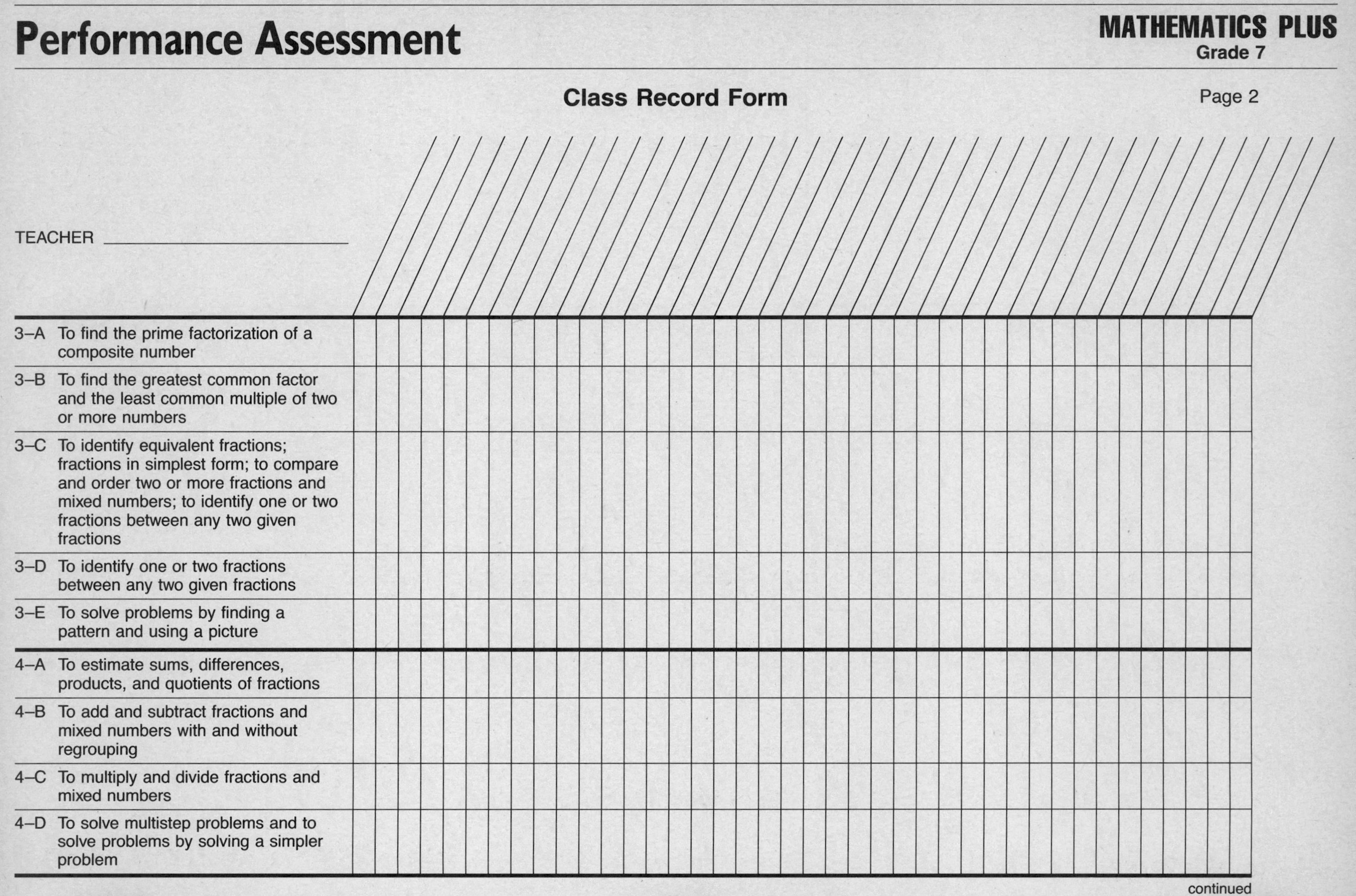

TEACHER ____________________

Objective																																								
3–A To find the prime factorization of a composite number																																								
3–B To find the greatest common factor and the least common multiple of two or more numbers																																								
3–C To identify equivalent fractions; fractions in simplest form; to compare and order two or more fractions and mixed numbers; to identify one or two fractions between any two given fractions																																								
3–D To identify one or two fractions between any two given fractions																																								
3–E To solve problems by finding a pattern and using a picture																																								
4–A To estimate sums, differences, products, and quotients of fractions																																								
4–B To add and subtract fractions and mixed numbers with and without regrouping																																								
4–C To multiply and divide fractions and mixed numbers																																								
4–D To solve multistep problems and to solve problems by solving a simpler problem																																								

continued

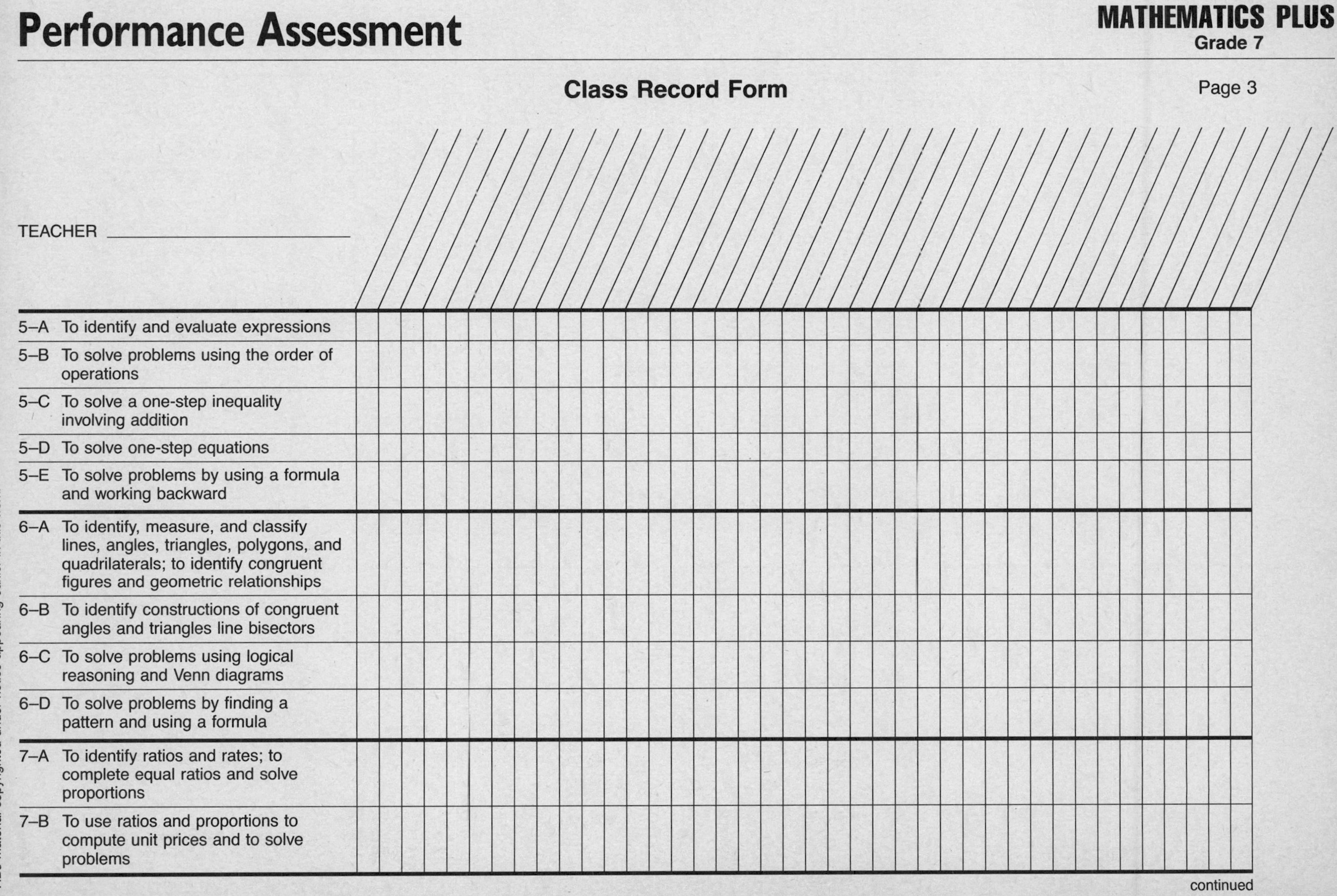

Class Record Form

TEACHER ____________________

Objective	
5–A	To identify and evaluate expressions
5–B	To solve problems using the order of operations
5–C	To solve a one-step inequality involving addition
5–D	To solve one-step equations
5–E	To solve problems by using a formula and working backward
6–A	To identify, measure, and classify lines, angles, triangles, polygons, and quadrilaterals; to identify congruent figures and geometric relationships
6–B	To identify constructions of congruent angles and triangles line bisectors
6–C	To solve problems using logical reasoning and Venn diagrams
6–D	To solve problems by finding a pattern and using a formula
7–A	To identify ratios and rates; to complete equal ratios and solve proportions
7–B	To use ratios and proportions to compute unit prices and to solve problems

continued

Class Record Form

TEACHER ______________________

Objective																																								
7–C To use scale drawings to solve problems																																								
7–D To identify corresponding parts of similar figures; to find the ratio of corresponding sides; to use proportions to find the missing measures of similar figures																																								
7–E To solve problems by using a map and using a formula																																								
8–A To identify fraction, decimal, ratio, and percent equivalencies																																								
8–B To estimate and find the percent of a number, the percent one number is of another; to find a number when a percent of it is known by using estimation, proportions, and equations																																								
8–C To analyze a circle graph																																								
8–D To solve sales tax, discount, and simple interest problems																																								
8–E To solve problems by using a graph and guessing and checking																																								

continued

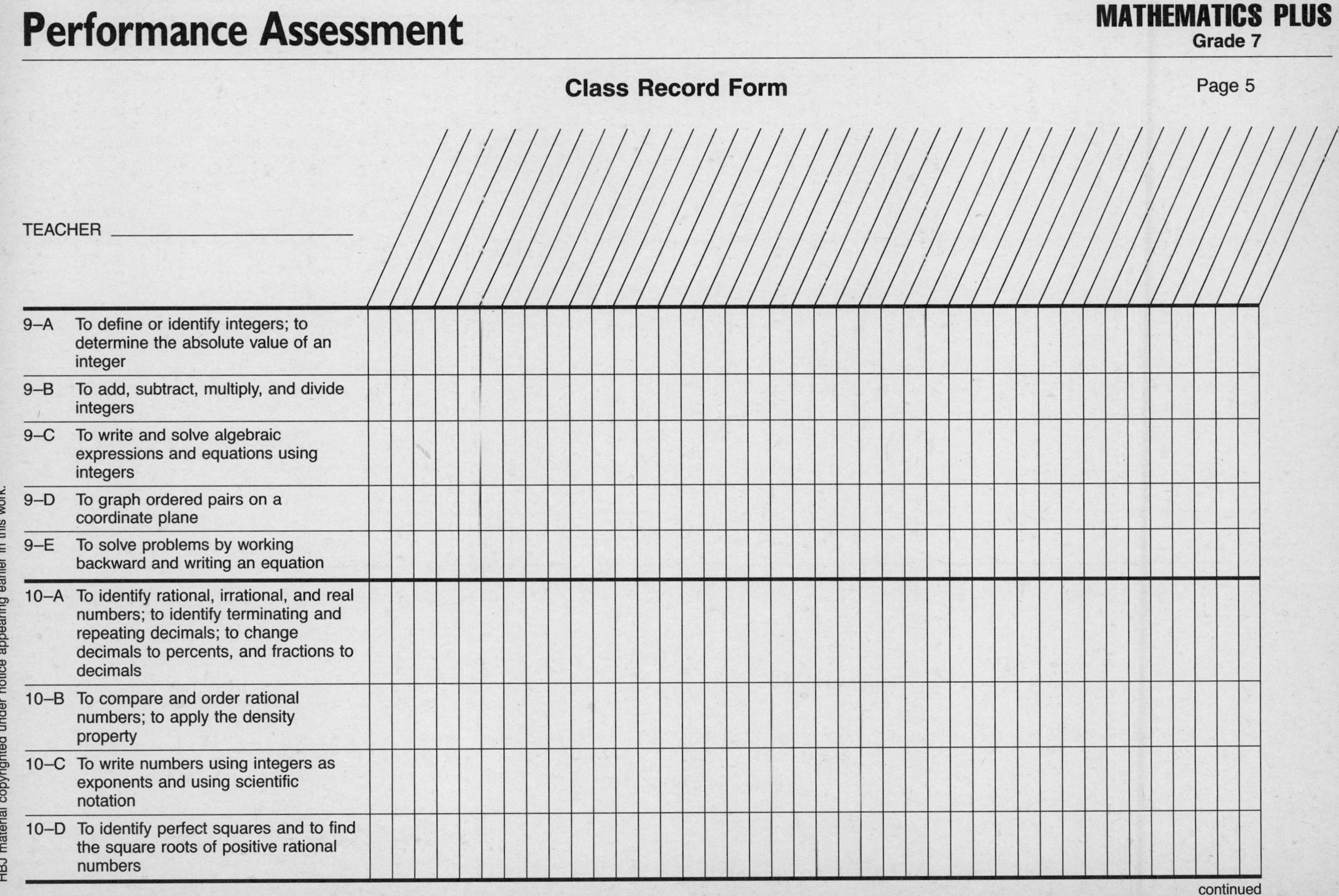

Class Record Form

TEACHER ______________________

9–A	To define or identify integers; to determine the absolute value of an integer
9–B	To add, subtract, multiply, and divide integers
9–C	To write and solve algebraic expressions and equations using integers
9–D	To graph ordered pairs on a coordinate plane
9–E	To solve problems by working backward and writing an equation
10–A	To identify rational, irrational, and real numbers; to identify terminating and repeating decimals; to change decimals to percents, and fractions to decimals
10–B	To compare and order rational numbers; to apply the density property
10–C	To write numbers using integers as exponents and using scientific notation
10–D	To identify perfect squares and to find the square roots of positive rational numbers

continued

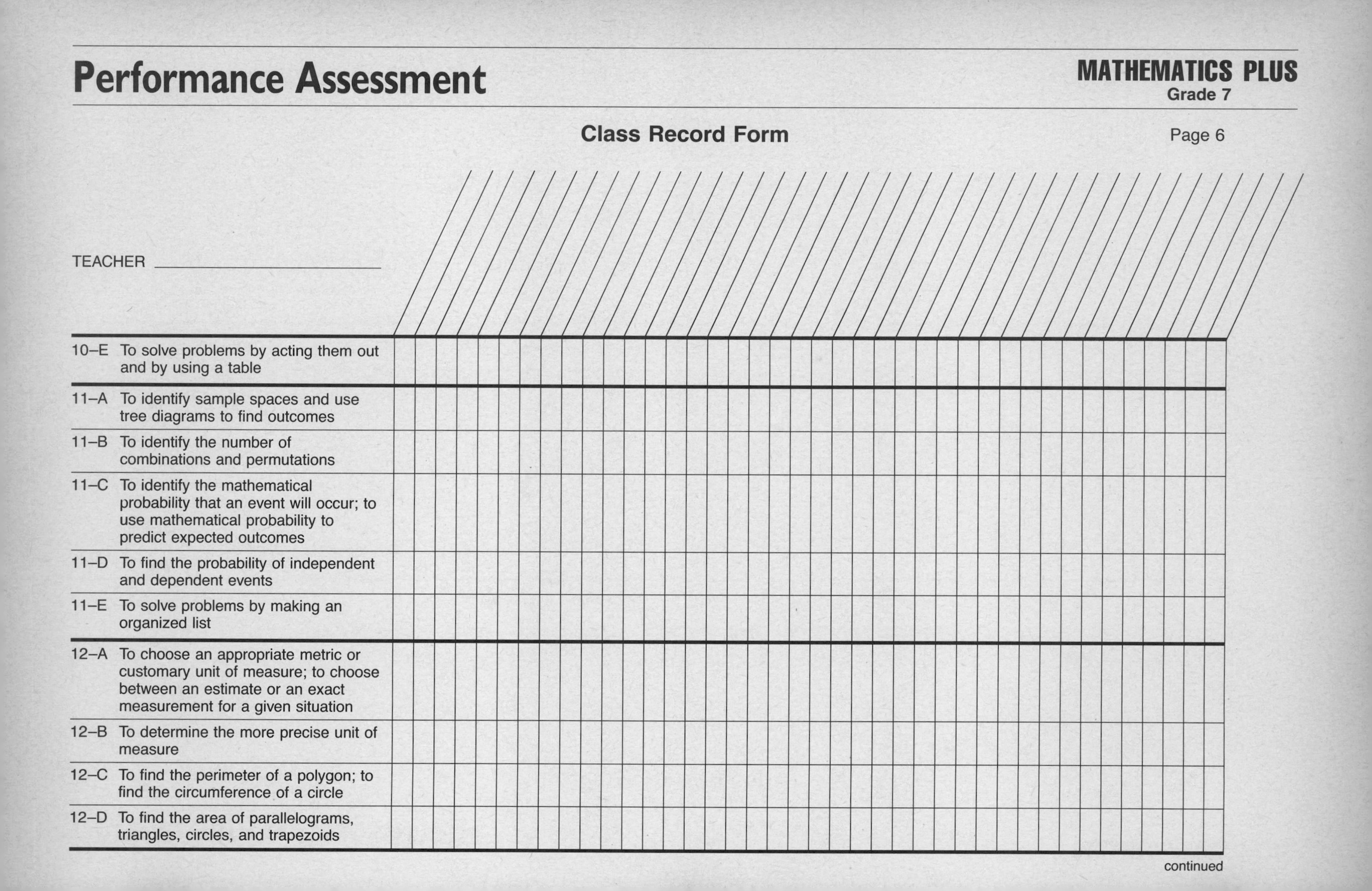

Class Record Form

TEACHER ______________________

10–E	To solve problems by acting them out and by using a table
11–A	To identify sample spaces and use tree diagrams to find outcomes
11–B	To identify the number of combinations and permutations
11–C	To identify the mathematical probability that an event will occur; to use mathematical probability to predict expected outcomes
11–D	To find the probability of independent and dependent events
11–E	To solve problems by making an organized list
12–A	To choose an appropriate metric or customary unit of measure; to choose between an estimate or an exact measurement for a given situation
12–B	To determine the more precise unit of measure
12–C	To find the perimeter of a polygon; to find the circumference of a circle
12–D	To find the area of parallelograms, triangles, circles, and trapezoids

continued

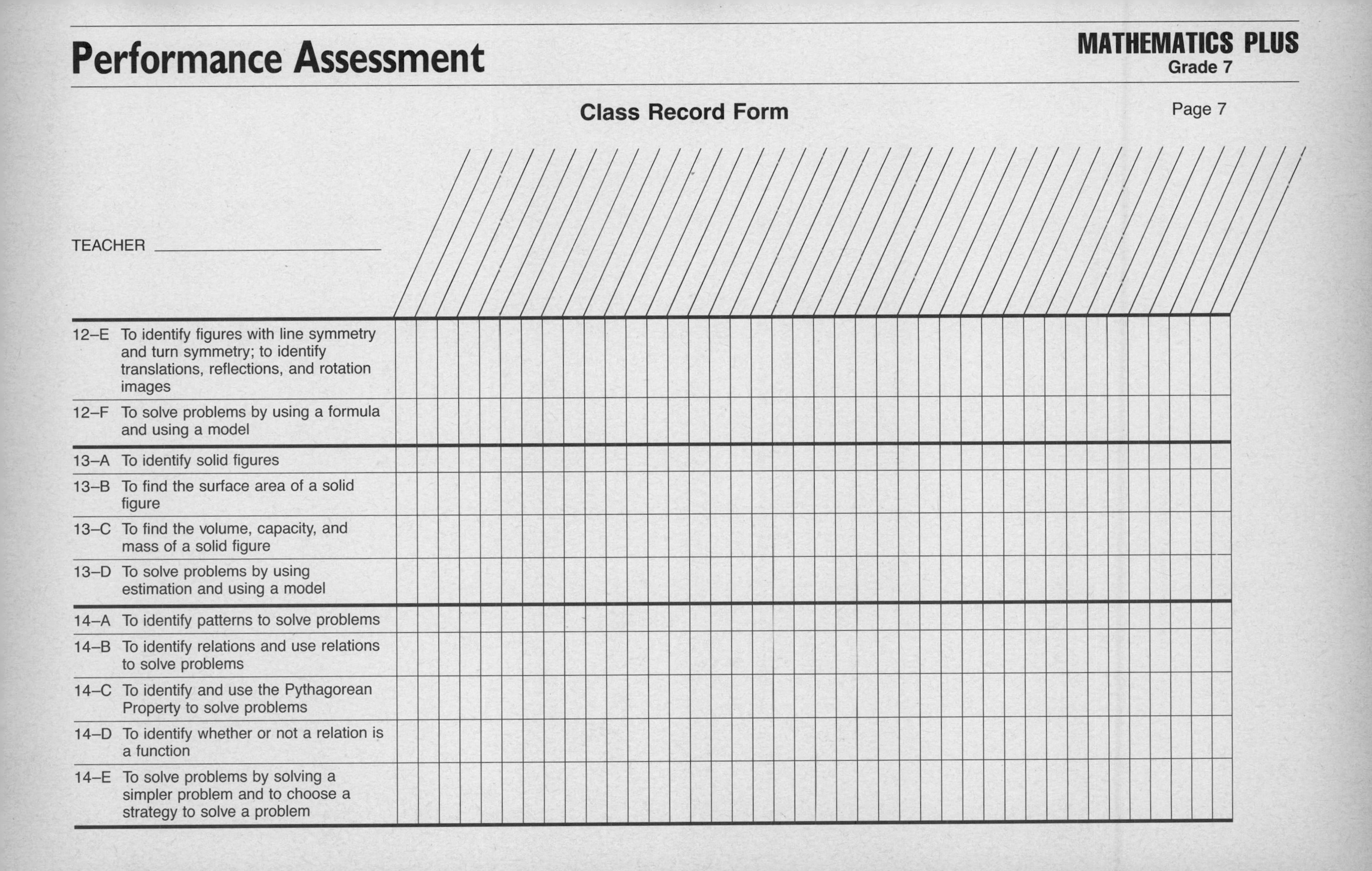

Class Record Form

TEACHER ______________________

12–E	To identify figures with line symmetry and turn symmetry; to identify translations, reflections, and rotation images
12–F	To solve problems by using a formula and using a model
13–A	To identify solid figures
13–B	To find the surface area of a solid figure
13–C	To find the volume, capacity, and mass of a solid figure
13–D	To solve problems by using estimation and using a model
14–A	To identify patterns to solve problems
14–B	To identify relations and use relations to solve problems
14–C	To identify and use the Pythagorean Property to solve problems
14–D	To identify whether or not a relation is a function
14–E	To solve problems by solving a simpler problem and to choose a strategy to solve a problem

Performance Checklists

Two types of checklists are included in this section: observation by the teacher and self-evaluation by the student. Observation checklists help evaluate student participation in cooperative learning groups, oral presentations, and projects. Student self-evaluation checklists give students a chance to reflect upon their own work in a group and their attitude about what they are learning.

These checklists give you information about students' confidence, flexibility, willingness to persevere, interest, curiosity, and inventiveness, inclination to monitor and reflect on their own thinking and doing, and appreciation of the role of mathematics in our culture.

Student's Name ______________________________

Cooperative Learning Checklist

Ring the response that best describes each student's behavior.

Never — Behavior is not observable.

Sometimes — Behavior is sometimes, but not always, observable when appropriate.

Always — Behavior is observable throughout the activity or whenever appropriate.

The student

• is actively involved in the activity.	Never	Sometimes	Always
• shares materials with others.	Never	Sometimes	Always
• helps others in the group.	Never	Sometimes	Always
• seeks the teacher's help only when all group members need help.	Never	Sometimes	Always
• fulfills their assigned role in the group.	Never	Sometimes	Always
• dominates the activity of the group.	Never	Sometimes	Always
• shares ideas with others.	Never	Sometimes	Always
• tolerates different views within the group about how to solve problems.	Never	Sometimes	Always

Use this checklist to discuss each student's successful cooperative learning experiences and ways in which he or she can become a more effective group member.

Student's Name ______________________________

Oral Presentation Checklist

This checklist describes the main features of a successful oral presentation. An "oral presentation" does not necessarily mean a presentation made in front of a full class or for a large group of fellow students, although these could be included. An oral presentation may be made for a small group of fellow students—explaining an idea (e.g., the difference between prime and composite numbers), demonstrating a way of solving a problem, or offering an argument to show why an idea is mathematically true.

Use this checklist to discuss each student's oral presentation and ways in which effective presentations are made. The checklist may also be used by the student as a self-checking guide during the planning stages of the oral presentation.

Rate each presentation on each item. Use a scale from 1—*strongly disagree*; 2—*disagree*; 3—*questionable*; 4—*agree*; 5—*strongly agree*; NA—*not applicable*.

The presenter

_____ clearly states what the presentation is going to be about.

_____ uses models and diagrams where appropriate.

_____ makes sure all members of the audience understand all main points.

_____ accepts questions comfortably.

_____ answers questions clearly.

_____ offers another way of looking at a topic or problem when an earlier explanation is not clear.

_____ demonstrates various ways to think about a problem.

_____ demonstrates a positive problem-solving attitude by searching and using appropriate strategies.

_____ demonstrates an understanding of the relationships among various mathematical ideas and makes connections among them.

_____ delivers the presentation clearly.

_____ paces the presentation appropriately for the audience.

Student's Name ______________________________

Projects Checklist

This checklist describes the main features of a successful project. A project, by its very nature, is constructed by the student to illustrate one or more mathematical ideas. It goes beyond regular and routine assignments and can be created by an individual or by a group. Usually open-ended, a project may include one or more of these experiences: research, model building, problem solving, mathematical art, gathering data, simulations, computer demonstrations, or writing reports.

Use this checklist as a guide to discuss each student's project. The checklist may also be used by the student as a self-checking guide during the planning stages of the project.

Rate each project on each item. Use a scale from 1—*strongly disagree*; 2—*disagree*; 3—*questionable*; 4—*agree*; 5—*strongly agree*; NA—*not applicable*.

The project

_____ demonstrates understanding of mathematical ideas or themes.

_____ has a unique aspect to it.

_____ demonstrates originality.

_____ demonstrates creativity.

_____ communicates its goals clearly.

_____ achieves its stated goals.

_____ shows evidence of careful planning.

_____ stimulates further interest in one or more mathematical topics.

_____ stimulates students' interest in mathematical learning.

_____ connects the math to the real world.

_____ integrates math with other subjects.

Name ______________________________

Attitude Survey

Ring the number that tells how you feel about each statement.

1 — strongly disagree
2 — disagree
3 — not sure
4 — agree
5 — strongly agree

1.	I enjoy solving problems.	**1**	**2**	**3**	**4**	**5**
2.	I enjoy working with others to solve problems.	**1**	**2**	**3**	**4**	**5**
3.	I want to complete math assignments as fast as possible.	**1**	**2**	**3**	**4**	**5**
4.	I think math is easy.	**1**	**2**	**3**	**4**	**5**
5.	I enjoy helping others in class and in groups.	**1**	**2**	**3**	**4**	**5**
6.	I would like extra assignments in math.	**1**	**2**	**3**	**4**	**5**
7.	I like to use math in other subjects.	**1**	**2**	**3**	**4**	**5**
8.	I enjoy challenges that require thinking.	**1**	**2**	**3**	**4**	**5**
9.	I enjoy writing about solutions.	**1**	**2**	**3**	**4**	**5**
10.	I enjoy talking about how to find solutions.	**1**	**2**	**3**	**4**	**5**
11.	I try very hard in math.	**1**	**2**	**3**	**4**	**5**
12.	Math is one of my favorite subjects.	**1**	**2**	**3**	**4**	**5**
13.	I like to work alone in math.	**1**	**2**	**3**	**4**	**5**
14.	I like to read about math outside of class.	**1**	**2**	**3**	**4**	**5**
15.	I like to solve problems and puzzles in my spare time.	**1**	**2**	**3**	**4**	**5**

How Well Did I Work in My Group?

Circle *yes* if you agree. Circle *no* if you disagree.

1. I solve problems better when I work with a group than when I work alone.	yes	no
2. A group is the best place for me to learn the type of math I learned today.	yes	no
3. I shared my ideas with my group today.	yes	no
4. I listened to the ideas of others in my group.	yes	no
5. I was able to ask questions of my group.	yes	no
6. I encouraged others in my group to share their ideas.	yes	no
7. I was able to discuss opposite ideas with my group.	yes	no
8. I understood the problem my group worked on today.	yes	no
9. I understood the solution to the problem my group worked on today.	yes	no
10. I can explain the problem my group worked on and its solution to others.	yes	no

II

Formal Assessment

Two types of formal assessment are provided in this section. The multiple-choice format is provided to assess mastery of the broad objectives of the program. These tests assess concepts, skills, and problem solving. The two forms of the test can be used as Pretest/Posttest or as two forms of the Posttest. The use of this test format helps prepare students for the standardized achievement tests.

The free-response tests, also given in two forms, are useful diagnostic tools. The work the student performs provides information about what this student understands about the concepts and/or procedures so that appropriate reteaching can be chosen from the many options in the program.

me ___

nventory Test

oose the letter of the correct answer.

What is the value of the 7 in 147,523,968?

A. 7,000 **B.** 70,000
C. 700,000 **D.** 7,000,000

What is two ten-thousandths in standard form?

A. 0.00002 **B.** 0.0002
C. 0.002 **D.** not here

$18 - 4 \times 3 + 2 =$ ___?___

A. 4 **B.** 8 **C.** 44 **D.** 70

$$\begin{array}{r} 2.471 \\ -0.69 \\ \hline \end{array}$$

A. 1.781 **B.** 2.221
C. 2.402 **D.** 3.161

$$\begin{array}{r} 3,219 \\ \times \quad 42 \\ \hline \end{array}$$

A. 19,314 **B.** 124,198
C. 135,198 **D.** 145,298

$26.163 \div 5.7 =$ ___?___

A. 0.459 **B.** 4.581
C. 4.59 **D.** 5.432

There are 9 chairs in a row. How many rows are needed to seat 387 people?

A. 41 rows **B.** 43 rows
C. 48 rows **D.** 3,483 rows

8. Lisa wants to compare the numbers of cars sold by two sales representatives during a six-month period. What kind of graph should she use?

A. circle graph
B. line graph
C. double bar graph
D. stem-and-leaf plot

Use the stem-and-leaf plot below to answer question 9.

Stem	Leaves
4	0 1 2 3 3 6
5	1 2 3 4 5 6 6 6
6	0 1 2 2 3 5

9. What is the mode?

A. 33 **B.** 55 **C.** 56 **D.** 65

10. What is the range of 1, 3, 5, 7, 9, 11, 13?

A. 7 **B.** 12 **C.** 13 **D.** not here

11. How many choices are possible for 9 pairs of socks and 5 pairs of shoes?

A. 4 choices **B.** 14 choices
C. 28 choices **D.** 45 choices

12. What is the prime factorization of 14?

A. 1×4 **B.** 2×7
C. 1×14 **D.** $2 \times 5 + 2 \times 2$

13. What is the greatest common factor (GCF) of 8 and 16?

A. 2 **B.** 4 **C.** 8 **D.** 16

14. $\frac{1}{4} + \frac{1}{5} =$ ___?___

A. $\frac{5}{20}$ **B.** $\frac{1}{9}$ **C.** $\frac{2}{9}$ **D.** $\frac{9}{20}$

Go on to the next page.

Name ______________________________

Inventory Test

15. $1\frac{1}{6} + 4\frac{3}{6} =$ ___?___

A. $5\frac{1}{3}$ **B.** $5\frac{2}{3}$

C. $6\frac{1}{3}$ **D.** 9

16. $\frac{3}{8} \times \frac{4}{5} =$ ___?___

A. $\frac{7}{40}$ **B.** $\frac{1}{4}$

C. $\frac{3}{10}$ **D.** $\frac{15}{32}$

17. Jack had $1\frac{1}{4}$ quarts of milk. He used $\frac{1}{3}$ of it. How much milk did he use?

A. $\frac{5}{12}$ qt **B.** $\frac{7}{8}$ qt

C. $\frac{11}{12}$ qt **D.** $1\frac{1}{12}$ qt

18. Compare. $\frac{2}{5} \div \frac{1}{7} \bigcirc 1$

A. $>$ **B.** $=$ **C.** $<$

19. $\frac{3}{4} \div \frac{1}{6} =$ ___?___

A. $\frac{1}{8}$ **B.** $\frac{7}{12}$

C. $\frac{11}{12}$ **D.** not here

20. What is $\frac{25}{100}$ written as a decimal?

A. 0.025 **B.** 0.25

C. 0.4 **D.** 4.0

21.
  2 hr 10 min
−1 hr 55 min

A. 15 min **B.** 45 min

C. 1 hr 45 min **D.** 4 hr 5 min

22. Joel has to compute the total weight of 3 objects. The first one weighs 2.41 g, the second 1.94 g, and the third 0.83 g. What is the total weight of the objects?

A. 4.18 g **B.** 5.15 g

C. 5.18 g **D.** 12.65 g

23. Which of the following is a ratio?

A. $4 + 5 = 8 + 1$ **B.** 9.374

C. $5 - 4$ **D.** 4:5

24. Which term makes these ratios equivalent?

$$\frac{1}{6} = \frac{6}{\blacksquare}$$

A. 1 **B.** 11 **C.** 12 **D.** 36

25. What are the cross products?

$$\frac{4}{5} = \frac{y}{35}$$

A. $4 \times y = 5 \times 35$

B. $5 \times y = 4 \times 35$

C. $4 \times 5 = y \times 35$

D. not here

26. A scale model of a new theater has a parking lot 15 in. long and a lawn 10 in. long. If the actual parking lot is 240 ft long, how long is the lawn?

A. 160 ft **B.** 180 ft

C. 245 ft **D.** 360 ft

27. If 40% of the fruit in a jar is apples, what fraction of the fruit is apples?

A. $\frac{1}{40}$ **B.** $\frac{1}{5}$

C. $\frac{1}{4}$ **D.** $\frac{2}{5}$

Go on to the next page.

Name ____________________

Inventory Test

28. Write 8% as a fraction in simplest form.

A. $\frac{1}{25}$ **B.** $\frac{2}{25}$

C. $\frac{1}{8}$ **D.** not here

29. 30% of 90 = ___?___

A. 3 **B.** 18 **C.** 27 **D.** 30

30. A store advertises sneakers for 20% off the regular price of $29.95. About how much is the sale price?

A. $6 **B.** $21 **C.** $24 **D.** $27

31. What is an angle called that measures less than 90°?

A. acute **B.** complementary

C. right **D.** obtuse

32. Angles *A* and *B* are complementary. If Angle *A* measures 10°, what is the measure of Angle *B*?

A. 35° **B.** 80° **C.** 90° **D.** 170°

33. Find the perimeter of this triangle.

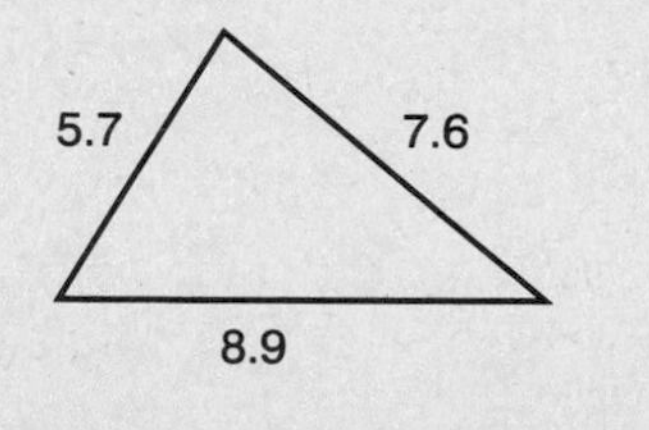

A. 11.1 units **B.** 22.2 units

C. 16.5 units **D.** not here

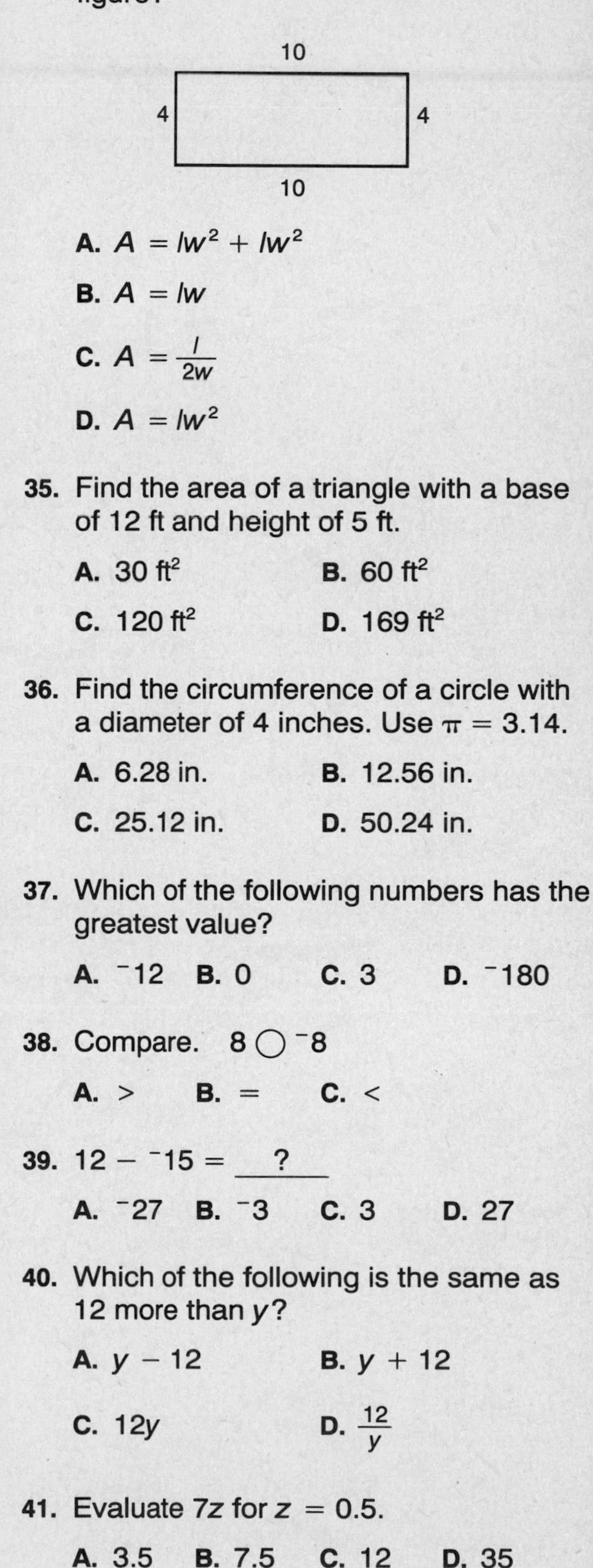

34. Which formula will find the area of this figure?

A. $A = lw^2 + lw^2$

B. $A = lw$

C. $A = \frac{l}{2w}$

D. $A = lw^2$

35. Find the area of a triangle with a base of 12 ft and height of 5 ft.

A. 30 ft² **B.** 60 ft²

C. 120 ft² **D.** 169 ft²

36. Find the circumference of a circle with a diameter of 4 inches. Use $\pi = 3.14$.

A. 6.28 in. **B.** 12.56 in.

C. 25.12 in. **D.** 50.24 in.

37. Which of the following numbers has the greatest value?

A. $^{-}12$ **B.** 0 **C.** 3 **D.** $^{-}180$

38. Compare. 8 ◯ $^{-}8$

A. $>$ **B.** $=$ **C.** $<$

39. $12 - {}^{-}15 =$ ___?___

A. $^{-}27$ **B.** $^{-}3$ **C.** 3 **D.** 27

40. Which of the following is the same as 12 more than *y*?

A. $y - 12$ **B.** $y + 12$

C. $12y$ **D.** $\frac{12}{y}$

41. Evaluate $7z$ for $z = 0.5$.

A. 3.5 **B.** 7.5 **C.** 12 **D.** 35

Go on to the next page.

Name ______________________________

Inventory Test

42. Which number line shows the solution for $p > {}^{-}2$?

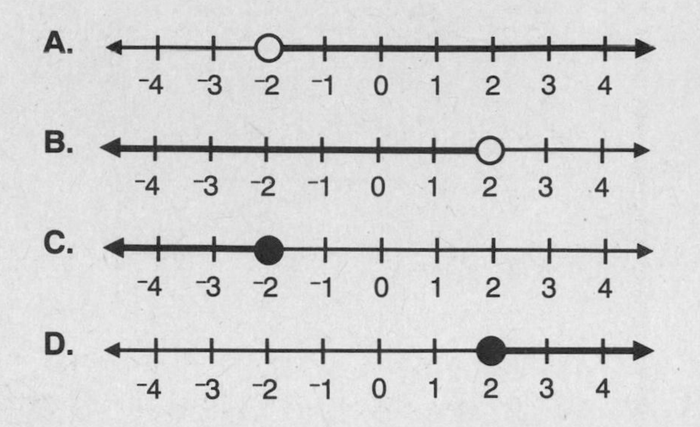

43. Peter spent $\frac{1}{2}$ of his money to buy a tennis racket and $\frac{1}{4}$ for shorts. He had $10 left to buy tennis balls. How much money did he have to start with?

A. $25 **B.** $35 **C.** $40 **D.** $50

44. Mike saved $10.50 in January, $8.75 in February, and $12.60 in March. On April 1, he had $50.00 in his bank. How much was in his bank before January?

A. $8.15 **B.** $18.15

C. $19.85 **D.** $31.85

45. Lee is twice as old as Marty. Bridget is twice as old as Lee, and Heidi is twice as old as Bridget. If Marty is 3, how old is Heidi?

A. 12 years old **B.** 18 years old

C. 21 years old **D.** 24 years old

Use the table below to answer questions 46–47.

Dry Cleaning Rates		
Day	First Pound	Each Additiona Pound
Mon–Fri	$2.50	$1.95
Sat–Sun	2.75	2.25
Holiday	3.00	2.50

46. What is the cost of dry cleaning 5 pounds on a Wednesday?

A. $9.75 **B.** $10.30

C. $12.25 **D.** $12.50

47. What is the cost of dry cleaning 3 pounds on a holiday?

A. $7.50 **B.** $7.75

C. $8.00 **D.** $10.50

48. The difference between two numbers 15. Their sum is 125. What is the larg number?

A. 60 **B.** 70 **C.** 75 **D.** 85

49. A contractor wants to put a 3-foot-wi door in the middle of a 14-foot-wide wall. If the door is centered, how far it be from each end of the wall?

A. 3 ft **B.** $5\frac{1}{2}$ ft

C. $8\frac{1}{2}$ ft **D.** 11 ft

50. Jed is making a bird feeder. He boug 3 boards for $2.19 each, 35 nails for $0.03 each, and 6 sheets of sandpa for $0.21 each. What was the total c of supplies?

A. $3.45 **B.** $5.51

C. $7.83 **D.** $8.88

Name ____________________

Pretest

Choose the letter of the correct answer.

1. Choose the *best* estimate.

3,198 + 2,025

A. less than 4,000
B. less than 5,000
C. greater than 5,000

2. Choose the *best* estimate.

982 − 386

A. less than 600
B. greater than 600
C. greater than 800

3. Choose the *best* estimate.

3.1 × 15.02

A. less than 40
B. greater than 45
C. greater than 50

4. Choose the *best* estimate.

49.97 ÷ 2.11

A. less than 20
B. less than 25
C. greater than 25

5. Estimate. 355 + 1,650 + 422

A. 2,100 **B.** 2,200
C. 2,500 **D.** 2,600

6. Estimate. 1,153.2 − 166.5

A. 700 **B.** 800
C. 1,000 **D.** 1,100

7. Estimate. 59.95 × 1,000

A. 6,000 **B.** 58,000
C. 60,000 **D.** 599,000

8. Estimate. 2,005 ÷ 9.57

A. 200 **B.** 950
C. 2,000 **D.** 18,000

9. Jane wants to buy 3 pens that cost $1.19 each and 2 pads that cost $1.15 each. *Estimate* the amount of money that Jane needs.

A. $3.00 **B.** $4.00
C. $4.50 **D.** $6.00

10. Bill had $18.99 to spend. He spent $3.09 on Friday, $2.10 on Saturday, and $2.95 on Sunday. *Estimate* the amount of money he had left.

A. $5.00 **B.** $8.00
C. $11.00 **D.** $14.00

11. Which quotient is greatest?

A. 99.2 ÷ 3
B. 99.2 ÷ 4
C. 99.2 ÷ 5
D. 99.2 ÷ 10

12. Which difference is greatest?

A. 142.5 − 12
B. 1,425.0 − 12
C. 14.25 − 12
D. 1.1425 − 12

13. Which product is greatest?

A. 4 × 1,001 **B.** 5 × 101
C. 7 × 10.1 **D.** 9 × 1.01

Go on to the next page.

Name ______________________________

Pretest

14. Which quotient is greatest?

A. $50 \div 0.1$ B. $50 \div 0.01$
C. $50 \div 0.001$ D. $50 \div 0.0001$

15. $7.86 + 2.99 + 3.07 =$ ___?___

A. 12.92 B. 13.92
C. 14.02 D. not here

16. $8.07 - 2.09 =$ ___?___

A. 5.08 B. 5.98
C. 6.02 D. not here

17. $9.90 \times 11.20 =$ ___?___

A. 11.088 B. 110.08
C. 110.88 D. not here

18. $16.4 \div 4.1 =$ ___?___

A. 0.25 B. 0.4
C. 3.50 D. not here

19. Julio bought 5.5 pounds of cheese for $13.75. How much did he pay per pound of cheese?

A. $2.25 per pound
B. $2.50 per pound
C. $2.75 per pound
D. $2.95 per pound

20. Myra runs 2.5 mi every weekday, 3.0 mi each Saturday, and 3.5 mi each Sunday. How many mi does she run in a 2-week period?

A. 31.5 mi B. 35.0 mi
C. 35.5 mi D. 38.0 mi

21. $5^3 =$ ___?___

A. 5 B. 25
C. 125 D. 625

22. $9^0 =$ ___?___

A. 0 B. 1
C. 9 D. 90

23. What is the value of 4 to the third power?

A. 4 B. 12
C. 16 D. 64

24. What is the value of 10^1?

A. 1 B. 10
C. 20 D. 100

25. Express 605,000,000 in scientific notation.

A. 6.05×10^8 B. 6.50×10^8
C. 65×10^6 D. 65×10^8

26. Express 4.5×10^3 in standard form.

A. 45 B. 450
C. 4,500 D. 45,000

27. Express 8,203 in scientific notation.

A. 8.23×10^2 B. 8.23×10^3
C. 8.203×10^2 D. 8.203×10^3

28. Express 3.5×10^1 in standard form.

A. 0.035 B. 0.35
C. 3.5 D. 35

Go on to the next page.

Name ______________________

Pretest

Use this information to answer questions 29–30.

The menu at Joe's Restaurant lists three main dishes: hamburger, hot dog, and pizza. It lists four salads: tossed, macaroni, potato, and cole slaw.

29. In how many combinations could Natasha order one main dish and one salad?

A. 3 combinations

B. 4 combinations

C. 7 combinations

D. 12 combinations

30. If Natasha orders a hamburger and two different salads, how many meals are possible for her to order?

A. 4 meals

B. 6 meals

C. 7 meals

D. 12 meals

31. A total of 56 boys and girls are in the school chorus. There are 20 more girls than boys. How many boys are in the chorus?

A. 16 boys

B. 18 boys

C. 20 boys

D. 36 boys

32. The Andersons went to a concert. They spent $39 for tickets. The tickets cost $12 for each adult and $5 for each child. How many children went to the concert?

A. 1 child

B. 2 children

C. 3 children

D. 4 children

Stop!

Name ______________________________

Posttest

Choose the letter of the correct answer.

1. Choose the *best* estimate.

   4,125 + 3,002

   A. less than 6,000
   B. less than 7,000
   C. greater than 7,000

2. Choose the *best* estimate.

   655 − 462

   A. less than 100
   B. less than 200
   C. less than 300

3. Choose the *best* estimate.

   4.1 × 25.05

   A. less than 90
   B. less than 100
   C. greater than 100

4. Choose the *best* estimate.

   99.98 ÷ 2.05

   A. less than 33
   B. less than 50
   C. greater than 60

5. Estimate. 544 + 2,499 + 351

   A. 2,800 B. 3,100
   C. 3,400 D. 3,500

6. Estimate. 2,742.1 − 653.5

   A. 1,900 B. 1,950
   C. 2,100 D. 2,300

7. Estimate. 69.97 × 10,000

   A. 70,000 B. 650,000
   C. 700,000 D. 750,000

8. Estimate. 5,005 ÷ 9.67

   A. 500 B. 700
   C. 2,500 D. 45,000

9. Mario wants to buy 2 suits that cost $149.95 each and 4 shirts that cost $19.95 each. *Estimate* the amount of money that Mario needs.

   A. $320.00 B. $340.00
   C. $360.00 D. $380.00

10. Doris had $25.92 to spend. She spent $9.65 on dinner, $5.00 on a movie, and $2.95 to park her car. *Estimate* the amount of money she had left.

    A. $5.00 B. $6.00
    C. $8.00 D. $11.00

11. Which quotient is greatest?

    A. 33.3 ÷ 6
    B. 33.3 ÷ 7
    C. 33.3 ÷ 8
    D. 33.3 ÷ 10

12. Which difference is greatest?

    A. 162 − 11.92
    B. 162 − 1.192
    C. 162 − 119.2
    D. 162 − 0.1192

13. Which product is greatest?

    A. 4 × 2,001 B. 6 × 201
    C. 8 × 20.1 D. 10 × 2.01

Go on to the next page.

Name ______

Posttest

14. Which quotient is greatest?

A. $800 \div 0.3$ **B.** $800 \div 0.03$
C. $800 \div 0.003$ **D.** $800 \div 0.0003$

15. $3.92 + 6.14 + 11.90 =$ ___?___

A. 21.06 **B.** 21.96
C. 22.96 **D.** not here

16. $8.68 - 4.79 =$ ___?___

A. 3.11 **B.** 3.89
C. 4.89 **D.** not here

17. $8.80 \times 10.2 =$ ___?___

A. 8.976 **B.** 88.76
C. 89.76 **D.** not here

18. Solve. $5.1\overline{)25.5}$

A. 0.2 **B.** 0.50
C. 4.50 **D.** not here

19. Juan bought 7.25 ounces of steak for $5.80. How much did he pay per ounce of steak?

A. $0.75 per ounce
B. $0.80 per ounce
C. $1.25 per ounce
D. $1.45 per ounce

20. Betty works 8 hr per day from Monday through Friday. She also works 2 hr per evening on Monday and Wednesday. How many hr does she work per week?

A. 40 hr **B.** 42 hr
C. 44 hr **D.** 50 hr

21. $2^4 =$ ___?___

A. 4 **B.** 8
C. 16 **D.** 24

22. $7^0 =$ ___?___

A. 0 **B.** 1
C. 7 **D.** 70

23. What is the value of 5 to the third power?

A. 5 **B.** 15
C. 25 **D.** 125

24. What is the value of 11^1?

A. 1 **B.** 11
C. 111 **D.** 121

25. Express 105,000 in scientific notation.

A. 1.05×10^5
B. 1.50×10^5
C. 15×10^4
D. 15×10^5

26. Express 5.6×10^4 in standard form.

A. 56 **B.** 560
C. 5,600 **D.** 56,000

27. Express 7,015 in scientific notation.

A. 7.015×10^2
B. 7.015×10^3
C. 7.15×10^2
D. 7.15×10^3

28. Express 0.25×10^1 in standard form.

A. 0.025 **B.** 1
C. 2.5 **D.** 25

Go on to the next page.

Posttest

Use this information to answer questions 29–30.

Sonya decides to spend New Year's Day watching football games. There are 3 games on television in the early afternoon, 3 in the late afternoon, and 2 at night.

29. Sonya decides to watch one game in the early afternoon, one in the late afternoon, and one at night. In how many different ways can she choose games to watch?

A. 3 ways **B.** 8 ways

C. 16 ways **D.** 18 ways

30. If Sonya decides to watch one game in the early afternoon, one in the late afternoon, but none at night, in how many different ways can she choose games to watch?

A. 3 ways **B.** 6 ways

C. 8 ways **D.** 9 ways

31. A total of 108 parents attended the seventh-grade open house. There were 20 more mothers present than fathers. How many parents that attended were fathers?

A. 40 fathers

B. 44 fathers

C. 60 fathers

D. 64 fathers

32. The Masons went to a play. They spent $34 for tickets. The tickets cost $9 for adults and $4 for children. How many children went to the play?

A. 1 child

B. 2 children

C. 3 children

D. 4 children

Stop!

Name ______________________________

Pretest

Choose the letter of the correct answer.

Use the bar graph below to answer questions 1–2.

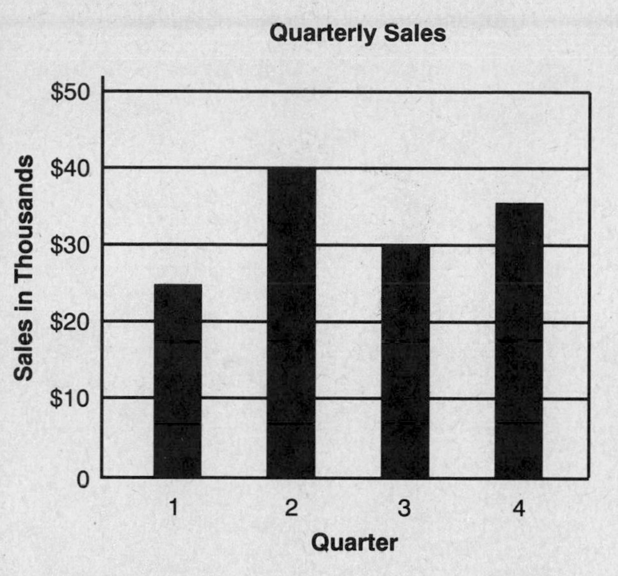

1. Which quarter had the greatest sales?

   A. Quarter 1

   B. Quarter 2

   C. Quarter 3

   D. Quarter 4

2. How much did sales drop between Quarters 2 and 3?

   A. \$2,000 **B.** \$10,000

   C. \$30,000 **D.** not here

Use the circle graph below to answer questions 3–4.

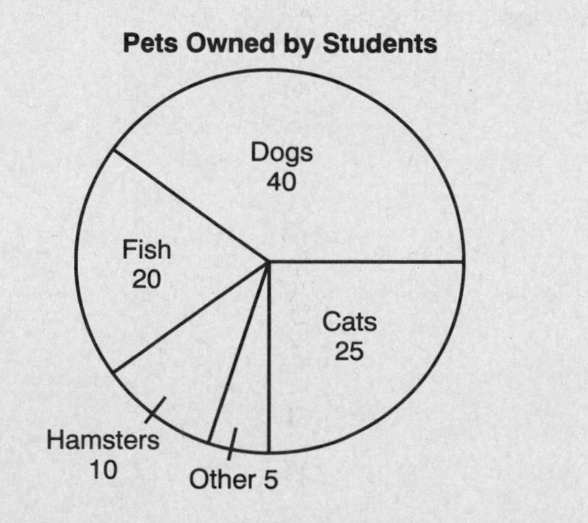

3. How many of the students own a cat or a hamster?

   A. 10

   B. 25

   C. 35

   D. 40

4. How many of the students own a dog, a cat, or fish?

   A. 20

   B. 25

   C. 40

   D. 85

Use the stem-and-leaf plot below to answer questions 5–6.

Number of Points Scored by a Football Team in 16 Games

Stem	Leaves
1	4 4 7 7
2	0 1 4 4 7 7 8 8
3	1 5
4	1 2

5. What was the least number of points scored by the team?

   A. 10 points

   B. 12 points

   C. 14 points

   D. 20 points

6. In how many of its games did the team score 30 or more points?

   A. 0

   B. 2

   C. 4

   D. 12

Go on to the next page.

Name ______________________________

Pretest

Use the information and the double-line graph below to answer questions 7–10.

Each December, a store predicts its sales (in millions) for the coming year. The double-line graph shows the predicted and actual sales for a five-year period.

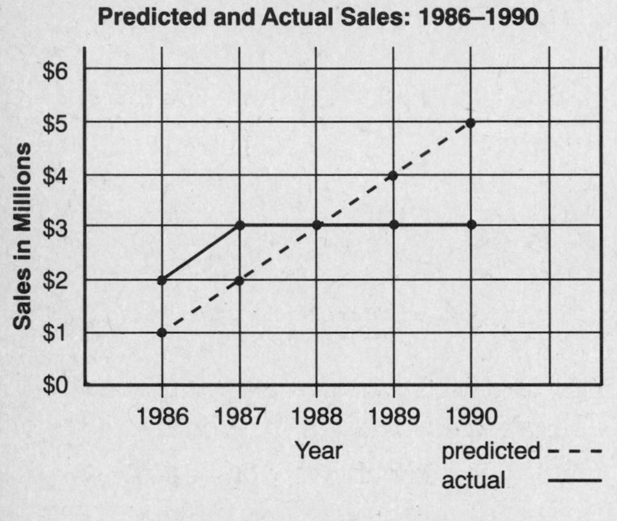

7. For which year did predicted sales equal actual sales?

   A. 1986 **B.** 1987
   C. 1988 **D.** not here

8. For which year were actual sales *most* behind predicted sales?

   A. 1986 **B.** 1988
   C. 1989 **D.** 1990

9. How much did actual sales increase between 1986 and 1989?

   A. $0 **B.** $1,000,000
   C. $2,000,000 **D.** $3,000,000

10. For how many years were actual sales at $3,000,000?

    A. 2 years **B.** 3 years
    C. 4 years **D.** not here

Use the table below to answer questions 11–14.

Grades Earned by English Students		
Grade	**Frequency**	**Cumulative Frequency**
A	6	6
B	10	16
C	8	24
D	7	31
F	1	32

11. Which grade was received by the most students?

    A. B **B.** C **C.** D **D.** A

12. How many students received a C?

    A. 1 **B.** 7 **C.** 8 **D.** 24

13. How many students received a grade of B or better?

    A. 6 **B.** 10 **C.** 16 **D.** 26

14. How many students received a grade of D?

    A. 1 **B.** 7 **C.** 8 **D.** 31

Use the information below to answer questions 15–18.

These are the scores of 10 students on a spelling test.

10, 13, 13, 13, 15, 15, 17, 18, 19, 20

15. What is the mode for the scores?

    A. 13 **B.** 14 **C.** 15 **D.** not here

16. What is the range of the scores?

    A. 10 **B.** 13 **C.** 15 **D.** 20

Go on to the next page.

Pretest

17. What is the mean of the scores?

    A. 13 B. 15 C. 15.3 D. 153

18. What is the median score?

    A. 10 B. 13 C. 17 D. not here

Use the information and the box-and-whisker graph below to answer questions 19–20.

Liz recorded the number of customers in her store each day for 20 days. The box-and-whisker graph below summarizes the information.

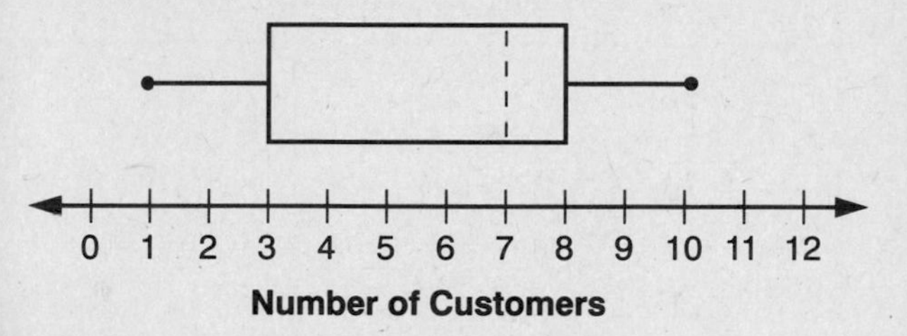

19. What was the median number of customers in the store?

    A. 3 customers

    B. 5 customers

    C. 7 customers

    D. 8 customers

20. What was the greatest number of customers in the store on any of the 20 days?

    A. 7 customers

    B. 8 customers

    C. 10 customers

    D. 12 customers

Use the table below to answer questions 21–23.

Senior Class Elections

Candidate	Girls' Votes	Boys' Votes
John	85	70
Thomas	40	85
Darlene	100	50
Vera	95	75
Total	320	280

21. Which candidate received the most votes?

    A. John B. Thomas

    C. Darlene D. Vera

22. How many of the boys' votes did John receive?

    A. 70 B. 85 C. 100 D. 280

23. If half of the boys who voted for Darlene had voted for John, who would have received the most votes?

    A. John B. Thomas

    C. Darlene D. Vera

24. A manager wants to draw a graph comparing the monthly sales of male and female sales staff. Which type of graph should she draw?

    A. double-bar graph

    B. circle graph

    C. box-and-whisker graph

    D. stem-and-leaf plot

Stop!

Name ______________________________

Posttest

Choose the letter of the correct answer.

Use the bar graph below to answer questions 1–2.

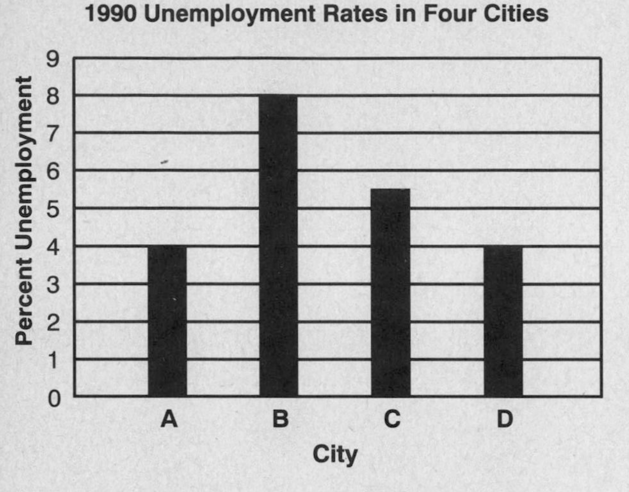

1. Which city had the highest rate of unemployment?

   A. City A **B.** City B

   C. City C **D.** City D

2. The percent of unemployment in City A was how much less than the percent of unemployment in City B?

   A. 2 percent **B.** 4 percent

   C. 6 percent **D.** 8 percent

Use the circle graph below to answer questions 3–4.

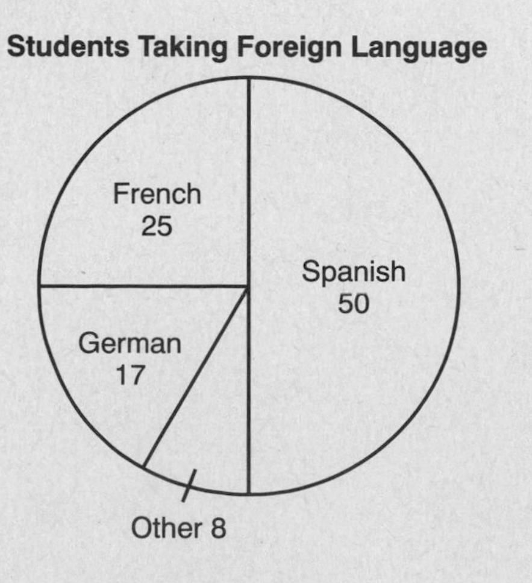

3. How many students are taking French?

   A. 8

   B. 17

   C. 25

   D. 50

4. How many students are taking Spanish or German?

   A. 17

   B. 25

   C. 67

   D. 75

Use the stem-and-leaf plot below to answer questions 5–6.

Number of Points Scored by a Basketball Team in 25 Games

Stem	Leaves
6	0 1 3 5 7
7	2 4 5 6 7 8 9 9
8	1 3 3 3 5 6 7
9	1 4 5 8 9

5. What was the greatest number of points scored by the team?

   A. 89 points

   B. 91 points

   C. 98 points

   D. 99 points

6. In how many of its games did the team score 95 or more points?

   A. 3

   B. 5

   C. 23

   D. not here

Go on to the next page.

Name ____________________

Use the information and the double-line graph below to answer questions 7–10.

Each month, a real-estate agent predicts how many homes his office will sell that month. The double-line graph below shows the predicted and actual sales for a five-month period.

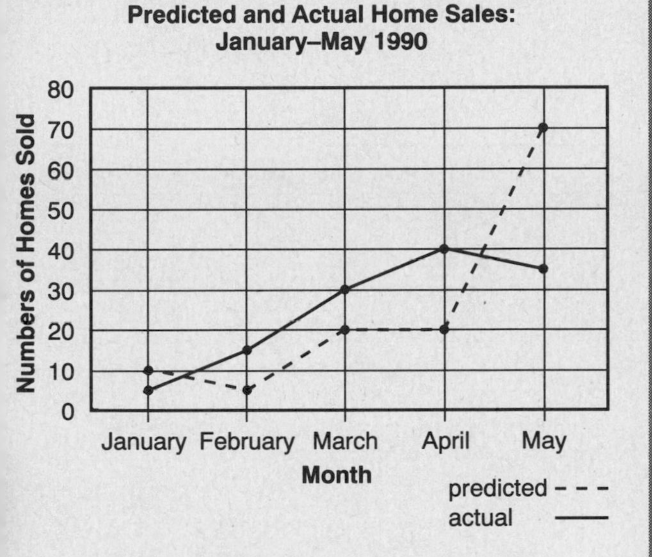

7. For which month were actual sales greatest?

A. February **B.** March

C. April **D.** May

8. For which month were actual sales most behind predicted sales?

A. January **B.** March

C. April **D.** May

9. By how many homes did actual sales increase between March and April?

A. 0 homes **B.** 10 homes

C. 20 homes **D.** 30 homes

10. During April, how many more homes were sold than predicted?

A. 10 **B.** 20 **C.** 30 **D.** 40

Use the table below to answer questions 11–14.

Weight Loss in Pounds During First Week in Diet Class

Weight Loss	Frequency	Cumulative Frequency
0	2	2
1	4	6
2	6	12
3	8	20
4	12	32
5	5	37
6	3	40

11. Which weight loss was most frequent?

A. 3 pounds

B. 4 pounds

C. 5 pounds

D. 6 pounds

12. How many of the dieters lost 2 pounds?

A. 2

B. 4

C. 6

D. 12

13. How many of the dieters lost 4 or fewer pounds?

A. 4

B. 8

C. 12

D. 32

14. How many dieters lost no weight?

A. 0 dieters

B. 1 dieter

C. 2 dieters

D. 4 dieters

Go on to the next page.

Name ______________________________

Posttest

Use the information below to answer questions 15–18.

These are the scores of 9 students on a math test.

15, 16, 16, 19, 21, 23, 24, 25, 30

15. What is the mode for the scores?

A. 15 **B.** 16 **C.** 19 **D.** 21

16. What is the range of the scores?

A. 14 **B.** 15 **C.** 30 **D.** not here

17. What is the mean of the scores?

A. 20.0 **B.** 21.0

C. 22.0 **D.** 22.1

18. What is the median score?

A. 16 **B.** 19 **C.** 21 **D.** not here

Use the information and the box-and-whisker graph below to answer questions 19–20.

Helen recorded the number of suits she sold each day for 30 days. The box-and-whisker graph below summarizes the sales.

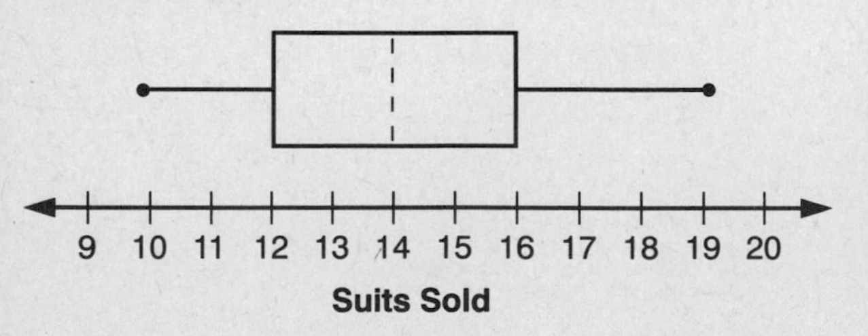

19. What was the median number of suits sold?

A. 10 suits

B. 12 suits

C. 14 suits

D. 16 suits

20. What was the greatest number of suits sold on any of the 30 days?

A. 14 suits **B.** 18 suits

C. 19 suits **D.** 20 suits

Use the table below to answer questions 21–23.

Music Preference Among Taft High School Students

Type of Music	Boys	Girls
Classical	30	90
Country	20	20
Jazz	40	60
Rock	150	100
Total	240	270

21. Which music received more votes from boys than from girls?

A. classical **B.** country

C. jazz **D.** rock

22. How many of the girls preferred classical music?

A. 30 **B.** 90 **C.** 240 **D.** 270

23. What music was *least* preferred by both groups?

A. classical **B.** country

C. jazz **D.** rock

24. A writer wants to draw a graph showing the favorite sports of teenage girls. Which type of graph should she draw?

A. bar graph

B. line graph

C. box-and-whisker graph

D. stem-and-leaf plot

Stop!

Name ______________________________

Pretest

Choose the letter of the correct answer.

1. Which is a prime number?

   A. 5 **B.** 10 **C.** 15 **D.** 35

2. Which is a composite number?

   A. 11 **B.** 17 **C.** 19 **D.** 21

3. Which is a prime number?

   A. 26 **B.** 29 **C.** 35 **D.** 51

4. How many factors does 37 have?

   A. 1 **B.** 2 **C.** 3 **D.** 37

5. What are the prime factors of 30?

   A. 5, 6 **B.** 2, 15
   C. 2, 3, 5 **D.** 3, 10

6. What are the prime factors of 52?

   A. 2, 2, 13 **B.** 2, 26
   C. 4, 13 **D.** 50, 2

7. What is the prime factorization of 24?

   A. $2^3 \times 3$ **B.** $2^2 \times 6$
   C. $4 \times 2 \times 3$ **D.** not here

8. What is the prime factorization of 1,000?

   A. $2^3 \times 25^2$ **B.** $2^3 \times 5^3$
   C. $2^2 \times 250$ **D.** 2×500

9. What is the greatest common factor of 24 and 36?

   A. 6 **B.** 12 **C.** 18 **D.** 24

10. What is the greatest common factor of 32 and 72?

    A. 8 **B.** 9 **C.** 16 **D.** 24

11. What is the least common multiple of 5 and 4?

    A. 4 **B.** 5 **C.** 10 **D.** 20

12. What is the least common multiple of 6 and 9?

    A. 6 **B.** 12 **C.** 18 **D.** 54

13. Which fraction is expressed in simplest form?

    A. $\frac{3}{15}$ **B.** $\frac{5}{15}$
    C. $\frac{55}{100}$ **D.** $\frac{99}{100}$

14. Which fraction is *not* expressed in simplest form?

    A. $\frac{1}{2}$ **B.** $\frac{2}{3}$ **C.** $\frac{4}{8}$ **D.** $\frac{11}{12}$

15. Which fraction is greater than $1\frac{3}{4}$?

    A. $\frac{27}{16}$ **B.** $\frac{32}{20}$
    C. $\frac{173}{100}$ **D.** $\frac{352}{200}$

16. Which group is ordered from least to greatest?

    A. $\frac{2}{5}, \frac{1}{2}, \frac{3}{10}$ **B.** $\frac{1}{3}, \frac{1}{2}, \frac{3}{4}$
    C. $\frac{3}{4}, \frac{3}{5}, \frac{3}{6}$ **D.** $\frac{2}{3}, \frac{4}{5}, \frac{7}{12}$

17. Compare. $\frac{9}{27} \bigcirc \frac{1}{3}$

    A. $<$ **B.** $=$ **C.** $>$

18. Compare. $3\frac{1}{7} \bigcirc 3\frac{9}{10}$

    A. $<$ **B.** $=$ **C.** $>$

19. $\frac{75}{90} =$ ___?___

    A. $\frac{5}{6}$ **B.** $\frac{15}{16}$ **C.** $\frac{5}{18}$ **D.** $\frac{10}{18}$

Go on to the next page.

Name ______________________________

Pretest

20. $\frac{2}{3} =$ ___?___

A. $\frac{3}{4}$ B. $\frac{5}{7}$

C. $\frac{7}{9}$ D. not here

Use the information below to answer questions 21–24.

A taxi charges $1.00 for the first mile and $0.20 for each additional $\frac{1}{6}$ mile.

21. How much does it cost to ride $2\frac{1}{2}$ miles?

A. $1.80 B. $2.80

C. $3.00 D. $3.80

22. How far can you travel for $5.00?

A. $3\frac{1}{3}$ miles B. $4\frac{1}{3}$ miles

C. 5 miles D. 10 miles

23. Jo rode 2 miles and Sam rode 3 miles. How much more did Sam pay?

A. $1.20 B. $2.00

C. $2.20 D. $2.40

24. Rashawn and a friend take a taxi for $1\frac{1}{2}$ miles. The two riders split the fare evenly. How much does each pay?

A. $0.80 B. $1.20

C. $1.60 D. $1.80

25. Which fraction falls between $\frac{4}{10}$ and $\frac{5}{10}$?

A. $\frac{9}{20}$ B. $\frac{11}{20}$ C. $\frac{16}{40}$ D. $\frac{9}{10}$

26. Which fraction falls between $\frac{3}{8}$ and $\frac{4}{8}$?

A. $\frac{6}{16}$ B. $\frac{8}{16}$ C. $\frac{14}{32}$ D. $\frac{17}{32}$

27. Which fraction falls between $\frac{1}{5}$ and $\frac{2}{5}$?

A. $\frac{3}{10}$ B. $\frac{4}{10}$

C. $\frac{1}{20}$ D. $\frac{9}{20}$

28. Which fraction falls between $\frac{1}{3}$ and $\frac{1}{2}$?

A. $\frac{3}{6}$ B. $\frac{2}{6}$

C. $\frac{5}{12}$ D. $\frac{7}{12}$

Use the information below to answer questions 29–32.

At Jones Pier the lake is 20 feet deep. Every $\frac{1}{4}$ mile from the pier it becomes 1 foot deeper until it reaches a maximum depth of 30 feet.

29. How deep is the water $1\frac{1}{2}$ miles from the pier?

A. 24 feet B. 26 feet

C. 28 feet D. 30 feet

30. How many miles from the pier does the lake reach its maximum depth?

A. $1\frac{1}{4}$ miles B. 2 miles

C. $2\frac{1}{4}$ miles D. $2\frac{1}{2}$ miles

31. How much deeper is the water 2 miles from the pier than it is 1 mile out?

A. 2 ft B. 4 ft

C. 6 ft D. 8 ft

32. How far from the pier does the water reach 22 feet in depth?

A. $\frac{1}{4}$ mile B. $\frac{1}{2}$ mile

C. $\frac{3}{4}$ mile D. 1 mile

Stop!

Name ____________________

Posttest

Choose the letter of the correct answer.

1. Which is a prime number?

   A. 3 B. 6 C. 10 D. 12

2. Which is a composite number?

   A. 13 B. 23 C. 33 D. 43

3. Which is a prime number?

   A. 42 B. 59 C. 65 D. not here

4. How many factors does 41 have?

   A. 1 B. 2 C. 4 D. 59

5. What are the prime factors of 42?

   A. 7, 6 B. 2, 21
   C. 2, 3, 7 D. not here

6. What are the prime factors of 68?

   A. 2, 2, 17 B. 2, 34
   C. 4, 17 D. 1, 68

7. What is the prime factorization of 48?

   A. $2^4 \times 3$ B. $2^3 \times 6$
   C. $4 \times 4 \times 4$ D. 8×6

8. What is the prime factorization of 100?

   A. $2^2 \times 25^2$ B. $2^2 \times 5^2$
   C. $2^2 \times 25$ D. $2 \times 5 \times 10$

9. What is the greatest common factor of 32 and 40?

   A. 4 B. 8 C. 10 D. 16

10. What is the greatest common factor of 16 and 24?

    A. 8 B. 12 C. 16 D. 24

11. What is the least common multiple of 3 and 4?

    A. 3 B. 4 C. 9 D. 12

12. What is the least common multiple of 9 and 12?

    A. 18 B. 24 C. 36 D. 72

13. Which fraction is expressed in simplest form?

    A. $\frac{4}{18}$ B. $\frac{5}{20}$
    C. $\frac{55}{200}$ D. $\frac{97}{102}$

14. Which fraction is *not* expressed in simplest form?

    A. $\frac{1}{3}$ B. $\frac{2}{5}$ C. $\frac{4}{10}$ D. $\frac{5}{11}$

15. Which fraction is greater than $1\frac{1}{4}$?

    A. $\frac{24}{20}$ B. $\frac{32}{30}$
    C. $\frac{123}{100}$ D. $\frac{257}{200}$

16. Which group is ordered from least to greatest?

    A. $\frac{1}{5}, \frac{2}{3}, \frac{4}{10}$ B. $\frac{1}{4}, \frac{1}{2}, \frac{3}{4}$
    C. $\frac{3}{6}, \frac{3}{7}, \frac{3}{8}$ D. $\frac{3}{4}, \frac{7}{8}, \frac{6}{7}$

17. Compare. $\frac{13}{18} \bigcirc \frac{3}{8}$

    A. < B. = C. >

18. Compare. $2\frac{7}{8} \bigcirc 2\frac{21}{24}$

    A. < B. = C. >

19. $\frac{7}{28} =$ ___?___

    A. $\frac{1}{3}$ B. $\frac{1}{4}$ C. $1\frac{3}{4}$ D. $\frac{4}{1}$

Go on to the next page.

Name ______________________________

Posttest

20. $\frac{3}{4}$ = ___?___

A. $\frac{4}{5}$ B. $\frac{5}{7}$ C. $\frac{6}{9}$ D. $\frac{6}{8}$

Use the information below to answer questions 21–24.

A parking lot charges $3.00 for the first hour and $1.00 for each additional $\frac{1}{2}$ hour.

21. How much does it cost to park for $2\frac{1}{2}$ hours?

A. $4.50 B. $6.00
C. $7.50 D. $9.00

22. For how long can you park for $4.00?

A. $1\frac{1}{2}$ hours B. $1\frac{3}{4}$ hours
C. 2 hours D. $2\frac{1}{4}$ hours

23. Alan parks for 4 hours and Alexa parks for 6 hours. How much more does Alexa pay than Alan?

A. $1.00 B. $2.00
C. $3.00 D. $4.00

24. Justin and Julie drive to work together. They split the cost of parking. If they park for $7\frac{1}{2}$ hours, how much will each pay?

A. $4.75 B. $8.00
C. $9.50 D. not here

25. Which fraction falls between $\frac{4}{12}$ and $\frac{5}{12}$?

A. $\frac{9}{24}$ B. $\frac{10}{24}$
C. $\frac{16}{48}$ D. $\frac{20}{48}$

26. Which fraction falls between $\frac{3}{10}$ and $\frac{4}{10}$?

A. $\frac{6}{20}$ B. $\frac{8}{20}$ C. $\frac{13}{40}$ D. $\frac{16}{40}$

27. Which fraction falls between $\frac{1}{8}$ and $\frac{2}{8}$?

A. $\frac{3}{16}$ B. $\frac{4}{16}$
C. $\frac{15}{32}$ D. $\frac{17}{32}$

28. Which fraction falls between $\frac{1}{4}$ and $\frac{1}{3}$?

A. $\frac{3}{12}$ B. $\frac{4}{12}$
C. $\frac{7}{24}$ D. $\frac{6}{24}$

Use the information below to answer questions 29–32.

Bob and Peg began a running program. Bob ran 1 mile on day 1 and increased the distance by $\frac{1}{10}$ mile each day. Peg ran 2 miles on day 1 and increased the distance by $\frac{1}{20}$ mile each day.

29. How far did Bob run on day 5?

A. $1\frac{3}{10}$ miles B. $1\frac{2}{5}$ miles
C. $1\frac{1}{2}$ miles D. $1\frac{3}{5}$ miles

30. How far did Peg run on day 6?

A. $2\frac{1}{10}$ mi B. $2\frac{3}{20}$ mi
C. $2\frac{1}{5}$ mi D. $2\frac{1}{4}$ mi

31. How far did Bob run on day 11?

A. $1\frac{9}{10}$ mi B. 2 mi
C. $2\frac{1}{10}$ mi D. $2\frac{1}{5}$ mi

32. How far did Peg run on day 15?

A. $2\frac{3}{5}$ mi B. $2\frac{13}{20}$ mi
C. $2\frac{7}{10}$ mi D. $2\frac{3}{4}$ mi

Stop!

Name ________________________________

Pretest

Choose the letter of the correct answer.

1. To estimate the sum of $6\frac{4}{9} + 3\frac{1}{8}$, what should $6\frac{4}{9}$ be rounded to?

   A. 5 **B.** $6\frac{1}{4}$ **C.** $6\frac{1}{2}$ **D.** 8

2. To estimate the quotient of $5\frac{7}{8} \div 2\frac{1}{8}$, what should $2\frac{1}{8}$ be rounded to?

   A. 2 **B.** $2\frac{3}{4}$ **C.** 3 **D.** 4

3. Estimate. $13\frac{5}{8} + 9\frac{2}{5} =$ ___?___

   A. 20 **B.** 21 **C.** 23 **D.** 24

4. Estimate. $10\frac{7}{8} - 1\frac{3}{5} =$ ___?___

   A. 8 **B.** $8\frac{1}{2}$ **C.** $9\frac{1}{2}$ **D.** 11

5. Estimate. $\frac{7}{6} \times \frac{4}{9} =$ ___?___

   A. $\frac{1}{8}$ **B.** $\frac{1}{2}$ **C.** 1 **D.** $1\frac{1}{2}$

6. Estimate. $12\frac{1}{9} \div 3\frac{9}{10} =$ ___?___

   A. 3 **B.** $4\frac{1}{2}$ **C.** 5 **D.** 9

7. Emilio bought two melons weighing $2\frac{7}{8}$ pounds and $3\frac{9}{10}$ pounds. Estimate the weight of the two melons combined.

   A. 4 lb **B.** 6 lb
   C. 7 lb **D.** 8 lb

8. Three farmers split $14\frac{8}{10}$ gallons of gas equally. Estimate the number of gallons each received.

   A. 3 gal **B.** 4 gal
   C. 5 gal **D.** 7 gal

9. To find the sum of $\frac{7}{8} + \frac{1}{3}$, $\frac{7}{8}$ should be renamed to which equivalent fraction?

   A. $\frac{14}{16}$ **B.** $\frac{21}{24}$
   C. $\frac{28}{32}$ **D.** $\frac{35}{40}$

10. To find the difference of $\frac{3}{4} - \frac{2}{5}$, $\frac{2}{5}$ should be renamed to which equivalent fraction?

    A. $\frac{4}{10}$ **B.** $\frac{6}{15}$
    C. $\frac{8}{20}$ **D.** $\frac{12}{30}$

11. $\frac{5}{8} + \frac{7}{8} =$ ___?___

    A. $\frac{3}{4}$ **B.** $1\frac{1}{3}$
    C. $1\frac{1}{2}$ **D.** $3\frac{9}{16}$

12. $3\frac{1}{3} + 4\frac{3}{4} =$ ___?___

    A. $6\frac{11}{12}$ **B.** $7\frac{4}{7}$
    C. $8\frac{11}{12}$ **D.** not here

13. $\frac{1}{2} - \frac{1}{6} =$ ___?___

    A. $\frac{1}{6}$ **B.** $\frac{1}{3}$
    C. $\frac{2}{3}$ **D.** not here

14. $12\frac{2}{5} - 11\frac{3}{5} =$ ___?___

    A. $\frac{1}{5}$ **B.** $\frac{4}{5}$
    C. $1\frac{1}{5}$ **D.** $1\frac{4}{5}$

Go on to the next page.

Name ______________________________

Pretest

15. Susan watched three short TV shows. The first lasted $\frac{1}{4}$ hour; the other two lasted $\frac{1}{2}$ hour each. For how long did she watch TV?

A. $\frac{3}{4}$ hr　　**B.** 1 hr

C. $1\frac{1}{8}$ hr　　**D.** $1\frac{1}{4}$ hr

16. Jose bought a bottle of juice. He drank $\frac{1}{3}$ of the bottle of juice, and his sister drank $\frac{1}{6}$ of the bottle of juice. How much of the bottle of juice was left?

A. $\frac{1}{6}$ bottle　　**B.** $\frac{1}{3}$ bottle

C. $\frac{1}{2}$ bottle　　**D.** $\frac{2}{3}$ bottle

17. Compare. $\frac{7}{8} \times \frac{3}{4} \bigcirc \frac{3}{4}$

A. <　　**B.** =　　**C.** >

18. Compare. $\frac{11}{10} \times \frac{2}{3} \bigcirc \frac{2}{3}$

A. <　　**B.** =　　**C.** >

19. What is $\frac{2}{3} \div \frac{1}{3}$ rewritten as a multiplication problem?

A. $\frac{3}{2} \times \frac{1}{3}$　　**B.** $\frac{2}{3} \times \frac{2}{3}$

C. $\frac{3}{2} \times \frac{3}{1}$　　**D.** $\frac{2}{3} \times \frac{3}{1}$

20. What is $8\frac{1}{2} \div 2\frac{3}{4}$ rewritten as a multiplication problem?

A. $\frac{17}{2} \times \frac{11}{4}$　　**B.** $\frac{17}{2} \times \frac{4}{11}$

C. $\frac{2}{17} \times \frac{11}{4}$　　**D.** $\frac{2}{17} \times \frac{4}{11}$

21. $\frac{3}{7} \times \frac{1}{9} =$ __?__

A. $\frac{1}{21}$　　**B.** $\frac{3}{16}$

C. $\frac{1}{4}$　　**D.** not here

22. $1\frac{3}{4} \times 2\frac{1}{3} =$ __?__

A. $\frac{7}{12}$　　**B.** $2\frac{1}{4}$

C. $3\frac{7}{12}$　　**D.** not here

23. $\frac{1}{3} \div \frac{1}{4} =$ __?__

A. $\frac{1}{12}$　　**B.** $\frac{3}{4}$

C. $1\frac{1}{3}$　　**D.** not here

24. $3\frac{1}{2} \div 1\frac{3}{4} =$ __?__

A. $\frac{1}{2}$　　**B.** $\frac{11}{14}$

C. 2　　**D.** $6\frac{1}{8}$

25. For $\frac{3}{4}$ of an hour, Rick practiced basketball. He spent $\frac{1}{2}$ of that time shooting foul shots. What part of an hour did he spend shooting foul shots?

A. $\frac{1}{8}$ hr　　**B.** $\frac{3}{16}$ hr

C. $\frac{1}{2}$ hr　　**D.** $\frac{3}{8}$ hr

26. Doreen bought $\frac{1}{3}$ yard of ribbon for crafts. She used it to make two bows. How much ribbon was needed for each bow?

A. $\frac{1}{12}$ yd　　**B.** $\frac{1}{6}$ yd

C. $\frac{1}{5}$ yd　　**D.** $\frac{2}{3}$ yd

Go on to the next page.

27. A carton containing 6 identical boxes weighs $12\frac{1}{2}$ pounds (lb). How many pounds does each box weigh?

A. $2\frac{1}{12}$ lb **B.** $2\frac{1}{6}$ lb

C. $2\frac{1}{3}$ lb **D.** $2\frac{1}{2}$ lb

28. Each floor tile is $\frac{10}{12}$ foot long. How many tiles are along one wall of a room that is 25 feet long?

A. 21 tiles **B.** 27 tiles

C. 30 tiles **D.** 300 tiles

29. Liza mowed lawns for $2\frac{1}{3}$ hours each day on Monday, Tuesday, and Wednesday. She raked leaves for $4\frac{1}{2}$ hours on Friday and $3\frac{2}{3}$ hours on Saturday. For how many hours did she work this week?

A. $10\frac{1}{2}$ hr **B.** $14\frac{1}{2}$ hr

C. $15\frac{1}{6}$ hr **D.** $51\frac{1}{2}$ hr

30. There are 12 teams in a baseball league. If each team plays each of the others once, how many games will be played?

A. 48 games **B.** 66 games

C. 132 games **D.** 144 games

Use the information below to answer questions 31–32.

Nancy must walk 5 blocks to school. It takes her $2\frac{1}{2}$ minutes to walk each block. Pat walks 7 blocks to school. It takes her only $1\frac{1}{2}$ minutes to walk each block.

31. How much longer does it take Nancy to walk to school than it takes Pat?

A. 1 min **B.** 2 min

C. $2\frac{1}{2}$ min **D.** 3 min

32. How long does it take Nancy to walk $2\frac{1}{2}$ blocks?

A. 5 min **B.** $5\frac{1}{2}$ min

C. 6 min **D.** $6\frac{1}{4}$ min

Stop!

Name ____________________

Posttest

Choose the letter of the correct answer.

1. To estimate the sum of $7\frac{5}{9} + 4\frac{1}{8}$, what should $7\frac{5}{9}$ be rounded to?

   A. 7 **B.** $7\frac{1}{8}$ **C.** $7\frac{1}{2}$ **D.** 9

2. To estimate the product of $5\frac{1}{6} \times 1\frac{1}{10}$, what should $1\frac{1}{10}$ be rounded to?

   A. 1 **B.** $1\frac{1}{6}$ **C.** $1\frac{1}{2}$ **D.** 2

3. Estimate. $4\frac{4}{7} + 9\frac{4}{9} =$ ___?___

   A. 11 **B.** $12\frac{1}{2}$

   C. $13\frac{1}{2}$ **D.** 14

4. Estimate. $10\frac{6}{7} - 2\frac{2}{5} =$ ___?___

   A. $7\frac{1}{2}$ **B.** $8\frac{1}{2}$ **C.** $9\frac{1}{2}$ **D.** 10

5. Estimate. $\frac{10}{9} \times \frac{4}{9} =$ ___?___

   A. $\frac{1}{2}$ **B.** $\frac{3}{4}$ **C.** $1\frac{1}{2}$ **D.** 2

6. Estimate. $15\frac{1}{8} \div 4\frac{9}{11} =$ ___?___

   A. 2 **B.** 3 **C.** 4 **D.** 5

7. Lucy bought two bags of peanuts. The first weighed $1\frac{7}{8}$ pounds, and the second weighed $1\frac{1}{16}$ pounds. Estimate the weight of the two bags combined.

   A. $1\frac{1}{2}$ lb **B.** 3 lb

   C. $3\frac{1}{2}$ lb **D.** 4 lb

8. Three friends divided $3\frac{1}{4}$ pints of yogurt equally. Estimate the amount of yogurt that each person had.

   A. $\frac{1}{2}$ pt **B.** 1 pt

   C. $1\frac{1}{2}$ pt **D.** 2 pt

9. To find the sum of $\frac{7}{8} + \frac{1}{5}$, $\frac{7}{8}$ should be renamed to which equivalent fraction?

   A. $\frac{14}{16}$ **B.** $\frac{21}{24}$

   C. $\frac{15}{40}$ **D.** $\frac{35}{40}$

10. To find the difference of $\frac{2}{3} - \frac{2}{5}$, $\frac{2}{5}$ should be renamed to which equivalent fraction?

    A. $\frac{4}{10}$ **B.** $\frac{6}{15}$

    C. $\frac{8}{20}$ **D.** $\frac{12}{30}$

11. $\frac{5}{12} + \frac{5}{6} =$ ___?___

    A. $\frac{5}{9}$ **B.** $\frac{5}{6}$ **C.** $1\frac{1}{4}$ **D.** $2\frac{1}{12}$

12. $4\frac{2}{3} + 4\frac{3}{4} =$ ___?___

    A. $8\frac{5}{12}$ **B.** $8\frac{5}{7}$

    C. $9\frac{7}{12}$ **D.** not here

13. $\frac{3}{4} - \frac{1}{2} =$ ___?___

    A. $\frac{1}{12}$ **B.** $\frac{1}{4}$

    C. 1 **D.** not here

Go on to the next page.

14. $11\frac{1}{6} - 10\frac{1}{3} =$ ___?___

A. $\frac{1}{6}$ **B.** $\frac{5}{6}$

C. $1\frac{1}{16}$ **D.** not here

15. Julie played three video games. The first lasted $\frac{1}{2}$ hour; the other two lasted $\frac{1}{4}$ hour each. For how long did she play video games?

A. $\frac{3}{4}$ hr **B.** 1 hr

C. $1\frac{1}{8}$ hr **D.** $1\frac{1}{4}$ hr

16. Beth and her two sisters each ate $\frac{1}{4}$ of a pizza. What fraction of the whole pizza was left?

A. $\frac{1}{8}$ pizza

B. $\frac{1}{6}$ pizza

C. $\frac{1}{4}$ pizza

D. $\frac{3}{8}$ pizza

17. Compare. $\frac{9}{10} \times \frac{2}{3} \bigcirc \frac{2}{3}$

A. $<$ **B.** $=$ **C.** $>$

18. Compare. $\frac{21}{20} \times \frac{4}{5} \bigcirc \frac{4}{5}$

A. $<$ **B.** $=$ **C.** $>$

19. What is $\frac{3}{5} \div \frac{1}{5}$ rewritten as a multiplication problem?

A. $\frac{5}{3} \times \frac{1}{5}$ **B.** $\frac{3}{5} \times \frac{3}{5}$

C. $\frac{5}{3} \times \frac{5}{1}$ **D.** $\frac{3}{5} \times \frac{5}{1}$

20. What is $9\frac{1}{3} \div 2\frac{2}{3}$ rewritten as a multiplication problem?

A. $\frac{28}{3} \times \frac{8}{3}$ **B.** $\frac{28}{3} \times \frac{3}{8}$

C. $\frac{3}{28} \times \frac{8}{3}$ **D.** $\frac{3}{28} \times \frac{3}{8}$

21. $\frac{3}{11} \times \frac{1}{3} =$ ___?___

A. $\frac{1}{11}$ **B.** $\frac{3}{11}$

C. $\frac{2}{7}$ **D.** not here

22. $2\frac{1}{2} \times 1\frac{1}{4} =$ ___?___

A. $\frac{5}{8}$ **B.** $2\frac{1}{8}$

C. $2\frac{7}{8}$ **D.** not here

23. $\frac{1}{2} \div \frac{1}{3} =$ ___?___

A. $\frac{1}{6}$ **B.** $\frac{2}{3}$

C. $1\frac{1}{2}$ **D.** not here

24. $6\frac{1}{2} \div 3\frac{1}{4} =$ ___?___

A. $1\frac{2}{15}$ **B.** 2

C. $2\frac{4}{5}$ **D.** $28\frac{1}{6}$

25. Paul spends $1\frac{1}{2}$ hours on his music every day. He spends $\frac{1}{3}$ of that time composing music. How much time does he spend composing daily?

A. $\frac{1}{6}$ hr **B.** $\frac{1}{3}$ hr

C. $\frac{1}{2}$ hr **D.** $1\frac{5}{6}$ hr

Go on to the next page.

Name ____________________

Posttest

26. Heidi watches a television show that lasts $\frac{1}{2}$ hour. Commercials make up $\frac{1}{6}$ of the show. What part of an hour does she spend watching commercials?

A. $\frac{1}{12}$ hr   **B.** $\frac{1}{6}$ hr

C. $\frac{1}{3}$ hr   **D.** $\frac{2}{3}$ hr

27. A carton containing 15 identical books weighs $22\frac{1}{2}$ pounds. How many pounds does each book weigh?

A. $1\frac{1}{2}$ lb   **B.** $1\frac{3}{4}$ lb

C. $1\frac{7}{8}$ lb   **D.** $2\frac{1}{2}$ lb

28. Ramon works for $3\frac{1}{4}$ hours on Saturday, $5\frac{1}{4}$ on Sunday, and $2\frac{1}{2}$ hours each weekday. For how many hours does he work in 2 weeks?

A. 23 hr   **B.** $33\frac{1}{2}$ hr

C. 42 hr   **D.** 77 hr

29. There are 12 girls in the state tennis tournament. Each girl will play each of the other girls once. How many matches will be played?

A. 36 matches   **B.** 45 matches

C. 66 matches   **D.** 78 matches

30. Jeri worked for 14 hours at $4.50 per hour. Alyce worked for $20\frac{1}{2}$ hours at $3.50 per hour. Who earned more, and by how much?

A. Alyce earned $8.75 more.

B. Alyce earned $18.75 more.

C. Jeri earned $1.25 more.

D. Jeri earned $63.00 more.

Use the information below to answer questions 31–32.

Justin can bicycle one mile in $\frac{1}{4}$ hour. Don can bicycle one mile in $\frac{1}{3}$ hour.

31. If both boys ride for two hours, how much farther will Justin have traveled?

A. $\frac{1}{12}$ mi   **B.** 1 mi

C. 2 mi   **C.** 4 mi

32. How long will it take Justin to ride $2\frac{1}{2}$ miles?

A. $\frac{3}{8}$ hr   **B.** $\frac{1}{2}$ hr

C. $\frac{5}{8}$ hr   **D.** $1\frac{1}{8}$ hr

Stop!

Name ______________________________

Cumulative Test

Choose the letter of the correct answer.

1. Choose the best estimate.
   2,196 + 1,023 = ___?___
   A. less than 2,000
   B. greater than 2,000
   C. less than 3,000
   D. greater than 3,000

2. Choose the best estimate.
   79.98 ÷ 2.05 = ___?___
   A. less than 25
   B. less than 40
   C. greater than 40
   D. greater than 45

3. 9 + 13 + 19 = ___?___
   A. 31 **B.** 41 **C.** 51 **D.** 122

4. 6.07 − 3.09 = ___?___
   A. 2.88 **B.** 2.98
   C. 3.02 **D.** 3.98

5. 19 × 17 = ___?___
   A. 152 **B.** 223
   C. 323 **D.** not here

6. 20.4 ÷ 5.1 = ___?___
   A. 0.25 **B.** 0.40
   C. 4.50 **D.** not here

7. Jason bought 7.5 pounds of cheese for $26.25. How much did he pay per pound of cheese?
   A. $3.25 **B.** $3.50
   C. $3.75 **D.** $3.95

8. Mary runs 1.5 miles every weekday, 2.5 miles each Saturday, and 3.0 miles each Sunday. How many miles does she run in two weeks?
   A. 20.5 mi **B.** 25.0 mi
   C. 25.5 mi **D.** 26.0 mi

9. What is the value of 4^3?
   A. 12 **B.** 16 **C.** 64 **D.** 81

10. What is the value of 2^3?
    A. 6 **B.** 8 **C.** 9 **D.** 18

11. Express 303,000,000 in scientific notation.
    A. 3.03×10^6 **B.** 3.30×10^6
    C. 3.03×10^8 **D.** 3.30×10^8

12. Express 2.4×10^4 in standard form.
    A. 240 **B.** 2,400
    C. 24,000 **D.** 240,000

Use the bar graph below to answer questions 13–14.

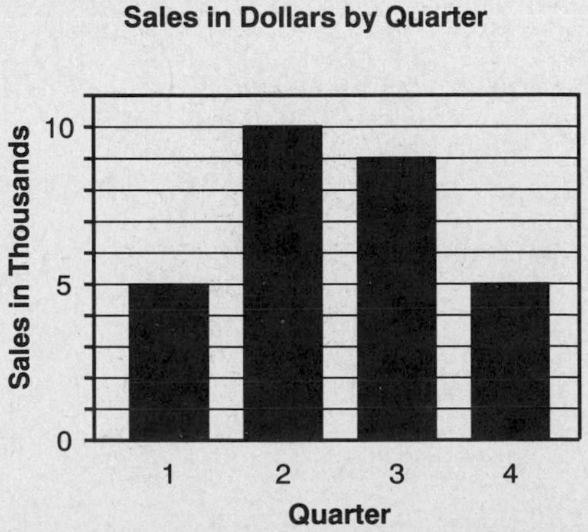

Sales in Dollars by Quarter

13. In which quarter were sales greatest?
    A. Quarter 1 **B.** Quarter 2
    C. Quarter 3 **D.** Quarter 4

Go on to the next page.

Cumulative Test

14. How much did sales drop between Quarter 3 and Quarter 4?

A. $400 B. $3,000
C. $4,000 D. $5,000

Use the circle graph to answer questions 15–16.

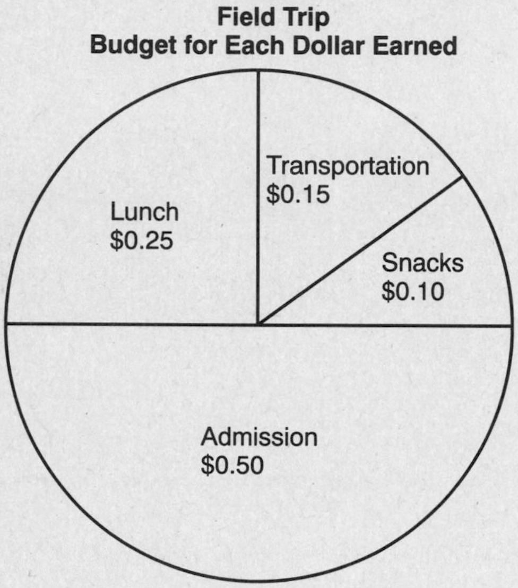

Field Trip
Budget for Each Dollar Earned

Transportation $0.15
Lunch $0.25
Snacks $0.10
Admission $0.50

15. If the students earn $350.00, how much can they spend for lunch?

A. $25.00 B. $87.50
C. $90.00 D. $117.50

16. If the students earn $375.00, how much can they spend for snacks?

A. $10.00 B. $37.50
C. $75.50 D. $100.00

Use the information below to answer questions 17–20.

The scores of 12 students on a math quiz were: 8, 10, 13, 13, 13, 15, 15, 17, 17, 21, 24, 26.

17. What is the mode for the above scores?

A. 13 B. 15 C. 17 D. not here

18. What is the range of these scores?

A. 8 B. 18 C. 26 D. 34

19. What is the mean of these scores?

A. 14.5 B. 15.0 C. 15.5 D. 16.0

20. What is the median of these scores?

A. 14 B. 15 C. 16 D. 17

21. What is the prime factorization of 48?

A. $2^4 \times 3$ B. $2^3 \times 5 + 4 \times 2$
C. $8 \times 2 \times 3$ D. not here

22. What is the prime factorization of 500?

A. $2^2 \times 25^2$ B. $2^2 \times 5^3$
C. $2^2 \times 125$ D. $5 \times 10 \times 10$

23. What is the greatest common factor of 48 and 60?

A. 4 B. 6 C. 12 D. 24

24. What is the least common multiple of 4 and 6?

A. 2 B. 8 C. 12 D. 24

25. Which fraction is expressed in simplest form?

A. $\frac{3}{21}$ B. $\frac{6}{27}$ C. $\frac{18}{32}$ D. $\frac{97}{100}$

26. Which fraction is greater than $1\frac{1}{4}$?

A. $\frac{19}{16}$ B. $\frac{30}{25}$
C. $\frac{123}{100}$ D. $\frac{255}{200}$

27. $\frac{35}{50} =$ ___?___

A. $\frac{3}{5}$ B. $\frac{7}{10}$ C. $\frac{5}{7}$ D. $\frac{20}{25}$

Go on to the next page.

Name ____________________

Cumulative Test

28. Compare. $\frac{3}{9} \bigcirc \frac{9}{27}$

A. $>$ **B.** $=$ **C.** $<$

29. Which fraction falls between $\frac{4}{20}$ and $\frac{5}{20}$?

A. $\frac{2}{9}$ **B.** $\frac{11}{40}$

C. $\frac{16}{80}$ **D.** $\frac{20}{80}$

30. Which fraction falls between $\frac{1}{3}$ and $\frac{1}{4}$?

A. $\frac{3}{12}$ **B.** $\frac{4}{12}$

C. $\frac{7}{24}$ **D.** not here

31. Which fraction falls between $\frac{1}{10}$ and $\frac{2}{10}$?

A. $\frac{3}{20}$ **B.** $\frac{4}{20}$

C. $\frac{5}{40}$ **D.** $\frac{7}{40}$

32. Which fraction falls between $\frac{1}{5}$ and $\frac{1}{6}$?

A. $\frac{5}{30}$ **B.** $\frac{6}{30}$ **C.** $\frac{11}{60}$ **D.** $\frac{12}{60}$

33. $\frac{3}{6} + \frac{4}{6} =$ ___?___

A. $\frac{6}{7}$ **B.** $1\frac{1}{6}$

C. $1\frac{1}{3}$ **D.** not here

34. $3\frac{2}{3} + 5\frac{1}{4} =$ ___?___

A. $8\frac{3}{7}$ **B.** $8\frac{11}{12}$

C. $9\frac{11}{12}$ **D.** $11\frac{6}{7}$

35. $\frac{1}{4} - \frac{1}{5} =$ ___?___

A. $\frac{1}{40}$ **B.** $\frac{1}{20}$

C. $\frac{1}{5}$ **D.** not here

36. $11\frac{1}{4} - 10\frac{3}{4} =$ ___?___

A. $\frac{3}{4}$ **B.** $1\frac{1}{4}$

C. $1\frac{1}{2}$ **D.** not here

37. Susan watched three movies. The first lasted $\frac{1}{2}$ hour; the other two lasted $\frac{3}{4}$ hour each. How long did she watch movies?

A. $1\frac{1}{4}$ hr **B.** $1\frac{3}{4}$ hr

C. 2 hr **D.** $2\frac{1}{4}$ hr

38. Tim had a bottle of juice. He drank $\frac{1}{4}$ of the bottle and his sister drank $\frac{1}{2}$ of the bottle. How much was left?

A. $\frac{1}{16}$ bottle **B.** $\frac{1}{8}$ bottle

C. $\frac{1}{4}$ bottle **D.** $\frac{3}{8}$ bottle

39. $\frac{3}{5} \times \frac{1}{7} =$ ___?___

A. $\frac{1}{3}$ **B.** $\frac{3}{35}$

C. $4\frac{1}{5}$ **D.** not here

40. $1\frac{1}{4} \times 1\frac{1}{3} =$ ___?___

A. $1\frac{5}{12}$ **B.** $1\frac{1}{2}$

C. $1\frac{7}{12}$ **D.** $1\frac{2}{3}$

Go on to the next page.

Name ______________________________________

Cumulative Test

41. $\frac{1}{2} \div \frac{1}{5} =$ ___?___

A. $\frac{1}{10}$ **B.** $\frac{2}{5}$ **C.** 2 **D.** $2\frac{1}{2}$

42. $1\frac{3}{4} \div 1\frac{1}{2} =$ ___?___

A. $\frac{6}{7}$ **B.** $1\frac{1}{6}$

C. $1\frac{1}{4}$ **D.** not here

43. The lake at Jones Pier is 15 feet deep. Every $\frac{1}{5}$ mile from the pier it becomes 1 foot deeper until it reaches its maximum depth of 30 feet. How many miles from the pier does the lake reach its maximum depth?

A. $2\frac{1}{2}$ mi **B.** 3 mi

C. 5 mi **D.** 15 mi

Use the table below to answer questions 44–45.

Senior Class Elections		
Candidate	**Girls' Votes**	**Boys' Votes**
Jack	50	80
Carol	60	84
Darla	100	51
Bill	90	65
	300	280

44. Who received the most total votes?

A. Jack **B.** Carol

C. Darla **D.** Bill

45. How many more votes did Bill receive than Carol?

A. 11 votes **B.** 19 votes

C. 30 votes **D.** 31 votes

46. A principal wants a graph to compare the monthly absences of boys and girls for the year. Which type of graph should be used?

A. double-bar graph

B. histogram

C. stem-and-leaf plot

D. circle graph

47. A family wants a graph to show how they spend their monthly income. Which type of graph should they use?

A. line graph

B. double-bar graph

C. box-and-whisker graph

D. circle graph

48. Tom gave Ben \$1.50 to buy stamps. Tom gave Ben only quarters and dimes. He gave Ben 9 coins in all. How many quarters did Tom give Ben?

A. 6 quarters **B.** 5 quarters

C. 4 quarters **D.** 3 quarters

49. There are 10 players in a tennis league. If each player plays one game with each of the other players, how many games will be played?

A. 20 games **B.** 30 games

C. 45 games **D.** 90 games

50. Nancy must walk 4 blocks to school. It takes her $2\frac{1}{2}$ minutes to walk each block. Pat walks 5 blocks to school. It takes her only $1\frac{1}{2}$ minutes to walk each block. How much longer does it take Nancy to walk to school than it takes Pat?

A. 1 minute **B.** $2\frac{1}{2}$ minutes

C. 3 minutes **D.** 10 minutes

Stop!

Name ____________________

Pretest

Choose the letter of the correct answer.

1. Choose the algebraic expression for the product of 9 and a number, *y*.

   A. $9y$ B. $\frac{9}{y}$ C. $\frac{y}{9}$ D. $y - 9$

2. Choose the algebraic expression for a number, *b*, decreased by 2.

   A. $2b$ B. $b + 2$
   C. $b - 2$ D. $2 - b$

For questions 3–6, let $a = 15$ and $b = 6$.

3. Evaluate. $4a - 6$

   A. 13 B. 36 C. 54 D. 66

4. Evaluate. $3b - 10$

   A. 3 B. 6 C. 8 D. 28

5. Evaluate. $2b - 12$

   A. 0 B. 3 C. 6 D. 12

6. Evaluate. $3a + 12$

   A. 48 B. 57 C. 105 D. not here

7. Which expression shows the number of minutes in *h* hours?

   A. $\frac{h}{60}$ B. $\frac{60}{h}$
   C. $h + 60$ D. $60h$

8. Mary and June scored a total of 25 points in a basketball game. Let m = the number of points scored by Mary. Which expression shows how many points June scored?

   A. $m + 25$ B. $m - 25$
   C. $25m$ D. $25 - m$

9. Find the value.

   $(3 + 5) \cdot (3 - 1) =$ ___?___

   A. 16 B. 17 C. 24 D. not here

10. Find the value.

    $(2 + 3) + (5 - 1)^2 =$ ___?___

    A. 9 B. 20 C. 81 D. not here

11. Find the value.

    $(2 + 1 + 3) \div (3 - 2) =$ ___?___

    A. 0 B. 2 C. 5 D. 6

12. Find the value.

    $5 \times (9 - 3) =$ ___?___

    A. 25 B. 30 C. 42 D. 56

13. Which inequality shows that a number, *b*, increased by 5 is less than 100?

    A. $5b > 100$ B. $b + 5 > 100$
    C. $5b < 100$ D. $b + 5 < 100$

14. Which inequality shows that a number, *c*, decreased by 10 is greater than 10?

    A. $10 - c > 10$ B. $c - 10 > 10$
    C. $c - 10 < 10$ D. $c + 10 > 10$

15. Solve the inequality. $x + 1 > 5$

    A. $x < 1$ B. $x < 4$
    C. $x > 4$ D. $x < 6$

16. Solve the inequality. $x + 5 < 14$

    A. $x > 9$ B. $x < 9$
    C. $x < 19$ D. $x > 19$

17. To solve the equation $y + 3 = 10$, what must be done to both sides of the equation?

    A. add 3 B. subtract 3
    C. add 10 D. subtract 10

18. To solve the equation $x - 5 = 7$, what must be done to both sides of the equation?

    A. add 5 B. subtract 5
    C. add 7 D. subtract 7

Go on to the next page.

Name ____________________

Pretest

19. To solve the equation $3x = 15$, what must be done to both sides of the equation?

A. multiply by 3 **B.** divide by 3

C. multiply by 15 **D.** divide by 15

20. To solve the equation $\frac{y}{20} = 2$, what must be done to both sides of the equation?

A. multiply by 2 **B.** divide by 2

C. multiply by 20 **D.** divide by 20

21. Solve. $p + 9 = 12$

A. $p = 3$ **B.** $p = 9$

C. $p = 21$ **D.** $p = 108$

22. Solve. $p - 10 = 0$

A. $p = 0$ **B.** $p = 10$

C. $p = 11$ **D.** $p = 100$

23. Solve. $3x = 30$

A. $x = 10$ **B.** $x = 30$

C. $x = 27$ **D.** $x = 90$

24. Solve. $\frac{x}{5} = 25$

A. $x = 5$ **B.** $x = 20$

C. $x = 30$ **D.** $x = 125$

25. After Joe gave 10 of his baseball cards to Bob, he had 92 cards left. How many cards did Joe have before giving cards to Bob?

A. 82 cards **B.** 92 cards

C. 102 cards **D.** 112 cards

26. Li earned $11.00 baby-sitting. That brought her savings to $51.50. How much money did Li have before baby-sitting?

A. $40.00 **B.** $40.50

C. $61.50 **D.** $62.50

27. Jon has 3 times as many stamps as Ali. If Jon has 15 stamps, how many does Ali have?

A. 5 stamps **B.** 18 stamps

C. 30 stamps **D.** 45 stamps

28. Ed's salary is $\frac{1}{4}$ of Bill's salary. Ed earns $20,000. How much does Bill earn?

A. $5,000 **B.** $10,000

C. $40,000 **D.** $80,000

29. Mr. Hynes withdrew money from the bank. He spent $30 on shoes and $10 on a tie. He gave $\frac{1}{2}$ of what was left to his wife. He then had $45. How much money did Mr. Hynes withdraw?

A. $85 **B.** $90 **C.** $130 **D.** $140

30. Betty joined a running club. After one month she doubled the distance she ran each day. After two months she doubled the distance again. She then was running 4 miles a day. How many miles a day did she run at the start?

A. 1 mi **B.** $1\frac{1}{2}$ mi

C. 2 mi **D.** 16 mi

31. Over the past 5 weeks, Tom lost 3 pounds per week. He now weighs 200 pounds. How much did Tom weigh before losing the weight?

A. 185 pounds **B.** 203 pounds

C. 212 pounds **D.** 215 pounds

32. Alicia collects stamps. On Monday she added 5 stamps to her collection. On Tuesday she doubled the number of stamps in her collection. She then had 90 stamps. How many stamps did she have on Sunday?

A. 35 stamps **B.** 40 stamps

C. 50 stamps **D.** 185 stamps

Stop!

Name ________________________________

Posttest

Choose the letter of the correct answer.

1. Choose the algebraic expression for the product of 5 and a number, c.

   A. $5c$ **B.** $\frac{5}{c}$ **C.** $\frac{c}{5}$ **D.** $c - 5$

2. Choose the algebraic expression for a number, x, increased by 3.

   A. $3x$ **B.** $x + 3$
   C. $x - 3$ **D.** $3 - x$

For questions 3–6, let $a = 20$ and $b = 5$.

3. Evaluate. $4b + 1$

   A. 5 **B.** 20 **C.** 21 **D.** not here

4. Evaluate. $2a - 27$

   A. 4 **B.** 13 **C.** 17 **D.** not here

5. Evaluate. $3b - 15$

   A. 0 **B.** 3 **C.** 12 **D.** 45

6. Evaluate. $4b - 20$

   A. 0 **B.** 1 **C.** 2 **D.** 4

7. Which expression shows the number of hours in d days?

   A. $\frac{d}{24}$ **B.** $\frac{24}{d}$
   C. $d + 24$ **D.** $24d$

8. Hank and Bobby scored a total of 6 goals in a hockey game. Let h = the number of goals scored by Hank. Which expression shows how many goals Bobby scored?

   A. $h + 6$ **B.** $h - 6$
   C. $6h$ **D.** $6 - h$

9. Find the value.

   $(2 + 6) \cdot (4 - 2) =$ ___?___

   A. 16 **B.** 24 **C.** 30 **D.** not here

10. Find the value.

    $(1 + 2) + (4 - 1)^2 =$ ___?___

    A. 6 **B.** 9 **C.** 36 **D.** not here

11. Find the value.

    $(3 + 2 + 7) \div (7 - 6) =$ ___?___

    A. 0 **B.** 1 **C.** 2 **D.** 12

12. Find the value.

    $10 \times (8 - 4) =$ ___?___

    A. 14 **B.** 40 **C.** 76 **D.** 104

13. Which inequality shows that a number, a, increased by 7 is less than 99?

    A. $7a > 99$ **B.** $a + 77 > 99$
    C. $7a < 99$ **D.** $a + 7 < 99$

14. Which inequality shows that a number, y, increased by 100 is greater than 100?

    A. $100y > 100$ **B.** $y + 100 > 100$
    C. $100y < 100$ **D.** $y + 100 < 100$

15. Solve the inequality. $x + 5 > 12$

    A. $x < 1$ **B.** $x > 1$
    C. $x < 5$ **D.** $x > 7$

16. Solve the inequality. $x + 4 < 9$

    A. $x > 4$ **B.** $x < 5$
    C. $x > 5$ **D.** $x > 13$

17. To solve the equation $y + 5 = 15$, what must be done to both sides of the equation?

    A. add 5 **B.** subtract 5
    C. add 15 **D.** subtract 15

18. To solve the equation $x - 3 = 6$, what must be done to both sides of the equation?

    A. add 3 **B.** subtract 3
    C. add 6 **D.** subtract 6

Go on to the next page.

Posttest

19. To solve the equation $4x = 20$, what must be done to both sides of the equation?

    A. multiply by 4 B. divide by 4
    C. multiply by 20 D. divide by 20

20. To solve the equation $\frac{y}{30} = 3$, what must be done to both sides of the equation?

    A. multiply by 3 B. divide by 3
    C. multiply by 30 D. divide by 30

21. Solve. $m + 12 = 14$

    A. $m = 2$ B. $m = 12$
    C. $m = 14$ D. $m = 26$

22. Solve. $m - 50 = 10$

    A. $m = {}^{-}40$ B. $m = 40$
    C. $m = 50$ D. $m = 60$

23. Solve. $5y = 100$

    A. $y = 2$ B. $y = 20$
    C. $y = 95$ D. $y = 105$

24. Solve. $\frac{x}{7} = 7$

    A. $x = 1$ B. $x = 7$
    C. $x = 49$ D. $x = 77$

25. Ben sells newspapers. One day he sold 110 papers. He had 45 left. How many papers did he have at the beginning of the day?

    A. 65 papers B. 75 papers
    C. 155 papers D. 165 papers

26. Ann is a waitress. One day she earned $12.50 in tips. Added to her salary, she made $35.00 that day. What was Ann's salary for the day?

    A. $22.50 B. $23.50
    C. $47.50 D. $642.50

27. In a basketball game, Nat scored 3 times as many points as Phil. Nat scored 18 points. How many points did Phil score?

    A. 6 points B. 15 points
    C. 21 points D. 54 points

28. Ed's weight is $\frac{3}{4}$ of Tony's weight. Ed weighs 120 pounds. How many pounds does Tony weigh?

    A. 60 lb B. 135 lb
    C. 160 lb D. 200 lb

29. Linda withdrew money from the bank. She gave $\frac{1}{2}$ of the money to her husband and then spent $50 on groceries. She then had $50. How much money did she withdraw?

    A. $50 B. $100 C. $200 D. $250

30. Betty joined an exercise club. In one month she doubled the number of sit ups she could do. After two months she doubled the number of sit-ups again. She then could do 60. How many sit-ups could she do at the start?

    A. 10 sit-ups B. 15 sit-ups
    C. 20 sit-ups D. 30 sit-ups

31. Over the past 6 weeks, Tom burned 3 logs per week. Tom now has 22 logs. How many logs did Tom have 6 weeks ago?

    A. 4 logs B. 13 logs
    C. 18 logs D. 40 logs

32. Natasha invested in the stock market. The first week she lost $\frac{1}{2}$ of her money. The second week she lost $\frac{1}{2}$ of the remainder. She was left with $1,200. How much money did she invest?

    A. $600 B. $1,200
    C. $3,600 D. $4,800

Stop!

Name ______________________________

Pretest

Choose the letter of the correct answer.

For questions 1–4, choose whether each statement is *always, sometimes,* or *never* true.

1. Intersecting lines are perpendicular.

A. always **B.** sometimes **C.** never

2. A right triangle is isosceles.

A. always **B.** sometimes **C.** never

3. Adjacent angles have a common vertex and a common ray.

A. always **B.** sometimes **C.** never

4. A regular hexagon has 6 congruent angles.

A. always **B.** sometimes **C.** never

5. What is the total angle measure of the interior angles of a quadrilateral?

A. 90° **B.** 180° **C.** 360° **D.** not here

6. What is the complement of a 65° angle?

A. 25° **B.** 35° **C.** 115° **D.** not here

7. Two angles of a triangle measure 40° and 60°. What is the measure of the third angle?

A. 50° **B.** 80° **C.** 90° **D.** 100°

8. One acute angle of a right triangle is twice as large as the other. What are the measures of the two acute angles?

A. 15°, 30° **B.** 45°, 45°

C. 25°, 50° **D.** 30°, 60°

9. If a triangle has angle measures of 45°, 50°, and 85°, what type of triangle is it?

A. isosceles **B.** equilateral

C. right **D.** acute

10. Classify the following triangle according to the lengths of its sides: 4 in., 5 in., 6 in.

A. isosceles **B.** equilateral

C. scalene **D.** regular

11. What is the measure of the *reflex* central angle formed by the hands of a clock at 12:40?

A. 60° **B.** 100°

C. 120° **D.** 240°

12. The measures of two sides of an isosceles triangle are 9 cm and 4 cm. What is the measure of the third side?

A. 4 cm **B.** 5 cm

C. 9 cm **D.** 13 cm

13. Figure 2 below represents what construction?

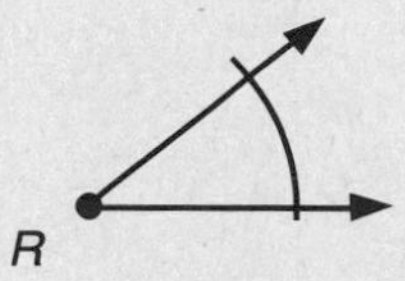

R X Y Z

Figure 1 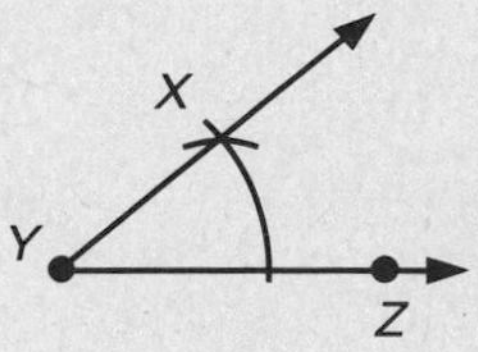 Figure 2

A. congruent line segment

B. angle bisector

C. line bisector

D. congruent angle

Go on to the next page.

Pretest

14. The figure below represents what construction?

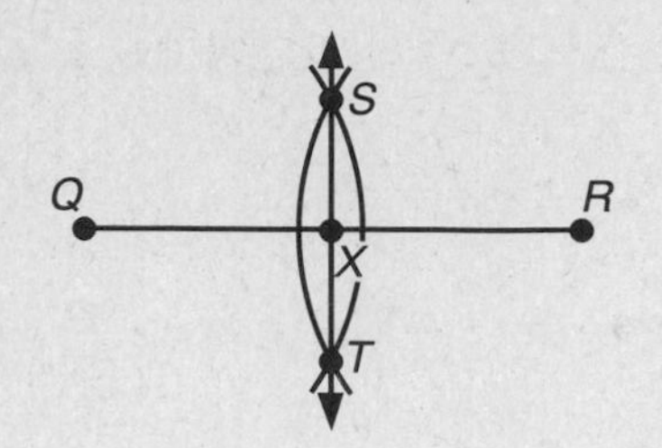

A. bisected line segment

B. bisected angle

C. congruent angle

D. congruent triangle

15. Figure 2 below represents what construction?

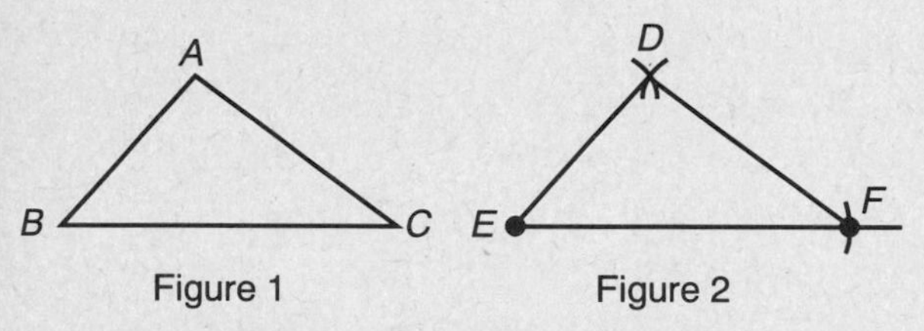

Figure 1 Figure 2

A. congruent line segment

B. congruent triangle

C. congruent angle

D. angle bisector

16. The figure below represents what construction?

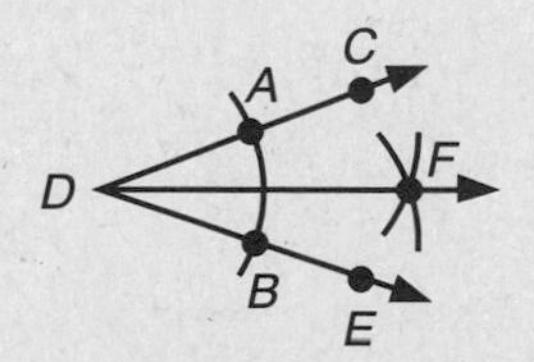

A. congruent angle

B. congruent line segment

C. bisected angle

D. bisected line segment

Use the Venn diagram below to answer questions 17–20.

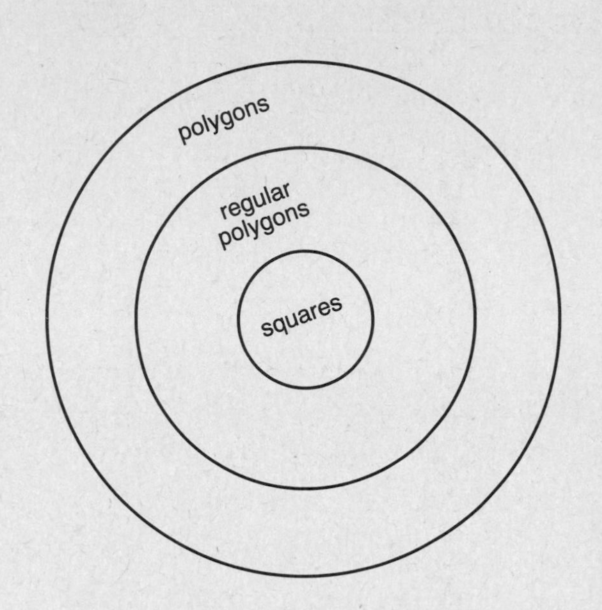

17. Does the diagram show that a polygon is always a square?

A. yes **B.** no

18. Does the diagram show that all polygons are regular polygons?

A. yes **B.** no

19. Does the diagram show that all squares are regular polygons?

A. yes **B.** no

20. Does the diagram show that some polygons are not squares?

A. yes **B.** no

Go on to the next page.

The table below shows the total angle measure of the interior angles in regular polygons. Use the table to answer questions 21–22.

Regular Polygons	
Number of Angles	**Total Angle Measure**
3	180°
4	360°
5	540°
6	720°
7	900°
8	?

21. What is the total interior angle measure of a regular 8-sided polygon?

A. 980°

B. 1,000°

C. 1,080°

D. 1,800°

22. What is the measure of each interior angle in a regular 10-sided polygon?

A. 90°

B. 126°

C. 144°

D. 180°

Use the information below to answer questions 23–24.

The taxi fare in a city is given by the formula $1.50 + 0.75*m*, where *m* = the distance traveled in miles.

23. Jan took a taxi for a distance of 4.2 miles. What was her cab fare?

A. $3.15 **B.** $3.51

C. $4.65 **D.** $31.50

24. What would be the cost of a 15.5-mile taxi ride?

A. $1.16 **B.** $11.63

C. $13.13 **D.** $31.30

Stop!

Name ______________________________

Posttest

Choose the letter of the correct answer.

For questions 1–4, choose whether each statement is *always, sometimes,* or *never* true.

1. Parallel lines intersect at one point.

A. always **B.** sometimes **C.** never

2. Two line segments with the same length are congruent.

A. always **B.** sometimes **C.** never

3. An acute triangle is equilateral.

A. always **B.** sometimes **C.** never

4. A square is a rhombus.

A. always **B.** sometimes **C.** never

5. Which quadrilateral has exactly one pair of parallel sides?

A. rhombus **B.** trapezoid

C. rectangle **D.** parallelogram

6. Angle *A* is congruent to angle *X*. If angle *A* measures 50°, what is the measure of angle *X*?

A. 40° **B.** 50° **C.** 130° **D.** 310°

7. What is the supplement of a 30° angle?

A. 60° **B.** 150°

C. 180° **D.** not here

8. Two angles of a triangle measure 40° and 50°. What is the measure of the third angle?

A. 10° **B.** 50°

C. 80° **D.** 90°

9. One acute angle of a right triangle is four times as large as the other. What are the measures of the two acute angles?

A. 10°, 40° **B.** 30°, 60°

C. 18°, 72° **D.** 20°, 80°

10. Classify the following triangle according to the measures of its sides: 4 cm, 4 cm, and 3 cm.

A. isosceles **B.** equilateral

C. right **D.** scalene

11. The measure of $\angle DEF$ is 70°. The measure of adjacent $\angle FEG$ is 10°. What is the measure of $\angle DEG$?

A. 10° **B.** 60°

C. 80° **D.** 100°

12. What is the measure of the *acute* central angle formed by the hands of a clock at 12:05?

A. 30° **B.** 60°

C. 300° **D.** 330°

13. The figure below represents what construction?

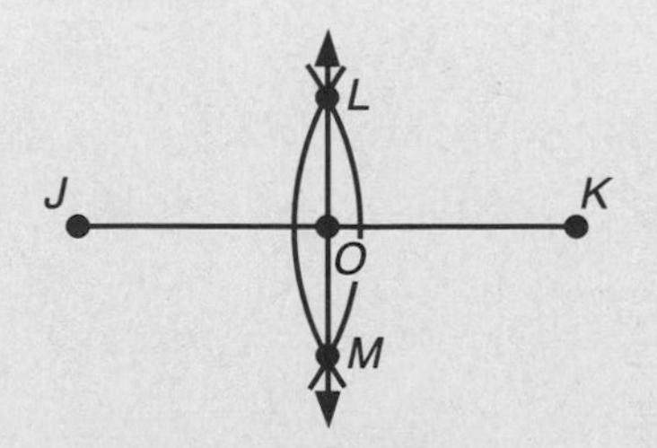

A. bisected angle

B. bisected line segment

C. congruent angle

D. congruent line segment

Go on to the next page.

Name ______________________________

Posttest

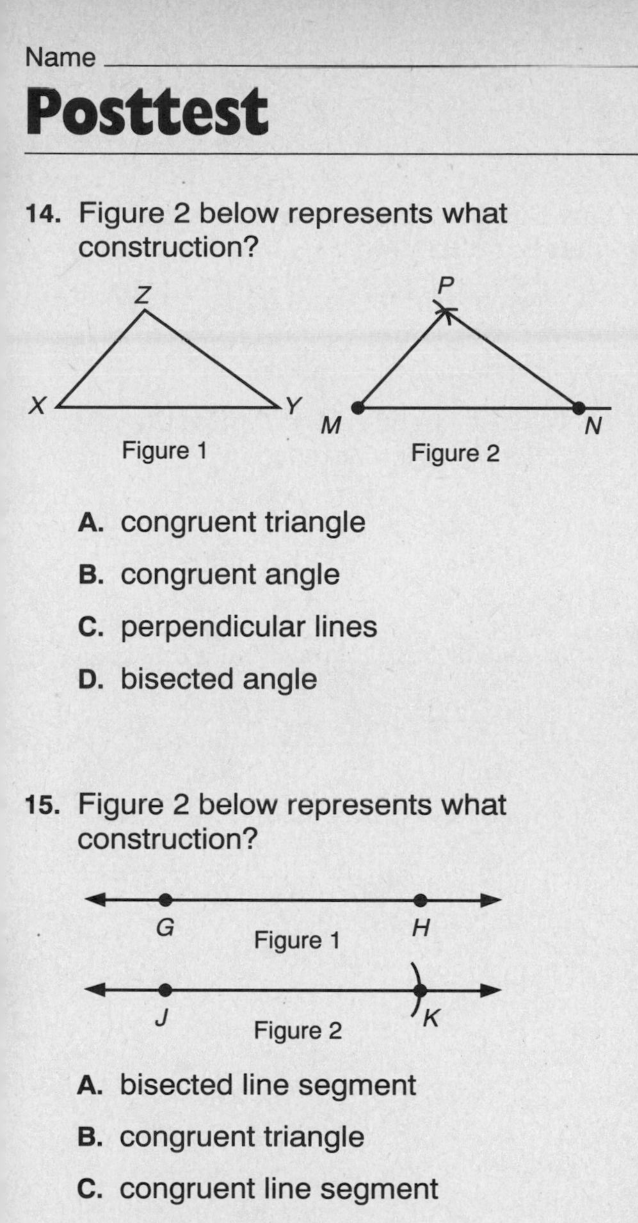

14. Figure 2 below represents what construction?

Figure 1 Figure 2

A. congruent triangle

B. congruent angle

C. perpendicular lines

D. bisected angle

15. Figure 2 below represents what construction?

Figure 1

Figure 2

A. bisected line segment

B. congruent triangle

C. congruent line segment

D. congruent angle

16. The figure below represents what construction?

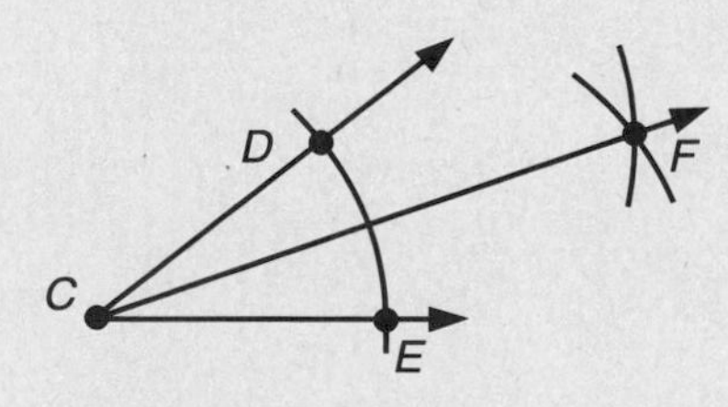

A. bisected line segment

B. bisected angle

C. congruent angle

D. congruent triangle

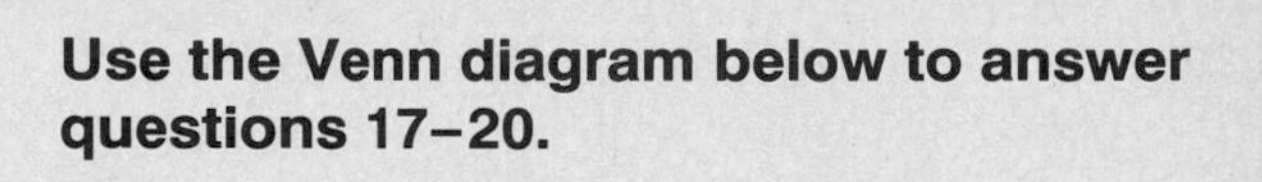

Use the Venn diagram below to answer questions 17–20.

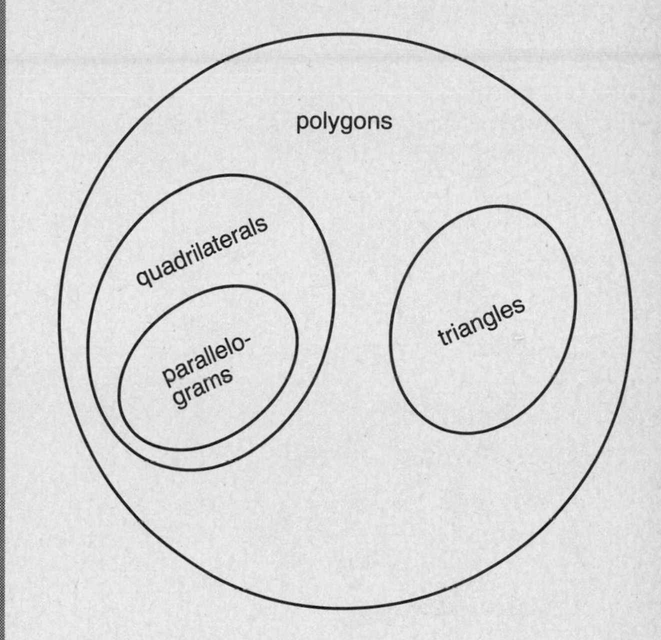

17. Does the diagram show that a parallelogram is always a quadrilateral?

A. yes **B.** no

18. Does the diagram show that all triangles are polygons?

A. yes **B.** no

19. Does the diagram show that some parallelograms are triangles?

A. yes **B.** no

20. Does the diagram show that all polygons are quadrilaterals?

A. yes **B.** no

Go on to the next page.

Name ____________________

Posttest

The table below shows the number of diagonals in regular polygons. Use it to answer questions 21–22.

Regular Polygons	
Number of Sides	**Number of Diagonals**
3	0
4	2
5	5
6	9
7	14
8	?

21. How many diagonals are in a regular polygon with 8 sides?

A. 19 diagonals

B. 20 diagonals

C. 23 diagonals

D. 28 diagonals

22. How many diagonals are in a regular 10-sided polygon?

A. 20 diagonals

B. 24 diagonals

C. 32 diagonals

D. 35 diagonals

Use this information to answer questions 23–24.

The formula for the total angle measure of a regular polygon is $(n - 2) \times 180°$, where $n =$ the number of angles.

23. What is the total angle measure of a 15-sided regular polygon?

A. 1,800° **B.** 2,340°

C. 2,700° **D.** 3,060°

24. What is the measure of each interior angle of a regular 10-sided polygon?

A. 144° **B.** 180°

C. 360° **D.** 1,800°

Stop!

Name ____________________

Pretest

Choose the letter of the correct answer.

1. A box contains 9 red marbles and 3 blue marbles. Which of the following expresses the ratio of red to blue marbles?

   A. 9 to 3 **B.** 9 to 12
   C. 9 to 1 **D.** 1 to 3

2. Which ratio does *not* have the same value as the other three?

   A. 1:5 **B.** 2:10 **C.** 4:20 **D.** 5:30

3. Which of the following is *not* a true proportion?

   A. $\frac{2}{3} = \frac{4}{6}$ **B.** $\frac{4}{5} = \frac{80}{100}$
   C. $\frac{5}{6} = \frac{155}{186}$ **D.** $\frac{7}{8} = \frac{75}{80}$

4. Which value of *a* makes the following a true proportion?

   $$\frac{a}{10} = \frac{4}{20}$$

   A. $a = 2$ **B.** $a = 4$
   C. $a = 8$ **D.** $a = 20$

5. Solve. $\frac{x}{4} = \frac{3}{6}$

   A. $x = 1$ **B.** $x = 2$
   C. $x = 3$ **D.** $x = 8$

6. Solve. $\frac{10}{y} = \frac{25}{7.5}$

   A. $y = 2.0$ **B.** $y = 2.5$
   C. $y = 3.0$ **D.** $y = 5.0$

7. Solve. $\frac{3a}{9} = \frac{8}{6}$

   A. $a = 4$ **B.** $a = 4\frac{1}{2}$
   C. $a = 5$ **D.** not here

8. Solve for *a*. $\frac{5}{a} = \frac{25}{55}$

   A. $a = 5$ **B.** $a = 11$
   C. $a = 16$ **D.** $a = 55$

9. Betty drives 10 miles in 15 minutes. At this rate how far will she drive in 75 minutes?

   A. 30 mi **B.** 40 mi
   C. 50 mi **D.** 112 mi

10. Jose's car runs 26 miles on 1 gallon of gas. At this rate how many gallons of gas will he use to travel 117 miles?

    A. $3\frac{1}{2}$ gal **B.** 4 gal
    C. $4\frac{1}{2}$ gal **D.** 5 gal

11. Mary is paid $7 for every 2 hours of work. How much does she earn for working 14 hours?

    A. $4 **B.** $23 **C.** $49 **D.** $98

12. Sean drives $\frac{9}{10}$ mile in one minute. At this rate how far will he drive in 1 hour?

    A. 45 mi **B.** 54 mi
    C. 60 mi **D.** 90 mi

13. A package of 8 batteries costs $7.60. What is the unit price?

    A. $0.90 **B.** $0.95
    C. $1.10 **D.** $9.50

14. If 8 batteries cost $7.60, how much would 20 batteries cost?

    A. $19.00 **B.** $22.80
    C. $26.60 **D.** $152.00

Go on to the next page.

15. A box of soap powder weighing 4 pounds (lb) costs \$3.96; a 6-lb box costs \$6.30. Which package has the lower unit price, and by how much per pound?

A. 4-lb, by \$0.06

B. 6-lb, by \$1.05

C. 4-lb, by \$0.58

D. 6-lb, by \$0.06

16. If 4 rolls cost \$2.16, how many rolls can be bought for \$4.32?

A. 6 rolls **B.** 8 rolls

C. 12 rolls **D.** 16 rolls

17. On a scale drawing, 1 inch is used to represent 4 feet. Which ratio expresses that relationship?

A. 1 : 8 **B.** 1 : 12 **C.** 1 : 48 **D.** 4 : 48

18. On Map A, 1 inch = 10 miles. On Map B, 1 inch = 20 miles. A road is 4 inches long on Map A. How long will the road be on Map B?

A. 2 in. **B.** 4 in.

C. 6 in. **D.** 8 in.

19. Use the scale 6 cm = 25 cm. What is the actual length of an object when the scale length is 24 cm?

A. 50 cm **B.** 75 cm

C. 100 cm **D.** 125 cm

20. Use the scale 1 m = 50 km. What is the actual length of an object when the scale length is $\frac{1}{4}$ m?

A. 2.5 km **B.** 12.5 km

C. 125 km **D.** 200 km

21. Lupe is making a scale drawing on which each centimeter represents 8 meters. One scale building measures 12 cm by 15 cm. What are the actual dimensions of the building?

A. 12 m $\times$ 15 m

B. 20 m $\times$ 23 m

C. 96 m $\times$ 120 m

D. 12 m $\times$ 120 m

22. On a map 1 cm = 25 km. Two cities are 12 cm apart on the map. How far are they actually apart?

A. 120 km **B.** 250 km

C. 300 km **D.** 370 km

For questions 23–24, choose whether the statement is *always, sometimes,* or *never* true.

23. Congruent triangles are similar.

A. always **B.** sometimes **C.** never

24. Similar figures have the same shape.

A. always **B.** sometimes **C.** never

25. Two rectangles are similar. The first rectangle is 6 in. wide and 10 in. long. The width of the second rectangle is 30 in. Find its length.

A. 18 in. **B.** 30 in.

C. 34 in. **D.** 50 in.

26. Two rectangles are similar. The first rectangle has a width of 2 cm and a length of 5 cm. The width of the second rectangle is 0.2 cm. Find the length.

A. 0.5 cm **B.** 3.8 cm

C. 10 cm **D.** 50 cm

Go on to the next page.

Pretest

27. A triangle has sides measuring 3 ft, 4 ft, 5 ft. Which of the following could be the lengths of the sides of a similar triangle?

A. 4 yd, 6 yd, 8 yd

B. 4 ft, 5 ft, 6 ft

C. 31 ft, 41 ft, 51 ft

D. 6 ft, 8 ft, 10 ft

28. These two rectangles are similar. Find x.

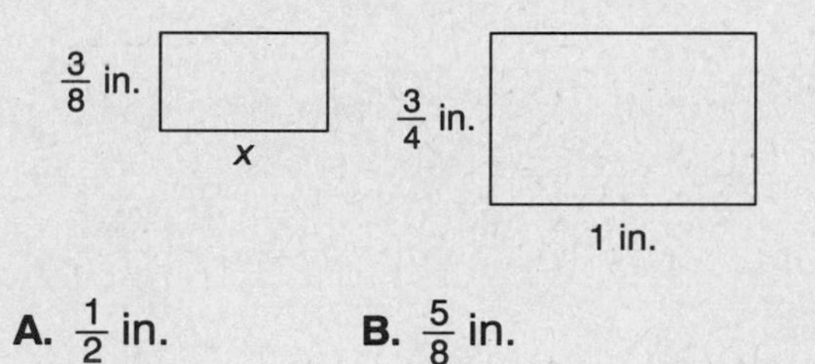

A. $\frac{1}{2}$ in. **B.** $\frac{5}{8}$ in.

C. $\frac{7}{8}$ in. **D.** $1\frac{1}{8}$ in.

29. A map scale is 2 cm = 7 km. Mt. Snow is 14 cm from Rayville on the map. How far is that?

A. 20 km

B. 30 km

C. 49 km

D. 100 km

30. The scale on a map is 1 in. = 200 mi. Two cities are $2\frac{1}{2}$ in. apart. If Toni averages 50 mi per hr, how long will it take her to drive from one city to the other?

A. $8\frac{1}{2}$ hr **B.** 9 hr

C. $9\frac{1}{2}$ hr **D.** 10 hr

Use the information below to answer questions 31–32.

The formula for finding the circumference of a circle is $C = 2\pi r$, where C = the circumference and r = the radius.

31. Use 3.14 as an approximate value of π. Find the circumference of a circle with a radius of 6 in.

A. 18.84 in. **B.** 37.68 in.

C. 113.04 in. **D.** 138.84 in.

32. Use $\frac{22}{7}$ as an approximate value of π. Find the radius of a circle with a circumference of 11 cm.

A. $1\frac{3}{4}$ cm

B. $3\frac{1}{2}$ cm

C. $17\frac{1}{2}$ cm

D. $69\frac{1}{10}$ cm

Stop!

Name ______________________________

Posttest

Choose the letter of the correct answer.

1. At the zoo there are 3 elephants and 8 tigers. Which of the following expresses the ratio of elephants to tigers?

   A. 3 : 11 **B.** 1 : 5 **C.** 8 : 3 **D.** 3 : 8

2. Which ratio does *not* have the same value as the other three?

   A. $\frac{2}{9}$ **B.** $\frac{3}{18}$ **C.** $\frac{6}{27}$ **D.** $\frac{10}{45}$

3. Which of the following is *not* a true proportion?

   A. $\frac{3}{4} = \frac{6}{8}$ **B.** $\frac{1}{5} = \frac{20}{100}$
   C. $\frac{7}{21} = \frac{1}{3}$ **D.** $\frac{5}{7} = \frac{55}{70}$

4. Which value of *a* makes the following a true proportion?

   $$\frac{a}{12} = \frac{3}{4}$$

   A. $a = 3$ **B.** $a = 4$
   C. $a = 6$ **D.** $a = 9$

5. Solve. $\frac{x}{8} = \frac{5}{10}$

   A. $x = 2$ **B.** $x = 4$
   C. $x = 16$ **D.** $x = 40$

6. Solve. $\frac{20}{y} = \frac{25}{2.5}$

   A. $y = 1.0$ **B.** $y = 1.5$
   C. $y = 2.0$ **D.** $y = 2.5$

7. Solve. $\frac{2a}{3} = \frac{4}{6}$

   A. $a = 1$ **B.** $a = 1\frac{1}{2}$
   C. $a = 2$ **D.** $a = 2\frac{1}{2}$

8. Solve for *a*. $\frac{4}{a} = \frac{1}{12}$

   A. $a = 3$ **B.** $a = 16$
   C. $a = 24$ **D.** $a = 48$

9. Martha drives 40 miles in 30 minutes. At this rate how far will she drive in 90 minutes?

   A. 68 mi **B.** 80 mi
   C. 100 mi **D.** 120 mi

10. Jim's car runs 32 miles on 1 gallon of gas. At this rate how many gallons of gas will he use to travel 112 miles?

    A. $2\frac{1}{2}$ gal **B.** 3 gal
    C. $3\frac{1}{2}$ gal **D.** 4 gal

11. Mario is paid $5 for every 2 hours of work. How much does he earn for working 10 hours?

    A. $4 **B.** $20 **C.** $25 **D.** $50

12. Jon drives $\frac{6}{10}$ mile in one minute. At this rate how far will he drive in 1 hour?

    A. 30 mi **B.** 36 mi
    C. 60 mi **D.** 100 mi

13. A package of 4 light bulbs costs $3.96. What is the unit price?

    A. $0.93 **B.** $0.99
    C. $1.02 **D.** $1.98

14. If 4 light bulbs cost $3.96, how much would 20 light bulbs cost?

    A. $19.80 **B.** $23.76
    C. $39.60 **D.** $316.80

Go on to the next page.

Posttest

15. A package of crackers weighing 10 ounces (oz) costs \$1.90; a 16-oz package costs \$2.40. Which package has the lower unit price, and by how much per ounce?

   A. 10-oz, by \$0.04
   B. 16-oz, by \$0.15
   C. 10-oz, by \$0.19
   D. 16-oz, by \$0.04

16. If 6 oranges cost \$1.98, how many oranges can be bought for \$5.94?

   A. 12 oranges B. 14 oranges
   C. 16 oranges D. 18 oranges

17. On a scale drawing, 1 inch is used to represent 2 feet. Which ratio expresses that relationship?

   A. 1 : 4 B. 1 : 12 C. 1 : 24 D. 2 : 24

18. On Map A, 1 inch = 5 miles. On Map B, 1 inch = 10 miles. A road is 3 inches long on Map A. How long will the road be on Map B?

   A. $1\frac{1}{2}$ in. B. 3 in.
   C. $4\frac{1}{2}$ in. D. 6 in.

19. Use the scale 3 cm = 8 m. What is the actual length of an object when the scale length is 27 cm?

   A. 9 m B. 32 m C. 72 m D. 216 m

20. Use the scale 1 cm = 25 cm. What is the actual length of an object when the scale length is 0.5 cm?

   A. 5 cm B. 10 cm
   C. 12.5 cm D. 37.5 cm

21. John is making a scale drawing on which each centimeter represents 5 meters. A field drawn to scale measures 15 cm by 20 cm. What are the actual dimensions of the field?

   A. 3 m × 4 m
   B. 20 m × 25 m
   C. 75 m × 100 m
   D. 20 m × 100 m

22. On a map 1 inch = 10 miles. Two cities are 10 inches apart on the map. How many miles are they actually apart?

   A. 10 mi B. 100 mi
   C. 200 mi D. 1,000 mi

For questions 23–24, choose whether the statement is *always, sometimes,* or *never* true.

23. Similar triangles are congruent.

   A. always B. sometimes C. never

24. Similar figures have different shapes.

   A. always B. sometimes C. never

25. Two rectangles are similar. The first rectangle has a width of 8 m and a length of 12 m. The width of the second rectangle is 24 m. Find its length.

   A. 20 m B. 24 m C. 28 m D. 36 m

26. Two rectangles are similar. The first rectangle has a width of 3 cm and a length of 6 cm. The width of the second rectangle is 0.5 cm. Find its length.

   A. 0.25 cm B. 1 cm
   C. 1.5 cm D. 3.5 cm

Go on to the next page.

Name ______________________

Posttest

27. A triangle has sides measuring 6 cm, 8 cm, and 10 cm. Which of the following could be the lengths of the sides of a similar triangle?

A. 6 km, 12 km, 15 km

B. 7 cm, 9 cm, 11 cm

C. 9 cm, 12 cm, 20 cm

D. 9 m, 12 m, 15 m

28. These two triangles are similar. Find *a*.

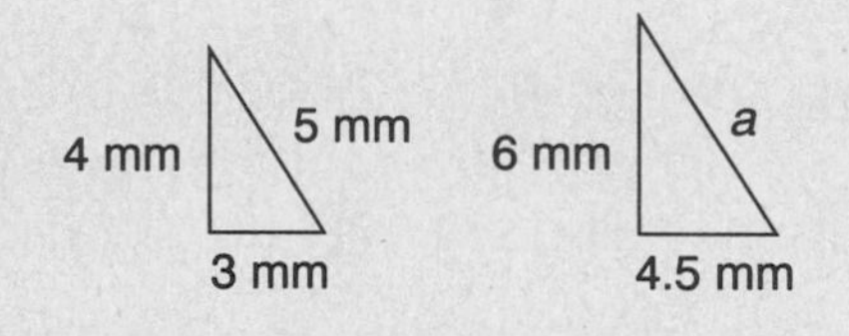

A. 5.5 mm **B.** 6.5 mm

C. 7.5 mm **D.** not here

29. A map scale is 2 cm = 9 km. The map distance from Redville to Leetown is 18 cm. The map distance from Leetown to Spring is 3 cm. How far is it from Redville to Spring through Leetown?

A. 21 km **B.** 42 km

C. 94.5 km **D.** 189 km

30. The scale on a map is 1 in. = 150 mi. Two cities are 3 in. apart. If Tom averages 50 mi per hr and makes no stops, how long will it take him to drive from one city to the other?

A. 3 hr **B.** $7\frac{1}{2}$ hr

C. $8\frac{1}{2}$ hr **D.** 9 hr

Use the information below to answer questions 31–32.

The formula for finding the circumference of a circle is $C = \pi \cdot d$, where C = the circumference and d = the diameter.

31. Use $\frac{22}{7}$ as an approximate value of π. Find the circumference of a circle with a diameter of 28 in.

A. 9 in. **B.** $7\frac{1}{7}$ in.

C. 88 in. **D.** $\frac{6}{16}$ in.

32. Use 3.14 as an approximate value of π. Find the diameter of a tree with a circumference of 8.32 m.

A. 0.38 m **B.** 2.65 m

C. 5.18 m **D.** 26.12 m

Stop!

Name ____________________

Pretest

Choose the letter of the correct answer.

1. Which of the following is not equivalent to the others?

   A. 18% **B.** 1.8 **C.** $\frac{9}{5}$ **D.** 180%

2. Which fraction can be most easily expressed as a percent?

   A. $\frac{7}{22}$ **B.** $\frac{3}{7}$ **C.** $\frac{2}{10}$ **D.** $\frac{1}{16}$

3. What is $\frac{90}{1,000}$ as a percent?

   A. 0.9% **B.** 9% **C.** 90% **D.** 99%

4. What is 0.7 as a percent?

   A. 0.7% **B.** 7% **C.** 70% **D.** 700%

5. What is 96% as a fraction in simplest form?

   A. $\frac{3}{20}$ **B.** $\frac{24}{25}$

   C. $\frac{49}{50}$ **D.** not here

6. What is $\frac{4}{10,000}$ as a decimal?

   A. 0.04 **B.** 0.004

   C. 0.0004 **D.** not here

7. A bookstore's sales increased by 25% over last year's sales. By what fraction did the store's sales increase?

   A. $\frac{1}{5}$ **B.** $\frac{1}{4}$ **C.** $\frac{2}{5}$ **D.** $\frac{3}{4}$

8. In 1992 attendance at a movie theater increased by $\frac{3}{20}$ in comparison to the attendance in 1991. By what percent did attendance increase?

   A. 3% **B.** 15% **C.** 30% **D.** 32%

9. Estimate. 19% of 397 = ___?___

   A. 50 **B.** 80 **C.** 90 **D.** 100

10. Which percent is the best estimate of 6 out of 17?

    A. 16% **B.** 20% **C.** 25% **D.** 33%

11. In a class of 30 students, 8 were absent. Which equation shows the percent who were absent?

    A. $n\% \times 30 = 8$ **B.** $n\% \times 8 = 30$

    C. $n\% = 8 \times 30$ **D.** $n\% \times 30 \div 8$

12. John saves $1.50 per week. That is equal to 20% of his allowance. Which is the correct equation to find John's allowance?

    A. $n = 20\% \times 1.50$

    B. $n = 20\% \div 1.50$

    C. $n \times 20\% = 1.50$

    D. $n \times 1.50 = 20\%$

13. 160% of 200 = ___?___

    A. 32 **B.** 120 **C.** 160 **D.** 320

14. $\frac{1}{2}$% of 400 = ___?___

    A. 2 **B.** 4 **C.** 20 **D.** 200

15. 25 is what percent of 125?

    A. 20% **B.** 25% **C.** 30% **D.** 50%

16. 20% of what number is 6?

    A. 12 **B.** 30 **C.** 40 **D.** 60

Go on to the next page.

Name ______________________________

Pretest

17. The Jacksons make a down payment of $15,000 on a house. The down payment is 20% of the total price. What is the price of the house?

A. $30,000 **B.** $60,000
C. $70,000 **D.** $75,000

18. Sarah bought a used car for $3,500 plus sales tax. The tax was $210. What was the percent of sales tax?

A. 3% **B.** 6% **C.** 9% **D.** 21%

19. Mara's Hardware Store usually sells a toolbox for $19.00. Today the store has a sale in which all prices are reduced by 20%. What is the sale price of the toolbox?

A. $15.20 **B.** $15.80
C. $16.20 **D.** $18.80

20. Justin's meal at the restaurant costs $15.00. He wants to leave a 15% tip. How much should he leave for the tip?

A. $1.50 **B.** $1.65
C. $1.75 **D.** $2.25

21. What is the sum of the degrees in the sections of a circle graph?

A. 180° **B.** 270°
C. 300% **D.** 360°

22. Rob made a circle graph to show how he spends his leisure time. A central angle of 72° represents what percent of his time?

A. 10% **B.** 20% **C.** 40% **D.** 72%

Use the circle graph below to answer questions 23–24.

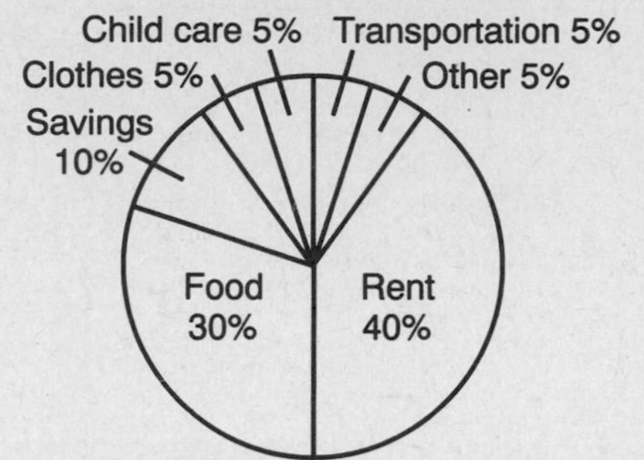

23. The McDonald family income is $1,500 per month. How much do they spend per month on food?

A. $50 **B.** $100 **C.** $450 **D.** $1,000

24. How much do the McDonalds spend on clothes and child care combined?

A. $25 **B.** $150 **C.** $175 **D.** $225

25. Sales tax is 6%. What is the tax on an item that costs $32?

A. $0.65 **B.** $1.92
C. $2.80 **D.** not here

26. An item that costs $80 is reduced by 25%. What is the sale price?

A. $20.00 **B.** $28.00
C. $54.00 **D.** $60.00

27. Juan buys a sweater that was discounted 25%. He pays $15.00. What was the original price of the sweater?

A. $18.00 **B.** $18.75
C. $20.00 **D.** $35.00

Go on to the next page.

Name ____________________

Pretest

28. Alexa borrows $800 for 6 months. The interest rate is 11% per year. How much interest must Alexa pay?

A. $40 **B.** $44
C. $82 **D.** $88

29. Two girls' ages add to 20. One girl is three times as old as the other. How old is the older girl?

A. 12 **B.** 14
C. 15 **D.** 18

30. Two numbers add to 52. One of the numbers is 10 more than the other. What are the numbers?

A. 16, 26 **B.** 22, 30
C. 21, 31 **D.** 26, 36

Use the circle graph below to answer questions 31–32.

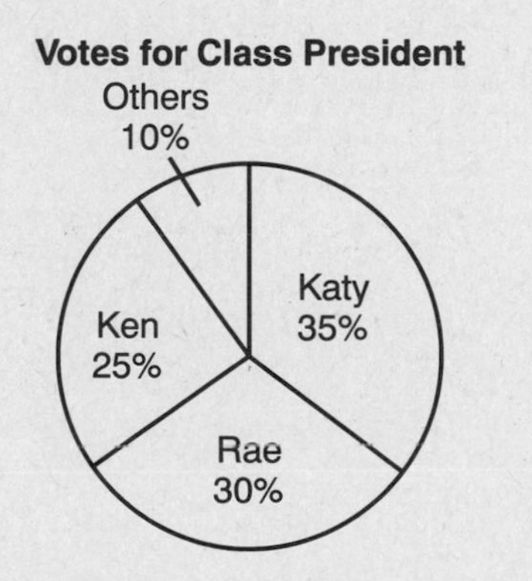

31. If 440 students voted, how many votes did Katy get?

A. 10 votes **B.** 22 votes
C. 44 votes **D.** 154 votes

32. Only two students other than Katy, Rae, and Ken received any of the 440 votes. If Sy received 18 of those votes, how many did Julia receive?

A. 5 votes **B.** 18 votes
C. 26 votes **D.** 44 votes

Stop!

Name ______________________________

Posttest

Choose the letter of the correct answer.

1. Which of the following is not equivalent to the others?

   A. 0.9% B. $\frac{9}{1,000}$

   C. 0.009 D. 0.09%

2. Which can be most easily expressed as a percent?

   A. $\frac{7}{26}$ B. $\frac{5}{7}$ C. $\frac{5}{10}$ D. $\frac{1}{12}$

3. What is 0.17 as a percent?

   A. 0.17% B. 1.7%

   C. 17% D. 170%

4. What is 0.15 as a percent?

   A. 0.15% B. 1.5%

   C. 15% D. 150%

5. What is 85% as a fraction in simplest form?

   A. $\frac{17}{20}$ B. $\frac{85}{100}$

   C. $\frac{19}{20}$ D. not here

6. What is 150% as a decimal?

   A. 1.05 B. 1.50

   C. 15.0 D. not here

7. A company's January sales dropped by $\frac{1}{5}$ compared to December sales. January sales were what percent of December sales?

   A. 20% B. 75% C. 80% D. 120%

8. In 1992 a magazine's subscribers doubled in number. What was the percent increase in subscribers?

   A. 20% B. 50% C. 100% D. 200%

9. Estimate. 30% of 198 = ___?___

   A. 35 B. 60 C. 90 D. 700

10. Which percent is the best estimate of 26 out of 36?

    A. 50% B. 60% C. 75% D. 90%

11. Danielle had $18.50. She spent $15.00 on clothes. Which is the correct equation to find what percent of her money Danielle spent?

    A. $n\% = \frac{15}{18.5}$

    B. $n\% = \frac{18.5}{15}$

    C. $n\% \times 15 = 18.5$

    D. $n\% = 15 \times 18.5$

12. Phil invited 12 friends to a party. Only 50% of them could come. Which equation shows how many came to the party?

    A. $n = \frac{12}{50}$

    B. $n \times 0.5 = 12$

    C. $n = \frac{0.5}{12}$

    D. $n = 0.5 \times 12$

13. 125% of 300 = ___?___

    A. 75 B. 325 C. 350 D. 375

14. 30% of 4 = ___?___

    A. 0.12 B. 1.2 C. 1.22 D. 12.0

15. 2.5 is what percent of 50?

    A. 0.5% B. 2% C. 5% D. 50%

16. 15% of what number is 60?

    A. 9 B. 90 C. 360 D. 400

Go on to the next page.

Name ______________________________

Posttest

17. Mr. Mosconi receives an 8% raise. The amount of the raise is $2,800. What was his salary before the raise?

    A. $17,500 **B.** $22,400
    C. $30,000 **D.** $35,000

18. Sandy bought a ring on sale for $560. Its original price was $700. What was the percent of discount?

    A. 14% **B.** 20%
    C. 25% **D.** 28%

19. Anna's Dress Shop is having a sale with all prices reduced by 20%. What is the sale price of a sweater that originally cost $35?

    A. $7.00 **B.** $15.00
    C. $28.00 **D.** $31.50

20. George buys a concert ticket through the mail. He pays $25.00 plus a 5% service charge. How much does George pay for the ticket?

    A. $26.25 **B.** $26.75
    C. $27.50 **D.** $30.00

21. A family makes a circle graph to show its budget. They draw a central angle of 54°. What percent of the budget is represented by the angle?

    A. 15% **B.** 20%
    C. 25% **D.** not here

22. How many degrees are there in 75% of a circle graph?

    A. 240° **B.** 270°
    C. 330° **D.** 360°

Use the circle graph below to answer questions 23–24.

Ethnic Breakdown at Lincoln High

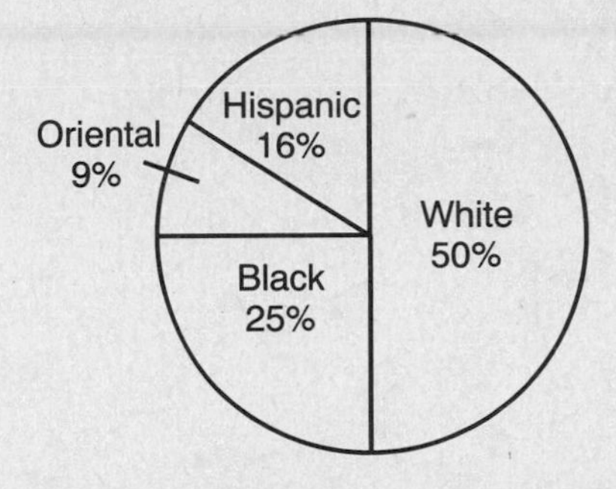

23. There are 1,050 black students at Lincoln High. What is the total number of white students at the school?

    A. 2,100 students **B.** 2,500 students
    C. 3,150 students **D.** 4,200 students

24. One half of the Hispanic students at Lincoln High take a biology class. If there are 1,200 students in the school, how many Hispanic students take a biology class?

    A. 32 students **B.** 96 students
    C. 160 students **D.** 192 students

25. The sales tax is 5%. What is the tax on an item that costs $24?

    A. $0.55 **B.** $1.20
    C. $1.42 **D.** $2.40

26. An item that costs $90 is discounted by 60%. What is the sale price?

    A. $30.00 **B.** $33.50
    C. $36.00 **D.** $54.00

27. Betty buys a bicycle that was discounted 20%. She pays $100. What was the original price of the bicycle?

    A. $105 **B.** $110 **C.** $120 **D.** $125

Go on to the next page.

Posttest

28. Alex borrows $500 for 24 months. The interest rate is 9% per year. How much interest must Alex pay?

A. $80.00 **B.** $85.00

C. $90.00 **D.** $900.00

29. The total number of points scored in a football game was 30. The winning team scored 4 times as many points as the losing team. How many points did the winning team score?

A. 18 points

B. 20 points

C. 24 points

D. 25 points

30. Henry and Anne went fishing. For every 2 fish that Henry caught, Anne caught 3. Together, they caught 15 fish. How many fish did Anne catch?

A. 6 fish **B.** 8 fish

C. 9 fish **D.** 10 fish

Use the circle graph below to answer questions 31–32.

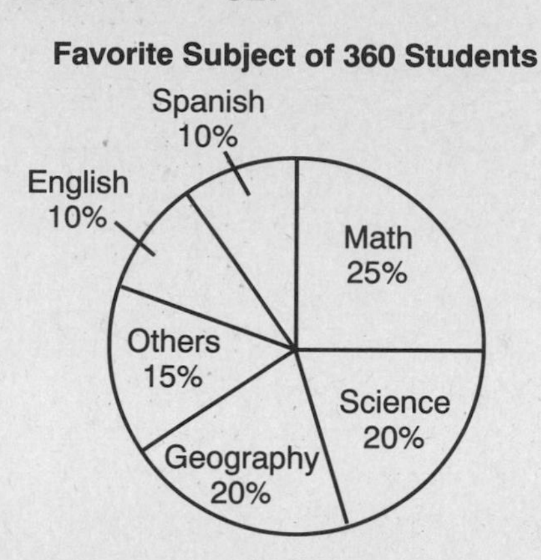

31. How many students chose science as their favorite subject?

A. 10 students **B.** 36 students

C. 54 students **D.** 72 students

32. If 20 students chose health as their favorite subject, what is the largest number of students who could have chosen French?

A. 5 students **B.** 34 students

C. 35 students **D.** 54 students

Stop!

Name ____________________

Cumulative Test

Choose the letter of the correct answer.

1. Juan bought 2 books for \$1.95 each, 2 pads at \$1.05 each, and a pen for \$1.95. *Estimate* how many dollars Juan spent.

   A. \$5.00 **B.** \$6.00
   C. \$7.00 **D.** \$8.00

2. Donna had \$31.00. She spent \$7.50 for dinner, \$4.00 for a movie, and \$1.50 for a drink at the movie. How much money did Donna have left?

   A. \$13.00 **B.** \$17.50
   C. \$18.00 **D.** \$19.00

3. What is the value of 5^1?

   A. 0 **B.** 1 **C.** 5 **D.** 50

4. Express 7,302 in scientific notation.

   A. 7.302×10^2 **B.** 7.32×10^2
   C. 7.302×10^3 **D.** 7.32×10^3

Use the stem-and-leaf plot below to answer questions 5–6.

Number of Points Scored by a Football Team in 20 Games

Stem	Leaves
1	0 4 4 4 7 7 9
2	0 0 1 1 1 1 4 4 7
3	1 5
4	2 4

5. What was the least number of points scored by the team in any game?

   A. 0 points **B.** 4 points
   C. 10 points **D.** 14 points

6. In what percentage of its games did the team score 25 or more points?

   A. 5% **B.** 25% **C.** 30% **D.** 40%

Use the double-bar graph to answer questions 7–8.

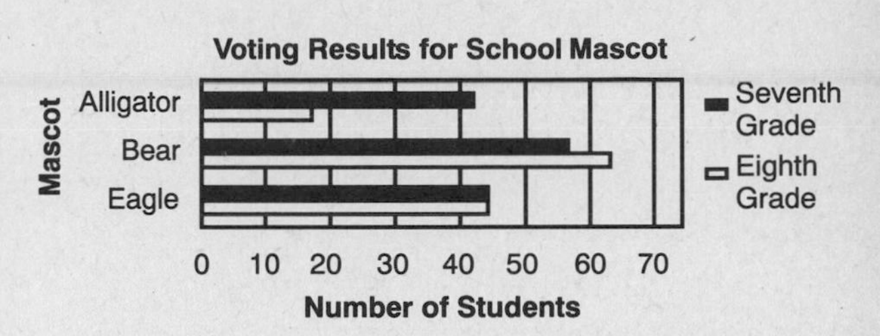

7. Which mascot was chosen by the greatest number of seventh graders?

   A. Alligator **B.** Bear **C.** Eagle

8. Which mascot was chosen by more seventh graders than eighth graders?

   A. Alligator **B.** Bear **C.** Eagle

Use the information below to answer questions 9–10.

A college basketball player scored the following numbers of points in the first 10 games of the season: 19, 15, 15, 22, 17, 20, 17, 19, 20, 20.

9. What is the mode?

   A. 15 points **B.** 17 points
   C. 19 points **D.** 20 points

10. What is the mean number of points scored?

    A. 18.0 points **B.** 18.2 points
    C. 18.4 points **D.** 19.0 points

11. Which fraction is *not* expressed in simplest form?

    A. $\frac{1}{13}$ **B.** $\frac{2}{9}$ **C.** $\frac{6}{15}$ **D.** $\frac{10}{11}$

12. Which group is ordered from least to greatest?

    A. $\frac{1}{6}, \frac{1}{2}, \frac{5}{11}$ **B.** $\frac{1}{7}, \frac{1}{2}, \frac{3}{5}$
    C. $\frac{3}{8}, \frac{3}{9}, \frac{3}{10}$ **D.** $\frac{5}{6}, \frac{4}{5}, \frac{3}{4}$

Go on to the next page.

Name ____________________

Cumulative Test

13. Which fraction is midway between $\frac{1}{50}$ and $\frac{2}{50}$?

A. $\frac{2}{100}$ B. $\frac{5}{200}$
C. $\frac{3}{100}$ D. $\frac{7}{200}$

14. $\frac{5}{12} + \frac{5}{8} =$ ___?___

A. $1\frac{1}{24}$ B. $1\frac{1}{12}$
C. $1\frac{1}{8}$ D. not here

15. $7\frac{1}{5} - 6\frac{3}{5} =$ ___?___

A. $\frac{2}{5}$ B. $\frac{3}{5}$ C. $1\frac{1}{5}$ D. not here

16. $1\frac{1}{3} \times 1\frac{1}{3} =$ ___?___

A. 1 B. $1\frac{1}{3}$ C. $1\frac{2}{3}$ D. $1\frac{7}{9}$

17. $\frac{1}{5} \div \frac{1}{10} =$ ___?___

A. $\frac{1}{50}$ B. $\frac{1}{2}$ C. 2 D. not here

18. Evaluate $\frac{2(a-6)}{3}$ for $a = 15$

A. 3 B. 6 C. 28 D. not here

19. Find the value of $(5 + 4) \times (4 - 1)$.

A. 17 B. 20 C. 27 D. 35

20. Find the value of $3 + 6 + (4 - 1)^2$.

A. 12 B. 15 C. 144 D. not here

21. Which inequality shows that a number, b, increased by 9 is less than 200?

A. $b - 9 < 200$ B. $b - 9 > 200$
C. $9b < 200$ D. $b + 9 < 200$

22. Solve the inequality. $x + 10 < 10$

A. $x < 0$ B. $x < 1$
C. $x < 10$ D. $x < 20$

23. Solve. $a - 20 = 0$

A. $a = 0$ B. $a = 10$
C. $a = 20$ D. $a = 40$

24. Solve. $\frac{x}{12} = 6$

A. $x = \frac{1}{2}$ B. $x = 2$
C. $x = 18$ D. $x = 72$

25. Are right triangles isosceles?

A. always B. sometimes C. never

26. What is the complement of a 75° angle?

A. 15° B. 25° C. 105° D. not here

27. Which ratio does *not* have the same meaning as the other three?

A. $\frac{2}{15}$ B. $\frac{5}{30}$ C. $\frac{6}{45}$ D. $\frac{8}{60}$

28. Which values of a and b make the following a true proportion?

$$\frac{a}{20} = \frac{6}{b}$$

A. $a = 4, b = 30$ B. $a = 4, b = 25$
C. $a = 3, b = 30$ D. $a = 12, b = 12$

29. A package of 6 cassette tapes costs $7.80. What is the unit price?

A. $1.10 B. $1.20
C. $1.30 D. $1.80

30. If 8 cakes cost $3.20 and 12 cakes cost $4.68, which offer has the lower unit price?

A. 8 cakes for $3.20
B. 12 cakes for $4.68

Go on to the next page.

Cumulative Test

31. On a scale drawing, 1 inch is used to represent 6 feet. Which ratio expresses that relationship?

A. 1:12 **B.** 5:12

C. 1:6 **D.** 6:1

32. On Map A, 1 inch = 10 miles. On Map B, 1 inch = 30 miles. A road between two cities is 3 inches long on Map A. How long is the same road on Map B?

A. 1 in. **B.** $1\frac{1}{2}$ in.

C. 2 in. **D.** 9 in.

33. Do similar figures have the same shape and same dimensions?

A. always **B.** sometimes **C.** never

34. A triangle has sides of 7.5 cm, 10 cm, and 12.5 cm. To which of the following triangles is it similar?

A. 2 cm, 3 cm, 4 cm

B. 6 cm, 8 cm, 12 cm

C. 1 cm, 3.5 cm, 6 cm

D. 9 cm, 12 cm, 15 cm

35. Express $\frac{88}{1,000}$ as a decimal.

A. 0.0088 **B.** 0.088

C. 0.880 **D.** 0.888

36. Express 0.78 as a fraction in simplest form.

A. $\frac{39}{500}$ **B.** $\frac{39}{100}$

C. $\frac{34}{50}$ **D.** $\frac{39}{50}$

37. Estimate. 15% of 498 = ___?___

A. 55 **B.** 60 **C.** 75 **D.** 90

38. 40% of what number is 8?

A. 3.2 **B.** 20 **C.** 32 **D.** 50

39. Bob made a circle graph to show how he spends his leisure time. A central angle of 90° represents what percent of his time?

A. $12\frac{1}{2}$% **B.** 15%

C. 25% **D.** not here

Use the circle graph below to answer question 40.

Club Membership at Lincoln School

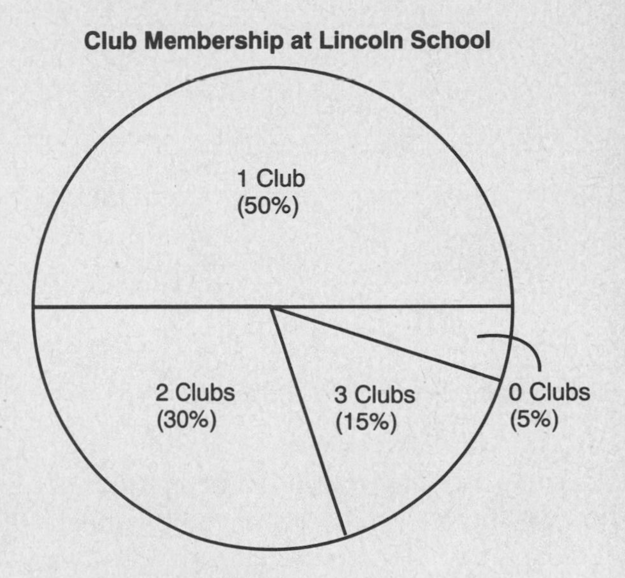

40. A total of 450 students at Lincoln School are members of two clubs. What is the total number of students at the school?

A. 750 students

B. 1,350 students

C. 1,500 students

D. 3,000 students

41. An item that regularly costs $200 is reduced by 35%. What is the sale price?

A. $35.00 **B.** $70.00

C. $130.00 **D.** $135.00

Go on to the next page.

42. Lee borrows $1,200. The interest rate is 10.5% per year. The payoff period is 6 months. How much interest will Lee pay? Use the formula $I = prt$.

A. $63.00 **B.** $66.00

C. $120.00 **D.** $126.00

Use the table below to answer question 43.

Students' Music Preferences		
Type of Music	Boys	Girls
Classical	30	70
Country	40	40
Jazz	60	35
Rock	110	135
	240	280

43. What percentage of boys prefer jazz?

A. $12\frac{1}{2}$% **B.** 25%

C. 35% **D.** 60%

44. Tony and Jen began a running program. Tony ran 2 miles on day 1 and increased the distance by $\frac{1}{4}$ mile each day. Jen ran 3 miles on day 1 and increased the distance by $\frac{1}{8}$ mile each day. On what day will Tony and Jen run the same distance?

A. day 5 **B.** day 8

C. day 9 **D.** day 13

45. Justin can bicycle one mile in $\frac{1}{4}$ hour. Don can bicycle one mile in $\frac{1}{2}$ hour. If both boys ride one hour, how much farther will Justin ride?

A. $\frac{1}{2}$ mi **B.** 1 mi

C. 2 mi **D.** $2\frac{1}{2}$ mi

Use the information below to answer question 46.

Taxi fare is given by the formula $1.25 + $0.90 m + $1.00 b, where m = the distance traveled in miles and b = the number of pieces of luggage carried.

46. Hal took a taxi for a distance of 4 miles. He had 2 pieces of luggage. If he tipped the driver 15%, how much did Hal pay in all?

A. about $5.58

B. about $6.44

C. about $6.73

D. about $7.88

47. Sid withdrew money from his bank account. He spent $50 for shoes and $10 for a tie. He gave $\frac{2}{3}$ of what was left to his son. He then had $45. How much money did Sid withdraw?

A. $120 **B.** $150

C. $195 **D.** $200

48. The scale on a map is 1 inch = 100 miles. Two cities are 2 inches apart. If Tom averages 50 miles per hour and makes no stops, how long will it take him to drive from one city to the other?

A. 2 hr **B.** $2\frac{1}{2}$ hr

C. 3 hr **D.** 4 hr

49. Two numbers add up to 72. One of the numbers is 10 more than the other. What are the numbers?

A. 26 and 36

B. 32 and 40

C. 31 and 41

D. 36 and 46

Go on to the next page.

Name ___________________________________

Cumulative Test

Use the information and double-line graph below to answer question 50.

Each December, a company predicts its sales (in millions) for the coming year. The double-line graph below shows the predicted and actual sales for the years 1989 through 1992.

Predicted and Actual Sales: 1989–1992

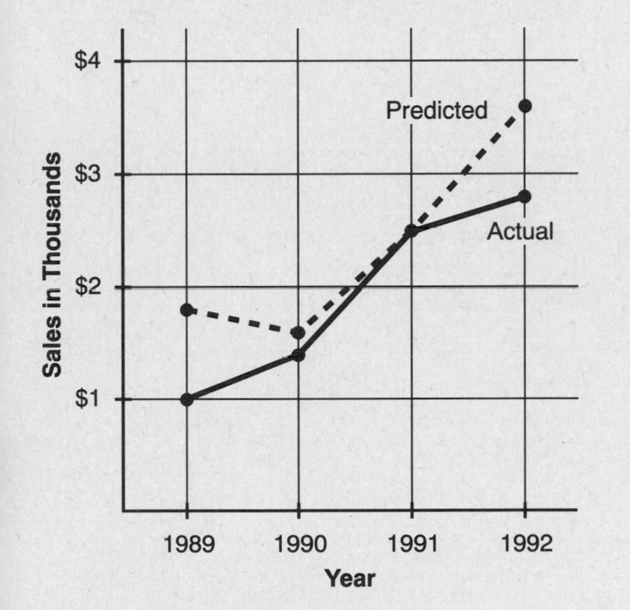

50. During which year did predicted sales equal actual sales?

    A. 1989

    B. 1990

    C. 1991

    D. 1992

Stop!

Name ________________________________

Pretest

Choose the letter of the correct answer.

1. Is the absolute value of a negative integer positive?

   A. always **B.** sometimes **C.** never

2. Is the opposite of a negative integer a positive integer?

   A. always **B.** sometimes **C.** never

3. Is the absolute value of a negative integer less than the absolute value of its opposite?

   A. always **B.** sometimes **C.** never

4. Which of the following has the greatest value?

   A. 6 **B.** $|^{-}9|$ **C.** 8 **D.** $^{-}100$

For questions 5–6, dark counters are positive integers, and light counters are negative integers.

5. What does the model represent?

   A. $^{-}7$ **B.** $^{-}1$

   C. 1 **D.** 7

6. What does the model represent?

   A. $2 - {^{-}5} = {^{-}7}$ **B.** $2 - {^{-}5} = {^{-}3}$

   C. $2 - {^{-}5} = 3$ **D.** $2 - {^{-}5} = 7$

7. When one negative integer is divided by another negative integer, the quotient is always:

   A. positive **B.** negative

   C. zero **D.** greater than 1

8. When an integer is subtracted from a greater integer, the result is always:

   A. a negative integer **B.** zero

   C. a positive integer **D.** less than 2

9. $^{-}9 + {^{-}6} =$ ___?___

   A. $^{-}15$ **B.** $^{-}3$

   C. 3 **D.** not here

10. $^{-}6 \div {^{-}2} =$ ___?___

    A. $^{-}3$ **B.** $\frac{1}{3}$

    C. 3 **D.** 12

11. $^{-}5 \times {^{-}7} =$ ___?___

    A. $^{-}35$ **B.** 35

    C. 57 **D.** not here

12. $3 - {^{-}4} =$ ___?___

    A. $^{-}7$ **B.** $^{-}1$

    C. 1 **D.** not here

13. One cold night, the temperature dropped from 3°F to $^{-}18$°F. How many degrees did the temperature drop?

    A. 3° **B.** 15°

    C. 18° **D.** 21°

14. Josie had $118 on May 1. During the month, she spent $214 and earned $108. How much did she have at the end of the month?

    A. $^{-}$\$12 **B.** $12 **C.** $224 **D.** $440

15. Alicia arrived 25 min late for a meeting. Cindy arrived 12 min earlier. How late was Cindy?

    A. 12 min **B.** 13 min

    C. 25 min **D.** 37 min

Go on to the next page.

Pretest

16. Five men went on a diet. Two men lost 10 pounds each. Two men lost 6 pounds each. The other man gained 3 pounds. How much did the men lose altogether?

A. 13 pounds **B.** 16 pounds
C. 29 pounds **D.** 35 pounds

17. Jo studied 4 hr less than 3 times the number of hours Moe studied. Which expression shows the number of hours Jo studied? Let m = the number of hours Moe studied.

A. $4 - 3m$ **B.** $3m - 4$
C. $4m - 3$ **D.** $3 - 4m$

18. To solve the equation $x - 7 = 22$, what must you do to both sides of the equation?

A. add 7 **B.** subtract 7
C. divide by 7 **D.** subtract 22

19. Solve for y. $y - 3 = {}^-6$

A. $y = {}^-9$ **B.** $y = {}^-3$
C. $y = 3$ **D.** $y = 9$

20. Solve for x. $x + 2 = 1$

A. $x = {}^-3$ **B.** $x = {}^-1$
C. $x = 1$ **D.** not here

21. Solve for n. $\frac{n}{3} = {}^-7$

A. $n = {}^-21$ **B.** $n = \frac{^-7}{3}$
C. $n = 10$ **D.** $n = 21$

22. Solve for p. $p \times {}^-5 = {}^-10$

A. $p = {}^-2$ **B.** $p = \frac{^-1}{2}$
C. $p = 2$ **D.** $p = 50$

23. Gary spent \$6.00 in a restaurant and \$4.00 at a movie. He had \$11.00 left. How much money did he start with?

A. \$15.00 **B.** \$17.00
C. \$20.00 **D.** \$21.00

24. A baseball team won $\frac{1}{3}$ of its games. The team won 15 games. How many games did the team play?

A. 5 games **B.** 30 games
C. 45 games **D.** 60 games

Use the graph below to answer questions 25–26.

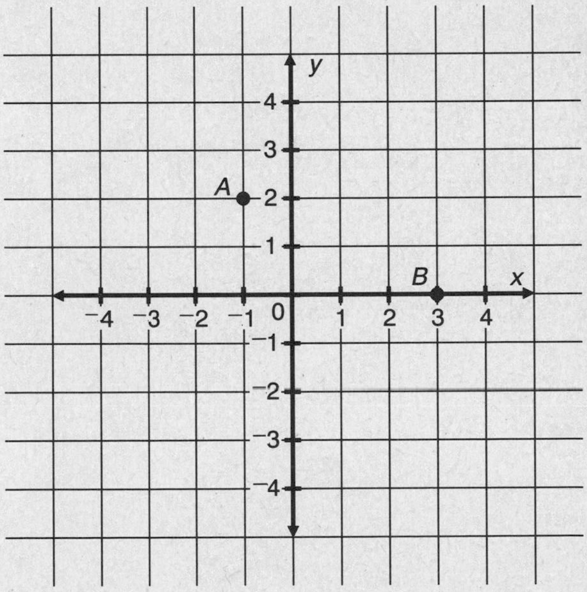

25. What are the coordinates of point A?

A. $({}^-1,{}^-2)$ **B.** $({}^-1,2)$
C. $(1,{}^-2)$ **D.** $(2,{}^-1)$

26. What are the coordinates of point B?

A. (3,0) **B.** (3,1)
C. $({}^-3,0)$ **D.** (0,3)

Go on to the next page.

Name ___

Pretest

Use the table below to answer questions 27–28.

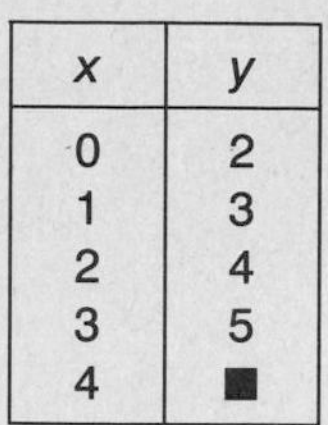

x	y
0	2
1	3
2	4
3	5
4	■

27. Which equation describes the pattern that relates y to x?

A. $y = 2x$ **B.** $y = x + 2$

C. $y = \frac{x}{2}$ **D.** $y = x - 2$

28. Which ordered pair would be graphed when $x = 4$?

A. (4,6) **B.** (4,8)

C. (6,4) **D.** (8,4)

29. Helena and Ruth bought some apples. Helena bought 2 more than 3 times the number of apples Ruth bought. If Helena bought 8 apples, how many did Ruth buy?

A. 1 apple **B.** 2 apples

C. 3 apples **D.** 26 apples

30. Boris receives three bills in the mail. The electric bill is 3 times as much as the phone bill. The gas bill is $\frac{1}{2}$ the amount of the phone bill. The gas bill is \$20. How much is the electric bill?

A. \$30 **B.** \$40

C. \$80 **D.** \$120

31. Nadia spent \$10.00 on books. She gave $\frac{1}{2}$ of the money she had left to her sister. Nadia then had \$1.50. How much money did she have before buying books?

A. \$3.50 **B.** \$11.00

C. \$11.50 **D.** \$13.00

32. Tricia got 42 questions correct on a test. Her score was 2 less than $\frac{1}{2}$ of Nan's score. How many questions did Nan get correct?

A. 19 questions

B. 84 questions

C. 86 questions

D. 88 questions

Stop!

Name ______________________

Posttest

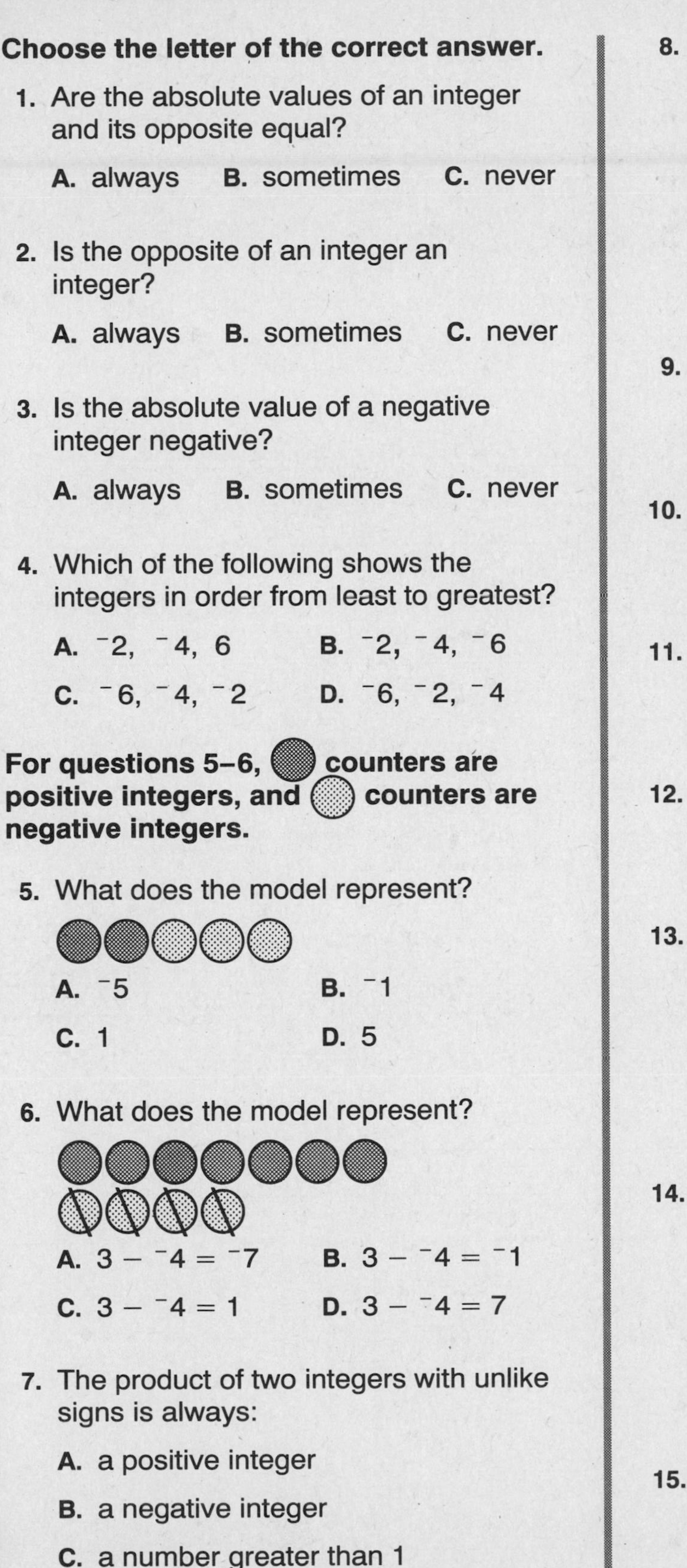

Choose the letter of the correct answer.

1. Are the absolute values of an integer and its opposite equal?

   A. always **B.** sometimes **C.** never

2. Is the opposite of an integer an integer?

   A. always **B.** sometimes **C.** never

3. Is the absolute value of a negative integer negative?

   A. always **B.** sometimes **C.** never

4. Which of the following shows the integers in order from least to greatest?

   A. $^{-}2$, $^{-}4$, 6 **B.** $^{-}2$, $^{-}4$, $^{-}6$
   C. $^{-}6$, $^{-}4$, $^{-}2$ **D.** $^{-}6$, $^{-}2$, $^{-}4$

For questions 5–6, (dark) counters are positive integers, and (light) counters are negative integers.

5. What does the model represent?

   A. $^{-}5$ **B.** $^{-}1$
   C. 1 **D.** 5

6. What does the model represent?

   A. $3 - {}^{-}4 = {}^{-}7$ **B.** $3 - {}^{-}4 = {}^{-}1$
   C. $3 - {}^{-}4 = 1$ **D.** $3 - {}^{-}4 = 7$

7. The product of two integers with unlike signs is always:

   A. a positive integer
   B. a negative integer
   C. a number greater than 1
   D. a number less than 1

8. When an integer is subtracted from a greater integer, the result is always:

   A. a positive integer
   B. a negative integer
   C. greater than 1
   D. zero

9. $^{-}15 + {}^{-}20 =$ __?__

   A. $^{-}5$ **B.** 5 **C.** 35 **D.** not here

10. $^{-}5 \div {}^{-}5 =$ __?__

    A. $^{-}5$ **B.** $^{-}1$ **C.** 1 **D.** 5

11. $^{-}50 \times {}^{-}2 =$ __?__

    A. $^{-}100$ **B.** 25 **C.** 100 **D.** not here

12. $100 - {}^{-}99 =$ __?__

    A. $^{-}1$ **B.** 1 **C.** 99 **D.** 199

13. The temperature increased from $^{-}9$°F to $^{-}7$°F. How many degrees did the temperature change?

    A. $^{-}2$° **B.** 2°
    C. 9° **D.** 16°

14. A club had \$140 on March 1. During the month, the members spent \$270 and earned \$181. How much did they have at the end of the month?

    A. $^{-}$\$311 **B.** \$51
    C. \$229 **D.** \$591

15. Roni's watch runs 7 sec slow each day. How slow will it be after one week?

    A. 14 sec **B.** 35 sec
    C. 49 sec **D.** 70 sec

Go on to the next page.

Name ______________________________

Posttest

16. The 10 members of the Garden Club went on diets. One half of the members lost 8 pounds each. Three members lost 5 pounds each. Two members gained 2 pounds each. What was the total amount lost?

A. 11 pounds **B.** 51 pounds
C. 91 pounds **D.** 109 pounds

17. Pat works 3 hr less per week than $\frac{1}{2}$ the time Lyn works. Which expression shows the number of hours per week Pat works? Let h = the number of hours Lyn works.

A. $3 - \frac{1}{2}h$ **B.** $\frac{1}{2}h - 3$
C. $3h - \frac{1}{2}$ **D.** $\frac{1}{2} - 3h$

18. To solve the equation $4x = 32$, what must you do to both sides of the equation?

A. divide by 4
B. subtract 4
C. multiply by 4
D. multiply by 32

19. Solve for y. $y - 11 = {}^-12$

A. $y = {}^-23$ **B.** $y = {}^-1$
C. $y = 1$ **D.** $y = 23$

20. Solve for x. $x + 12 = 2$

A. $x = {}^-14$ **B.** $x = {}^-10$
C. $x = 10$ **D.** $x = 14$

21. Solve for n. $\frac{n}{10} = {}^-3$

A. $n = {}^-30$ **B.** $n = \frac{{}^-10}{3}$
C. $n = 30$ **D.** not here

22. Solve for p. $p \times {}^-1 = {}^-99$

A. $p = {}^-1$ **B.** $p = \frac{{}^-1}{99}$
C. $p = \frac{1}{99}$ **D.** $p = 99$

23. Hal gave \$15 to his brother and \$10 to each of his 2 sisters. He was left with \$44. How much money did he start with?

A. \$9 **B.** \$35 **C.** \$69 **D.** \$79

24. A soccer team lost $\frac{3}{4}$ of its games. The team won 8 games. How many games did the team play?

A. 6 games **B.** 12 games
C. 24 games **D.** 32 games

Use the graph below to answer questions 25–26.

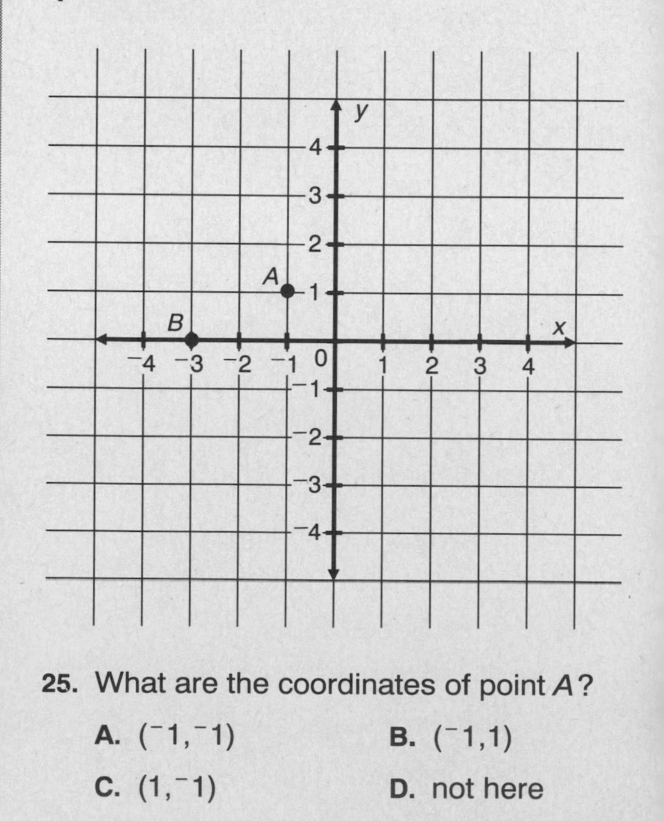

25. What are the coordinates of point A?

A. $({}^-1, {}^-1)$ **B.** $({}^-1, 1)$
C. $(1, {}^-1)$ **D.** not here

Go on to the next page.

Name

Posttest

26. What are the coordinates of point B?

A. $(0,{}^{-}3)$ **B.** $(1,3)$
C. $(0,3)$ **D.** $({}^{-}3,0)$

Use the table below to answer questions 27–28.

x	y
0	0
1	2
2	4
3	6
4	■

27. Which equation describes the pattern that relates y to x?

A. $y = 2x$ **B.** $y = x + 2$
C. $y = \frac{x}{2}$ **D.** $y = x - 2$

28. Which ordered pair would be graphed when $x = 4$?

A. (4,7) **B.** (4,8)
C. (7,4) **D.** (8,4)

29. Helen played a game three times. Her second score was twice as high as her first score. Her third score was 5 points higher than her second score. Helen's third score was 225. What was her first score?

A. 100 points **B.** 110 points
C. 115 points **D.** 120 points

30. Mike got 5 words wrong on a spelling test. His sister got 3 more than twice as many words wrong. How many words did his sister get wrong?

A. 7 words **B.** 10 words
C. 13 words **D.** 18 words

31. Emile spent $50.00 on groceries. He spent $\frac{1}{5}$ of the money on fruit, $\frac{1}{2}$ on dairy items, and the rest on meat. How much did he spend on meat?

A. $3.00 **B.** $15.00
C. $30.00 **D.** $35.00

32. Bill weighs twice as much as Phil. John weighs $\frac{2}{3}$ as much as Phil. Bill weighs 240 pounds. How much does John weigh?

A. 60 pounds **B.** 80 pounds
C. 120 pounds **D.** 160 pounds

Stop!

Pretest

Choose the letter of the correct answer.

1. Are mixed numbers rational numbers?

   A. always **B.** sometimes **C.** never

2. Are the square roots of numbers that are not perfect squares irrational numbers?

   A. always **B.** sometimes **C.** never

3. Does a fraction whose denominator has only 2 and 5 as prime factors result in a repeating decimal?

   A. always **B.** sometimes **C.** never

4. Are irrational numbers real numbers?

   A. always **B.** sometimes **C.** never

5. Express $\frac{39}{50}$ as a decimal.

   A. 0.075 **B.** 0.078
   C. 0.395 **D.** 0.78

6. Express 0.825 as a percent.

   A. 0.825% **B.** 8.25%
   C. 82.5% **D.** not here

7. Express 1.75 as a percent.

   A. 1.75% **B.** 17.5%
   C. 175% **D.** not here

8. Which of the following fractions can be expressed as a repeating decimal?

   A. $\frac{1}{16}$ **B.** $\frac{6}{9}$ **C.** $\frac{3}{20}$ **D.** $\frac{12}{40}$

9. Which of the following fractions can be expressed as a terminating decimal?

   A. $\frac{3}{7}$ **B.** $\frac{6}{18}$ **C.** $\frac{1}{12}$ **D.** $\frac{3}{25}$

10. Which square root is an irrational number?

    A. $\sqrt{4}$ **B.** $\sqrt{8}$ **C.** $\sqrt{16}$ **D.** $\sqrt{81}$

11. Compare. $^{-}1{,}000 \bigcirc 0.0001$

    A. $<$ **B.** $=$ **C.** $>$

12. Compare. $\frac{^{-}1}{50} \bigcirc {}^{-}0.5$

    A. $<$ **B.** $=$ **C.** $>$

13. Which is a rational number between 0.870 and $\frac{7}{8}$?

    A. 0.869 **B.** 0.871
    C. $\frac{13}{16}$ **D.** 0.876

14. Which is a rational number between $^{-}0.50$ and $^{-}0.51$?

    A. $^{-}0.499$ **B.** $^{-}0.501$
    C. $^{-}0.511$ **D.** not here

15. Order from least to greatest.

    $\frac{6}{7}, \frac{7}{8}, \frac{^{-}5}{6}, \frac{^{-}9}{8}$

    A. $\frac{^{-}5}{6}, \frac{6}{7}, \frac{7}{8}, \frac{^{-}9}{8}$ **B.** $\frac{^{-}5}{6}, \frac{^{-}9}{8}, \frac{6}{7}, \frac{7}{8}$
    C. $\frac{^{-}9}{8}, \frac{^{-}5}{6}, \frac{6}{7}, \frac{7}{8}$ **D.** $\frac{^{-}9}{8}, \frac{^{-}5}{6}, \frac{7}{8}, \frac{6}{7}$

16. Order from least to greatest.

    $\frac{1}{6}, \frac{1}{7}, 0.17, 0.14$

    A. $0.14, \frac{1}{7}, \frac{1}{6}, 0.17$
    B. $\frac{1}{7}, 0.14, 0.17, \frac{1}{6}$
    C. $\frac{1}{7}, \frac{1}{6}, 0.14, 0.17$
    D. $0.14, 0.17, \frac{1}{7}, \frac{1}{6}$

Go on to the next page.

Name ______________________________

Pretest

17. Compare. $\frac{1}{4^3} \bigcirc \frac{1}{2^5}$

A. > B. = C. <

18. Compare. $\frac{1}{3^7} \bigcirc \frac{1}{4^8}$

A. < B. = C. >

19. $3^4 =$ ___?___

A. 34 B. 64 C. 81 D. 243

20. What is 3^{-4} expressed with a positive exponent?

A. $\frac{1}{3^4}$ B. $\frac{1}{81^2}$

C. $\frac{3}{3^3}$ D. $\frac{3^3}{33}$

21. What is 4.2×10^3 in standard form?

A. 420 B. 4,200

C. 42,000 D. 420,000

22. What is 502,000,000 in scientific notation?

A. 5.02×10^6 B. 5.02×10^8

C. 5.2×10^7 D. 5.20×10^8

23. Which is a perfect square?

A. 24 B. 44 C. 99 D. 144

24. Which is the best estimate of $\sqrt{83}$?

A. 8.9 B. 9.1 C. 10.1 D. 11.2

25. $\sqrt{121} =$ ___?___

A. 11 B. 12

C. 21 D. not here

26. $\sqrt{\frac{9}{16}} =$ ___?___

A. $\frac{3}{16}$ B. $\frac{3}{4}$

C. $\frac{9}{4}$ D. not here

27. $\sqrt{0.0049} =$ ___?___

A. 0.0007 B. 0.007

C. 0.07 D. 0.7

28. $\sqrt{10{,}000} =$ ___?___

A. 10 B. 100 C. 500 D. 1,000

Use the table below to answer questions 29–32.

Students Attending Grant High

Year	Number
1970	1,050
1975	1,100
1980	1,200
1985	1,350
1990	1,550

29. How many students attended Grant High in 1980?

A. 1,100 students B. 1,150 students

C. 1,200 students D. 1,550 students

30. If the trend continues, how many students will attend Grant High in 1995?

A. 1,700 students B. 1,750 students

C. 1,800 students D. 1,850 students

31. If the trend continues, how many students will attend Grant High in 2005?

A. 2,100 students B. 2,400 students

C. 2,450 students D. 2,500 students

32. How much did the student population increase from 1970 to 1990?

A. 50 students B. 300 students

C. 450 students D. 500 students

Stop!

Name ______________________________

Posttest

Choose the letter of the correct answer.

1. Are fractions rational numbers?

   A. always **B.** sometimes **C.** never

2. Are the square roots of numbers that are perfect squares irrational numbers?

   A. always **B.** sometimes **C.** never

3. Can fractions with a denominator of 15 be changed to terminating decimals?

   A. always **B.** sometimes **C.** never

4. Are rational numbers real numbers?

   A. always **B.** sometimes **C.** never

5. Express 0.905 as a percent.

   A. 0.905% **B.** 9.05%
   C. 90.5% **D.** not here

6. Express $\frac{19}{20}$ as a decimal.

   A. 0.095 **B.** 0.098
   C. 0.95 **D.** 0.98

7. Express 2.5 as a percent.

   A. 0.25% **B.** 2.50%
   C. 25.0% **D.** 250%

8. Express $\frac{9}{10{,}000}$ as a decimal.

   A. 0.00009 **B.** 0.0009
   C. 0.009 **D.** not here

9. Which of the following fractions can be expressed as a repeating decimal?

   A. $\frac{1}{8}$ **B.** $\frac{1}{16}$ **C.** $\frac{1}{24}$ **D.** $\frac{1}{32}$

10. Which square root is an irrational number?

    A. $\sqrt{64}$ **B.** $\sqrt{95}$
    C. $\sqrt{121}$ **D.** $\sqrt{144}$

11. Compare. $^-0.003 \bigcirc \frac{^-2}{1{,}000}$

    A. $<$ **B.** $=$ **C.** $>$

12. Compare. $\frac{^-1}{99} \bigcirc \frac{^-2}{99}$

    A. $<$ **B.** $=$ **C.** $>$

13. Which is a rational number between $\frac{99}{100}$ and 0.991?

    A. 0.9899 **B.** 0.9911
    C. $\frac{992}{1{,}000}$ **D.** not here

14. Which is a rational number between $^-0.001$ and $\frac{^-1}{500}$?

    A. $^-0.0009$ **B.** $^-0.0011$
    C. $^-0.0021$ **D.** not here

15. Order from least to greatest.

    $\frac{1}{3}, \frac{1}{4}, \frac{^-1}{5}, \frac{^-1}{6}$

    A. $\frac{^-1}{6}, \frac{^-1}{5}, \frac{1}{4}, \frac{1}{3}$ **B.** $\frac{^-1}{6}, \frac{^-1}{5}, \frac{1}{3}, \frac{1}{4}$
    C. $\frac{^-1}{5}, \frac{^-1}{6}, \frac{1}{4}, \frac{1}{3}$ **D.** $\frac{^-1}{5}, \frac{^-1}{6}, \frac{1}{3}, \frac{1}{4}$

16. Order from least to greatest.

    $\frac{1}{9}, \frac{1}{11}, 0.100, \frac{1}{8}$

    A. $\frac{1}{11}, 0.100, \frac{1}{9}, \frac{1}{8}$
    B. $0.100, \frac{1}{8}, \frac{1}{11}, \frac{1}{9}$
    C. $\frac{1}{11}, \frac{1}{9}, 0.100, \frac{1}{8}$
    D. $\frac{1}{11}, 0.100, \frac{1}{8}, \frac{1}{9}$

Go on to the next page.

Posttest

17. What is another way to write $7 \times 7 \times 7 \times 7$?

A. $4 \cdot 7^2$ **B.** 4×7

C. 4^7 **D.** 7^4

18. Compare. $\frac{1}{2^3} \bigcirc \frac{1}{3^2}$

A. $<$ **B.** $=$ **C.** $>$

19. What is 5^{-3} expressed with a positive exponent?

A. 3^5 **B.** $\frac{5^3}{25}$

C. $\frac{1}{53}$ **D.** $\frac{1}{5^3}$

20. $\frac{1,000}{10^4} = \underline{\quad ? \quad}$

A. $\frac{1}{100}$ **B.** $\frac{1}{10}$

C. $\frac{3}{4}$ **D.** 10

21. What is 1.2×10^{-3} in standard form?

A. 0.00012 **B.** 0.0012

C. 0.012 **D.** 0.12

22. What is 32,000 in scientific notation?

A. 3.20×10^3 **B.** 3.02×10^5

C. 3.2×10^4 **D.** 3.20×10^5

23. Which is a perfect square?

A. 200 **B.** 300 **C.** 400 **D.** 500

24. Which is the best estimate of $\sqrt{124}$?

A. 10.9 **B.** 11.1

C. 12.4 **D.** 13.1

25. $\sqrt{144} = \underline{\quad ? \quad}$

A. 12 **B.** 12.5

C. 13 **D.** not here

26. $\sqrt{\frac{4}{25}} = \underline{\quad ? \quad}$

A. $\frac{2}{25}$ **B.** $\frac{2}{5}$

C. $2\frac{1}{2}$ **D.** not here

27. $\sqrt{0.09} = \underline{\quad ? \quad}$

A. 0.003 **B.** 0.03

C. 0.3 **D.** 0.81

28. $\sqrt{1,000,000} = \underline{\quad ? \quad}$

A. 100 **B.** 500

C. 1,000 **D.** 10,000

Use the table below to answer questions 29–32.

Town Population

Year	Population
1986	2,000
1987	2,100
1988	2,300
1989	2,600
1990	3,000
1991	3,500

29. What was the town population in 1988?

A. 230 **B.** 2,100

C. 2,300 **D.** 23,000

30. If the trend continues, what will be the town population in 1992?

A. 4,000 **B.** 4,100

C. 4,200 **D.** 41,000

31. If the trend continues, what will be the town population in 1995?

A. 5,600 **B.** 6,400

C. 6,500 **D.** 65,000

32. By how many people did the population increase from 1986 to 1991?

A. 1,000 **B.** 1,500 **C.** 2,000 **D.** 2,500

Stop!

Name ____________________

Pretest

Choose the letter of the correct answer.

1. Jan picks one month at random from a calendar. What is the number of possible outcomes?

   A. 6 outcomes　**B.** 9 outcomes
   C. 12 outcomes　**D.** 24 outcomes

2. Al has a spinner with five equal sections, labeled 3, 5, 7, 9, and 10. What is the number of possible outcomes of one spin?

   A. 3 outcomes　**B.** 5 outcomes
   C. 9 outcomes　**D.** 19 outcomes

3. A restaurant has 3 choices for soup, 6 possible main courses, and 5 desserts. If Sandra orders soup, a main course, and dessert, how many possible selections can she make?

   A. 14 selections　**B.** 18 selections
   C. 30 selections　**D.** 90 selections

4. What is the number of possible three-digit area codes? Assume that zero *cannot* be used as the first digit.

   A. 29 codes　**B.** 90 codes
   C. 900 codes　**D.** 1,000 codes

5. Can a greater number of 2-member committees be selected from a group of 6 people or from a group of 8 people?

   A. from a group of 6 people
   B. from a group of 8 people
   C. The same number of committees can be selected from either group.

6. Which of these words provides the *least* number of possible letter arrangements?

   A. *dog*　**B.** *bird*
   C. *horse*　**D.** *flower*

7. How many different combinations of 2 items can be selected from a set of 8 items?

   A. 16 combinations
   B. 28 combinations
   C. 56 combinations
   D. not here

8. How many different arrangements of the letters *A, B, C, D* are possible?

   A. 12 arrangements
   B. 16 arrangements
   C. 24 arrangements
   D. 120 arrangements

9. A tennis team has 6 members. How many possible teams of 2 can be formed?

   A. 12 teams　**B.** 15 teams
   C. 30 teams　**D.** 120 teams

10. A basketball league has 4 teams. In how many different ways can the teams be arranged in the final standings?

    A. 12 ways　**B.** 16 ways
    C. 20 ways　**D.** 24 ways

11. A coin is flipped 3 times. What is the probability of obtaining a head on any *one* of the flips?

    A. $\frac{1}{8}$　**B.** $\frac{1}{4}$　**C.** $\frac{1}{2}$　**D.** $\frac{3}{4}$

12. A coin is flipped 4 times. Which of the following is the most likely outcome?

    A. 4 heads, 0 tails
    B. 3 heads, 1 tail
    C. 2 heads, 2 tails
    D. 1 head, 3 tails

Go on to the next page.

Pretest

13. Hector randomly picks an integer between 0 and 9. Which of the following is the most likely outcome?

A. The number is 5.

B. The number is 9.

C. The number is odd.

D. The number is not 1.

14. On one roll of a number cube marked 1–6, which of these has the highest probability?

A. a number greater than 4

B. a 4

C. a number less than 4

D. a 1 or a 2

15. A coin is flipped 4 times. What is the probability of obtaining 4 heads?

A. $\frac{1}{64}$ **B.** $\frac{1}{32}$ **C.** $\frac{1}{8}$ **D.** not here

Use the information below to answer questions 16–18.

A spinner has 10 equal sections, numbered from 1 to 10.

16. What is the probability of spinning an odd number?

A. $\frac{1}{5}$ **B.** $\frac{4}{10}$ **C.** $\frac{5}{10}$ **D.** $\frac{6}{10}$

17. What is the probability of spinning a number greater than 6?

A. $\frac{4}{10}$ **B.** $\frac{5}{10}$ **C.** $\frac{6}{10}$ **D.** not here

18. What is the best prediction of how many times the number 10 will be obtained in 900 spins?

A. 45 times **B.** 90 times

C. 100 times **D.** 180 times

Use the information below to answer questions 19–20.

A large jar is filled with red beans and blue beans. It holds 100 beans in all. Bea shakes the jar and removes a handful. Her handful contains 12 red beans and 18 blue beans.

19. What ratio should Bea use to predict the number of blue beans in the jar?

A. $\frac{18}{30}$ **B.** $\frac{12}{18}$

C. $\frac{12}{30}$ **D.** $\frac{12}{100}$

20. What is the best prediction of how many red beans are in the jar?

A. 36 beans

B. 40 beans

C. 50 beans

D. 60 beans

21. How should you compute the probability of two independent events both happening?

A. add the two probabilities

B. subtract the lower probability from the higher

C. multiply the two probabilities

D. divide the higher probability by the lower

22. Which of the following does *not* describe independent events?

A. flipping a coin 2 times

B. rolling a number cube 2 times

C. spinning a spinner 2 times

D. drawing 2 balls from a box without replacing the first ball

Go on to the next page.

Name ______________________________

Pretest

To answer questions 23–24, assume that a jar contains 3 red balls and 3 black balls. The first ball drawn is *not* replaced in the jar.

23. Helen takes 2 balls from the jar. What is the probability that both balls are black?

A. $\frac{1}{10}$ **B.** $\frac{1}{5}$ **C.** $\frac{1}{2}$ **D.** $\frac{2}{3}$

24. Mike takes 2 balls from the jar. What is the probability that they are the same color?

A. $\frac{1}{10}$ **B.** $\frac{2}{10}$ **C.** $\frac{2}{5}$ **D.** $\frac{1}{2}$

25. Juan flips a coin and rolls a number cube. What is the probability that the coin lands on heads and the number cube shows a 1?

A. $\frac{1}{24}$ **B.** $\frac{1}{12}$ **C.** $\frac{1}{6}$ **D.** $\frac{1}{2}$

26. Olga flips a coin 2 times. What is the probability that both flips come up heads?

A. $\frac{1}{16}$ **B.** $\frac{1}{8}$ **C.** $\frac{1}{4}$ **D.** $\frac{1}{2}$

27. Two baseball teams play 2 games. Since Team A is the better team, its chance of winning each game is $\frac{2}{3}$. What is the probability that Team A will win both games?

A. $\frac{1}{3}$ **B.** $\frac{4}{9}$ **C.** $\frac{2}{3}$ **D.** $\frac{4}{3}$

28. A jar contains 5 red and 5 white marbles. Jo takes 1 marble from the jar, does not replace it, and then removes another marble. What is the probability that both marbles are white?

A. $\frac{4}{18}$ **B.** $\frac{1}{5}$ **C.** $\frac{1}{4}$ **D.** $\frac{17}{18}$

29. Sam, Tina, and Ray each have a collection. Ray and the stamp collector are best friends. The coin collector lives next to Tina, who enjoys trading bells from her collection. Who has the coin collection?

A. Ray **B.** Sam **C.** Tina

Use the information below to answer questions 30–32.

Joe and Edna must take music, health, and art courses during a two-year period. Each student is randomly assigned one of the courses for the current semester. Assume that the probability of assignment to each course is $\frac{1}{3}$.

30. What is the probability that both students will be assigned the art course?

A. $\frac{3}{9}$ **B.** $\frac{2}{9}$ **C.** $\frac{1}{6}$ **D.** $\frac{1}{9}$

31. What is the probability that both students will be assigned any of the three courses together?

A. $\frac{2}{3}$ **B.** $\frac{3}{9}$ **C.** $\frac{2}{9}$ **D.** $\frac{1}{9}$

32. What is the probability that neither student will be assigned the music course?

A. $\frac{2}{3}$ **B.** $\frac{4}{9}$ **C.** $\frac{1}{3}$ **D.** $\frac{4}{81}$

Stop!

Name ________________________________

Posttest

Choose the letter of the correct answer.

1. Lisa picks one day of the week from the calendar. What is the number of possible outcomes?

   A. 1 outcome **B.** 3 outcomes

   C. 7 outcomes **D.** 12 outcomes

2. John has a spinner with 8 equal sections, labeled 1, 2, 3, 4, 5, 6, 7, and 8. What is the number of possible outcomes?

   A. 2 outcomes **B.** 6 outcomes

   C. 8 outcomes **D.** 12 outcomes

3. Len is at a fruit stand. He has 6 kinds of fruit and 5 kinds of vegetables to choose from. If he buys 1 fruit and 1 vegetable, how many outcomes are possible?

   A. 6 outcomes

   B. 11 outcomes

   C. 25 outcomes

   D. 30 outcomes

4. What is the number of possible four-digit numbers? Do *not* use zero in the first place.

   A. 4,000 numbers

   B. 9,000 numbers

   C. 10,000 numbers

   D. 100,000 numbers

5. Can a greater number of 2-member teams be selected from a group of 4 girls or from a group of 5 girls?

   A. from a group of 4 girls

   B. from a group of 5 girls

   C. The same number of teams can be selected from either group.

6. Which of these words provides the *greatest* number of possible letter arrangements?

   A. *hat* **B.** *noon*

   C. *seven* **D.** *jacket*

7. How many different combinations of 2 items can be selected from a set of 6 items?

   A. 12 combinations

   B. 15 combinations

   C. 36 combinations

   D. not here

8. How many different arrangements of the letters *A, B, C, D, E* are possible?

   A. 12 arrangements

   B. 24 arrangements

   C. 120 arrangements

   D. 720 arrangements

9. A baseball team has 12 members. The team needs 2 co-captains. How many different pairs of cocaptains can be selected from the 12 members?

   A. 66 pairs

   B. 120 pairs

   C. 121 pairs

   D. 132 pairs

10. A basketball league has 7 teams. In how many different ways can the teams be arranged in the final standings?

    A. 49 ways

    B. 210 ways

    C. 720 ways

    D. 5,040 ways

Go on to the next page.

Posttest

11. What is the probability of an event that is impossible?

A. a negative number

B. 0

C. $\frac{1}{2}$

D. 1

12. A coin is flipped 2 times. Which of the following is the most likely outcome?

A. 2 heads, 0 tails

B. 1 head, 1 tail

C. 0 heads, 2 tails

D. All of these are equally likely.

13. Lupe randomly picks one letter of the alphabet. Which of the following is the most likely outcome?

A. The letter is *A.*

B. The letter is *Z.*

C. The letter is a vowel.

D. The letter is a consonant.

14. On one roll of a number cube marked 1–6, which of these has the highest probability?

A. an odd number

B. an even number

C. a number less than 5

D. a number greater than 5

15. A coin is flipped 3 times. What is the probability of obtaining 3 heads?

A. $\frac{1}{9}$ **B.** $\frac{1}{8}$

C. $\frac{1}{3}$ **D.** not here

Use the information below to answer questions 16–18.

A spinner has 10 equal sections, numbered from 1 to 10.

16. What is the probability of spinning 1, 4, or 9?

A. $\frac{1}{10}$ **B.** $\frac{2}{10}$ **C.** $\frac{3}{10}$ **D.** $\frac{1}{3}$

17. What is the probability of spinning a number greater than 2?

A. $\frac{2}{10}$ **B.** $\frac{6}{10}$ **C.** $\frac{7}{10}$ **D.** not here

18. What is the best prediction of how many times the number 2 will be obtained in 1,000 spins?

A. 25 times **B.** 50 times

C. 100 times **D.** 200 times

Use the information below to answer questions 19–20.

A large jar is filled with black candies and white candies. It holds 200 candies in all. Hal shakes the jar and removes a handful. His handful contains 10 white candies and 15 black candies.

19. What ratio should Hal use to predict the number of white candies in the jar?

A. $\frac{10}{25}$ **B.** $\frac{10}{15}$ **C.** $\frac{15}{200}$ **D.** $\frac{10}{200}$

20. What is the best prediction of the total number of black candies in the jar?

A. 40 candies **B.** 80 candies

C. 120 candies **D.** 160 candies

21. When the outcome of an event is *not* affected by the outcome of an earlier event, what are the events called?

A. dependent **B.** independent

C. equally likely **D.** impossible

Go on to the next page.

Posttest

22. Which of the following pairs of events are most likely to be independent?

A. It rains heavily.
You carry an umbrella.

B. You are tall.
You get an A in math.

C. You are sick.
You do not go to school.

D. You study hard in science.
You get an A in science.

To answer questions 23–24, assume that a jar contains 3 red balls and 2 black balls. The first ball drawn is *not* replaced in the jar.

23. Anne takes 2 balls from the jar. What is the probability that both balls are black?

A. $\frac{2}{20}$ **B.** $\frac{2}{10}$ **C.** $\frac{4}{25}$ **D.** $\frac{2}{5}$

24. Jake takes 2 balls from the jar. What is the probability that he picks a red ball and then a black ball?

A. $\frac{3}{20}$ **B.** $\frac{3}{15}$ **C.** $\frac{3}{10}$ **D.** $\frac{3}{30}$

25. Donna flips a coin 3 times. What is the probability of obtaining 3 heads?

A. $\frac{1}{3}$ **B.** $\frac{1}{4}$ **C.** $\frac{1}{8}$ **D.** $\frac{1}{16}$

26. Lupe rolls a number cube, marked 1–6, twice. What is the probability of rolling a 6 both times?

A. $\frac{1}{36}$ **B.** $\frac{1}{18}$ **C.** $\frac{1}{12}$ **D.** $\frac{2}{6}$

27. Tom and Mae play chess against each other. Tom has a $\frac{2}{3}$ probability of beating Mae in each game. What is the probability of Tom winning 2 games in a row?

A. $\frac{1}{9}$ **B.** $\frac{4}{9}$ **C.** $\frac{5}{9}$ **D.** $\frac{6}{9}$

28. A jar contains 7 blue marbles and 3 black marbles. Rae removes 1 marble, does not replace it, and then removes another marble. What is the probability that both marbles are blue?

A. $\frac{1}{49}$ **B.** $\frac{2}{7}$ **C.** $\frac{42}{90}$ **D.** $\frac{49}{100}$

29. Bo, Cara, and Don are neighbors. They each have one pet: a dog, a hamster, or a turtle. Bo and the turtle owner are the same age. Bo walks his pet every morning. Don is afraid of turtles. Who owns the hamster?

A. Bo **B.** Cara **C.** Don

Use the information below to answer questions 30–32.

Bob and Cari must take health and music during a two-year period. Each student is randomly assigned one of the courses for the current semester. Assume that the probability of assignment to each course is $\frac{1}{2}$.

30. What is the probability that both students will be assigned the health course?

A. $\frac{3}{4}$ **B.** $\frac{1}{2}$ **C.** $\frac{1}{4}$ **D.** $\frac{1}{8}$

31. What is the probability that neither student will be assigned the music course?

A. $\frac{3}{4}$ **B.** $\frac{1}{2}$ **C.** $\frac{1}{4}$ **D.** $\frac{1}{8}$

32. What is the probability that Bob and Cari will be assigned different classes?

A. $\frac{3}{4}$ **B.** $\frac{1}{2}$ **C.** $\frac{1}{4}$ **D.** $\frac{1}{8}$

Stop!

Name ____________________

Cumulative Test

Choose the letter of the correct answer.

1. Bill wants to buy 2 suits that cost \$199.95 each and 3 shirts that cost \$14.95 each. *Estimate* how much money Bill needs.

   A. \$245.00 **B.** \$430.00
   C. \$435.00 **D.** \$445.00

2. What is the value of 3^4?

   A. 12 **B.** 64 **C.** 81 **D.** 243

3. Express 506,000 in scientific notation.

   A. 5.06×10^5 **B.** 5.6×10^5
   C. 5.06×10^6 **D.** 5.60×10^6

Use the bar graph below to answer questions 4–5.

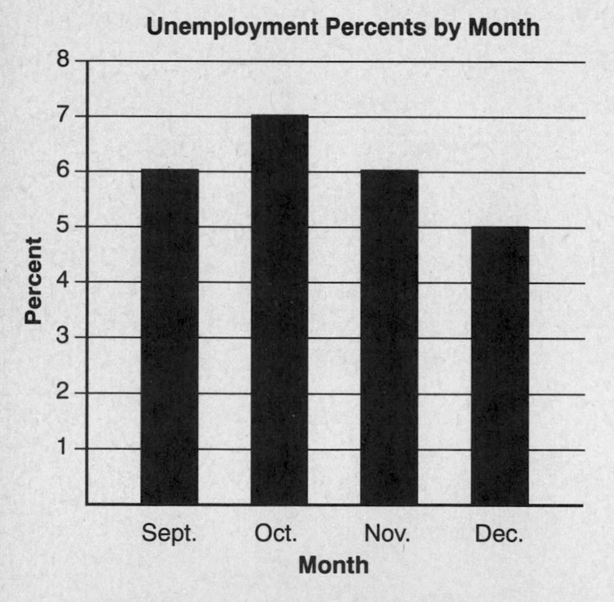

4. Which month had the greatest unemployment percent?

   A. September **B.** October
   C. November **D.** December

5. The unemployment percent in December was what fraction of the unemployment percent in November?

   A. $\frac{1}{5}$ **B.** $\frac{2}{3}$ **C.** $\frac{5}{7}$ **D.** $\frac{5}{6}$

Use the information and box-and-whisker graph below to answer questions 6–7.

A clothing-store owner records the number of suits sold in his store each day for 30 days. He summarizes the information in the box-and-whisker graph below.

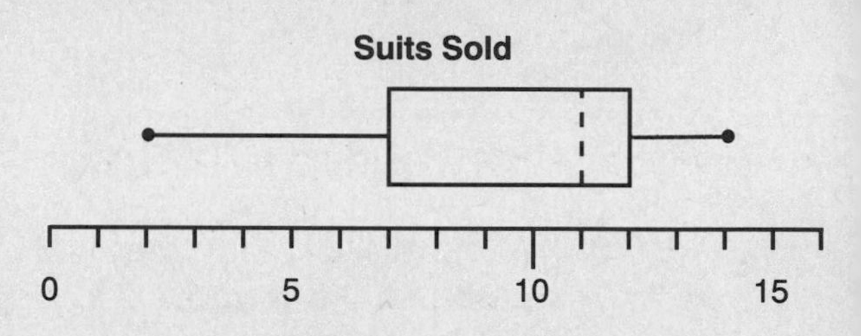

6. What was the median number of suits sold in his store each day?

   A. 7 suits **B.** 11 suits
   C. 12 suits **D.** 14 suits

7. What was the greatest number of suits sold on any of the 30 days?

   A. 12 suits **B.** 14 suits
   C. 15 suits **D.** 16 suits

Use the information below to answer questions 8–9.

Li's bowling scores for her last 10 games were 85, 90, 90, 90, 92, 92, 95, 97, 99, and 105.

8. What is the range of these scores?

   A. 20 **B.** 85 **C.** 105 **D.** 190

9. What is the median of these scores?

   A. 90 **B.** 92 **C.** 93 **D.** 93.5

10. Which fractions are ordered from least to greatest?

    A. $\frac{5}{6}, \frac{4}{5}, \frac{3}{4}$ **B.** $\frac{1}{4}, \frac{1}{3}, \frac{1}{2}$
    C. $\frac{3}{9}, \frac{2}{9}, \frac{1}{9}$ **D.** $\frac{2}{3}, \frac{2}{4}, \frac{2}{5}$

Go on to the next page.

Cumulative Test

11. Which fraction falls between $\frac{8}{10}$ and $\frac{9}{10}$?

    A. $\frac{15}{20}$ B. $\frac{31}{40}$

    C. $\frac{17}{20}$ D. $\frac{7}{10}$

12. $13\frac{1}{4} - 10\frac{1}{6} = $ ___?___

    A. $2\frac{1}{12}$ B. $2\frac{1}{6}$

    C. $3\frac{1}{12}$ D. not here

13. $3\frac{1}{2} \times 1\frac{1}{3} = $ ___?___

    A. $3\frac{1}{6}$ B. $3\frac{2}{3}$

    C. $4\frac{2}{3}$ D. not here

14. $\frac{7}{8} \div \frac{8}{7} = $ ___?___

    A. 1 B. $1\frac{1}{8}$

    C. $1\frac{1}{7}$ D. not here

15. Find the value of $(2 + 4) \times (7 - 2)$.

    A. 22 B. 28 C. 30 D. 40

16. Solve the inequality. $x + 9 < 11$

    A. $x < 1$ B. $x < \frac{11}{9}$

    C. $x < 2$ D. $x < 20$

17. Rectangles *A* and *B* are congruent. Rectangle *A* is 10 cm long by 20 cm wide. Which of the following are possible dimensions of Rectangle *B*?

    A. 1 cm × 2 cm

    B. 5 m × 15 cm

    C. 20 cm × 40 cm

    D. 10 cm × 20 cm

18. A box of 5 cakes costs $4.20. What is the unit price?

    A. $0.21 B. $0.82

    C. $0.84 D. $0.94

19. Use the scale 2 cm = 100 cm. What is the actual length of an object when the scale length is 5 cm?

    A. 20 cm B. 40 cm

    C. 250 cm D. 500 cm

20. Two rectangles are similar. The first rectangle has a width of 8 ft and a length of 20 ft. The width of the second rectangle is 10 ft. Find its length.

    A. 4 ft B. 16 ft

    C. 22 ft D. 25 ft

21. Express the ratio 9:12 as a percent.

    A. 0.75% B. 7.5%

    C. 75% D. 750%

22. Express 85% as a fraction.

    A. $\frac{5}{8}$ B. $\frac{17}{20}$

    C. $\frac{8}{5}$ D. not here

23. Phil invited 28 friends to a party. Only 25% of them could come. Which is the correct equation to find how many came to the party?

    A. $n = \frac{28}{0.25}$ B. $n \times 0.25 = 28$

    C. $n = \frac{0.25}{28}$ D. $n = 0.25 \times 28$

24. 3.5 is what percent of 70?

    A. 0.5% B. 5% C. 20% D. 50%

25. 15% of what number is 120?

    A. 18 B. 80 C. 720 D. 800

Go on to the next page.

Name ______________________________

Cumulative Test

26. Jan is buying a bicycle that is marked down 20%. What will be the sale price of a $200 bicycle?

A. $160 **B.** $220

C. $225 **D.** $250

27. Dick borrows $600. The interest rate is 9.5% per year. The payoff period is 24 months. How much interest will Dick pay? Use the formula $I = prt$.

A. $54 **B.** $57 **C.** $108 **D.** $114

28. ${}^{-}4 + {}^{-}3 = \underline{\quad ? \quad}$

A. ${}^{-}7$ **B.** ${}^{-}1$

C. 1 **D.** not here

29. $\frac{{}^{-}8}{{}^{-}4} = \underline{\quad ? \quad}$

A. ${}^{-}2$ **B.** $\frac{1}{2}$

C. 2 **D.** not here

30. To solve the equation $x - 9 = 22$, what must you do to both sides of the equation?

A. add 9 **B.** subtract 9

C. add 22 **D.** subtract 22

31. Solve for y. $y + 6 = {}^{-}12$

A. $y = {}^{-}18$ **B.** $y = {}^{-}6$

C. $y = 6$ **D.** $y = 18$

32. Are mixed numbers rational?

A. always **B.** sometimes **C.** never

33. Which square root is an irrational number?

A. $\sqrt{4}$ **B.** $\sqrt{6}$ **C.** $\sqrt{9}$ **D.** $\sqrt{16}$

34. Compare. ${}^{-}500 \bigcirc 0.0006$

A. $>$ **B.** $=$ **C.** $<$

35. Compare. $2.30 \times 10^3 \bigcirc 2.03 \times 10^4$

A. $>$ **B.** $=$ **C.** $<$

36. $\frac{10^4}{10^2} = \underline{\quad ? \quad}$

A. 2 **B.** 10 **C.** 100 **D.** 1,000

37. Compare. $\sqrt{144} \bigcirc 13$

A. $>$ **B.** $=$ **C.** $<$

38. How many different combinations of 2 can be selected from a set of 6 items?

A. 12 combinations

B. 20 combinations

C. 30 combinations

D. not here

39. On one roll of a number cube numbered from 1 to 6, which has the highest probability?

A. a 1

B. a 5

C. a number less than 3

D. a number greater than 3

40. A spinner has 8 equal-sized sections numbered from 1 to 8. What is the probability of spinning an even number?

A. $\frac{1}{8}$ **B.** $\frac{2}{8}$ **C.** $\frac{1}{2}$ **D.** $\frac{6}{8}$

To answer questions 41–42, assume that there is a vase containing 3 red balls and 3 black balls.

41. Helen takes 2 balls from the vase without replacement. What is the probability that both balls are black?

A. $\frac{1}{5}$ **B.** $\frac{1}{4}$ **C.** $\frac{3}{5}$ **D.** $\frac{2}{3}$

Go on to the next page.

Name ____________________

Cumulative Test

42. Helen takes 1 ball from the vase, replaces it, and then takes a second ball. What is the probability that both balls are red?

A. $\frac{1}{5}$ **B.** $\frac{1}{4}$ **C.** $\frac{1}{2}$ **D.** $\frac{3}{5}$

43. The number of points scored by two football teams totals 50. One team scored 4 times as many points as the other. How many points did the winning team score?

A. 24 points **B.** 32 points
C. 40 points **D.** 44 points

44. Nadia spent $20 buying books. She gave $\frac{1}{4}$ of the remaining money to her sister. Nadia then had $15. How much money did she have before buying her books?

A. $35 **B.** $40
C. $60 **D.** $80

Use the table below to answer question 45.

Students Graduating from Grant High	
Year	**Number**
1986	1,250
1987	1,300
1988	1,400
1989	1,550
1990	1,750

45. Assuming the trend continued, how many students graduated from Grant High in 1991?

A. 1,900 students
B. 1,950 students
C. 2,000 students
D. 2,050 students

46. Tricia had 32 questions correct on a math test. Her score was 2 items more than $\frac{3}{4}$ of her twin sister's score. How many questions did Tricia's sister have correct?

A. 23 questions **B.** 26 questions
C. 40 questions **D.** 42 questions

47. Joel and Gary begin a swimming program. Joel swims 4 laps on day 1 and increases the distance by 1 lap each day. Gary swims 4 laps on day 1 and increases the distance by $1\frac{1}{2}$ laps each day. On day 7, how much farther will Gary swim than Joel?

A. 3 laps **B.** $4\frac{1}{2}$ laps
C. 6 laps **D.** 13 laps

48. There are 8 teams in a softball league. If each team plays each of the other teams twice, how many total games will be played?

A. 28 games **B.** 56 games
C. 112 games **D.** 128 games

49. Heidi watches a television program that lasts $\frac{1}{2}$ hour. Commercials make up $\frac{1}{6}$ of the show. What part of an hour does she spend watching the actual program?

A. $\frac{1}{12}$ hr **B.** $\frac{1}{3}$ hr
C. $\frac{5}{12}$ hr **D.** $\frac{5}{6}$ hr

50. Mrs. Chan drove at an average speed of 48 miles per hour. She drove a distance of 120 miles. How long did the drive take? (distance = rate × time)

A. $2\frac{1}{4}$ hr **B.** $2\frac{1}{2}$ hr
C. $2\frac{3}{4}$ hr **D.** 3 hr

Stop!

Name ____________________

Pretest

Choose the letter of the correct answer.

1. To measure the width of a closet, which is the most appropriate unit of measure?

   A. inch **B.** foot **C.** yard **D.** mile

2. In which of the following situations would precise measurements be necessary?

   A. determining your car's gas mileage
   B. determining how much paint is needed to paint your house
   C. planning how far you will drive on a vacation
   D. measuring the time to run a 50-yd dash at a track meet

3. Which might be the height of a man?

   A. 3,000 mm **B.** 2,000 cm
   C. 1.7 m **D.** 0.5 km

4. Which might be the weight of a banana?

   A. 4 oz **B.** 4 lb **C.** 4 g **D.** 4 kg

5. Which is the best estimate of the amount of water it takes to fill an eyedropper?

   A. 1 mL **B.** 0.5 L **C.** 1 qt **D.** $\frac{1}{4}$ pint

6. Which is the best estimate of the width of a desk?

   A. 15 in. **B.** 3 ft **C.** 4 yd **D.** 4 m

7. Which measurement is most precise?

   A. 36 in. **B.** 3 ft
   C. 1 yd **D.** $1\frac{1}{4}$ yd

8. Which measurement is most precise?

   A. 98 ml **B.** 1 ml
   C. 1,000 mL **D.** 1,000.0 mL

9. Which measurement is most precise?

   A. 4 qt **B.** $4\frac{1}{4}$ qt
   C. $1\frac{1}{2}$ gal **D.** $1\frac{3}{4}$ gal

10. Which measurement is most precise?

    A. 3,600 sec **B.** 60 min
    C. 1 hr **D.** 1.0 hr

11. What is the perimeter of a regular pentagon when $s = 4\frac{1}{2}$ cm?

    A. 18 cm **B.** $20\frac{1}{4}$ cm
    C. $22\frac{1}{2}$ cm **D.** not here

12. What is the circumference of a circle with a radius of 3 in.? Use $\pi = 3.14$.

    A. 9.14 in. **B.** 9.42 in.
    C. 18.84 in. **D.** 28.26 in.

13. Jim is making a basketball hoop from a metal bar. He wants the diameter of the hoop to be 50 cm. How long does the metal bar need to be? Use $\pi = 3.14$.

    A. 15.7 cm **B.** 157 cm
    C. 31.4 cm **D.** 314 cm

14. Doreen wants to put wooden trim along the top of the walls in her room. The room is 9 ft wide by 12 ft long. How many feet of trim does she need?

    A. 21 ft **B.** 42 ft
    C. 108 ft **D.** 225 ft

Go on to the next page.

Name ____________________

Pretest

15. Which figure has the greater area?

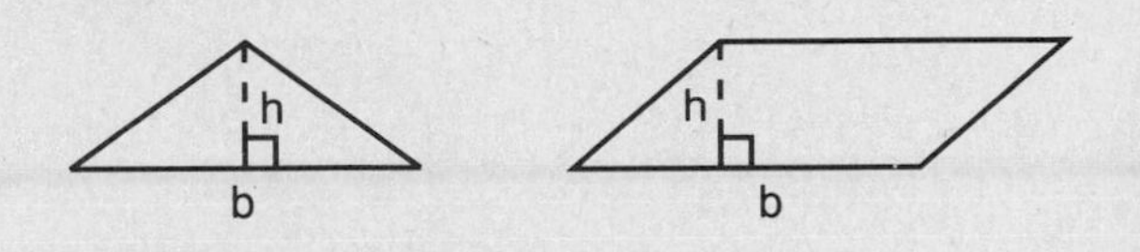

A. the triangle

B. the parallelogram

C. The areas are equal.

16. Do two trapezoids with the same height have the same area?

A. always **B.** sometimes **C.** never

17. What is the area of a parallelogram with a base of 10 cm and a height of 5 cm?

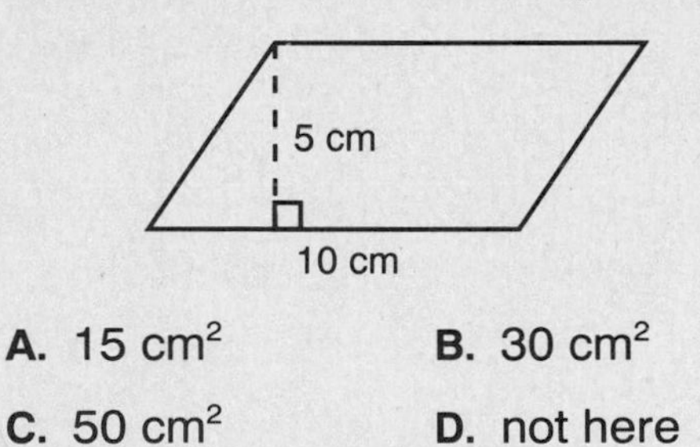

A. 15 cm^2 **B.** 30 cm^2

C. 50 cm^2 **D.** not here

18. What is the area of a triangle with a base of 9 ft and a height of 7 ft?

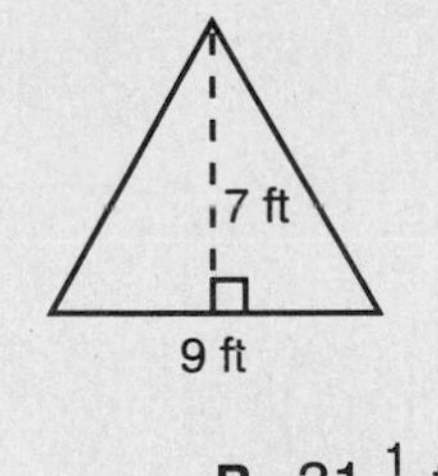

A. 16 ft^2 **B.** $31\frac{1}{2}$ ft^2

C. 63 ft^2 **D.** 130 ft^2

19. What is the area of a circle with a diameter of 7 m? Use $\pi = \frac{22}{7}$.

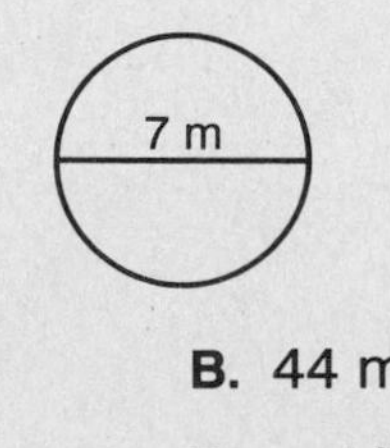

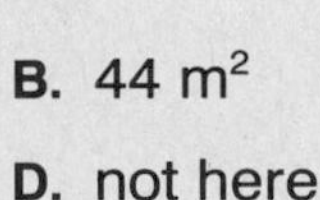

A. 22 m^2 **B.** 44 m^2

C. 154 m^2 **D.** not here

20. What is the area of a trapezoid with bases of 4 cm and 6 cm and a height of 10 cm?

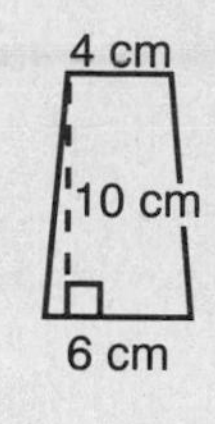

A. 20 cm^2 **B.** 30 cm^2

C. 100 cm^2 **D.** not here

21. Jack's garden is in the shape of a square. Its perimeter is 26 ft. What is the area of the garden?

A. 26 ft^2 **B.** 42.25 ft^2

C. 169 ft^2 **D.** 676 ft^2

22. Erica arranged a play space for her dog. She put a pole into the ground and attached a 10-ft leash. The dog can go exactly 10 ft in any direction from the pole. What is the area of the play space?

A. 31.4 ft^2 **B.** 62.8 ft^2

C. 100 ft^2 **D.** 314 ft^2

23. Does a figure with line symmetry also have turn symmetry?

A. always **B.** sometimes **C.** never

24. Does translation change a figure's size but not its shape?

A. always **B.** sometimes **C.** never

25. Which of the following letters has line symmetry?

A. P **B.** R **C.** F **D.** E

26. Which of the following letters has turn symmetry?

A. L **B.** E **C.** H **D.** K

Go on to the next page.

Pretest

27. How many lines of symmetry, does a square have?

A. 1 line **B.** 2 lines
C. 3 lines **D.** 4 lines

28. For turn symmetry, what is the angle measure of each turn for a regular six-sided polygon?

A. 30° **B.** 60°
C. 90° **D.** 120°

Use the formulas below to answer questions 29–31.

$F = \quad C + 32$ $C = \quad (F - 32)$

29. A mixture freezes at 10°C. At what Fahrenheit temperature does it freeze?

A. 22°F **B.** 42°F
C. 50°F **D.** 75.6°F

30. One day the temperature dropped from 60°F to 42°F in one hour. What was the temperature drop in degrees Celsius?

A. 2° **B.** 10° **C.** 18° **D.** 42°

31. Alice heated wax and observed its temperature with a Celsius thermometer. If the reading on her thermometer was 45°C, what was the temperature in degrees Fahrenheit?

A. 13°F **B.** 25°F
C. 81°F **D.** 113°F

32. Harry wants to make a wallpaper design from congruent regular polygons. Which of the following regular polygons do *not* tessellate a plane?

A. triangles **B.** squares
C. hexagons **D.** pentagons

Stop!

Name ______________________________

Posttest

Choose the letter of the correct answer.

1. To measure the distance between two cities, which is the most appropriate unit of measure?

A. millimeter **B.** centimeter
C. meter **D.** kilometer

2. In which situation would estimation be most appropriate?

A. pouring a dose of medicine
B. measuring your ring size for a jeweler
C. measuring the temperature of meat you are cooking
D. measuring a table leg that you must replace

3. Which might be the height of a room in a home?

A. 1,000 mm **B.** 5,000 cm
C. 3.5 m **D.** 0.1 km

4. Which might be the weight of an egg?

A. 2 oz **B.** 1 lb **C.** 2 g **D.** 1 kg

5. Which is the best estimate of the amount of water it takes to fill a cup?

A. 10 mL **B.** 1 L **C.** 8 oz **D.** 2 pt

6. Which is the best estimate of the width of a refrigerator?

A. 15 in. **B.** 3 ft **C.** 4 yd **D.** 5 m

7. Which measurement is most precise?

A. 35 in. **B.** $36\frac{1}{2}$ in.
C. 3 ft **D.** $3\frac{1}{2}$ ft

8. Which measurement is most precise?

A. 98 cm **B.** 98.1 cm
C. 1 m **D.** 1.13 m

9. Which measurement is most precise?

A. 128 fl oz **B.** 8 pt
C. 2 qt **D.** $\frac{1}{2}$ gal

10. Which measurement is most precise?

A. 2 sec **B.** 1.1 sec
C. $1\frac{1}{2}$ sec **D.** 1.001 sec

11. What is the perimeter of a regular hexagon when $s = 2.5$ cm?

A. 6.25 cm **B.** 12.5 cm
C. 15 cm **D.** 17.5 cm

12. What is the circumference of a circle with a radius of 2 ft? Use $\pi = 3.14$.

A. 6.28 ft **B.** 12.56 ft
C. 18.84 ft **D.** not here

13. Harold wants to build a hoop with a circumference of 16 ft. What will be the diameter? Use $\pi = 3.14$.

A. 2.55 ft **B.** 5.10 ft
C. 25.12 ft **D.** 50.24 ft

14. The perimeter of a rectangle is 10 cm. The shorter sides are each 2 cm. What is the length of each of the longer sides?

A. $2\frac{1}{2}$ cm **B.** 3 cm
C. 4 cm **D.** 5 cm

Go on to the next page.

Name ________________________________

Posttest

15. Which triangle has the greater area?

A. the equilateral triangle

B. the right triangle

C. The areas are equal.

16. Do two circles with the same circumference have the same area?

A. always **B.** sometimes **C.** never

17. What is the area of a parallelogram with a base of 20 cm and a height of 15 cm?

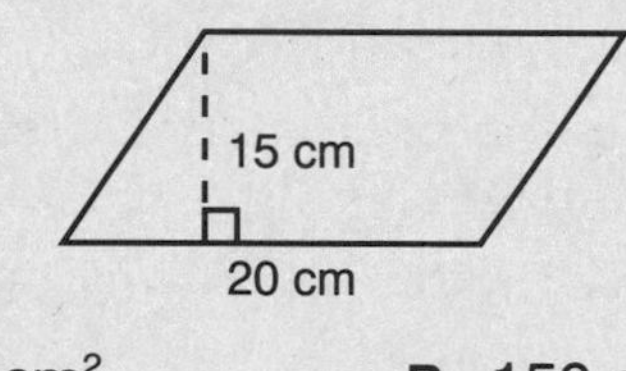

A. 70 cm^2 **B.** 150 cm^2

C. 300 cm^2 **D.** not here

18. What is the area of a triangle with a base of 3 ft and a height of 4 ft?

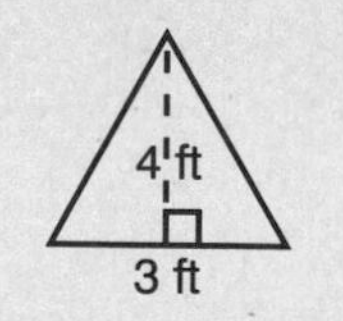

A. $3\frac{1}{2}$ ft^2 **B.** 6 ft^2

C. 12 ft^2 **D.** 49 ft^2

19. What is the area of a circle with a diameter of 10 m? Use $\pi = 3.14$.

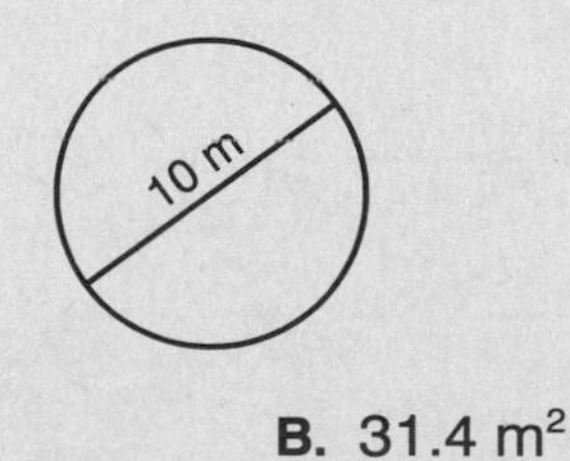

A. 15.7 m^2 **B.** 31.4 m^2

C. 314 m^2 **D.** not here

20. What is the area of a trapezoid with bases of 10 cm and 20 cm and a height of 10 cm?

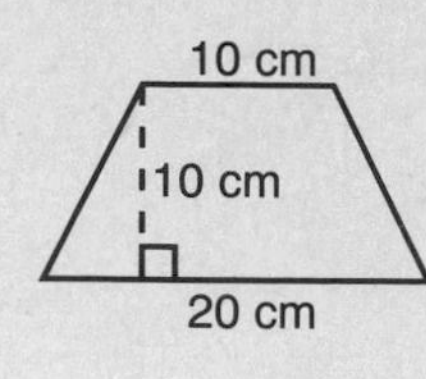

A. 50 cm^2 **B.** 150 cm^2

C. 200 cm^2 **D.** 300 cm^2

21. Jack's garden is in the shape of a square. Its area is 36 ft^2. What is the perimeter of the garden?

A. 6 ft **B.** 12 ft **C.** 18 ft **D.** 24 ft

22. Hilda's garden is circular. The area is 3.14 ft^2. What is the diameter?

A. $\frac{1}{2}$ ft **B.** 1 ft

C. $1\frac{1}{2}$ ft **D.** 2 ft

23. Does a circle have more than 4 lines of symmetry?

A. always **B.** sometimes **C.** never

24. Does reflection change both the size and the shape of a figure?

A. always **B.** sometimes **C.** never

25. Which of the following letters has line symmetry?

A. J **B.** Q **C.** G **D.** H

26. Which of the following letters has turn symmetry?

A. A **B.** P **C.** W **D.** X

27. How many lines of symmetry does an isosceles triangle have?

A. 1 line **B.** 2 lines

C. 3 lines **D.** 4 lines

Go on to the next page.

Name ____________________

Posttest

28. For turn symmetry, what is the angle measure of each turn for a square?

A. 30° **B.** 60° **C.** 90° **D.** 120°

Use the formulas below to answer questions 29–31.

$F = \frac{9}{5}C + 32$ $C = \frac{5}{9}(F - 32)$

29. Water boils at 100°C. At what Fahrenheit temperature does it boil?

A. 112°F **B.** 132°F

C. 180°F **D.** 212°F

30. One day the temperature increased from 70°F to 79°F in one hour. What was the temperature increase in degrees Celsius?

A. 1° **B.** 5° **C.** 9° **D.** 18°

31. Berto chilled water and observed its temperature with a Fahrenheit thermometer. If the reading on his thermometer was 45°F, what was the temperature in degrees Celsius?

A. 13°C **B.** 25°C

C. 77°C **D.** not here

32. John wants to make a quilt design from congruent regular polygons. He can use any regular polygon for which the measures of the angles meeting at a vertex have a sum of which of the following?

A. 90° **B.** 180°

C. 360° **D.** 720°

Stop!

Name ______________________________

Pretest

Choose the letter of the correct answer.

1. Paco is drawing a pattern for a solid figure. The base of the figure is a polygon. Each side of the base is the base of an isosceles triangle. Which figure is Paco drawing?

   A. cone **B.** cylinder
   C. pyramid **D.** triangular prism

2. Helen built a solid figure with 2 flat surfaces and 1 curved surface. Which figure did she build?

   A. cylinder **B.** sphere
   C. cone **D.** polyhedron

3. Which of these figures has no flat surfaces?

   A. cone **B.** cylinder
   C. pyramid **D.** sphere

4. What do prisms and pyramids have in common?

   A. Their lateral faces are all triangles.
   B. They are polyhedrons.
   C. They have congruent, parallel bases.
   D. None of their surfaces are polygons.

5. How many faces does a triangular prism have?

   A. 3 faces **B.** 4 faces
   C. 5 faces **D.** 6 faces

6. A polyhedron has 6 faces and 8 vertices. How many edges does it have?

   A. 12 edges **B.** 14 edges
   C. 16 edges **D.** 18 edges

7. How many faces does a rectangular pyramid have?

   A. 4 faces **B.** 5 faces
   C. 6 faces **D.** 8 faces

8. A solid figure has 6 faces: 5 triangles and 1 pentagon. What is the figure?

   A. pentagonal pyramid
   B. triangular prism
   C. pentagonal prism
   D. not here

9. What is the surface area of this cylinder? Use $\pi = 3.14$.

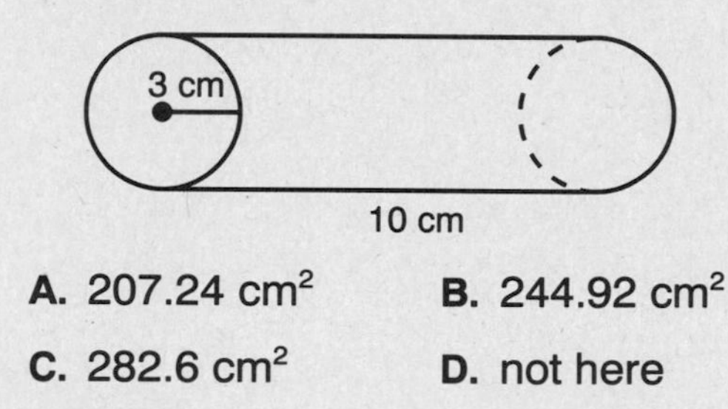

   A. 207.24 cm^2 **B.** 244.92 cm^2
   C. 282.6 cm^2 **D.** not here

10. The dimensions of a rectangular prism are 5 ft × 4 ft × 4 ft. What is its surface area?

    A. 80 ft^2 **B.** 104 ft^2
    C. 112 ft^2 **D.** 400 ft^2

11. A cube has a surface area of 600 cm^2. What is the length of each face?

    A. 50 cm **B.** 60 cm
    C. 100 cm **D.** not here

12. What is the surface area of this square pyramid?

    A. 120 $in.^2$
    B. 145 $in.^2$
    C. 240 $in.^2$
    D. 300 $in.^2$

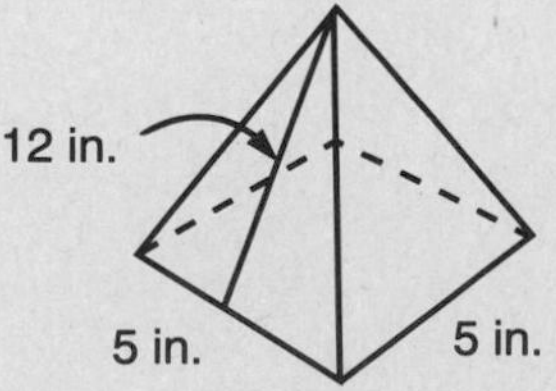

Go on to the next page.

Pretest

13. What is the volume of a cylinder whose radius is 2 m and whose height is 0.25 m? Use $\pi = 3.14$.

A. 3.14 m^3 **B.** 6.28 m^3
C. 9.42 m^3 **D.** 12.56 m^3

14. What is the volume of this rectangular pyramid?

A. 8 ft^3
B. 12 ft^3
C. 16 ft^3
D. 24 ft^3

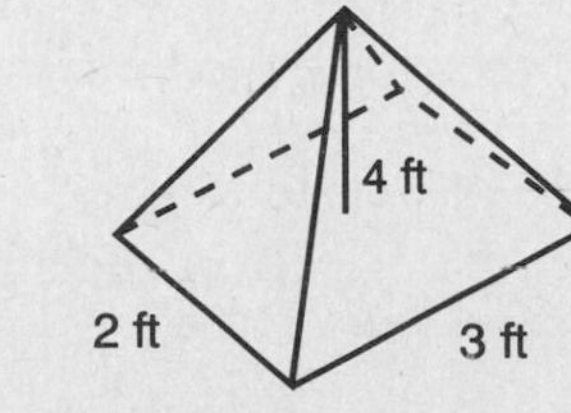

15. What is the volume of a rectangular prism with the dimensions 4 in. × 6 in. × 8 in.?

A. 18 $in.^3$ **B.** 72 $in.^3$
C. 144 $in.^3$ **D.** 192 $in.^3$

16. What volume of water is held by a spoon with a capacity of 10 mL?

A. 1 cm^3 **B.** 10 cm^3
C. 100 cm^3 **D.** not here

17. A swimming pool is 25 m long, 10 m wide, and 2 m deep. What is the capacity of the pool?

A. 50 kL **B.** 250 kL
C. 500 kL **D.** 5,000 kL

18. The volume of a pitcher is 1,500 cm^3. What mass of water will fill it?

A. 1.5 kg **B.** 15 kg
C. 150 kg **D.** 1,500 kg

19. Lena's aquarium is in the shape of a cylinder. Its radius is 5 cm and its height is 10 cm. What is the volume? Use $\pi = 3.14$.

A. 157 cm^3 **B.** 785 cm^3
C. 1,000 cm **D.** 1,570 cm^3

20. Harry packs yogurt into tubs that are in the shape of a cylinder 8 in. high and 8 in. in diameter. What are the dimensions of a tub that will hold 4 times as much yogurt?

A. $h = 16$ in.; $d = 8$ in.
B. $h = 8$ in.; $d = 16$ in.
C. $h = 12$ in.; $d = 12$ in.
D. $h = 16$ in.; $d = 16$ in.

21. *Estimate* how much money you need in order to buy 3 pairs of socks at $1.98 a pair and 4 cans of tennis balls at $2.99 a can.

A. $17.00 **B.** $18.00
C. $18.50 **D.** $19.00

22. *About* how many square feet of carpet are needed for a rectangular room 9 ft 11 in. by 17 ft 2 in.?

A. 150 ft^2 **B.** 170 ft^2
C. 180 ft^2 **D.** 190 ft^2

23. Joe is wrapping a box. *Estimate* how many square feet of wrapping paper he needs if the box measures $11\frac{1}{4}$ in. by $12\frac{1}{2}$ in. by 23 in.

A. 2 ft^2 **B.** 6 ft^2
C. 8 ft^2 **D.** 10 ft^2

24. Aquarium A and aquarium B are both cylinders. They have the same radius, but A is twice as high as B. What is the ratio of the volume of water held by A to the volume of water held by B?

A. $\frac{1}{2}$ to 1 **B.** 2 to 1
C. 4 to 1 **D.** 8 to 1

Stop!

Name ___

Posttest

Choose the letter of the correct answer.

1. Lorna is building a solid figure. The figure has 2 parallel bases that are congruent triangles. Its other faces are rectangles. What is the figure?

   A. rectangular pyramid
   B. rectangular prism
   C. triangular pyramid
   D. triangular prism

2. Sean built a solid figure with 1 flat surface and 1 curved surface. Which figure did he build?

   A. cylinder **B.** sphere
   C. cone **D.** polyhedron

3. Which of these figures is *not* a polyhedron?

   A. cylinder **B.** prism
   C. pyramid **D.** cube

4. What do prisms and cylinders have in common?

   A. Their bases are circles.
   B. They have parallel, congruent bases.
   C. Their bases are polygons.
   D. They have curved surfaces.

5. How many faces does a triangular pyramid have?

   A. 3 faces **B.** 4 faces
   C. 5 faces **D.** 6 faces

6. A polyhedron has 8 faces and 12 vertices. How many edges does it have?

   A. 16 edges **B.** 18 edges
   C. 20 edges **D.** 22 edges

7. How many faces does a rectangular prism have?

   A. 4 faces **B.** 5 faces
   C. 6 faces **D.** 8 faces

8. A solid figure has 6 rectangular surfaces. What is the figure?

   A. rectangular prism
   B. hexagonal prism
   C. rectangular pyramid
   D. not here

9. What is the surface area of this cylinder? Use $\pi = 3.14$.

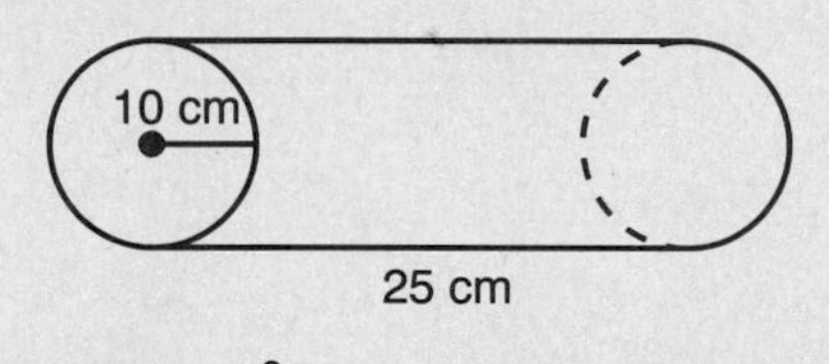

   A. 628 cm^2 **B.** 1,570 cm^2
   C. 2,198 cm^2 **D.** not here

10. What is the surface area of this rectangular prism?

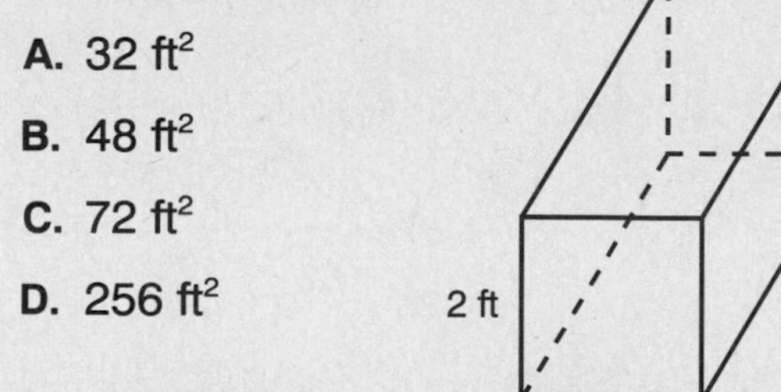

   A. 32 ft^2
   B. 48 ft^2
   C. 72 ft^2
   D. 256 ft^2

11. A cube has a surface area of 96 cm^2. What is the length of each face?

   A. 8 cm **B.** 16 cm
   C. 24 cm **D.** not here

12. The base of a square pyramid is 4 in. $\times$ 4 in. The height of each triangular face is 6 in. What is the surface area of the pyramid?

   A. 48 in.^{2} **B.** 64 in.^{2}
   C. 96 in.^{2} **D.** 112 in.^{2}

Go on to the next page.

Name ____________________

Posttest

13. What is the volume of a cylinder with a radius of 1 m and a height of 0.5 m? Use $\pi = 3.14$.

A. 0.785 m^3 **B.** 1.57 m^3
C. 3.14 m^3 **D.** not here

14. What is the volume of a rectangular pyramid that has a base 3 ft × 4 ft and a height of 5 ft?

A. 20 ft^3 **B.** 40 ft^3
C. 48 ft^3 **D.** 60 ft^3

15. What is the volume of this rectangular prism?

A. 60 $in.^3$
B. 120 $in.^3$
C. 540 $in.^3$
D. 600 $in.^3$

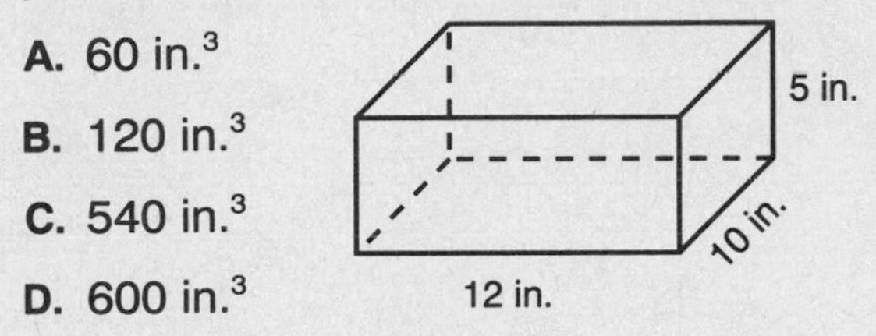

16. A cup with a volume of 50 cm^3 has what liquid capacity?

A. 5 mL **B.** 50 mL
C. 500 mL **D.** not here

17. A swimming pool is 30 m long, 15 m wide, and 1.5 m deep. What is the capacity of the pool?

A. 67.5 kL **B.** 450 kL
C. 675 kL **D.** 6,750 kL

18. The volume of a bowl is 3,500 cm^3. What mass of water will fill it?

A. 3.5 kg **B.** 35 kg
C. 350 kg **D.** 3,500 kg

19. Eli's aquarium is in the shape of a cylinder. It has a radius of 10 cm and height of 20 cm. What is the volume? Use $\pi = 3.14$.

A. 314 cm^3 **B.** 628 cm^3
C. 3,140 cm^3 **D.** 6,280 cm^3

20. Joel packs food into containers that are in the shape of a cone 6 cm high and 6 cm in diameter. What are the dimensions of a cone that will hold 2 times as much food?

A. $h = 12$ cm; $d = 6$ cm

B. $h = 6$ cm; $d = 12$ cm

C. $h = 9$ cm; $d = 9$ cm

D. $h = 12$ cm; $d = 12$ cm

21. *Estimate* how much money Lee needs in order to buy 2 shirts at $14.95 each, 2 pairs of shoes at $49.50 a pair, and 3 ties at $19.95 each.

A. $160.00 **B.** $175.00
C. $185.00 **D.** $190.00

22. *Estimate* the area of a rectangular garden that has the dimensions 15 ft 11 in. by 20 ft 2 in.

A. 300 ft^2 **B.** 315 ft^2
C. 320 ft^2 **D.** 350 ft^2

23. Emily is wrapping a box. *About* how many square feet of wrapping paper will she need if the dimensions of the box are $11\frac{1}{4}$ in. by $24\frac{1}{2}$ in. by 23 in.?

A. 4 ft^2 **B.** 10 ft^2
C. 12 ft^2 **D.** 16 ft^2

24. Aquarium A and aquarium B are both rectangular prisms. Their bases have the same dimensions, but A is twice as high as B. What is the ratio of the volume of water held by A to that held by B?

A. $\frac{1}{2}$ to 1 **B.** 2 to 1
C. 4 to 1 **D.** 8 to 1

Stop!

Name ____________________

Pretest

Choose the letter of the correct answer.

1. What are the first four square numbers?

   A. 1, 3, 6, 12 **B.** 1, 4, 9, 16

   C. 1, 5, 10, 15 **D.** 1, 2, 4, 8

2. How many dots would come next in this pattern?

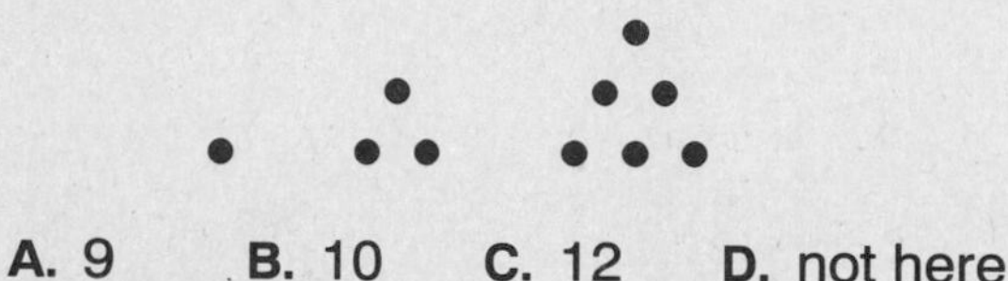

   A. 9 **B.** 10 **C.** 12 **D.** not here

3. What is the next number in the pattern?
   1, 3, 6, 10, 15

   A. 20 **B.** 21

   C. 22 **D.** not here

4. What is the next number in the pattern?
   243, 81, 27

   A. 3 **B.** 9 **C.** 18 **D.** not here

5. What are the first four numbers in this pattern? The first number is 7. Add 12 each time.

   A. 7, 12, 24, 36 **B.** 7, 19, 26, 38

   C. 7, 19, 31, 43 **D.** 12, 24, 36, 43

6. What are the first four numbers in this pattern? The first number is 2. Multiply by 3 and subtract 1 each time.

   A. 2, 3, 8, 23 **B.** 2, 5, 14, 41

   C. 2, 5, 15, 44 **D.** 2, 6, 5, 15

To answer questions 7–8, use the relation {(0,0), (1,2), (2,4), (3,6)}.

7. What is the domain of the relation?

   A. 0, 1 **B.** 2, 3

   C. 0, 1, 2, 3 **D.** 0, 2, 4, 6

8. What is the range of the relation?

   A. 0, 2 **B.** 4, 6

   C. 0, 1, 2, 3 **D.** 0, 2, 4, 6

9. Which ordered pairs are graphed?

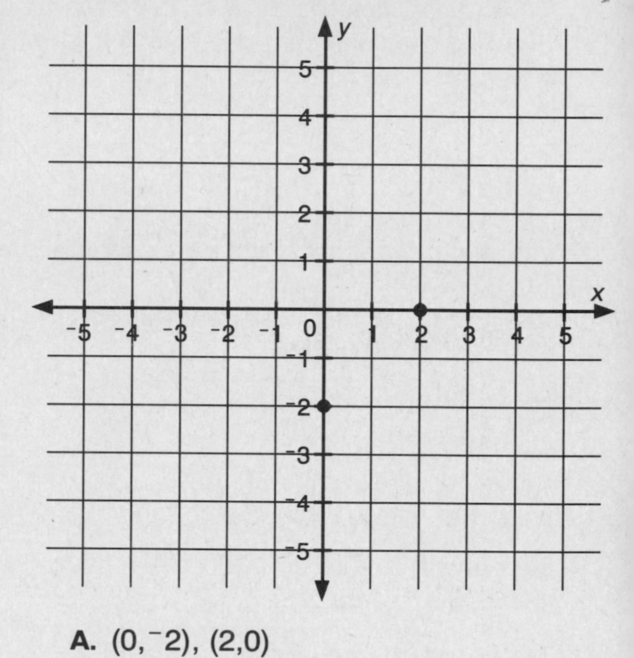

   A. $(0,{}^{-}2)$, $(2,0)$

   B. $({}^{-}2,0)$, $(2,0)$

   C. $(0,2)$, $(2,0)$

   D. $({}^{-}2,0)$, $(0,2)$

10. If the length and width of a rectangle both triple, what happens to the area of the rectangle?

    A. It doubles.

    B. It triples.

    C. It increases by 6 times.

    D. It increases by 9 times.

11. Which equation represents the relation {(0,0), (1,3), (2,6), (3,9)}?

    A. $y = x + 6$ **B.** $y = \frac{1}{3}x$

    C. $y = x + 3$ **D.** $y = 3x$

Go on to the next page.

Name ______________________________

Pretest

12. Which equation represents the relation presented in the table below?

Domain	Range
1	7
2	14
3	21
4	28

A. $y = x + 7$ **B.** $y = \frac{1}{7}x$

C. $y = 7x$ **D.** $y = 6x + 1$

13. What is the missing value in the following table?

Domain	Range
1	1
2	4
3	9
4	16
5	■

A. 23 **B.** 25 **C.** 32 **D.** not here

14. Which ordered pair fits the relation $y = 3x + 2$?

A. (5,1) **B.** (0,5)

C. (1,8) **D.** (0,2)

15. Which of the following gives the lengths of the sides of a right triangle?

A. 3 m, 4 m, 6 m

B. 6 m, 8 m, 10 m

C. 5 m, 6 m, 8 m

D. 6 m, 9 m, 12 m

16. To which triangles does the Pythagorean Property apply?

A. all triangles

B. equilateral triangles

C. isosceles triangles

D. right triangles

17. Find c.

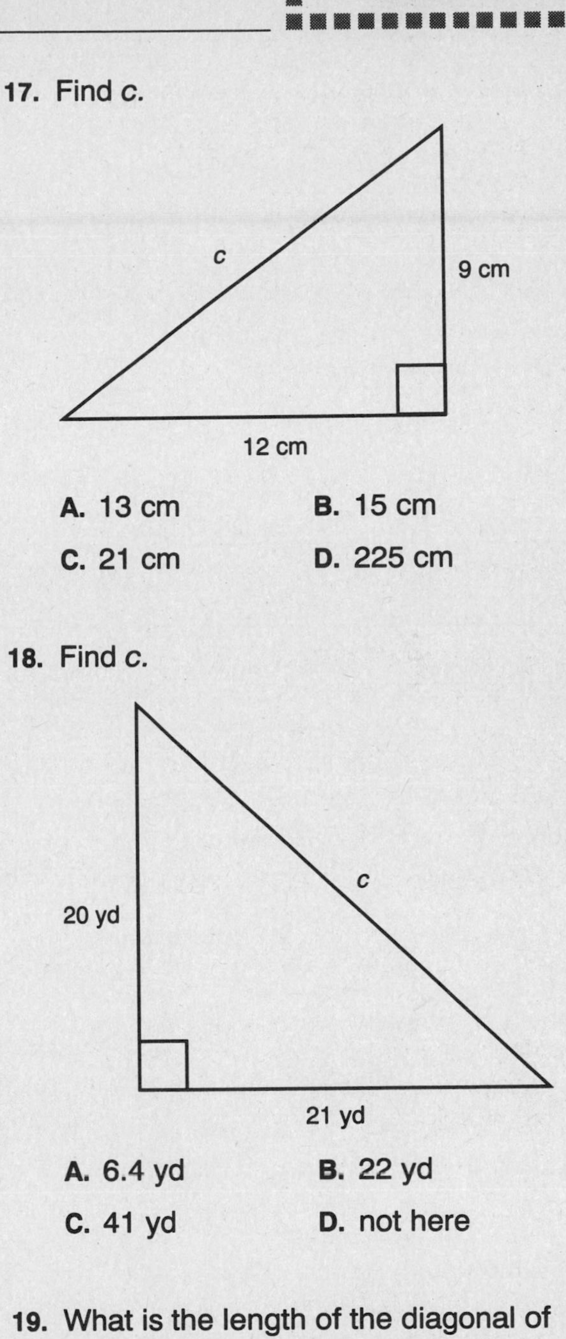

A. 13 cm **B.** 15 cm

C. 21 cm **D.** 225 cm

18. Find c.

A. 6.4 yd **B.** 22 yd

C. 41 yd **D.** not here

19. What is the length of the diagonal of this square, to the nearest inch?

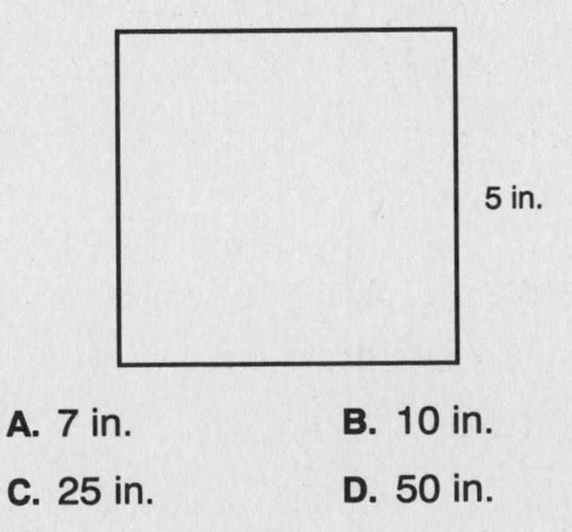

A. 7 in. **B.** 10 in.

C. 25 in. **D.** 50 in.

Go on to the next page.

20. What is the length of the diagonal of this rectangle, to the nearest tenth?

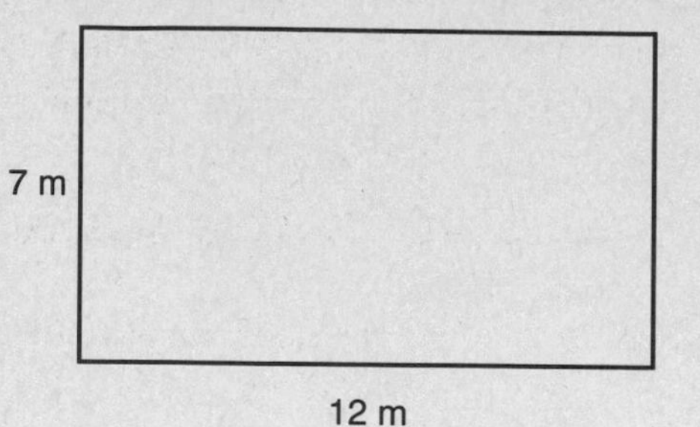

A. 13.9 m **B.** 14.0 m

C. 14.1 m **D.** 19.0 m

21. Is a relation a function?

A. always **B.** sometimes **C.** never

22. What is a relation called that has only one element of the range for each element of the domain?

A. a domain **B.** an equation

C. a function **D.** not here

23. Which ordered pair fits this equation?

$$y = 10x + 1$$

A. (1,11) **B.** (1,0)

C. (21,2) **D.** (1,12)

24. Which point on the graph is (2,0)?

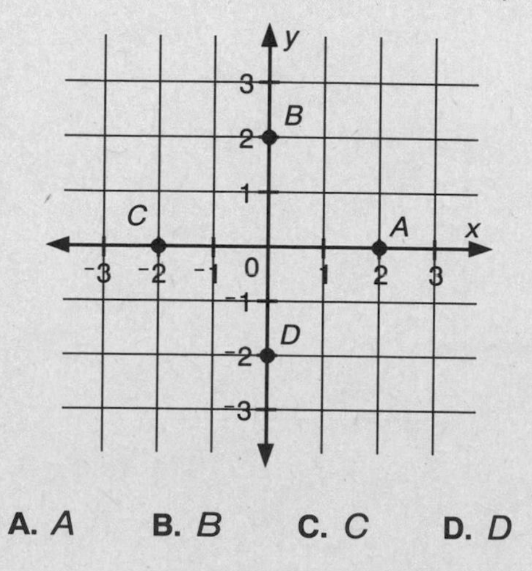

A. *A* **B.** *B* **C.** *C* **D.** *D*

25. Which of the following relations is a function?

A. {(1,6), (2,6), (3,6)}

B. {(1,6), (1,7)}

C. {(1,6), (2,6), (2,7)}

D. {(2,1), (2,2)}

26. Which of the following is a function?

A. {(1,1), (1,2)}

B. {(1,1), (2,1)}

C. {(2,1), (2,2)}

D. {(0,1), (0,2)}

27. Which of the following relations is *not* a function?

A. $y = x$

B. $y = 2x$

C. $y = 2x + 3$

D. $y = \sqrt{x^2 + 1}$

28. Which of the following graphs does *not* represent a function?

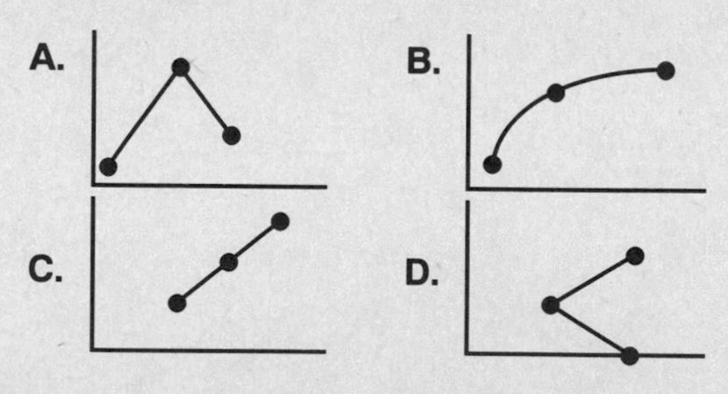

29. Otis sold 99 drinks. He sold only milk and juice. He sold twice as many cartons of milk as cartons of juice. How many cartons of milk did he sell?

A. 33 cartons

B. 60 cartons

C. 63 cartons

D. 66 cartons

Go on to the next page.

30. Ron sold 80 hamburgers and hot dogs. He sold 20 more hot dogs than hamburgers. Which equation could be used to find how many hamburgers he sold? (Let x = the number of hamburgers sold.)

A. $x + 20 = 80$

B. $2x = 80$

C. $x + x + 20 = 80$

D. $x + x + x = 60$

31. John took some money from his bank account. He spent $5 on lunch and $\frac{1}{2}$ of the remainder at the grocery store. He was left with $45. How much money did he take from his account?

A. $90 **B.** $95 **C.** $100 **D.** $105

32. The sum of two numbers is 32. One number is 6 greater than the other. What are the two numbers?

A. 12 and 18 **B.** 13 and 19

C. 14 and 18 **D.** 14 and 20

Stop!

Name ______________________________

Posttest

Choose the letter of the correct answer.

1. What are the first four triangular numbers?

   A. 1, 4, 9, 16 **B.** 1, 3, 6, 10
   C. 1, 3, 9, 27 **D.** 1, 5, 12, 22

2. How many dots would come next in this pattern?

   A. 23 **B.** 25 **C.** 32 **D.** not here

3. What is the next number in the pattern? 10, 15, 21, 28

   A. 35 **B.** 36 **C.** 37 **D.** not here

4. What is the next number in the pattern? 122, 100, 80, 62

   A. 44 **B.** 46 **C.** 52 **D.** 60

5. What are the first four numbers in this pattern? The first number is 99. Subtract 7 each time.

   A. 92, 85, 78, 71
   B. 99, 92, 85, 78
   C. 99, 97, 90, 83
   D. 99, 90, 83, 76

6. What are the first four numbers in this pattern? The first number is 0. Multiply by 5 and add 2 each time.

   A. 0, 2, 10, 12 **B.** 0, 2, 12, 62
   C. 0, 5, 15, 35 **D.** 0, 7, 37, 187

To answer questions 7–8, use the relation {(1,2), (2,4), (3,6), (4,8)}.

7. What is the domain of the relation?

   A. 1, 2 **B.** 3, 4
   C. 1, 2, 3, 4 **D.** 2, 4, 6, 8

8. What is the range of the relation?

   A. 2, 4
   B. 6, 8
   C. 1, 2, 3, 4
   D. 2, 4, 6, 8

9. Which ordered pairs are graphed?

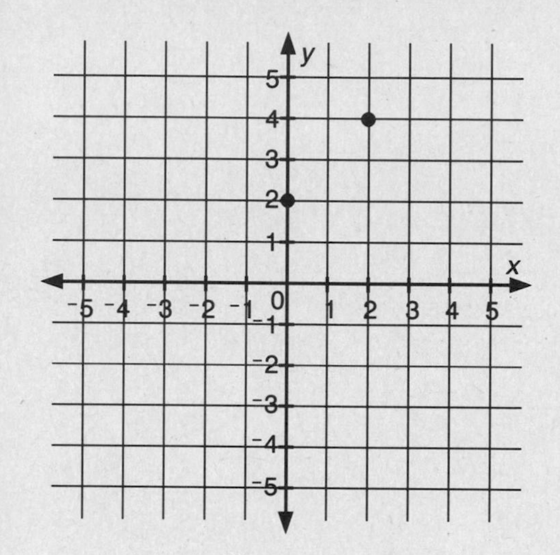

   A. (0,2), (2,4)
   B. (2,0), (2,4)
   C. (2,0), (4,2)
   D. (0,2), (4,2)

10. If the length of a rectangle triples and the width doubles, what happens to the area of the rectangle?

    A. It doubles.
    B. It increases by 5 times.
    C. It increases by 6 times.
    D. It increases by 10 times.

11. Which equation represents the relation {(0,0), (1,5), (2,10), (3,15)}?

    A. $y = x + 12$
    B. $y = 4x + 3$
    C. $y = x + 4$
    D. $y = 5x$

Go on to the next page.

Name ____________________

Posttest

12. Which equation represents the relation presented in the table below?

Domain	Range
1	4
2	8
3	12
4	16

A. $y = x + 3$

B. $y = x + 12$

C. $y = 4x$

D. $y = \frac{1}{4}x$

13. What is the missing value in the following table?

Domain	Range
1	4
2	7
3	10
4	13
5	■

A. 15 **B.** 16

C. 17 **D.** not here

14. Which ordered pair fits the relation $y = x - 2$?

A. (0,2)

B. ($^-2$,0)

C. (2,4)

D. (0,$^-2$)

15. Which of the following gives the lengths of the sides of a right triangle?

A. 9 cm, 12 cm, 16 cm

B. 9 cm, 12 cm, 15 cm

C. 3 cm, 6 cm, 9 cm

D. 5 cm, 9 cm, 11 cm

16. The square of the longest side of a triangle is greater than the sum of the squares of the two shorter sides. What do you know about the triangle?

A. It cannot be an isosceles triangle.

B. It must be a scalene triangle.

C. It must be an equilateral triangle.

D. It cannot be a right triangle.

17. Find *c*.

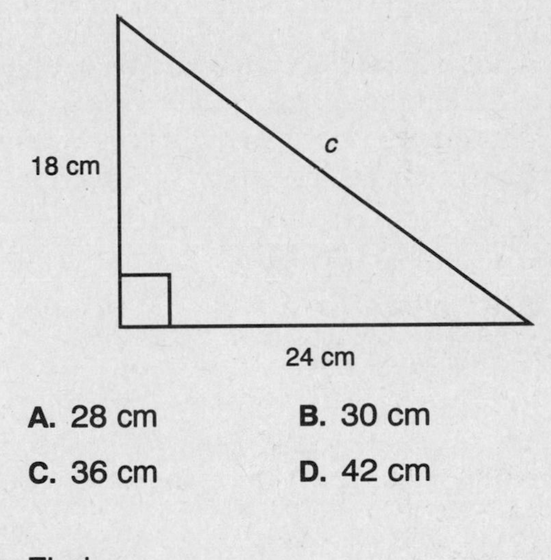

A. 28 cm **B.** 30 cm

C. 36 cm **D.** 42 cm

18. Find *c*.

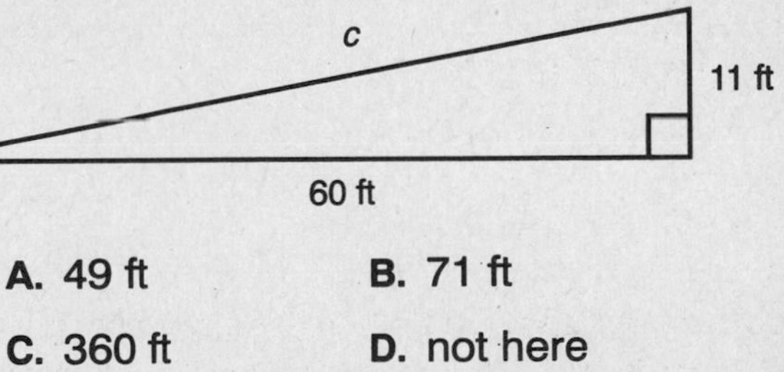

A. 49 ft **B.** 71 ft

C. 360 ft **D.** not here

19. What is the length of the diagonal of this square, to the nearest foot?

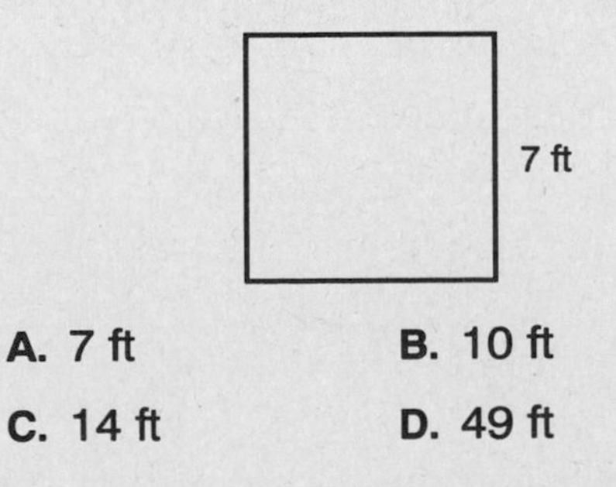

A. 7 ft **B.** 10 ft

C. 14 ft **D.** 49 ft

Go on to the next page.

20. What is the length of the diagonal of this rectangle, to the nearest tenth?

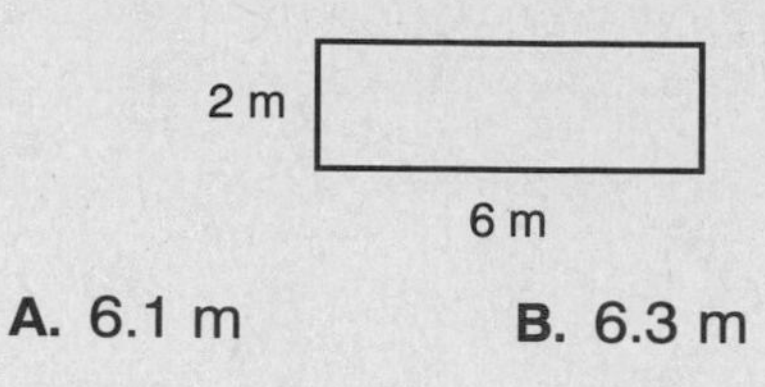

A. 6.1 m **B.** 6.3 m
C. 6.6 m **D.** 6.9 m

21. Is a function a relation?

A. always **B.** sometimes **C.** never

22. What is a relation called that has only one element of the range for each element of the domain?

A. a function
B. a figurate number
C. a domain
D. not here

23. Which ordered pair fits this equation?

$$y = 5x + 10$$

A. (20,2) **B.** (0,15)
C. (10,0) **D.** (1,15)

24. Which point on the graph is (2,3)?

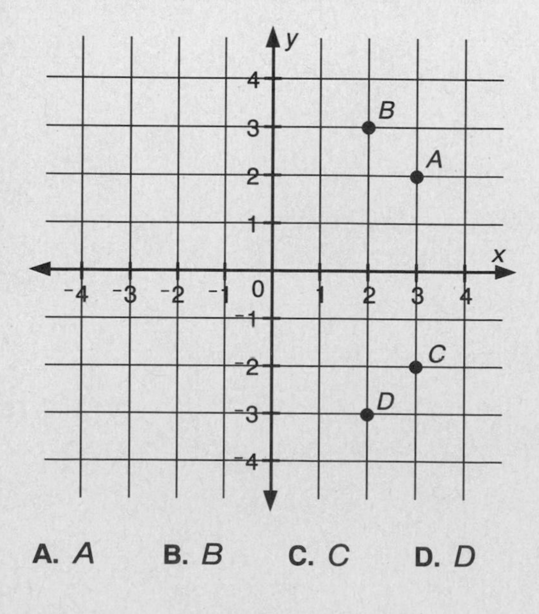

A. *A* **B.** *B* **C.** *C* **D.** *D*

25. Which of the following relations is a function?

A. {(1,3), (2,3), (3,3)}
B. {(1,3), (1,4)}
C. {(1,3), (2,3), (2,4)}
D. {(2,3), (2,4)}

26. Which of the following is a function?

A. {(0,1), (0,2)} **B.** {(0,1), (2,1)}
C. {(2,0), (2,1)} **D.** {(0,0), (0,2)}

27. Which of the following relations is *not* a function?

A. $y = x + 50$ **B.** $y = 2x$
C. $y = x$ **D.** $y = \sqrt{x}$

28. Which of the following graphs does *not* represent a function?

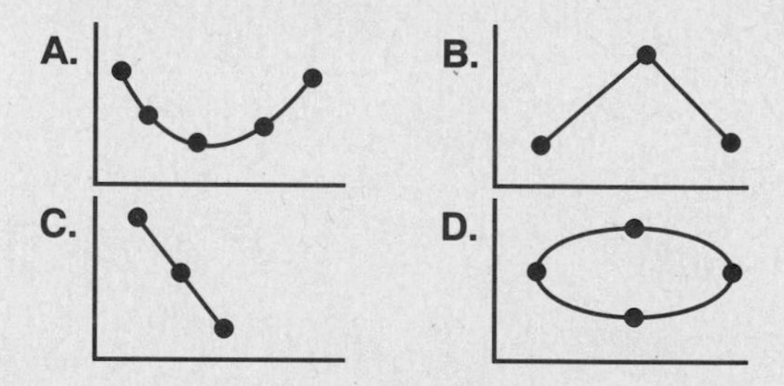

29. Hilda sold 200 hats. She sold 3 times as many men's hats as women's hats. How many men's hats did she sell?

A. 50 men's hats **B.** 125 men's hats
C. 150 men's hats **D.** 160 men's hats

30. John sold 100 hats. He sold 60 more women's hats than men's hats. Which equation could be used to find how many men's hats he sold? (Let x = the number of men's hats sold.)

A. $x + 60 = 100$
B. $2x = 100$
C. $x + x + 60 = 100$
D. $x + x + x = 100 + 60$

Go on to the next page.

Posttest

31. Bill took money from his bank account. He spent \$6 at a movie and $\frac{1}{3}$ of the remainder for groceries. He then had \$40 left. How much money did he take from his bank account?

A. \$60 **B.** \$66 **C.** \$96 **D.** \$126

32. The product of two numbers is 72. One number is 6 greater than the other. What are the two numbers?

A. 4 and 18 **B.** 6 and 12

C. 14 and 8 **D.** 7 and 13

Stop!

Name ____________

Cumulative Test

Choose the letter of the correct answer.

1. Choose the best estimate.

$$\frac{99.77}{2.07} = \underline{\quad ? \quad}$$

A. less than 45 B. less than 50

C. greater than 50 D. greater than 55

2. $9^0 = \underline{\quad ? \quad}$

A. 0 B. 1 C. 9 D. 90

3. What is 1.12×10^4 in standard form?

A. 1,102 B. 11,200

C. 112,000 D. 1,120,000

Use the line graph below to answer questions 4–5.

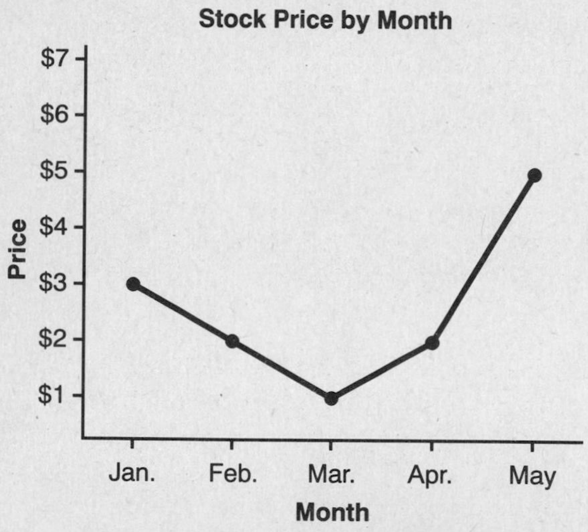

Stock Price by Month

Price: $7, $6, $5, $4, $3, $2, $1

Month: Jan., Feb., Mar., Apr., May

4. During what month was the stock selling at the lowest price?

A. January B. February

C. March D. May

5. How much did the price of the stock go up from April to May?

A. $1.50 B. $2.00

C. $3.00 D. $5.00

6. Seven students took a quiz. Their scores were 12, 6, 5, 8, 10, 11, 11. What was the median score?

A. 8 B. 9 C. 10 D. 11

7. Which group is ordered from least to greatest?

A. $1\frac{1}{5}, 1\frac{1}{2}, 1\frac{4}{10}$

B. $\frac{6}{5}, \frac{6}{4}, \frac{6}{3}$

C. $\frac{3}{8}, \frac{3}{10}, \frac{3}{12}$

D. $\frac{11}{5}, \frac{12}{6}, \frac{13}{7}$

8. $5\frac{2}{3} + 5\frac{3}{4} = \underline{\quad ? \quad}$

A. $10\frac{5}{12}$ B. $10\frac{5}{7}$

C. $11\frac{7}{12}$ D. not here

9. $8\frac{1}{2} \div 4\frac{1}{4} = \underline{\quad ? \quad}$

A. $\frac{1}{4}$ B. $\frac{1}{2}$ C. 2 D. $2\frac{1}{4}$

10. Find the value of $4 \times (8 - 2)$.

A. 18 B. 24 C. 30 D. not here

11. Solve the inequality. $x + 11 < 9$

A. $x < {}^{-}2$ B. $x < 1$

C. $x < 2$ D. $x < 20$

12. What is the measure of each angle in a regular pentagon? Use the formula $(n - 2) \times 180° \div n$.

A. 72° B. 90° C. 108° D. not here

13. If oranges are priced at 6 for $1.25 and 18 for $3.40, which has the lower unit price?

A. 6 for $1.25 B. 18 for $3.40

Go on to the next page.

Cumulative Test

14. On a map, 1 in. = 50 mi. Two cities are 14 inches apart on the map. How many miles are they actually apart?

A. 64 mi **B.** 70 mi
C. 280 mi **D.** 700 mi

15. Two rectangles are similar. The first rectangle has a width of 6 cm and length of 12 cm. The width of the second rectangle is 0.5 cm. Find its length.

A. 0.25 cm **B.** 1 cm
C. 2 cm **D.** 6 cm

16. Express $\frac{2}{10,000}$ as a decimal.

A. 0.02 **B.** 0.002
C. 0.0002 **D.** not here

17. Estimate. 79% of 82 = __?__

A. 64 **B.** 80 **C.** 90 **D.** 105

18. 30% of what number is 120?

A. 36 **B.** 40 **C.** 400 **D.** not here

19. Hilda's Dress Shop sells sweaters for $40.00. Yesterday, all prices were reduced by 20%. Today, prices are reduced by an additional 10% of the sale price. What is today's sale price of a sweater?

A. $11.20 **B.** $28.00
C. $28.80 **D.** $29.80

20. George bought a concert ticket. He paid $22.00 plus 5% sales tax. How much did George pay for the ticket?

A. $23.10 **B.** $24.10
C. $24.20 **D.** $27.00

21. $^{-}50 \times {}^{-}10 =$ __?__

A. $^{-}500$ **B.** 500
C. 5,000 **D.** not here

22. Solve for x. $x + 14 = 3$

A. $x = {}^{-}17$ **B.** $x = {}^{-}11$
C. $x = 11$ **D.** $x = 17$

23. Will a fraction whose denominator is 10 result in a repeating decimal?

A. always **B.** sometimes **C.** never

24. Express $\frac{37}{50}$ as a decimal.

A. 0.074 **B.** 0.375
C. 0.74 **D.** 7.4

25. Which is a rational number between $^{-}0.20$ and $^{-}0.21$?

A. $^{-}0.199$ **B.** $^{-}0.201$
C. $^{-}0.211$ **D.** $^{-}0.220$

26. Compare. $\left(\frac{1}{4}\right)^2 \bigcirc \left(\frac{1}{5}\right)^3$

A. > **B.** = **C.** <

27. How many different combinations of 2 can be selected from a set of 8 coins?

A. 4 combinations
B. 10 combinations
C. 16 combinations
D. 28 combinations

28. How many different arrangements are possible of the letters A, B, C, D, E?

A. 12 arrangements
B. 20 arrangements
C. 60 arrangements
D. 120 arrangements

Go on to the next page.

Cumulative Test

29. Harvey rolls a single number cube numbered from 1 to 6. Which event has a $\frac{1}{3}$ probability of occurrence?

A. a roll of 3

B. a roll of any odd number

C. a roll of 1 or 2

D. a roll of any number greater than 3

30. A coin is flipped 4 times. What is the probability of obtaining 4 heads?

A. $\frac{1}{32}$ B. $\frac{1}{16}$ C. $\frac{1}{8}$ D. $\frac{1}{4}$

31. A bowl has 4 oranges and 5 apples. If Amy chooses one piece of fruit at random, how likely is it that she will choose a banana?

A. certain

B. impossible

C. neither

32. To measure the height of a room, which is the most appropriate unit of measurement?

A. inches B. feet C. yards D. miles

33. In which of the following situations would a precise measurement be necessary?

A. determining your car's gas mileage

B. determining how much soap you need to wash your clothes

C. planning how many miles you will fly on a trip to Mexico

D. measuring the time to run a 100-yard dash at a track meet

34. What is the circumference of a circle with a radius of 2 inches? Use 3.14 for π.

A. 1.57 in. B. 6.28 in.

C. 12.56 in. D. not here

35. Find the area of a parallelogram with a base of 40 cm and height of 20 cm.

A. 60 cm^2 B. 80 cm^2

C. 400 cm^2 D. not here

36. Does a figure with line symmetry also have turn symmetry?

A. always B. sometimes C. never

37. Hank is drawing a pattern for a figure. The figure has a rectangle as its base. Each side of the base is the base of a triangle. Which figure is he drawing?

A. cone B. cylinder

C. pyramid D. rectangular prism

38. The dimensions of a rectangular prism are 6 ft $\times$ 4 ft $\times$ 4 ft. What is its surface area?

A. 96 ft^2 B. 112 ft^2

C. 128 ft^2 D. 144 ft^2

39. Find the area of a triangle with a base of 40 cm and a height of 20 cm.

A. 40 cm^2 B. 80 cm^2

C. 400 cm^2 D. 800 cm^2

40. Which ordered pair fits the relation $y = 4x + 2$?

A. (10,2) B. (8,2)

C. (2,8) D. (2,10)

Go on to the next page.

Name ________________________________

Cumulative Test

41. Find the hypotenuse of a right triangle with legs measuring 12 cm and 9 cm. The sum of the squares of the legs is equal to the square of the hypotenuse.

A. 13 cm **B.** 15 cm
C. 18 cm **D.** 19 cm

42. Does a graph that shows a circle describe a function?

A. always **B.** sometimes **C.** never

43. What is the total angle measure of a regular 7-sided polygon? Use the formula $(n - 2) \times 180°$.

A. 800° **B.** 880° **C.** 900° **D.** 1,080°

Use the information and the double-bar graph below to answer question 44.

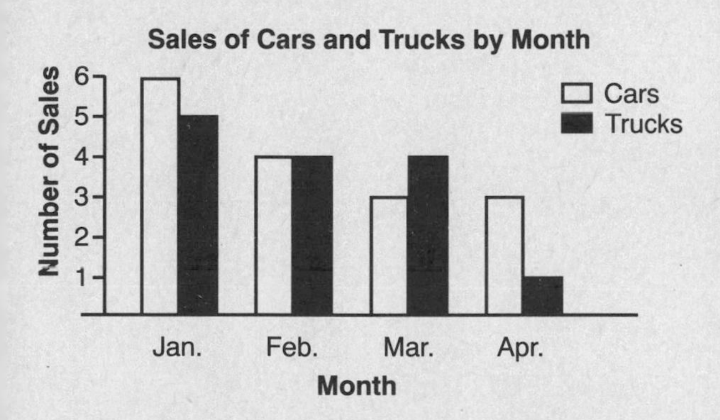

44. During which month did truck sales exceed car sales?

A. January **B.** February
C. March **D.** April

45. Two numbers add to 201. One of the numbers is 9 more than the other. What are the numbers?

A. 95 and 106 **B.** 94 and 107
C. 96 and 105 **D.** 98 and 103

46. Roger is paid $2.00 per hour for baby-sitting during the day and $2.50 per hour after 6 P.M. This week, Roger worked 6 hours Friday night and 3 hours Saturday afternoon. How much did he earn?

A. $19.50 **B.** $21.00
C. $23.00 **D.** $40.50

47. Joy travels 195 miles. The trip takes exactly 3 hours. What is her rate of speed? (distance = rate × time.)

A. 60 mph **B.** 65 mph
C. 75 mph **D.** 80 mph

48. Helen bowled 3 games. Her second game score was twice as high as her first game. Her third game was 15 points higher than her second game. Her score in game 3 was 255. What was Helen's score in game 1?

A. 100 **B.** 105 **C.** 110 **D.** 120

49. Mike missed 3 words on a spelling test. His sister missed 6 more than twice as many words as Mike missed. How many words did his sister miss?

A. 9 words **B.** 12 words
C. 15 words **D.** 18 words

50. Paula wants to make a scarf from congruent regular polygons that tessellate a plane. She can use any regular polygon for which the angles meeting at a vertex total to which of the following?

A. 180° **B.** 240° **C.** 320° **D.** 360°

Stop!

Name ____________________

End-of-Book Test

Choose the letter of the correct answer.

1. Elliot wants to buy 3 magazines that cost \$1.29 each and a newspaper that costs \$0.35. *Estimate* how much money Elliot needs.

   A. \$2.00 **B.** \$3.00
   C. \$4.00 **D.** \$5.00

2. $10^{-3} = \underline{\quad ? \quad}$

   A. $^{-}1{,}000$ **B.** $^{-}300$
   C. $\frac{^{-}1}{1{,}000}$ **D.** $\frac{1}{1{,}000}$

3. What is 3.009×10^3 in standard form?

   A. 300.9 **B.** 3,009
   C. 3,090 **D.** 3,900

Use the information and bar graph to answer questions 4–5.

Maxine used a bar graph to compare the performance of 4 classes on a math quiz.

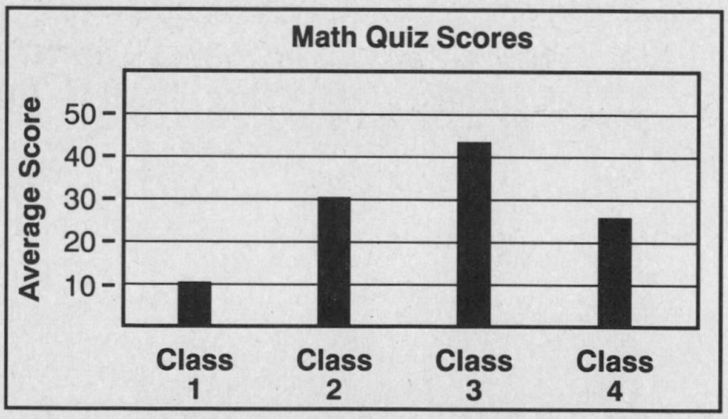

Math Quiz Scores

Average Score: 10, 20, 30, 40, 50

Class 1, Class 2, Class 3, Class 4

4. Which class had the highest average score on the quiz?

   A. Class 1 **B.** Class 2
   C. Class 3 **D.** Class 4

5. How much higher did Class 4 average than did Class 1?

   A. 2 points **B.** 10 points
   C. 15 points **D.** 25 points

6. Seven students took a quiz. Their scores were 42, 37, 38, 24, 24, 16, 29. What is the mean of these scores?

   A. 24 **B.** 29 **C.** 30 **D.** 35

7. Which fraction lies between $\frac{1}{11}$ and $\frac{2}{11}$?

   A. $\frac{1}{22}$ **B.** $\frac{3}{44}$
   C. $\frac{6}{44}$ **D.** $\frac{5}{22}$

8. $9\frac{1}{2} - 2\frac{2}{3} = \underline{\quad ? \quad}$

   A. $6\frac{5}{6}$ **B.** $7\frac{1}{6}$
   C. $7\frac{5}{6}$ **D.** not here

9. $\frac{3}{4} \div \frac{3}{8} = \underline{\quad ? \quad}$

   A. $\frac{9}{32}$ **B.** $\frac{1}{2}$
   C. 2 **D.** not here

10. Find the value of $(4 + 2) \times (3 + 4)$.

    A. 14 **B.** 18 **C.** 36 **D.** 42

11. Solve. $x + 20 < 25$

    A. $x < 5$ **B.** $x > 5$
    C. $x < 45$ **D.** $x > 45$

12. A triangle has angle measures of 40°, 50°, and 90°. What type of triangle is it?

    A. isosceles **B.** equilateral
    C. right **D.** acute

13. If juice is priced at 6 cans for \$2.70 and 8 cans for \$3.52, which has the lower unit price?

    A. 6 cans for \$2.70
    B. 8 cans for \$3.52

Go on to the next page.

Name ____________________

End-of-Book Test

14. On Map A, 1 cm = 50 km. On Map B, 1 cm = 75 km. Two cities are 10 cm apart on Map B. How far apart are they on Map A?

A. 15 cm **B.** 25 cm

C. 250 cm **D.** 375 cm

15. A triangle has sides measuring 6 cm, 8 cm, and 10 cm. Which of the following could be the lengths of the sides of a similar triangle?

A. 7 cm, 9 cm, 11 cm

B. 16 km, 18 km, 20 km

C. 9 yd, 12 yd, 20 yd

D. 9 m, 12 m, 15 m

16. Express 0.3% as a decimal.

A. 0.0003 **B.** 0.003

C. 0.03 **D.** not here

17. 27 is what percent of 108?

A. 19% **B.** 25%

C. 27% **D.** 400%

18. 20% of what number is 200?

A. 40 **B.** 400 **C.** 1,000 **D.** 10,000

19. Nikki bought a shirt for $22.00 plus 7% sales tax. How much did Nikki pay for the shirt?

A. $22.70 **B.** $23.44

C. $23.54 **D.** $24.54

20. Andy is buying a suit that is marked down 25%. What is the sale price of a $240 suit?

A. $175 **B.** $180 **C.** $215 **D.** $300

21. $^{-}25 \times 20 =$ ____?____

A. $^{-}250$ **B.** $^{-}50$

C. 500 **D.** not here

22. Solve for x. $x + 20 = 18$

A. $x = {}^{-}2$ **B.** $x = {}^{-}1$

C. $x = 2$ **D.** $x = 38$

23. Are mixed numbers rational numbers?

A. always **B.** sometimes **C.** never

24. Which of these fractions can be expressed as a terminating decimal?

A. $\frac{2}{18}$ **B.** $\frac{1}{7}$ **C.** $\frac{1}{5}$ **D.** $\frac{1}{3}$

25. Which is a rational number between 0.61 and 0.62?

A. 0.601 **B.** 0.609

C. 0.6110 **D.** 0.6201

26. Compare. $10^{-5} \bigcirc 10^{3}$

A. $>$ **B.** $=$ **C.** $<$

27. How many different combinations of 2 can be selected from a set of 20 items?

A. 90 combinations

B. 190 combinations

C. 380 combinations

D. not here

28. In how many different orders can 5 people arrange themselves in a line?

A. 24 orders **B.** 60 orders

C. 120 orders **D.** 360 orders

29. Harvey rolls a single number cube numbered from 1 to 6. What is the probability that he will roll an even number?

A. $\frac{1}{6}$ **B.** $\frac{1}{3}$ **C.** $\frac{1}{2}$ **D.** $\frac{3}{5}$

Go on to the next page.

Name ____________________

End-of-Book Test

30. A coin is flipped 3 times. What is the probability of obtaining 3 heads?

A. $\frac{1}{16}$ **B.** $\frac{1}{9}$ **C.** $\frac{1}{8}$ **D.** $\frac{1}{6}$

31. Jon and Ruth play 2 chess matches. Ruth is a better player than Jon. Her probability of winning each game is $\frac{3}{4}$. What is the probability that Ruth will win both games?

A. $\frac{3}{8}$ **B.** $\frac{9}{16}$ **C.** $\frac{2}{3}$ **D.** $\frac{3}{4}$

32. Which unit of measurement is the most appropriate for measuring the length of a book?

A. millimeter **B.** centimeter

C. meter **D.** kilometer

33. Which might be the weight of a peach?

A. 4 oz **B.** 4 lb

C. 4 g **D.** 4 kg

34. What is the perimeter of a regular pentagon with a side measuring 3.5 cm?

A. 14 cm **B.** 15.5 cm

C. 17.5 cm **D.** not here

35. Find the area of a triangle with a base of 20 in. and height of 35 in.

A. 55 in.^{2} **B.** 110 in.^{2}

C. 700 in.^{2} **D.** not here

36. Which of these letters has line symmetry?

A. the letter "J" **B.** the letter "R"

C. the letter "L" **D.** the letter "H"

37. Which of these figures is a polyhedron?

A. sphere **B.** cone

C. cylinder **D.** prism

38. The dimensions of a rectangular prism are 3 ft × 1 ft × 1 ft. What is its surface area?

A. 3 ft^2 **B.** 10 ft^2

C. 14 ft^2 **D.** 25 ft^2

39. Find the volume of a rectangular prism with the dimensions 2 in. × 4 in. × 8 in.

A. 14 in.^{3} **B.** 64 in.^{3}

C. 112 in.^{3} **D.** not here

40. Which ordered pair fits the relation $y = 3x + 2$?

A. (5,0) **B.** (0,5)

C. (2,0) **D.** (0,2)

41. Find the length of the hypotenuse of this right triangle. The sum of the squares of the legs is equal to the square of the hypotenuse.

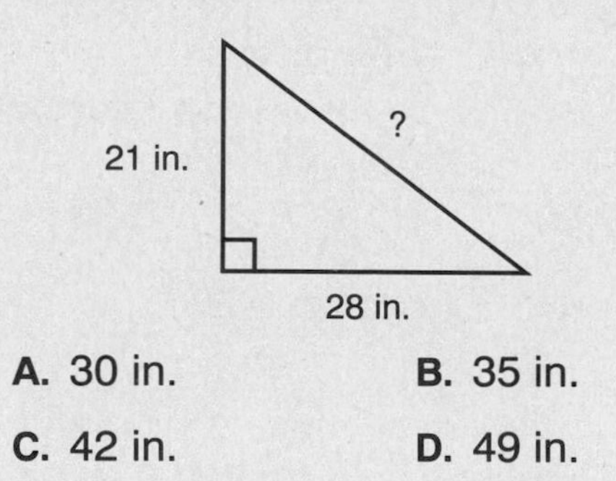

A. 30 in. **B.** 35 in.

C. 42 in. **D.** 49 in.

42. A relation that has only one element of the range for each element of the domain is called

A. a domain. **B.** a function.

C. an equation. **D.** a figurate number.

43. The scale of a road map is 1 inch = 100 miles. On the map, the highway distance from Budd Lake to Mount Tom is $\frac{3}{4}$ in. What is the distance in miles between the two points?

A. 3 miles **B.** 25 miles

C. 75 miles **D.** 125 miles

Go on to the next page.

Name

End-of-Book Test

Use the information and the double-bar graph to answer question 44.

Joan used a double-bar graph to compare the monthly sales of hatchbacks and sedans over a period of 4 months.

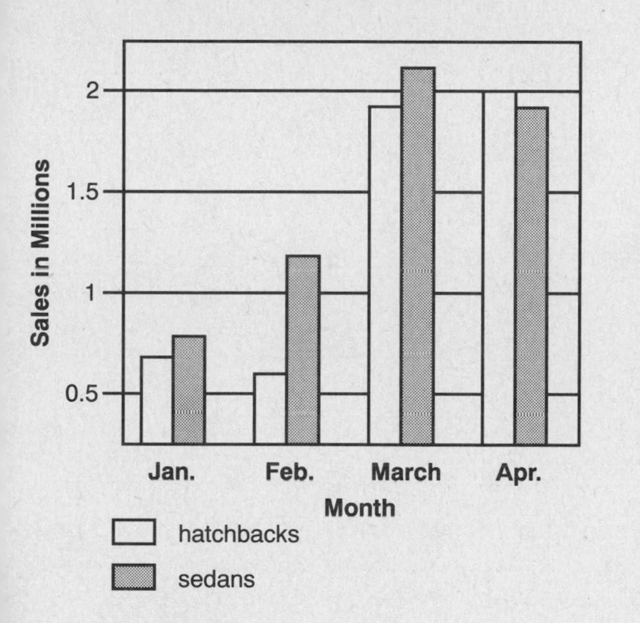

44. During which month was the difference in sales of hatchbacks and sedans the greatest?

    A. January **B.** February
    C. March **D.** April

45. Two numbers add to 360. One of the numbers is 3 times the other. What is the smaller number?

    A. 60 **B.** 80 **C.** 90 **D.** 120

46. Nancy walks 5 blocks to school. It takes her $2\frac{1}{2}$ minutes to walk each block. Pat walks 6 blocks to school. It takes her only $1\frac{1}{2}$ minutes to walk each block. How much longer does it take Nancy to walk to school than it takes Pat?

    A. $2\frac{1}{4}$ min **B.** $2\frac{1}{2}$ min
    C. $3\frac{1}{2}$ min **D.** $3\frac{3}{4}$ min

47. Art drove a distance of 275 miles in his car. His average speed was 50 miles per hour. How long did the drive take? (distance = rate × time)

    A. $5\frac{1}{4}$ hr **B.** $5\frac{1}{2}$ hr
    C. $5\frac{3}{4}$ hr **D.** $6\frac{1}{2}$ hr

48. Mimi spent $\frac{1}{2}$ of her savings on Monday. On Tuesday, after she spent $50.00, she had only $2.00 left. How much money did Mimi start with?

    A. $98.00 **B.** $102.00
    C. $104.00 **D.** $106.00

49. Helena and Mark went shopping. Mark spent $2.00 more than 3 times the amount Helena spent. Helena spent $5.50. How much did Mark spend?

    A. $16.50 **B.** $17.50
    C. $18.50 **D.** $19.50

Use the table to answer question 50.

Music Preferred by Students		
Music	**Frequency**	**Cumulative Frequency**
Rock	50	50
Jazz	20	70
Country	20	90
Rap	10	100

50. How many more students chose rock music than jazz music?

    A. 20 students **B.** 30 students
    C. 50 students **D.** 100 students

Stop!

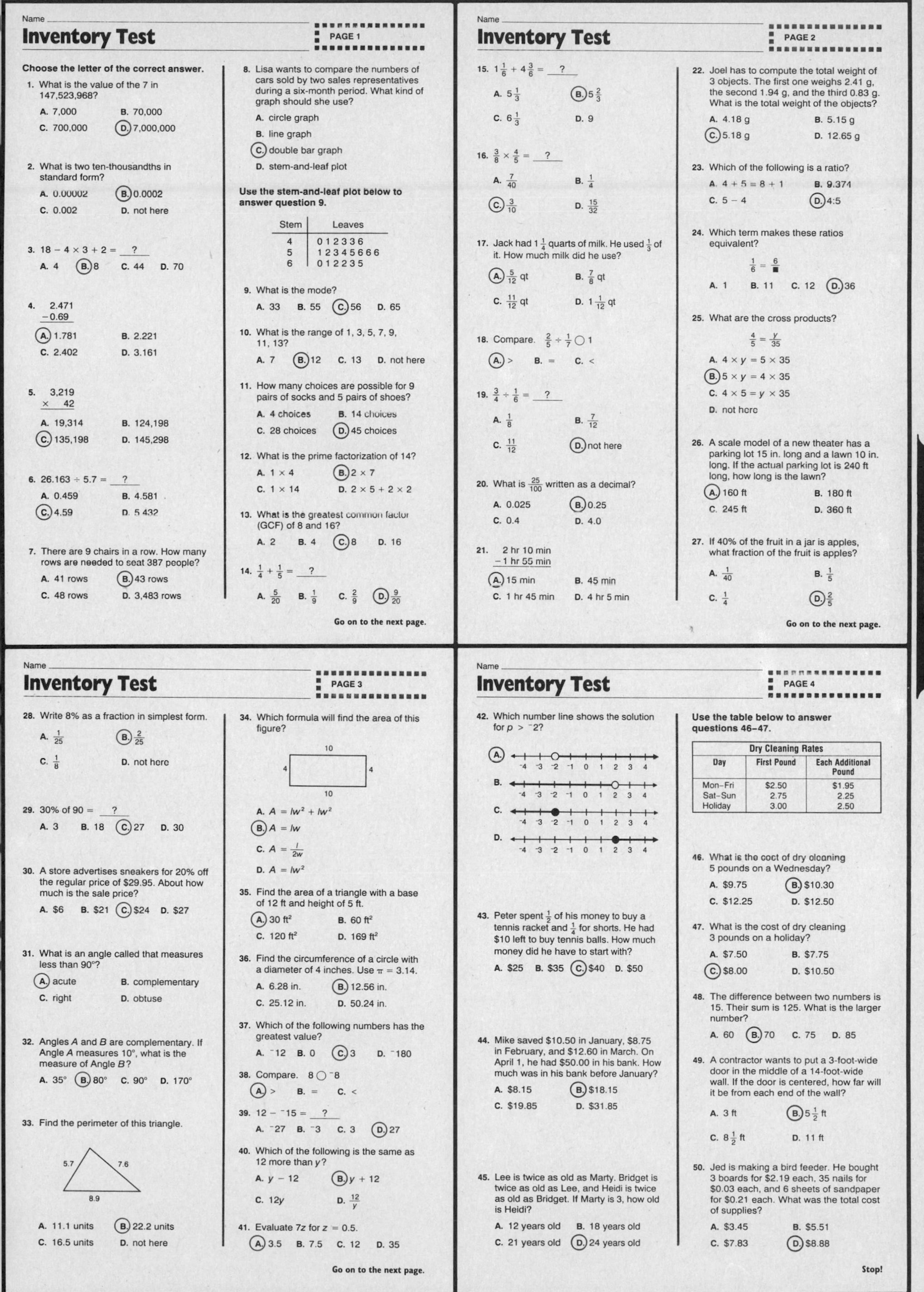

Name

Inventory Test

PAGE 1

Choose the letter of the correct answer.

1. What is the value of the 7 in 147,523,968?

A. 7,000 B. 70,000 C. 700,000 (D.) 7,000,000

2. What is two ten-thousandths in standard form?

A. 0.00002 (B.) 0.0002 C. 0.002 D. not here

3. $18 - 4 \times 3 + 2 =$?

A. 4 (B.) 8 C. 44 D. 70

4. $2.471 - 0.69$

(A.) 1.781 B. 2.221 C. 2.402 D. 3.161

5. $3{,}219 \times 42$

A. 19,314 B. 124,198 (C.) 135,198 D. 145,298

6. $26.163 \div 5.7 =$?

A. 0.459 B. 4.581 (C.) 4.59 D. 5.432

7. There are 9 chairs in a row. How many rows are needed to seat 387 people?

A. 41 rows (B.) 43 rows C. 48 rows D. 3,483 rows

8. Lisa wants to compare the numbers of cars sold by two sales representatives during a six-month period. What kind of graph should she use?

A. circle graph B. line graph (C.) double bar graph D. stem-and-leaf plot

Use the stem-and-leaf plot below to answer question 9.

Stem	Leaves
4	0 1 2 3 3 6
5	1 2 3 4 5 6 6 6
6	0 1 2 2 3 5

9. What is the mode?

A. 33 B. 55 (C.) 56 D. 65

10. What is the range of 1, 3, 5, 7, 9, 11, 13?

A. 7 (B.) 12 C. 13 D. not here

11. How many choices are possible for 9 pairs of socks and 5 pairs of shoes?

A. 4 choices B. 14 choices C. 28 choices (D.) 45 choices

12. What is the prime factorization of 14?

A. 1×4 (B.) 2×7 C. 1×14 D. $2 \times 5 + 2 \times 2$

13. What is the greatest common factor (GCF) of 8 and 16?

A. 2 B. 4 (C.) 8 D. 16

14. $\frac{1}{4} + \frac{1}{5} =$?

A. $\frac{5}{20}$ B. $\frac{1}{9}$ C. $\frac{2}{9}$ (D.) $\frac{9}{20}$

Go on to the next page.

Name

Inventory Test

PAGE 2

15. $1\frac{1}{6} + 4\frac{3}{6} =$?

A. $5\frac{1}{3}$ (B.) $5\frac{2}{3}$ C. $6\frac{1}{3}$ D. 9

16. $\frac{3}{8} \times \frac{4}{5} =$?

A. $\frac{7}{40}$ B. $\frac{1}{4}$ (C.) $\frac{3}{10}$ D. $\frac{15}{32}$

17. Jack had $1\frac{1}{4}$ quarts of milk. He used $\frac{1}{3}$ of it. How much milk did he use?

(A.) $\frac{5}{12}$ qt B. $\frac{7}{8}$ qt C. $\frac{11}{12}$ qt D. $1\frac{1}{12}$ qt

18. Compare. $\frac{2}{5} \div \frac{1}{7} \bigcirc 1$

(A.) > B. = C. <

19. $\frac{3}{4} \div \frac{1}{6} =$?

A. $\frac{1}{8}$ B. $\frac{7}{12}$ C. $\frac{11}{12}$ (D.) not here

20. What is $\frac{25}{100}$ written as a decimal?

A. 0.025 (B.) 0.25 C. 0.4 D. 4.0

21. 2 hr 10 min − 1 hr 55 min

(A.) 15 min B. 45 min C. 1 hr 45 min D. 4 hr 5 min

22. Joel has to compute the total weight of 3 objects. The first one weighs 2.41 g, the second 1.94 g, and the third 0.83 g. What is the total weight of the objects?

A. 4.18 g B. 5.15 g (C.) 5.18 g D. 12.65 g

23. Which of the following is a ratio?

A. $4 + 5 = 8 + 1$ B. 9.374 C. $5 - 4$ (D.) 4:5

24. Which term makes these ratios equivalent?

$\frac{1}{6} = \frac{6}{■}$

A. 1 B. 11 C. 12 (D.) 36

25. What are the cross products?

$\frac{4}{5} = \frac{y}{35}$

A. $4 \times y = 5 \times 35$ (B.) $5 \times y = 4 \times 35$ C. $4 \times 5 = y \times 35$ D. not here

26. A scale model of a new theater has a parking lot 15 in. long and a lawn 10 in. long. If the actual parking lot is 240 ft long, how long is the lawn?

(A.) 160 ft B. 180 ft C. 245 ft D. 360 ft

27. If 40% of the fruit in a jar is apples, what fraction of the fruit is apples?

A. $\frac{1}{40}$ B. $\frac{1}{5}$ C. $\frac{1}{4}$ (D.) $\frac{2}{5}$

Go on to the next page.

Name

Inventory Test

PAGE 3

28. Write 8% as a fraction in simplest form.

A. $\frac{1}{25}$ (B.) $\frac{2}{25}$ C. $\frac{1}{8}$ D. not here

29. 30% of 90 = ?

A. 3 B. 18 (C.) 27 D. 30

30. A store advertises sneakers for 20% off the regular price of $29.95. About how much is the sale price?

A. $6 B. $21 (C.) $24 D. $27

31. What is an angle called that measures less than 90°?

(A.) acute B. complementary C. right D. obtuse

32. Angles *A* and *B* are complementary. If Angle *A* measures 10°, what is the measure of Angle *B*?

A. 35° (B.) 80° C. 90° D. 170°

33. Find the perimeter of this triangle.

A. 11.1 units (B.) 22.2 units C. 16.5 units D. not here

34. Which formula will find the area of this figure?

A. $A = lw^2 + lw^2$ (B.) $A = lw$ C. $A = \frac{l}{2w}$ D. $A = lw^2$

35. Find the area of a triangle with a base of 12 ft and height of 5 ft.

(A.) 30 ft² B. 60 ft² C. 120 ft² D. 169 ft²

36. Find the circumference of a circle with a diameter of 4 inches. Use $\pi = 3.14$.

A. 6.28 in. (B.) 12.56 in. C. 25.12 in. D. 50.24 in.

37. Which of the following numbers has the greatest value?

A. $^-12$ B. 0 (C.) 3 D. $^-180$

38. Compare. $8 \bigcirc {}^-8$

(A.) > B. = C. <

39. $12 - {}^-15 =$?

A. $^-27$ B. $^-3$ C. 3 (D.) 27

40. Which of the following is the same as 12 more than *y*?

A. $y - 12$ (B.) $y + 12$ C. $12y$ D. $\frac{12}{y}$

41. Evaluate $7z$ for $z = 0.5$.

(A.) 3.5 B. 7.5 C. 12 D. 35

Go on to the next page.

Name

Inventory Test

PAGE 4

42. Which number line shows the solution for $p > {}^-2$?

(A.) B. C. D.

43. Peter spent $\frac{1}{2}$ of his money to buy a tennis racket and $\frac{1}{4}$ for shorts. He had $10 left to buy tennis balls. How much money did he have to start with?

A. $25 B. $35 (C.) $40 D. $50

44. Mike saved $10.50 in January, $8.75 in February, and $12.60 in March. On April 1, he had $50.00 in his bank. How much was in his bank before January?

A. $8.15 (B.) $18.15 C. $19.85 D. $31.85

45. Lee is twice as old as Marty. Bridget is twice as old as Lee, and Heidi is twice as old as Bridget. If Marty is 3, how old is Heidi?

A. 12 years old B. 18 years old C. 21 years old (D.) 24 years old

Use the table below to answer questions 46–47.

Dry Cleaning Rates		
Day	First Pound	Each Additional Pound
Mon–Fri	$2.50	$1.95
Sat–Sun	2.75	2.25
Holiday	3.00	2.50

46. What is the cost of dry cleaning 5 pounds on a Wednesday?

A. $9.75 (B.) $10.30 C. $12.25 D. $12.50

47. What is the cost of dry cleaning 3 pounds on a holiday?

A. $7.50 B. $7.75 (C.) $8.00 D. $10.50

48. The difference between two numbers is 15. Their sum is 125. What is the larger number?

A. 60 (B.) 70 C. 75 D. 85

49. A contractor wants to put a 3-foot-wide door in the middle of a 14-foot-wide wall. If the door is centered, how far will it be from each end of the wall?

A. 3 ft (B.) $5\frac{1}{2}$ ft C. $8\frac{1}{2}$ ft D. 11 ft

50. Jed is making a bird feeder. He bought 3 boards for $2.19 each, 35 nails for $0.03 each, and 6 sheets of sandpaper for $0.21 each. What was the total cost of supplies?

A. $3.45 B. $5.51 C. $7.83 (D.) $8.88

Stop!

A. TEST ANSWERS
STANDARDIZED FORMAT

Name

Pretest

Choose the letter of the correct answer.

1. Choose the *best* estimate.

   3,198 + 2,025

   A. less than 4,000
   B. less than 5,000
   C. greater than 5,000

2. Choose the *best* estimate.

   982 − 386

   A. less than 600
   B. greater than 600
   C. greater than 800

3. Choose the *best* estimate.

   3.1 × 15.02

   A. less than 40
   B. greater than 45
   C. greater than 50

4. Choose the *best* estimate.

   49.97 ÷ 2.11

   A. less than 20
   B. less than 25
   C. greater than 25

5. Estimate. 355 + 1,650 + 422

   A. 2,100 B. 2,200
   C. 2,500 D. 2,600

6. Estimate. 1,153.2 − 166.5

   A. 700 B. 800
   C. 1,000 D. 1,100

7. Estimate. 59.95 × 1,000

   A. 6,000 B. 58,000
   C. 60,000 D. 599,000

8. Estimate. 2,005 ÷ 9.57

   A. 200 B. 950
   C. 2,000 D. 18,000

9. Jane wants to buy 3 pens that cost $1.19 each and 2 pads that cost $1.15 each. *Estimate* the amount of money that Jane needs.

   A. $3.00 B. $4.00
   C. $4.50 **D.** $6.00

10. Bill had $18.99 to spend. He spent $3.09 on Friday, $2.10 on Saturday, and $2.95 on Sunday. *Estimate* the amount of money he had left.

    A. $5.00 B. $8.00
    C. $11.00 D. $14.00

11. Which quotient is greatest?

    A. 99.2 ÷ 3
    B. 99.2 ÷ 4
    C. 99.2 ÷ 5
    D. 99.2 ÷ 10

12. Which difference is greatest?

    A. 142.5 − 12
    B. 1,425.0 − 12
    C. 14.25 − 12
    D. 1.1425 − 12

13. Which product is greatest?

    A. 4 × 1,001 B. 5 × 101
    C. 7 × 10.1 D. 9 × 1.01

Go on to the next page.

Name

Pretest

14. Which quotient is greatest?

    A. 50 ÷ 0.1 B. 50 ÷ 0.01
    C. 50 ÷ 0.001 **D.** 50 ÷ 0.0001

15. 7.86 + 2.99 + 3.07 = __?__

    A. 12.92 **B.** 13.92
    C. 14.02 D. not here

16. 8.07 − 2.09 = __?__

    A. 5.08 **B.** 5.98
    C. 6.02 D. not here

17. 9.90 × 11.20 = __?__

    A. 11.088 B. 110.08
    C. 110.88 D. not here

18. 16.4 ÷ 4.1 = __?__

    A. 0.25 B. 0.4
    C. 3.50 **D.** not here

19. Julio bought 5.5 pounds of cheese for $13.75. How much did he pay per pound of cheese?

    A. $2.25 per pound
    B. $2.50 per pound
    C. $2.75 per pound
    D. $2.95 per pound

20. Myra runs 2.5 mi every weekday, 3.0 mi each Saturday, and 3.5 mi each Sunday. How many mi does she run in a 2-week period?

    A. 31.5 mi B. 35.0 mi
    C. 35.5 mi **D.** 38.0 mi

21. 5^3 = __?__

    A. 5 B. 25
    C. 125 D. 625

22. 9^0 = __?__

    A. 0 **B.** 1
    C. 9 D. 90

23. What is the value of 4 to the third power?

    A. 4 B. 12
    C. 16 **D.** 64

24. What is the value of 10^1?

    A. 1 **B.** 10
    C. 20 D. 100

25. Express 605,000,000 in scientific notation.

    A. 6.05×10^8 B. 6.50×10^8
    C. 65×10^6 D. 65×10^8

26. Express 4.5×10^3 in standard form.

    A. 45 B. 450
    C. 4,500 D. 45,000

27. Express 8,203 in scientific notation.

    A. 8.23×10^2 B. 8.23×10^3
    C. 8.203×10^2 **D.** 8.203×10^3

28. Express 3.5×10^1 in standard form.

    A. 0.035 B. 0.35
    C. 3.5 **D.** 35

Go on to the next page.

Name

Pretest

Use this information to answer questions 29–30.

The menu at Joe's Restaurant lists three main dishes: hamburger, hot dog, and pizza. It lists four salads: tossed, macaroni, potato, and cole slaw.

29. In how many combinations could Natasha order one main dish and one salad?

    A. 3 combinations
    B. 4 combinations
    C. 7 combinations
    D. 12 combinations

30. If Natasha orders a hamburger and two different salads, how many meals are possible for her to order?

    A. 4 meals
    B. 6 meals
    C. 7 meals
    D. 12 meals

31. A total of 56 boys and girls are in the school chorus. There are 20 more girls than boys. How many boys are in the chorus?

    A. 16 boys
    B. 18 boys
    C. 20 boys
    D. 36 boys

32. The Andersons went to a concert. They spent $39 for tickets. The tickets cost $12 for each adult and $5 for each child. How many children went to the concert?

    A. 1 child
    B. 2 children
    C. 3 children
    D. 4 children

Stop!

Name

Posttest

Choose the letter of the correct answer.

1. Choose the *best* estimate.

   4,125 + 3,002

   A. less than 6,000
   B. less than 7,000
   C. greater than 7,000

2. Choose the *best* estimate.

   655 − 462

   A. less than 100
   B. less than 200
   C. less than 300

3. Choose the *best* estimate.

   4.1 × 25.05

   A. less than 90
   B. less than 100
   C. greater than 100

4. Choose the *best* estimate.

   99.98 ÷ 2.05

   A. less than 33
   B. less than 50
   C. greater than 60

5. Estimate. 544 + 2,499 + 351

   A. 2,800 B. 3,100
   C. 3,400 D. 3,500

6. Estimate. 2,742.1 − 653.5

   A. 1,900 B. 1,950
   C. 2,100 D. 2,300

7. Estimate. 69.97 × 10,000

   A. 70,000 B. 650,000
   C. 700,000 D. 750,000

8. Estimate. 5,005 ÷ 9.67

   A. 500 B. 700
   C. 2,500 D. 45,000

9. Mario wants to buy 2 suits that cost $149.95 each and 4 shirts that cost $19.95 each. *Estimate* the amount of money that Mario needs.

   A. $320.00 B. $340.00
   C. $360.00 **D.** $380.00

10. Doris had $25.92 to spend. She spent $9.65 on dinner, $5.00 on a movie, and $2.95 to park her car. *Estimate* the amount of money she had left.

    A. $5.00 B. $6.00
    C. $8.00 D. $11.00

11. Which quotient is greatest?

    A. 33.3 ÷ 6
    B. 33.3 ÷ 7
    C. 33.3 ÷ 8
    D. 33.3 ÷ 10

12. Which difference is greatest?

    A. 162 − 11.02
    B. 162 − 1.192
    C. 162 − 119.2
    D. 162 − 0.1192

13. Which product is greatest?

    A. 4 × 2,001 B. 6 × 201
    C. 8 × 20.1 D. 10 × 2.01

Go on to the next page.

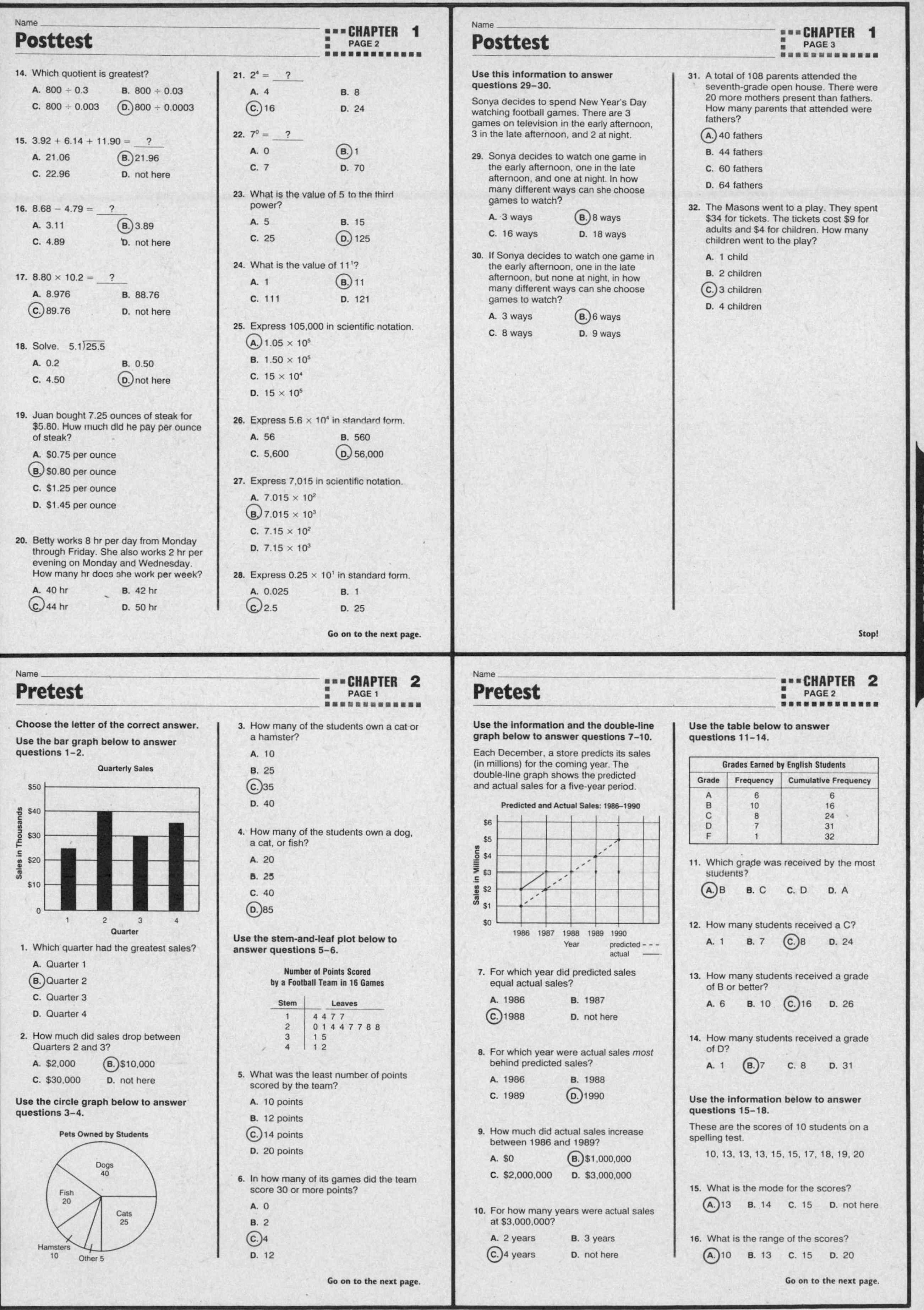

Name

Posttest

CHAPTER 1
PAGE 2

14. Which quotient is greatest?
A. 800 ÷ 0.3 B. 800 ÷ 0.03
C. 800 ÷ 0.003 **(D.) 800 ÷ 0.0003**

15. 3.92 + 6.14 + 11.90 = ___?___
A. 21.06 **(B.) 21.96**
C. 22.96 D. not here

16. 8.68 − 4.79 = ___?___
A. 3.11 **(B.) 3.89**
C. 4.89 D. not here

17. 8.80 × 10.2 = ___?___
A. 8.976 B. 88.76
(C.) 89.76 D. not here

18. Solve. 5.1)25.5
A. 0.2 B. 0.50
C. 4.50 **(D.) not here**

19. Juan bought 7.25 ounces of steak for $5.80. How much did he pay per ounce of steak?
A. $0.75 per ounce
(B.) $0.80 per ounce
C. $1.25 per ounce
D. $1.45 per ounce

20. Betty works 8 hr per day from Monday through Friday. She also works 2 hr per evening on Monday and Wednesday. How many hr does she work per week?
A. 40 hr B. 42 hr
(C.) 44 hr D. 50 hr

21. 2^4 = ___?___
A. 4 B. 8
(C.) 16 D. 24

22. 7^0 = ___?___
A. 0 **(B.) 1**
C. 7 D. 70

23. What is the value of 5 to the third power?
A. 5 B. 15
C. 25 **(D.) 125**

24. What is the value of 11^1?
A. 1 **(B.) 11**
C. 111 D. 121

25. Express 105,000 in scientific notation.
(A.) 1.05×10^5
B. 1.50×10^5
C. 15×10^4
D. 15×10^5

26. Express 5.6×10^4 in standard form.
A. 56 B. 560
C. 5,600 **(D.) 56,000**

27. Express 7,015 in scientific notation.
A. 7.015×10^2
(B.) 7.015×10^3
C. 7.15×10^2
D. 7.15×10^3

28. Express 0.25×10^1 in standard form.
A. 0.025 B. 1
(C.) 2.5 D. 25

Go on to the next page.

Name

Posttest

CHAPTER 1
PAGE 3

Use this information to answer questions 29–30.

Sonya decides to spend New Year's Day watching football games. There are 3 games on television in the early afternoon, 3 in the late afternoon, and 2 at night.

29. Sonya decides to watch one game in the early afternoon, one in the late afternoon, and one at night. In how many different ways can she choose games to watch?
A. 3 ways **(B.) 8 ways**
C. 16 ways D. 18 ways

30. If Sonya decides to watch one game in the early afternoon, one in the late afternoon, but none at night, in how many different ways can she choose games to watch?
A. 3 ways **(B.) 6 ways**
C. 8 ways D. 9 ways

31. A total of 108 parents attended the seventh-grade open house. There were 20 more mothers present than fathers. How many parents that attended were fathers?
(A.) 40 fathers
B. 44 fathers
C. 60 fathers
D. 64 fathers

32. The Masons went to a play. They spent $34 for tickets. The tickets cost $9 for adults and $4 for children. How many children went to the play?
A. 1 child
B. 2 children
(C.) 3 children
D. 4 children

Stop!

Name

Pretest

CHAPTER 2
PAGE 1

Choose the letter of the correct answer.

Use the bar graph below to answer questions 1–2.

1. Which quarter had the greatest sales?
A. Quarter 1
(B.) Quarter 2
C. Quarter 3
D. Quarter 4

2. How much did sales drop between Quarters 2 and 3?
A. $2,000 **(B.) $10,000**
C. $30,000 D. not here

Use the circle graph below to answer questions 3–4.

3. How many of the students own a cat or a hamster?
A. 10
B. 25
(C.) 35
D. 40

4. How many of the students own a dog, a cat, or fish?
A. 20
B. 25
C. 40
(D.) 85

Use the stem-and-leaf plot below to answer questions 5–6.

Number of Points Scored by a Football Team in 16 Games

Stem	Leaves
1	4 4 7 7
2	0 1 4 4 7 7 8 8
3	1 5
4	1 2

5. What was the least number of points scored by the team?
A. 10 points
B. 12 points
(C.) 14 points
D. 20 points

6. In how many of its games did the team score 30 or more points?
A. 0
B. 2
(C.) 4
D. 12

Go on to the next page.

Name

Pretest

CHAPTER 2
PAGE 2

Use the information and the double-line graph below to answer questions 7–10.

Each December, a store predicts its sales (in millions) for the coming year. The double-line graph shows the predicted and actual sales for a five-year period.

7. For which year did predicted sales equal actual sales?
A. 1986 B. 1987
(C.) 1988 D. not here

8. For which year were actual sales *most* behind predicted sales?
A. 1986 B. 1988
C. 1989 **(D.) 1990**

9. How much did actual sales increase between 1986 and 1989?
A. $0 **(B.) $1,000,000**
C. $2,000,000 D. $3,000,000

10. For how many years were actual sales at $3,000,000?
A. 2 years B. 3 years
(C.) 4 years D. not here

Use the table below to answer questions 11–14.

Grades Earned by English Students

Grade	Frequency	Cumulative Frequency
A	6	6
B	10	16
C	8	24
D	7	31
F	1	32

11. Which grade was received by the most students?
(A.) B B. C C. D D. A

12. How many students received a C?
A. 1 B. 7 **(C.) 8** D. 24

13. How many students received a grade of B or better?
A. 6 B. 10 **(C.) 16** D. 26

14. How many students received a grade of D?
A. 1 **(B.) 7** C. 8 D. 31

Use the information below to answer questions 15–18.

These are the scores of 10 students on a spelling test.

10, 13, 13, 13, 15, 15, 17, 18, 19, 20

15. What is the mode for the scores?
(A.) 13 B. 14 C. 15 D. not here

16. What is the range of the scores?
(A.) 10 B. 13 C. 15 D. 20

Go on to the next page.

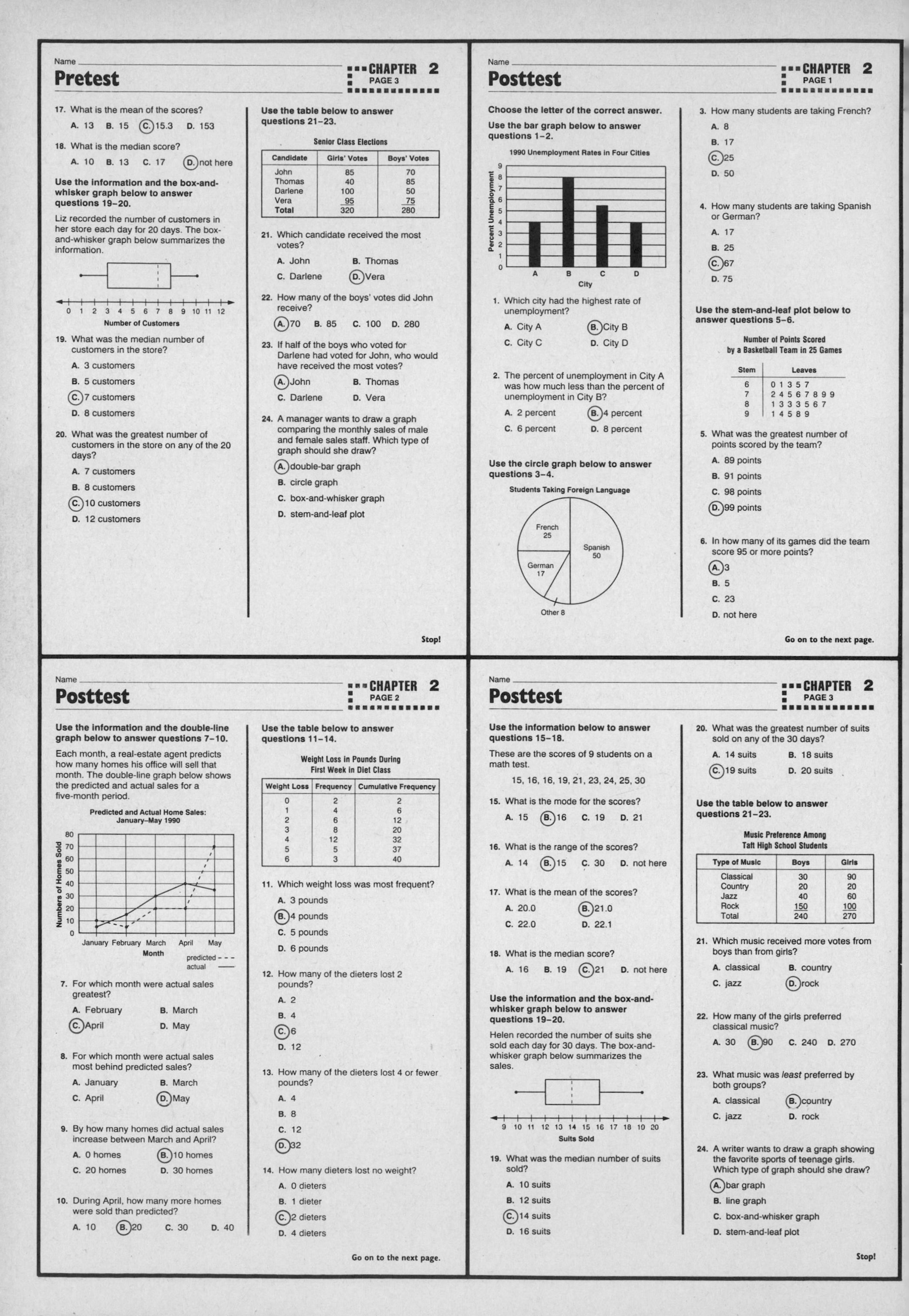

Name

Pretest

CHAPTER 2
PAGE 3

17. What is the mean of the scores?
A. 13 B. 15 (C.) 15.3 D. 153

18. What is the median score?
A. 10 B. 13 C. 17 (D.) not here

Use the information and the box-and-whisker graph below to answer questions 19–20.

Liz recorded the number of customers in her store each day for 20 days. The box-and-whisker graph below summarizes the information.

0 1 2 3 4 5 6 7 8 9 10 11 12
Number of Customers

19. What was the median number of customers in the store?
A. 3 customers
B. 5 customers
(C.) 7 customers
D. 8 customers

20. What was the greatest number of customers in the store on any of the 20 days?
A. 7 customers
B. 8 customers
(C.) 10 customers
D. 12 customers

Use the table below to answer questions 21–23.

Senior Class Elections

Candidate	Girls' Votes	Boys' Votes
John	85	70
Thomas	40	85
Darlene	100	50
Vera	95	75
Total	320	280

21. Which candidate received the most votes?
A. John B. Thomas
C. Darlene (D.) Vera

22. How many of the boys' votes did John receive?
(A.) 70 B. 85 C. 100 D. 280

23. If half of the boys who voted for Darlene had voted for John, who would have received the most votes?
(A.) John B. Thomas
C. Darlene D. Vera

24. A manager wants to draw a graph comparing the monthly sales of male and female sales staff. Which type of graph should she draw?
(A.) double-bar graph
B. circle graph
C. box-and-whisker graph
D. stem-and-leaf plot

Stop!

Name

Posttest

CHAPTER 2
PAGE 1

Choose the letter of the correct answer.

Use the bar graph below to answer questions 1–2.

1990 Unemployment Rates in Four Cities

Percent Unemployment
0 1 2 3 4 5 6 7 8 9
A B C D
City

1. Which city had the highest rate of unemployment?
A. City A (B.) City B
C. City C D. City D

2. The percent of unemployment in City A was how much less than the percent of unemployment in City B?
A. 2 percent (B.) 4 percent
C. 6 percent D. 8 percent

Use the circle graph below to answer questions 3–4.

Students Taking Foreign Language

French 25
Spanish 50
German 17
Other 8

3. How many students are taking French?
A. 8
B. 17
(C.) 25
D. 50

4. How many students are taking Spanish or German?
A. 17
B. 25
(C.) 67
D. 75

Use the stem-and-leaf plot below to answer questions 5–6.

Number of Points Scored by a Basketball Team in 25 Games

Stem	Leaves
6	0 1 3 5 7
7	2 4 5 6 7 8 9 9
8	1 3 3 3 5 6 7
9	1 4 5 8 9

5. What was the greatest number of points scored by the team?
A. 89 points
B. 91 points
C. 98 points
(D.) 99 points

6. In how many of its games did the team score 95 or more points?
(A.) 3
B. 5
C. 23
D. not here

Go on to the next page.

Name

Posttest

CHAPTER 2
PAGE 2

Use the information and the double-line graph below to answer questions 7–10.

Each month, a real-estate agent predicts how many homes his office will sell that month. The double-line graph below shows the predicted and actual sales for a five-month period.

Predicted and Actual Home Sales: January–May 1990

Numbers of Homes Sold
0 10 20 30 40 50 60 70 80
January February March April May
Month
predicted - - -
actual ——

7. For which month were actual sales greatest?
A. February B. March
(C.) April D. May

8. For which month were actual sales most behind predicted sales?
A. January B. March
C. April (D.) May

9. By how many homes did actual sales increase between March and April?
A. 0 homes (B.) 10 homes
C. 20 homes D. 30 homes

10. During April, how many more homes were sold than predicted?
A. 10 (B.) 20 C. 30 D. 40

Use the table below to answer questions 11–14.

Weight Loss in Pounds During First Week in Diet Class

Weight Loss	Frequency	Cumulative Frequency
0	2	2
1	4	6
2	6	12
3	8	20
4	12	32
5	5	37
6	3	40

11. Which weight loss was most frequent?
A. 3 pounds
(B.) 4 pounds
C. 5 pounds
D. 6 pounds

12. How many of the dieters lost 2 pounds?
A. 2
B. 4
(C.) 6
D. 12

13. How many of the dieters lost 4 or fewer pounds?
A. 4
B. 8
C. 12
(D.) 32

14. How many dieters lost no weight?
A. 0 dieters
B. 1 dieter
(C.) 2 dieters
D. 4 dieters

Go on to the next page.

Name

Posttest

CHAPTER 2
PAGE 3

Use the information below to answer questions 15–18.

These are the scores of 9 students on a math test.

15, 16, 16, 19, 21, 23, 24, 25, 30

15. What is the mode for the scores?
A. 15 (B.) 16 C. 19 D. 21

16. What is the range of the scores?
A. 14 (B.) 15 C. 30 D. not here

17. What is the mean of the scores?
A. 20.0 (B.) 21.0
C. 22.0 D. 22.1

18. What is the median score?
A. 16 B. 19 (C.) 21 D. not here

Use the information and the box-and-whisker graph below to answer questions 19–20.

Helen recorded the number of suits she sold each day for 30 days. The box-and-whisker graph below summarizes the sales.

9 10 11 12 13 14 15 16 17 18 19 20
Suits Sold

19. What was the median number of suits sold?
A. 10 suits
B. 12 suits
(C.) 14 suits
D. 16 suits

20. What was the greatest number of suits sold on any of the 30 days?
A. 14 suits B. 18 suits
(C.) 19 suits D. 20 suits

Use the table below to answer questions 21–23.

Music Preference Among Taft High School Students

Type of Music	Boys	Girls
Classical	30	90
Country	20	20
Jazz	40	60
Rock	150	100
Total	240	270

21. Which music received more votes from boys than from girls?
A. classical B. country
C. jazz (D.) rock

22. How many of the girls preferred classical music?
A. 30 (B.) 90 C. 240 D. 270

23. What music was *least* preferred by both groups?
A. classical (B.) country
C. jazz D. rock

24. A writer wants to draw a graph showing the favorite sports of teenage girls. Which type of graph should she draw?
(A.) bar graph
B. line graph
C. box-and-whisker graph
D. stem-and-leaf plot

Stop!

Name

Pretest

CHAPTER 3 PAGE 1

Choose the letter of the correct answer.

1. Which is a prime number?
   (A.) 5 B. 10 C. 15 D. 35
2. Which is a composite number?
   A. 11 B. 17 C. 19 (D.) 21
3. Which is a prime number?
   A. 26 (B.) 29 C. 35 D. 51
4. How many factors does 37 have?
   A. 1 (B.) 2 C. 3 D. 37
5. What are the prime factors of 30?
   A. 5, 6 B. 2, 15 (C.) 2, 3, 5 D. 3, 10
6. What are the prime factors of 52?
   (A.) 2, 2, 13 B. 2, 26 C. 4, 13 D. 50, 2
7. What is the prime factorization of 24?
   (A.) $2^3 \times 3$ B. $2^2 \times 6$ C. $4 \times 2 \times 3$ D. not here
8. What is the prime factorization of 1,000?
   A. $2^3 \times 25^2$ (B.) $2^3 \times 5^3$ C. $2^2 \times 250$ D. 2×500
9. What is the greatest common factor of 24 and 36?
   A. 6 (B.) 12 C. 18 D. 24
10. What is the greatest common factor of 32 and 72?
    (A.) 8 B. 9 C. 16 D. 24
11. What is the least common multiple of 5 and 4?
    A. 4 B. 5 C. 10 (D.) 20
12. What is the least common multiple of 6 and 9?
    A. 6 B. 12 (C.) 18 D. 54
13. Which fraction is expressed in simplest form?
    A. $\frac{3}{15}$ B. $\frac{5}{15}$ C. $\frac{55}{100}$ (D.) $\frac{99}{100}$
14. Which fraction is *not* expressed in simplest form?
    A. $\frac{1}{2}$ B. $\frac{2}{3}$ (C.) $\frac{4}{8}$ D. $\frac{11}{12}$
15. Which fraction is greater than $1\frac{3}{4}$?
    A. $\frac{27}{16}$ B. $\frac{32}{20}$ C. $\frac{173}{100}$ (D.) $\frac{352}{200}$
16. Which group is ordered from least to greatest?
    A. $\frac{2}{5}, \frac{1}{2}, \frac{3}{10}$ (B.) $\frac{1}{3}, \frac{1}{2}, \frac{3}{4}$ C. $\frac{3}{4}, \frac{3}{5}, \frac{3}{6}$ D. $\frac{2}{3}, \frac{4}{5}, \frac{7}{12}$
17. Compare. $\frac{9}{27} \bigcirc \frac{1}{3}$
    A. < (B.) = C. >
18. Compare. $3\frac{1}{7} \bigcirc 3\frac{9}{10}$
    (A.) < B. = C. >
19. $\frac{75}{90}$ = ___?___
    (A.) $\frac{5}{6}$ B. $\frac{15}{16}$ C. $\frac{5}{18}$ D. $\frac{10}{18}$

Go on to the next page.

Name

Pretest

CHAPTER 3 PAGE 2

20. $\frac{2}{3}$ = ___?___
    A. $\frac{3}{4}$ B. $\frac{5}{7}$ C. $\frac{7}{9}$ (D.) not here

Use the information below to answer questions 21–24.

A taxi charges $1.00 for the first mile and $0.20 for each additional $\frac{1}{8}$ mile.

21. How much does it cost to ride $2\frac{1}{2}$ miles?
    A. $1.80 (B.) $2.80 C. $3.00 D. $3.80
22. How far can you travel for $5.00?
    A. $3\frac{1}{3}$ miles (B.) $4\frac{1}{3}$ miles C. 5 miles D. 10 miles
23. Jo rode 2 miles and Sam rode 3 miles. How much more did Sam pay?
    (A.) $1.20 B. $2.00 C. $2.20 D. $2.40
24. Rashawn and a friend take a taxi for $1\frac{1}{2}$ miles. The two riders split the fare evenly. How much does each pay?
    (A.) $0.80 B. $1.20 C. $1.60 D. $1.80
25. Which fraction falls between $\frac{4}{10}$ and $\frac{5}{10}$?
    (A.) $\frac{9}{20}$ B. $\frac{11}{20}$ C. $\frac{16}{40}$ D. $\frac{9}{10}$
26. Which fraction falls between $\frac{3}{8}$ and $\frac{4}{8}$?
    A. $\frac{6}{16}$ B. $\frac{8}{16}$ (C.) $\frac{14}{32}$ D. $\frac{17}{32}$
27. Which fraction falls between $\frac{1}{5}$ and $\frac{2}{5}$?
    (A.) $\frac{3}{10}$ B. $\frac{4}{10}$ C. $\frac{1}{20}$ D. $\frac{9}{20}$
28. Which fraction falls between $\frac{1}{3}$ and $\frac{1}{2}$?
    A. $\frac{3}{6}$ B. $\frac{2}{6}$ (C.) $\frac{5}{12}$ D. $\frac{7}{12}$

Use the information below to answer questions 29–32.

At Jones Pier the lake is 20 feet deep. Every $\frac{1}{4}$ mile from the pier it becomes 1 foot deeper until it reaches a maximum depth of 30 feet.

29. How deep is the water $1\frac{1}{2}$ miles from the pier?
    A. 24 feet (B.) 26 feet C. 28 feet D. 30 feet
30. How many miles from the pier does the lake reach its maximum depth?
    A. $1\frac{1}{4}$ miles B. 2 miles C. $2\frac{1}{4}$ miles (D.) $2\frac{1}{2}$ miles
31. How much deeper is the water 2 miles from the pier than it is 1 mile out?
    A. 2 ft (B.) 4 ft C. 6 ft D. 8 ft
32. How far from the pier does the water reach 22 feet in depth?
    A. $\frac{1}{4}$ mile (B.) $\frac{1}{2}$ mile C. $\frac{3}{4}$ mile D. 1 mile

Stop!

Name

Posttest

CHAPTER 3 PAGE 1

Choose the letter of the correct answer.

1. Which is a prime number?
   (A.) 3 B. 6 C. 10 D. 12
2. Which is a composite number?
   A. 13 B. 23 (C.) 33 D. 43
3. Which is a prime number?
   A. 42 (B.) 59 C. 65 D. not here
4. How many factors does 41 have?
   A. 1 (B.) 2 C. 4 D. 59
5. What are the prime factors of 42?
   A. 7, 6 B. 2, 21 (C.) 2, 3, 7 D. not here
6. What are the prime factors of 68?
   (A.) 2, 2, 17 B. 2, 34 C. 4, 17 D. 1, 68
7. What is the prime factorization of 48?
   (A.) $2^4 \times 3$ B. $2^3 \times 6$ C. $4 \times 4 \times 4$ D. 8×6
8. What is the prime factorization of 100?
   A. $2^2 \times 25^2$ (B.) $2^2 \times 5^2$ C. $2^2 \times 25$ D. $2 \times 5 \times 10$
9. What is the greatest common factor of 32 and 40?
   A. 4 (B.) 8 C. 10 D. 16
10. What is the greatest common factor of 16 and 24?
    (A.) 8 B. 12 C. 16 D. 24
11. What is the least common multiple of 3 and 4?
    A. 3 B. 4 C. 9 (D.) 12
12. What is the least common multiple of 9 and 12?
    A. 18 B. 24 (C.) 36 D. 72
13. Which fraction is expressed in simplest form?
    A. $\frac{4}{18}$ B. $\frac{5}{20}$ C. $\frac{55}{200}$ (D.) $\frac{97}{102}$
14. Which fraction is *not* expressed in simplest form?
    A. $\frac{1}{3}$ B. $\frac{2}{5}$ (C.) $\frac{4}{10}$ D. $\frac{5}{11}$
15. Which fraction is greater than $1\frac{1}{4}$?
    A. $\frac{24}{20}$ B. $\frac{32}{30}$ C. $\frac{123}{100}$ (D.) $\frac{257}{200}$
16. Which group is ordered from least to greatest?
    A. $\frac{1}{5}, \frac{2}{3}, \frac{4}{10}$ (B.) $\frac{1}{4}, \frac{1}{2}, \frac{3}{4}$ C. $\frac{3}{6}, \frac{3}{7}, \frac{3}{8}$ D. $\frac{3}{4}, \frac{7}{8}, \frac{6}{7}$
17. Compare. $\frac{13}{18} \bigcirc \frac{3}{8}$
    A. < B. = (C.) >
18. Compare. $2\frac{7}{8} \bigcirc 2\frac{21}{24}$
    A. < (B.) = C. >
19. $\frac{7}{28}$ = ___?___
    A. $\frac{1}{3}$ (B.) $\frac{1}{4}$ C. $1\frac{3}{4}$ D. $\frac{4}{1}$

Go on to the next page.

Name

Posttest

CHAPTER 3 PAGE 2

20. $\frac{3}{4}$ = ___?___
    A. $\frac{4}{5}$ B. $\frac{5}{7}$ C. $\frac{6}{9}$ (D.) $\frac{6}{8}$

Use the information below to answer questions 21–24.

A parking lot charges $3.00 for the first hour and $1.00 for each additional $\frac{1}{2}$ hour.

21. How much does it cost to park for $2\frac{1}{2}$ hours?
    A. $4.50 (B.) $6.00 C. $7.50 D. $9.00
22. For how long can you park for $4.00?
    (A.) $1\frac{1}{2}$ hours B. $1\frac{3}{4}$ hours C. 2 hours D. $2\frac{1}{4}$ hours
23. Alan parks for 4 hours and Alexa parks for 6 hours. How much more does Alexa pay than Alan?
    A. $1.00 B. $2.00 C. $3.00 (D.) $4.00
24. Justin and Julie drive to work together. They split the cost of parking. If they park for $7\frac{1}{2}$ hours, how much will each pay?
    A. $4.75 (B.) $8.00 C. $9.50 D. not here
25. Which fraction falls between $\frac{4}{12}$ and $\frac{5}{12}$?
    (A.) $\frac{9}{24}$ B. $\frac{10}{24}$ C. $\frac{16}{48}$ D. $\frac{20}{48}$
26. Which fraction falls between $\frac{3}{10}$ and $\frac{4}{10}$?
    A. $\frac{6}{20}$ B. $\frac{8}{20}$ (C.) $\frac{13}{40}$ D. $\frac{16}{40}$
27. Which fraction falls between $\frac{1}{8}$ and $\frac{2}{8}$?
    (A.) $\frac{3}{16}$ B. $\frac{4}{16}$ C. $\frac{15}{32}$ D. $\frac{17}{32}$
28. Which fraction falls between $\frac{1}{4}$ and $\frac{1}{3}$?
    A. $\frac{3}{12}$ B. $\frac{4}{12}$ (C.) $\frac{7}{24}$ D. $\frac{6}{24}$

Use the information below to answer questions 29–32.

Bob and Peg began a running program. Bob ran 1 mile on day 1 and increased the distance by $\frac{1}{10}$ mile each day. Peg ran 2 miles on day 1 and increased the distance by $\frac{1}{20}$ mile each day.

29. How far did Bob run on day 5?
    A. $1\frac{3}{10}$ miles (B.) $1\frac{2}{5}$ miles C. $1\frac{1}{2}$ miles D. $1\frac{3}{5}$ miles
30. How far did Peg run on day 6?
    A. $2\frac{1}{10}$ mi B. $2\frac{3}{20}$ mi C. $2\frac{1}{5}$ mi (D.) $2\frac{1}{4}$ mi
31. How far did Bob run on day 11?
    A. $1\frac{9}{10}$ mi (B.) 2 mi C. $2\frac{1}{10}$ mi D. $2\frac{1}{5}$ mi
32. How far did Peg run on day 15?
    A. $2\frac{3}{5}$ mi B. $2\frac{13}{20}$ mi (C.) $2\frac{7}{10}$ mi D. $2\frac{3}{4}$ mi

Stop!

Name ______________________

Pretest

CHAPTER 4 — PAGE 1

Choose the letter of the correct answer.

1. To estimate the sum of $6\frac{4}{9} + 3\frac{1}{8}$, what should $6\frac{4}{9}$ be rounded to?
   A. 5 B. $6\frac{1}{4}$ **C. $6\frac{1}{2}$** D. 8

2. To estimate the quotient of $5\frac{7}{8} \div 2\frac{1}{8}$, what should $2\frac{1}{8}$ be rounded to?
   A. 2 B. $2\frac{3}{4}$ C. 3 D. 4

3. Estimate. $13\frac{5}{8} + 9\frac{2}{5} =$ ___?___
   A. 20 B. 21 **C. 23** D. 24

4. Estimate. $10\frac{7}{8} - 1\frac{3}{5} =$ ___?___
   A. 8 B. $8\frac{1}{2}$ **C. $9\frac{1}{2}$** D. 11

5. Estimate. $\frac{7}{6} \times \frac{4}{9} =$ ___?___
   A. $\frac{1}{8}$ **B. $\frac{1}{2}$** C. 1 D. $1\frac{1}{2}$

6. Estimate. $12\frac{1}{9} \div 3\frac{9}{10} =$ ___?___
   A. 3 B. $4\frac{1}{2}$ C. 5 D. 9

7. Emilio bought two melons weighing $2\frac{7}{8}$ pounds and $3\frac{9}{10}$ pounds. Estimate the weight of the two melons combined.
   A. 4 lb B. 6 lb **C. 7 lb** D. 8 lb

8. Three farmers split $14\frac{8}{10}$ gallons of gas equally. Estimate the number of gallons each received.
   A. 3 gal B. 4 gal **C. 5 gal** D. 7 gal

9. To find the sum of $\frac{7}{8} + \frac{1}{3}$, $\frac{7}{8}$ should be renamed to which equivalent fraction?
   A. $\frac{14}{16}$ **B. $\frac{21}{24}$** C. $\frac{28}{32}$ D. $\frac{35}{40}$

10. To find the difference of $\frac{3}{4} - \frac{2}{5}$, $\frac{2}{5}$ should be renamed to which equivalent fraction?
    A. $\frac{4}{10}$ B. $\frac{6}{15}$ **C. $\frac{8}{20}$** D. $\frac{12}{30}$

11. $\frac{5}{8} + \frac{7}{8} =$ ___?___
    A. $\frac{3}{4}$ B. $1\frac{1}{3}$ **C. $1\frac{1}{2}$** D. $3\frac{9}{16}$

12. $3\frac{1}{3} + 4\frac{3}{4} =$ ___?___
    A. $6\frac{11}{12}$ B. $7\frac{4}{7}$ C. $8\frac{11}{12}$ **D. not here**

13. $\frac{1}{2} - \frac{1}{6} =$ ___?___
    A. $\frac{1}{6}$ **B. $\frac{1}{3}$** C. $\frac{2}{3}$ D. not here

14. $12\frac{2}{5} - 11\frac{3}{5} =$ ___?___
    A. $\frac{1}{5}$ **B. $\frac{4}{5}$** C. $1\frac{1}{5}$ D. $1\frac{4}{5}$

Go on to the next page.

Name ______________________

Pretest

CHAPTER 4 — PAGE 2

15. Susan watched three short TV shows. The first lasted $\frac{1}{4}$ hour; the other two lasted $\frac{1}{2}$ hour each. For how long did she watch TV?
    A. $\frac{3}{4}$ hr B. 1 hr C. $1\frac{1}{8}$ hr **D. $1\frac{1}{4}$ hr**

16. Jose bought a bottle of juice. He drank $\frac{1}{3}$ of the bottle of juice, and his sister drank $\frac{1}{6}$ of the bottle of juice. How much of the bottle of juice was left?
    A. $\frac{1}{6}$ bottle B. $\frac{1}{3}$ bottle **C. $\frac{1}{2}$ bottle** D. $\frac{2}{3}$ bottle

17. Compare. $\frac{7}{8} \times \frac{3}{4} \bigcirc \frac{3}{4}$
    A. $<$ B. $=$ C. $>$

18. Compare. $\frac{11}{10} \times \frac{2}{3} \bigcirc \frac{2}{3}$
    A. $<$ B. $=$ **C. $>$**

19. What is $\frac{2}{3} \div \frac{1}{3}$ rewritten as a multiplication problem?
    A. $\frac{3}{2} \times \frac{1}{3}$ B. $\frac{2}{3} \times \frac{2}{3}$ C. $\frac{3}{2} \times \frac{3}{1}$ **D. $\frac{2}{3} \times \frac{3}{1}$**

20. What is $8\frac{1}{2} \div 2\frac{3}{4}$ rewritten as a multiplication problem?
    A. $\frac{17}{2} \times \frac{11}{4}$ **B. $\frac{17}{2} \times \frac{4}{11}$** C. $\frac{2}{17} \times \frac{11}{4}$ D. $\frac{2}{17} \times \frac{4}{11}$

21. $\frac{3}{7} \times \frac{1}{9} =$ ___?___
    A. $\frac{1}{21}$ B. $\frac{3}{16}$ C. $\frac{1}{4}$ D. not here

22. $1\frac{3}{4} \times 2\frac{1}{3} =$ ___?___
    A. $\frac{7}{12}$ B. $2\frac{1}{4}$ C. $3\frac{7}{12}$ **D. not here**

23. $\frac{1}{3} \div \frac{1}{4} =$ ___?___
    A. $\frac{1}{12}$ B. $\frac{3}{4}$ **C. $1\frac{1}{3}$** D. not here

24. $3\frac{1}{2} \div 1\frac{3}{4} =$ ___?___
    A. $\frac{1}{2}$ B. $\frac{11}{14}$ **C. 2** D. $6\frac{1}{8}$

25. For $\frac{3}{4}$ of an hour, Rick practiced basketball. He spent $\frac{1}{2}$ of that time shooting foul shots. What part of an hour did he spend shooting foul shots?
    A. $\frac{1}{8}$ hr B. $\frac{3}{16}$ hr C. $\frac{1}{2}$ hr **D. $\frac{3}{8}$ hr**

26. Doreen bought $\frac{1}{3}$ yard of ribbon for crafts. She used it to make two bows. How much ribbon was needed for each bow?
    A. $\frac{1}{12}$ yd **B. $\frac{1}{6}$ yd** C. $\frac{1}{5}$ yd D. $\frac{2}{3}$ yd

Go on to the next page.

Name ______________________

Pretest

CHAPTER 4 — PAGE 3

27. A carton containing 6 identical boxes weighs $12\frac{1}{2}$ pounds (lb). How many pounds does each box weigh?
    A. $2\frac{1}{12}$ lb B. $2\frac{1}{6}$ lb C. $2\frac{1}{3}$ lb D. $2\frac{1}{2}$ lb

28. Each floor tile is $\frac{10}{12}$ foot long. How many tiles are along one wall of a room that is 25 feet long?
    A. 21 tiles B. 27 tiles **C. 30 tiles** D. 300 tiles

29. Liza mowed lawns for $2\frac{1}{3}$ hours each day on Monday, Tuesday, and Wednesday. She raked leaves for $4\frac{1}{2}$ hours on Friday and $3\frac{2}{3}$ hours on Saturday. For how many hours did she work this week?
    A. $10\frac{1}{2}$ hr B. $14\frac{1}{2}$ hr **C. $15\frac{1}{6}$ hr** D. $51\frac{1}{2}$ hr

30. There are 12 teams in a baseball league. If each team plays each of the others once, how many games will be played?
    A. 48 games **B. 66 games** C. 132 games D. 144 games

Use the information below to answer questions 31–32.

Nancy must walk 5 blocks to school. It takes her $2\frac{1}{2}$ minutes to walk each block. Pat walks 7 blocks to school. It takes her only $1\frac{1}{2}$ minutes to walk each block.

31. How much longer does it take Nancy to walk to school than it takes Pat?
    A. 1 min **B. 2 min** C. $2\frac{1}{2}$ min D. 3 min

32. How long does it take Nancy to walk $2\frac{1}{2}$ blocks?
    A. 5 min B. $5\frac{1}{2}$ min C. 6 min **D. $6\frac{1}{4}$ min**

Stop!

Name ______________________

Posttest

CHAPTER 4 — PAGE 1

Choose the letter of the correct answer.

1. To estimate the sum of $7\frac{5}{9} + 4\frac{1}{8}$, what should $7\frac{5}{9}$ be rounded to?
   A. 7 B. $7\frac{1}{8}$ **C. $7\frac{1}{2}$** D. 9

2. To estimate the product of $5\frac{1}{6} \times 1\frac{1}{10}$, what should $1\frac{1}{10}$ be rounded to?
   A. 1 B. $1\frac{1}{6}$ C. $1\frac{1}{2}$ D. 2

3. Estimate. $4\frac{4}{7} + 9\frac{4}{9} =$ ___?___
   A. 11 B. $12\frac{1}{2}$ C. $13\frac{1}{2}$ **D. 14**

4. Estimate. $10\frac{6}{7} - 2\frac{2}{5} =$ ___?___
   A. $7\frac{1}{2}$ **B. $8\frac{1}{2}$** C. $9\frac{1}{2}$ D. 10

5. Estimate. $\frac{10}{9} \times \frac{4}{9} =$ ___?___
   A. $\frac{1}{2}$ B. $\frac{3}{4}$ C. $1\frac{1}{2}$ D. 2

6. Estimate. $15\frac{1}{8} \div 4\frac{9}{11} =$ ___?___
   A. 2 **B. 3** C. 4 D. 5

7. Lucy bought two bags of peanuts. The first weighed $1\frac{7}{8}$ pounds, and the second weighed $1\frac{1}{16}$ pounds. Estimate the weight of the two bags combined.
   A. $1\frac{1}{2}$ lb **B. 3 lb** C. $3\frac{1}{2}$ lb D. 4 lb

8. Three friends divided $3\frac{1}{4}$ pints of yogurt equally. Estimate the amount of yogurt that each person had.
   A. $\frac{1}{2}$ pt **B. 1 pt** C. $1\frac{1}{2}$ pt D. 2 pt

9. To find the sum of $\frac{7}{8} + \frac{1}{5}$, $\frac{7}{8}$ should be renamed to which equivalent fraction?
   A. $\frac{14}{16}$ B. $\frac{21}{24}$ C. $\frac{15}{40}$ **D. $\frac{35}{40}$**

10. To find the difference of $\frac{2}{3} - \frac{2}{5}$, $\frac{2}{5}$ should be renamed to which equivalent fraction?
    A. $\frac{4}{10}$ **B. $\frac{6}{15}$** C. $\frac{8}{20}$ D. $\frac{12}{30}$

11. $\frac{5}{12} + \frac{5}{6} =$ ___?___
    A. $\frac{5}{9}$ B. $\frac{5}{6}$ **C. $1\frac{1}{4}$** D. $2\frac{1}{12}$

12. $4\frac{2}{3} + 4\frac{3}{4} =$?
    A. $8\frac{5}{12}$ B. $8\frac{5}{7}$ C. $9\frac{7}{12}$ **D. not here**

13. $\frac{3}{4} - \frac{1}{2} =$ ___?___
    A. $\frac{1}{12}$ **B. $\frac{1}{4}$** C. 1 D. not here

Go on to the next page.

Name

Posttest

CHAPTER 4
PAGE 2

14. $11\frac{1}{6} - 10\frac{1}{3} =$ ___?___

A. $\frac{1}{6}$ (B.) $\frac{5}{6}$

C. $1\frac{1}{16}$ D. not here

15. Julie played three video games. The first lasted $\frac{1}{2}$ hour; the other two lasted $\frac{1}{4}$ hour each. For how long did she play video games?

A. $\frac{3}{4}$ hr (B.) 1 hr

C. $1\frac{1}{8}$ hr D. $1\frac{1}{4}$ hr

16. Beth and her two sisters each ate $\frac{1}{4}$ of a pizza. What fraction of the whole pizza was left?

A. $\frac{1}{8}$ pizza

B. $\frac{1}{6}$ pizza

(C.) $\frac{1}{4}$ pizza

D. $\frac{3}{8}$ pizza

17. Compare. $\frac{9}{10} \times \frac{2}{3} \bigcirc \frac{2}{3}$

(A.) < B. = C. >

18. Compare. $\frac{21}{20} \times \frac{4}{5} \bigcirc \frac{4}{5}$

A. < B. = (C.) >

19. What is $\frac{3}{5} \div \frac{1}{5}$ rewritten as a multiplication problem?

A. $\frac{5}{3} \times \frac{1}{5}$ B. $\frac{3}{5} \times \frac{3}{5}$

C. $\frac{5}{3} \times \frac{5}{1}$ (D.) $\frac{3}{5} \times \frac{5}{1}$

20. What is $9\frac{1}{3} \div 2\frac{2}{3}$ rewritten as a multiplication problem?

A. $\frac{28}{3} \times \frac{8}{3}$ (B.) $\frac{28}{3} \times \frac{3}{8}$

C. $\frac{3}{28} \times \frac{8}{3}$ D. $\frac{3}{28} \times \frac{3}{8}$

21. $\frac{3}{11} \times \frac{1}{3} =$ ___?___

(A.) $\frac{1}{11}$ B. $\frac{3}{11}$

C. $\frac{2}{7}$ D. not here

22. $2\frac{1}{2} \times 1\frac{1}{4} =$ ___?___

A. $\frac{5}{8}$ B. $2\frac{1}{8}$

C. $2\frac{7}{8}$ (D.) not here

23. $\frac{1}{2} \div \frac{1}{3} =$ ___?___

A. $\frac{1}{6}$ B. $\frac{2}{3}$

(C.) $1\frac{1}{2}$ D. not here

24. $6\frac{1}{2} \div 3\frac{1}{4} =$ ___?___

A. $1\frac{2}{15}$ (B.) 2

C. $2\frac{4}{5}$ D. $28\frac{1}{6}$

25. Paul spends $1\frac{1}{2}$ hours on his music every day. He spends $\frac{1}{3}$ of that time composing music. How much time does he spend composing daily?

A. $\frac{1}{6}$ hr B. $\frac{1}{3}$ hr

(C.) $\frac{1}{2}$ hr D. $1\frac{5}{6}$ hr

Go on to the next page.

Name

Posttest

CHAPTER 4
PAGE 3

26. Heidi watches a television show that lasts $\frac{1}{2}$ hour. Commercials make up $\frac{1}{6}$ of the show. What part of an hour does she spend watching commercials?

(A.) $\frac{1}{12}$ hr B. $\frac{1}{6}$ hr

C. $\frac{1}{3}$ hr D. $\frac{2}{3}$ hr

27. A carton containing 15 identical books weighs $22\frac{1}{2}$ pounds. How many pounds does each book weigh?

(A.) $1\frac{1}{2}$ lb B. $1\frac{3}{4}$ lb

C. $1\frac{7}{8}$ lb D. $2\frac{1}{2}$ lb

28. Ramon works for $3\frac{1}{4}$ hours on Saturday, $5\frac{1}{4}$ on Sunday, and $2\frac{1}{2}$ hours each weekday. For how many hours does he work in 2 weeks?

A. 23 hr B. $33\frac{1}{2}$ hr

(C.) 42 hr D. 77 hr

29. There are 12 girls in the state tennis tournament. Each girl will play each of the other girls once. How many matches will be played?

A. 36 matches B. 45 matches

(C.) 66 matches D. 78 matches

30. Jeri worked for 14 hours at \$4.50 per hour. Alyce worked for $20\frac{1}{2}$ hours at \$3.50 per hour. Who earned more, and by how much?

(A.) Alyce earned \$8.75 more.

B. Alyce earned \$18.75 more.

C. Jeri earned \$1.25 more.

D. Jeri earned \$63.00 more.

Use the information below to answer questions 31–32.

Justin can bicycle one mile in $\frac{1}{4}$ hour. Don can bicycle one mile in $\frac{1}{3}$ hour.

31. If both boys ride for two hours, how much farther will Justin have traveled?

A. $\frac{1}{12}$ mi B. 1 mi

(C.) 2 mi C. 4 mi

32. How long will it take Justin to ride $2\frac{1}{2}$ miles?

A. $\frac{3}{8}$ hr B. $\frac{1}{2}$ hr

(C.) $\frac{5}{8}$ hr D. $1\frac{1}{8}$ hr

Stop!

Name

Cumulative Test

CHAPTERS 1–4
PAGE 1

Choose the letter of the correct answer.

1. Choose the best estimate.
$2,196 + 1,023 =$ ___?___

A. less than 2,000

B. greater than 2,000

C. less than 3,000

(D.) greater than 3,000

2. Choose the best estimate.
$79.98 \div 2.05 =$ ___?___

A. less than 25

(B.) less than 40

C. greater than 40

D. greater than 45

3. $9 + 13 + 19 =$ ___?___

A. 31 (B.) 41 C. 51 D. 122

4. $6.07 - 3.09 =$ ___?___

A. 2.88 (B.) 2.98

C. 3.02 D. 3.98

5. $19 \times 17 =$ ___?___

A. 152 B. 223

(C.) 323 D. not here

6. $20.4 \div 5.1 =$ ___?___

A. 0.25 B. 0.40

C. 4.50 (D.) not here

7. Jason bought 7.5 pounds of cheese for \$26.25. How much did he pay per pound of cheese?

A. \$3.25 (B.) \$3.50

C. \$3.75 D. \$3.95

8. Mary runs 1.5 miles every weekday, 2.5 miles each Saturday, and 3.0 miles each Sunday. How many miles does she run in two weeks?

A. 20.5 mi B. 25.0 mi

C. 25.5 mi (D.) 26.0 mi

9. What is the value of 4^3?

A. 12 B. 16 (C.) 64 D. 81

10. What is the value of 2^3?

A. 6 (B.) 8 C. 9 D. 18

11. Express 303,000,000 in scientific notation.

A. 3.03×10^6 B. 3.30×10^6

(C.) 3.03×10^8 D. 3.30×10^8

12. Express 2.4×10^4 in standard form.

A. 240 B. 2,400

(C.) 24,000 D. 240,000

Use the bar graph below to answer questions 13–14.

13. In which quarter were sales greatest?

A. Quarter 1 (B.) Quarter 2

C. Quarter 3 D. Quarter 4

Go on to the next page.

Name

Cumulative Test

CHAPTERS 1–4
PAGE 2

14. How much did sales drop between Quarter 3 and Quarter 4?

A. \$400 B. \$3,000

(C.) \$4,000 D. \$5,000

Use the circle graph to answer questions 15–16.

15. If the students earn \$350.00, how much can they spend for lunch?

A. \$25.00 (B.) \$87.50

C. \$90.00 D. \$117.50

16. If the students earn \$375.00, how much can they spend for snacks?

A. \$10.00 (B.) \$37.50

C. \$75.50 D. \$100.00

Use the information below to answer questions 17–20.

The scores of 12 students on a math quiz were: 8, 10, 13, 13, 13, 15, 15, 17, 17, 21, 24, 26.

17. What is the mode for the above scores?

(A.) 13 B. 15 C. 17 D. not here

18. What is the range of these scores?

A. 8 (B.) 18 C. 26 D. 34

19. What is the mean of these scores?

A. 14.5 B. 15.0 C. 15.5 (D.) 16.0

20. What is the median of these scores?

A. 14 (B.) 15 C. 16 D. 17

21. What is the prime factorization of 48?

(A.) $2^4 \times 3$ B. $2^3 \times 5 + 4 \times 2$

C. $8 \times 2 \times 3$ D. not here

22. What is the prime factorization of 500?

A. $2^2 \times 25^2$ (B.) $2^2 \times 5^3$

C. $2^2 \times 125$ D. $5 \times 10 \times 10$

23. What is the greatest common factor of 48 and 60?

A. 4 B. 6 (C.) 12 D. 24

24. What is the least common multiple of 4 and 6?

A. 2 B. 8 (C.) 12 D. 24

25. Which fraction is expressed in simplest form?

A. $\frac{3}{21}$ B. $\frac{6}{27}$ C. $\frac{18}{32}$ (D.) $\frac{97}{100}$

26. Which fraction is greater than $1\frac{1}{4}$?

A. $\frac{19}{16}$ B. $\frac{30}{25}$

C. $\frac{123}{100}$ (D.) $\frac{255}{200}$

27. $\frac{35}{50} =$ ___?___

A. $\frac{3}{5}$ (B.) $\frac{7}{10}$ C. $\frac{5}{7}$ D. $\frac{20}{25}$

Go on to the next page

A. TEST ANSWERS
STANDARDIZED FORMAT

Name

Cumulative Test

CHAPTERS 1–4 PAGE 3

28. Compare. $\frac{3}{9} \bigcirc \frac{9}{27}$

A. > (B.) = C. <

29. Which fraction falls between $\frac{4}{20}$ and $\frac{5}{20}$?

(A.) $\frac{2}{9}$ B. $\frac{11}{40}$ C. $\frac{16}{80}$ D. $\frac{20}{80}$

30. Which fraction falls between $\frac{1}{3}$ and $\frac{1}{4}$?

A. $\frac{3}{12}$ B. $\frac{4}{12}$ (C.) $\frac{7}{24}$ D. not here

31. Which fraction falls between $\frac{1}{10}$ and $\frac{2}{10}$?

(A.) $\frac{3}{20}$ B. $\frac{4}{20}$ C. $\frac{5}{40}$ D. $\frac{7}{40}$

32. Which fraction falls between $\frac{1}{5}$ and $\frac{1}{6}$?

A. $\frac{5}{30}$ B. $\frac{6}{30}$ (C.) $\frac{11}{60}$ D. $\frac{12}{60}$

33. $\frac{3}{6} + \frac{4}{6} =$ ___?___

A. $\frac{6}{7}$ (B.) $1\frac{1}{6}$ C. $1\frac{1}{3}$ D. not here

34. $3\frac{2}{3} + 5\frac{1}{4} =$ ___?___

A. $8\frac{3}{7}$ (B.) $8\frac{11}{12}$ C. $9\frac{11}{12}$ D. $11\frac{6}{7}$

35. $\frac{1}{4} - \frac{1}{5} =$ ___?___

A. $\frac{1}{40}$ (B.) $\frac{1}{20}$ C. $\frac{1}{5}$ D. not here

36. $11\frac{1}{4} - 10\frac{3}{4} =$ ___?___

A. $\frac{3}{4}$ B. $1\frac{1}{4}$ C. $1\frac{1}{2}$ (D.) not here

37. Susan watched three movies. The first lasted $\frac{1}{2}$ hour; the other two lasted $\frac{3}{4}$ hour each. How long did she watch movies?

A. $1\frac{1}{4}$ hr B. $1\frac{3}{4}$ hr (C.) 2 hr D. $2\frac{1}{4}$ hr

38. Tim had a bottle of juice. He drank $\frac{1}{4}$ of the bottle and his sister drank $\frac{1}{2}$ of the bottle. How much was left?

A. $\frac{1}{16}$ bottle B. $\frac{1}{8}$ bottle (C.) $\frac{1}{4}$ bottle D. $\frac{3}{8}$ bottle

39. $\frac{3}{5} \times \frac{1}{7} =$ ___?___

A. $\frac{1}{3}$ (B.) $\frac{3}{35}$ C. $4\frac{1}{5}$ D. not here

40. $1\frac{1}{4} \times 1\frac{1}{3} =$ ___?___

A. $1\frac{5}{12}$ B. $1\frac{1}{2}$ C. $1\frac{7}{12}$ (D.) $1\frac{2}{3}$

Go on to the next page.

Name

Cumulative Test

CHAPTERS 1–4 PAGE 4

41. $\frac{1}{2} \div \frac{1}{5} =$ ___?___

A. $\frac{1}{10}$ B. $\frac{2}{5}$ C. 2 (D.) $2\frac{1}{2}$

42. $1\frac{3}{4} \div 1\frac{1}{2} =$ ___?___

A. $\frac{6}{7}$ (B.) $1\frac{1}{6}$ C. $1\frac{1}{4}$ D. not here

43. The lake at Jones Pier is 15 feet deep. Every $\frac{1}{5}$ mile from the pier it becomes 1 foot deeper until it reaches its maximum depth of 30 feet. How many miles from the pier does the lake reach its maximum depth?

A. $2\frac{1}{2}$ mi (B.) 3 mi C. 5 mi D. 15 mi

Use the table below to answer questions 44–45.

Senior Class Elections

Candidate	Girls' Votes	Boys' Votes
Jack	50	80
Carol	60	84
Darla	100	51
Bill	90	65
	300	280

44. Who received the most total votes?

A. Jack B. Carol C. Darla (D.) Bill

45. How many more votes did Bill receive than Carol?

(A.) 11 votes B. 19 votes C. 30 votes D. 31 votes

46. A principal wants a graph to compare the monthly absences of boys and girls for the year. Which type of graph should be used?

(A.) double-bar graph
B. histogram
C. stem-and-leaf plot
D. circle graph

47. A family wants a graph to show how they spend their monthly income. Which type of graph should they use?

A. line graph
B. double-bar graph
C. box-and-whisker graph
(D.) circle graph

48. Tom gave Ben $1.50 to buy stamps. Tom gave Ben only quarters and dimes. He gave Ben 9 coins in all. How many quarters did Tom give Ben?

A. 6 quarters B. 5 quarters (C.) 4 quarters D. 3 quarters

49. There are 10 players in a tennis league. If each player plays one game with each of the other players, how many games will be played?

A. 20 games B. 30 games (C.) 45 games D. 90 games

50. Nancy must walk 4 blocks to school. It takes her $2\frac{1}{2}$ minutes to walk each block. Pat walks 5 blocks to school. It takes her only $1\frac{1}{2}$ minutes to walk each block. How much longer does it take Nancy to walk to school than it takes Pat?

A. 1 minute (B.) $2\frac{1}{2}$ minutes C. 3 minutes D. 10 minutes

Stop!

Name

Pretest

CHAPTER 5 PAGE 1

Choose the letter of the correct answer.

1. Choose the algebraic expression for the product of 9 and a number, y.

(A.) $9y$ B. $\frac{9}{y}$ C. $\frac{y}{9}$ D. $y - 9$

2. Choose the algebraic expression for a number, b, decreased by 2.

A. $2b$ B. $b + 2$ (C.) $b - 2$ D. $2 - b$

For questions 3–6, let $a = 15$ and $b = 6$.

3. Evaluate. $4a - 6$

A. 13 B. 36 (C.) 54 D. 66

4. Evaluate. $3b - 10$

A. 3 B. 6 (C.) 8 D. 28

5. Evaluate. $2b - 12$

(A.) 0 B. 3 C. 6 D. 12

6. Evaluate. $3a + 12$

A. 48 (B.) 57 C. 105 D. not here

7. Which expression shows the number of minutes in h hours?

A. $\frac{h}{60}$ B. $\frac{60}{h}$ C. $h + 60$ (D.) $60h$

8. Mary and June scored a total of 25 points in a basketball game. Let m = the number of points scored by Mary. Which expression shows how many points June scored?

A. $m + 25$ B. $m - 25$ C. $25m$ (D.) $25 - m$

9. Find the value.

$(3 + 5) \cdot (3 - 1) =$ ___?___

(A.) 16 B. 17 C. 24 D. not here

10. Find the value.

$(2 + 3) + (5 - 1)^2 =$ ___?___

A. 9 B. 20 C. 81 (D.) not here

11. Find the value.

$(2 + 1 + 3) \div (3 - 2) =$ ___?___

A. 0 B. 2 C. 5 (D.) 6

12. Find the value.

$5 \times (9 - 3) =$ ___?___

A. 25 (B.) 30 C. 42 D. 56

13. Which inequality shows that a number, b, increased by 5 is less than 100?

A. $5b > 100$ B. $b + 5 > 100$ C. $5b < 100$ (D.) $b + 5 < 100$

14. Which inequality shows that a number, c, decreased by 10 is greater than 10?

A. $10 - c > 10$ (B.) $c - 10 > 10$ C. $c - 10 < 10$ D. $c + 10 > 10$

15. Solve the inequality. $x + 1 > 5$

A. $x < 1$ B. $x < 4$ (C.) $x > 4$ D. $x < 6$

16. Solve the inequality. $x + 5 < 14$

A. $x > 9$ (B.) $x < 9$ C. $x < 19$ D. $x > 19$

17. To solve the equation $y + 3 = 10$, what must be done to both sides of the equation?

A. add 3 (B.) subtract 3 C. add 10 D. subtract 10

18. To solve the equation $x - 5 = 7$, what must be done to both sides of the equation?

(A.) add 5 B. subtract 5 C. add 7 D. subtract 7

Go on to the next page.

Name

Pretest

CHAPTER 5 PAGE 2

19. To solve the equation $3x = 15$, what must be done to both sides of the equation?

A. multiply by 3 (B.) divide by 3 C. multiply by 15 D. divide by 15

20. To solve the equation $\frac{y}{20} = 2$, what must be done to both sides of the equation?

A. multiply by 2 B. divide by 2 (C.) multiply by 20 D. divide by 20

21. Solve. $p + 9 = 12$

(A.) $p = 3$ B. $p = 9$ C. $p = 21$ D. $p = 108$

22. Solve. $p - 10 = 0$

A. $p = 0$ (B.) $p = 10$ C. $p = 11$ D. $p = 100$

23. Solve. $3x = 30$

(A.) $x = 10$ B. $x = 30$ C. $x = 27$ D. $x = 90$

24. Solve. $\frac{x}{5} = 25$

A. $x = 5$ B. $x = 20$ C. $x = 30$ (D.) $x = 125$

25. After Joe gave 10 of his baseball cards to Bob, he had 92 cards left. How many cards did Joe have before giving cards to Bob?

A. 82 cards B. 92 cards (C.) 102 cards D. 112 cards

26. Li earned $11.00 baby-sitting. That brought her savings to $51.50. How much money did Li have before baby-sitting?

A. $40.00 (B.) $40.50 C. $61.50 D. $62.50

27. Jon has 3 times as many stamps as Ali. If Jon has 15 stamps, how many does Ali have?

(A.) 5 stamps B. 18 stamps C. 30 stamps D. 45 stamps

28. Ed's salary is $\frac{1}{4}$ of Bill's salary. Ed earns $20,000. How much does Bill earn?

A. $5,000 B. $10,000 C. $40,000 (D.) $80,000

29. Mr. Hynes withdrew money from the bank. He spent $30 on shoes and $10 on a tie. He gave $\frac{1}{2}$ of what was left to his wife. He then had $45. How much money did Mr. Hynes withdraw?

A. $85 B. $90 (C.) $130 D. $140

30. Betty joined a running club. After one month she doubled the distance she ran each day. After two months she doubled the distance again. She then was running 4 miles a day. How many miles a day did she run at the start?

(A.) 1 mi B. $1\frac{1}{2}$ mi C. 2 mi D. 16 mi

31. Over the past 5 weeks, Tom lost 3 pounds per week. He now weighs 200 pounds. How much did Tom weigh before losing the weight?

A. 185 pounds B. 203 pounds C. 212 pounds (D.) 215 pounds

32. Alicia collects stamps. On Monday she added 5 stamps to her collection. On Tuesday she doubled the number of stamps in her collection. She then had 90 stamps. How many stamps did she have on Sunday?

A. 35 stamps (B.) 40 stamps C. 50 stamps D. 185 stamps

Stop!

Name

Posttest

CHAPTER 5 PAGE 1

Choose the letter of the correct answer.

1. Choose the algebraic expression for the product of 5 and a number, c.

(A.) $5c$ B. $\frac{5}{c}$ C. $\frac{c}{5}$ D. $c - 5$

2. Choose the algebraic expression for a number, x, increased by 3.

A. $3x$ (B.) $x + 3$
C. $x - 3$ D. $3 - x$

For questions 3–6, let $a = 20$ and $b = 5$.

3. Evaluate. $4b + 1$

A. 5 B. 20 (C.) 21 D. not here

4. Evaluate. $2a - 27$

A. 4 (B.) 13 C. 17 D. not here

5. Evaluate. $3b - 15$

(A.) 0 B. 3 C. 12 D. 45

6. Evaluate. $4b - 20$

(A.) 0 B. 1 C. 2 D. 4

7. Which expression shows the number of hours in d days?

A. $\frac{d}{24}$ B. $\frac{24}{d}$
C. $d + 24$ (D.) $24d$

8. Hank and Bobby scored a total of 6 goals in a hockey game. Let h = the number of goals scored by Hank. Which expression shows how many goals Bobby scored?

A. $h + 6$ B. $h - 6$
C. $6h$ (D.) $6 - h$

9. Find the value.

$(2 + 6) \cdot (4 - 2) =$ ___?___

(A.) 16 B. 24 C. 30 D. not here

10. Find the value.

$(1 + 2) + (4 - 1)^2 =$ ___?___

A. 6 B. 9 C. 36 (D.) not here

11. Find the value.

$(3 + 2 + 7) \div (7 - 6) =$ ___?___

A. 0 B. 1 C. 2 (D.) 12

12. Find the value.

$10 \times (8 - 4) -$ ___?___

A. 14 (B.) 40 C. 76 D. 104

13. Which inequality shows that a number, a, increased by 7 is less than 99?

A. $7a > 99$ B. $a + 77 > 99$
C. $7a < 99$ (D.) $a + 7 < 99$

14. Which inequality shows that a number, y, increased by 100 is greater than 100?

A. $100y > 100$ (B.) $y + 100 > 100$
C. $100y < 100$ D. $y + 100 < 100$

15. Solve the inequality. $x + 5 > 12$

A. $x < 1$ B. $x > 1$
C. $x < 5$ (D.) $x > 7$

16. Solve the inequality. $x + 4 < 9$

A. $x > 4$ (B.) $x < 5$
C. $x > 5$ D. $x > 13$

17. To solve the equation $y + 5 = 15$, what must be done to both sides of the equation?

A. add 5 (B.) subtract 5
C. add 15 D. subtract 15

18. To solve the equation $x - 3 = 6$, what must be done to both sides of the equation?

(A.) add 3 B. subtract 3
C. add 6 D. subtract 6

Go on to the next page.

Name

Posttest

CHAPTER 5 PAGE 2

19. To solve the equation $4x = 20$, what must be done to both sides of the equation?

A. multiply by 4 (B.) divide by 4
C. multiply by 20 D. divide by 20

20. To solve the equation $\frac{y}{30} = 3$, what must be done to both sides of the equation?

A. multiply by 3 B. divide by 3
(C.) multiply by 30 D. divide by 30

21. Solve. $m + 12 = 14$

(A.) $m = 2$ B. $m = 12$
C. $m = 14$ D. $m = 26$

22. Solve. $m - 50 = 10$

A. $m = {}^-40$ B. $m = 40$
C. $m = 50$ (D.) $m = 60$

23. Solve. $5y = 100$

A. $y = 2$ (B.) $y = 20$
C. $y = 95$ D. $y = 105$

24. Solve. $\frac{x}{7} = 7$

A. $x = 1$ B. $x = 7$
(C.) $x = 49$ D. $x = 77$

25. Ben sells newspapers. One day he sold 110 papers. He had 45 left. How many papers did he have at the beginning of the day?

A. 65 papers B. 75 papers
(C.) 155 papers D. 165 papers

26. Ann is a waitress. One day she earned $12.50 in tips. Added to her salary, she made $35.00 that day. What was Ann's salary for the day?

(A.) $22.50 B. $23.50
C. $47.50 D. $642.50

27. In a basketball game, Nat scored 3 times as many points as Phil. Nat scored 18 points. How many points did Phil score?

(A.) 6 points B. 15 points
C. 21 points D. 54 points

28. Ed's weight is $\frac{3}{4}$ of Tony's weight. Ed weighs 120 pounds. How many pounds does Tony weigh?

A. 60 lb B. 135 lb
(C.) 160 lb D. 200 lb

29. Linda withdrew money from the bank. She gave $\frac{1}{2}$ of the money to her husband and then spent $50 on groceries. She then had $50. How much money did she withdraw?

A. $50 B. $100 (C.) $200 D. $250

30. Betty joined an exercise club. In one month she doubled the number of sit ups she could do. After two months she doubled the number of sit-ups again. She then could do 60. How many sit-ups could she do at the start?

A. 10 sit-ups (B.) 15 sit-ups
C. 20 sit-ups D. 30 sit-ups

31. Over the past 6 weeks, Tom burned 3 logs per week. Tom now has 22 logs. How many logs did Tom have 6 weeks ago?

A. 4 logs B. 13 logs
C. 18 logs (D.) 40 logs

32. Natasha invested in the stock market. The first week she lost $\frac{1}{2}$ of her money. The second week she lost $\frac{1}{2}$ of the remainder. She was left with $1,200. How much money did she invest?

A. $600 B. $1,200
C. $3,600 (D.) $4,800

Stop!

Name

Pretest

CHAPTER 6 PAGE 1

Choose the letter of the correct answer.

For questions 1–4, choose whether each statement is *always, sometimes,* or *never* true.

1. Intersecting lines are perpendicular.

A. always (B.) sometimes C. never

2. A right triangle is isosceles.

A. always (B.) sometimes C. never

3. Adjacent angles have a common vertex and a common ray.

(A.) always B. sometimes C. never

4. A regular hexagon has 6 congruent angles.

(A.) always B. sometimes C. never

5. What is the total angle measure of the interior angles of a quadrilateral?

A. 90° B. 180° (C.) 360° D. not here

6. What is the complement of a 65° angle?

(A.) 25° B. 35° C. 115° D. not here

7. Two angles of a triangle measure 40° and 60°. What is the measure of the third angle?

A. 50° (B.) 80° C. 90° D. 100°

8. One acute angle of a right triangle is twice as large as the other. What are the measures of the two acute angles?

A. 15°, 30° B. 45°, 45°
C. 25°, 50° (D.) 30°, 60°

9. If a triangle has angle measures of 45°, 50°, and 85°, what type of triangle is it?

A. isosceles B. equilateral
C. right (D.) acute

10. Classify the following triangle according to the lengths of its sides: 4 in., 5 in., 6 in.

A. isosceles B. equilateral
(C.) scalene D. regular

11. What is the measure of the *reflex* central angle formed by the hands of a clock at 12:40?

A. 60° B. 100°
C. 120° (D.) 240°

12. The measures of two sides of an isosceles triangle are 9 cm and 4 cm. What is the measure of the third side?

A. 4 cm B. 5 cm
(C.) 9 cm D. 13 cm

13. Figure 2 below represents what construction?

R Y X Z

Figure 1 Figure 2

A. congruent line segment
B. angle bisector
C. line bisector
(D.) congruent angle

Go on to the next page.

Name

Pretest

CHAPTER 6 PAGE 2

14. The figure below represents what construction?

S Q R X T

(A.) bisected line segment
B. bisected angle
C. congruent angle
D. congruent triangle

15. Figure 2 below represents what construction?

A B C D E F

Figure 1 Figure 2

A. congruent line segment
(B.) congruent triangle
C. congruent angle
D. angle bisector

16. The figure below represents what construction?

C A D F B E

A. congruent angle
B. congruent line segment
(C.) bisected angle
D. bisected line segment

Use the Venn diagram below to answer questions 17–20.

polygons
regular polygons
squares

17. Does the diagram show that a polygon is always a square?

A. yes (B.) no

18. Does the diagram show that all polygons are regular polygons?

A. yes (B.) no

19. Does the diagram show that all squares are regular polygons?

(A.) yes B. no

20. Does the diagram show that some polygons are not squares?

(A.) yes B. no

Go on to the next page.

A. TEST ANSWERS STANDARDIZED FORMAT

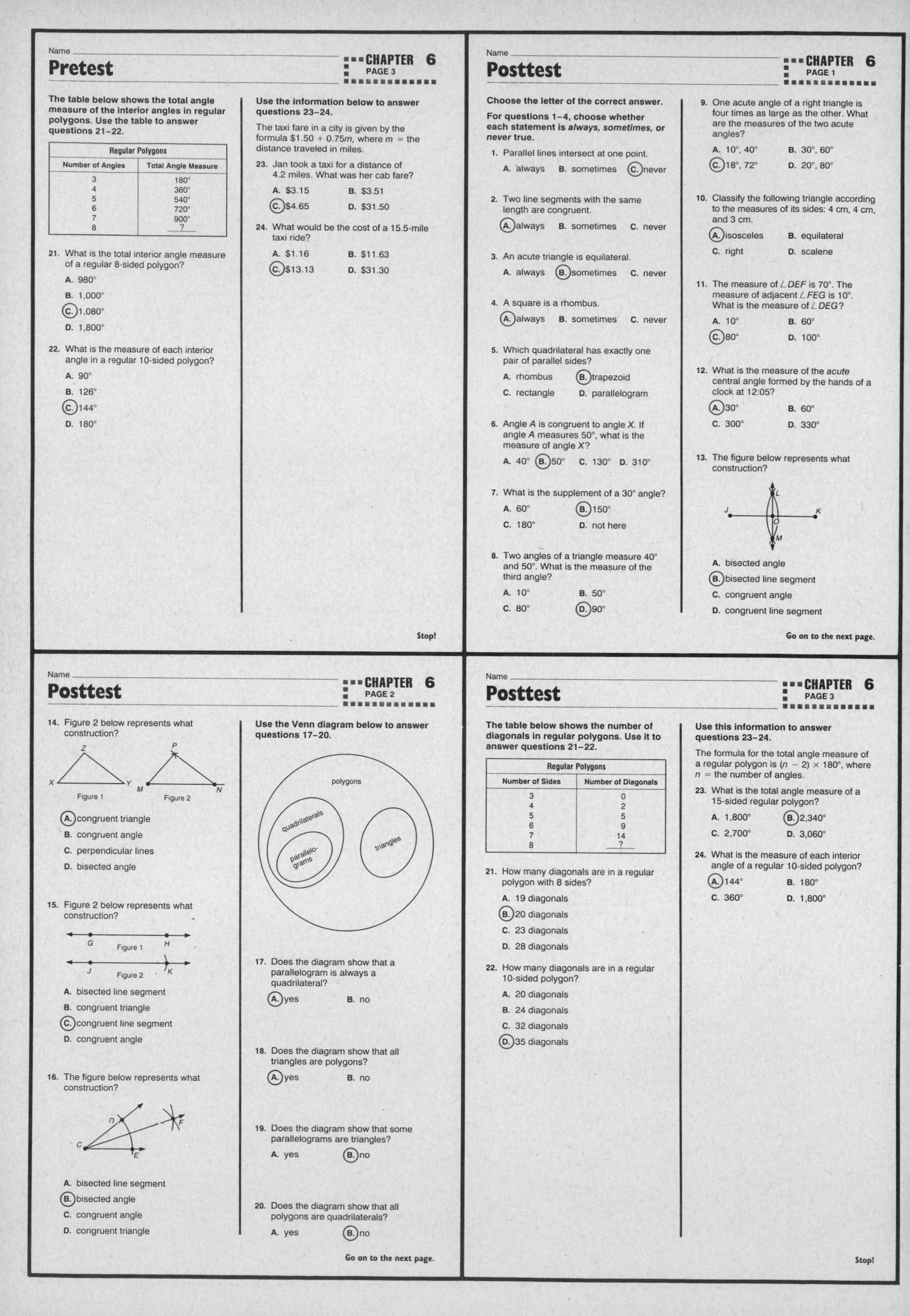

Name

Pretest

CHAPTER 6
PAGE 3

The table below shows the total angle measure of the interior angles in regular polygons. Use the table to answer questions 21–22.

Regular Polygons	
Number of Angles	Total Angle Measure
3	180°
4	360°
5	540°
6	720°
7	900°
8	?

21. What is the total interior angle measure of a regular 8-sided polygon?
    A. 980°
    B. 1,000°
    C. 1,080°
    D. 1,800°

22. What is the measure of each interior angle in a regular 10-sided polygon?
    A. 90°
    B. 126°
    C. 144°
    D. 180°

Use the information below to answer questions 23–24.

The taxi fare in a city is given by the formula $1.50 + 0.75*m*, where *m* = the distance traveled in miles.

23. Jan took a taxi for a distance of 4.2 miles. What was her cab fare?
    A. $3.15 B. $3.51
    C. $4.65 D. $31.50

24. What would be the cost of a 15.5-mile taxi ride?
    A. $1.16 B. $11.63
    C. $13.13 D. $31.30

Stop!

Name

Posttest

CHAPTER 6
PAGE 1

Choose the letter of the correct answer.

For questions 1–4, choose whether each statement is *always, sometimes,* or *never* true.

1. Parallel lines intersect at one point.
   A. always B. sometimes **C.** never

2. Two line segments with the same length are congruent.
   A. always B. sometimes C. never

3. An acute triangle is equilateral.
   A. always **B.** sometimes C. never

4. A square is a rhombus.
   A. always B. sometimes C. never

5. Which quadrilateral has exactly one pair of parallel sides?
   A. rhombus **B.** trapezoid
   C. rectangle D. parallelogram

6. Angle *A* is congruent to angle *X*. If angle *A* measures 50°, what is the measure of angle *X*?
   A. 40° **B.** 50° C. 130° D. 310°

7. What is the supplement of a 30° angle?
   A. 60° **B.** 150°
   C. 180° D. not here

8. Two angles of a triangle measure 40° and 50°. What is the measure of the third angle?
   A. 10° B. 50°
   C. 80° **D.** 90°

9. One acute angle of a right triangle is four times as large as the other. What are the measures of the two acute angles?
   A. 10°, 40° B. 30°, 60°
   C. 18°, 72° D. 20°, 80°

10. Classify the following triangle according to the measures of its sides: 4 cm, 4 cm, and 3 cm.
    A. isosceles B. equilateral
    C. right D. scalene

11. The measure of $\angle DEF$ is 70°. The measure of adjacent $\angle FEG$ is 10°. What is the measure of $\angle DEG$?
    A. 10° B. 60°
    C. 80° D. 100°

12. What is the measure of the *acute* central angle formed by the hands of a clock at 12:05?
    A. 30° B. 60°
    C. 300° D. 330°

13. The figure below represents what construction?

    A. bisected angle
    B. bisected line segment
    C. congruent angle
    D. congruent line segment

Go on to the next page.

Name

Posttest

CHAPTER 6
PAGE 2

14. Figure 2 below represents what construction?

    A. congruent triangle
    B. congruent angle
    C. perpendicular lines
    D. bisected angle

15. Figure 2 below represents what construction?

    A. bisected line segment
    B. congruent triangle
    C. congruent line segment
    D. congruent angle

16. The figure below represents what construction?

    A. bisected line segment
    B. bisected angle
    C. congruent angle
    D. congruent triangle

Use the Venn diagram below to answer questions 17–20.

17. Does the diagram show that a parallelogram is always a quadrilateral?
    A. yes B. no

18. Does the diagram show that all triangles are polygons?
    A. yes B. no

19. Does the diagram show that some parallelograms are triangles?
    A. yes **B.** no

20. Does the diagram show that all polygons are quadrilaterals?
    A. yes **B.** no

Go on to the next page.

Name

Posttest

CHAPTER 6
PAGE 3

The table below shows the number of diagonals in regular polygons. Use it to answer questions 21–22.

Regular Polygons	
Number of Sides	Number of Diagonals
3	0
4	2
5	5
6	9
7	14
8	?

21. How many diagonals are in a regular polygon with 8 sides?
    A. 19 diagonals
    B. 20 diagonals
    C. 23 diagonals
    D. 28 diagonals

22. How many diagonals are in a regular 10-sided polygon?
    A. 20 diagonals
    B. 24 diagonals
    C. 32 diagonals
    D. 35 diagonals

Use this information to answer questions 23–24.

The formula for the total angle measure of a regular polygon is $(n - 2) \times 180°$, where n = the number of angles.

23. What is the total angle measure of a 15-sided regular polygon?
    A. 1,800° **B.** 2,340°
    C. 2,700° D. 3,060°

24. What is the measure of each interior angle of a regular 10-sided polygon?
    A. 144° B. 180°
    C. 360° D. 1,800°

Stop!

Name ______________________

Pretest

CHAPTER 7 PAGE 1

Choose the letter of the correct answer.

1. A box contains 9 red marbles and 3 blue marbles. Which of the following expresses the ratio of red to blue marbles?

   (A.) 9 to 3 B. 9 to 12
   C. 9 to 1 D. 1 to 3

2. Which ratio does *not* have the same value as the other three?

   A. 1:5 B. 2:10 C. 4:20 **(D.)** 5:30

3. Which of the following is *not* a true proportion?

   A. $\frac{2}{3} = \frac{4}{6}$ B. $\frac{4}{5} = \frac{80}{100}$
   C. $\frac{5}{6} = \frac{155}{186}$ **(D.)** $\frac{7}{8} = \frac{75}{80}$

4. Which value of *a* makes the following a true proportion?

   $\frac{a}{10} = \frac{4}{20}$

   (A.) $a = 2$ B. $a = 4$
   C. $a = 8$ D. $a = 20$

5. Solve. $\frac{x}{4} = \frac{3}{6}$

   A. $x = 1$ **(B.)** $x = 2$
   C. $x = 3$ D. $x = 8$

6. Solve. $\frac{10}{y} = \frac{25}{7.5}$

   A. $y = 2.0$ B. $y = 2.5$
   (C.) $y = 3.0$ D. $y = 5.0$

7. Solve. $\frac{3a}{9} = \frac{8}{6}$

   (A.) $a = 4$ B. $a = 4\frac{1}{2}$
   C. $a = 5$ D. not here

8. Solve for *a*. $\frac{5}{a} = \frac{25}{55}$

   A. $a = 5$ **(B.)** $a = 11$
   C. $a = 16$ D. $a = 55$

9. Betty drives 10 miles in 15 minutes. At this rate how far will she drive in 75 minutes?

   A. 30 mi B. 40 mi
   (C.) 50 mi D. 112 mi

10. Jose's car runs 26 miles on 1 gallon of gas. At this rate how many gallons of gas will he use to travel 117 miles?

    A. $3\frac{1}{2}$ gal B. 4 gal
    (C.) $4\frac{1}{2}$ gal D. 5 gal

11. Mary is paid $7 for every 2 hours of work. How much does she earn for working 14 hours?

    A. $4 B. $23 **(C.)** $49 D. $98

12. Sean drives $\frac{9}{10}$ mile in one minute. At this rate how far will he drive in 1 hour?

    A. 45 mi **(B.)** 54 mi
    C. 60 mi D. 90 mi

13. A package of 8 batteries costs $7.60. What is the unit price?

    A. $0.90 **(B.)** $0.95
    C. $1.10 D. $9.50

14. If 8 batteries cost $7.60, how much would 20 batteries cost?

    (A.) $19.00 B. $22.80
    C. $26.60 D. $152.00

Go on to the next page.

Name ______________________

Pretest

CHAPTER 7 PAGE 2

15. A box of soap powder weighing 4 pounds (lb) costs $3.96; a 6-lb box costs $6.30. Which package has the lower unit price, and by how much per pound?

    (A.) 4-lb, by $0.06
    B. 6-lb, by $1.05
    C. 4-lb, by $0.58
    D. 6-lb, by $0.06

16. If 4 rolls cost $2.16, how many rolls can be bought for $4.32?

    A. 6 rolls **(B.)** 8 rolls
    C. 12 rolls D. 16 rolls

17. On a scale drawing, 1 inch is used to represent 4 feet. Which ratio expresses that relationship?

    A. 1 : 8 B. 1 : 12 **(C.)** 1 : 48 D. 4 : 48

18. On Map A, 1 inch = 10 miles. On Map B, 1 inch = 20 miles. A road is 4 inches long on Map A. How long will the road be on Map B?

    (A.) 2 in. B. 4 in.
    C. 6 in. D. 8 in.

19. Use the scale 6 cm = 25 cm. What is the actual length of an object when the scale length is 24 cm?

    A. 50 cm B. 75 cm
    (C.) 100 cm D. 125 cm

20. Use the scale 1 m = 50 km. What is the actual length of an object when the scale length is $\frac{1}{4}$ m?

    A. 2.5 km **(B.)** 12.5 km
    C. 125 km D. 200 km

21. Lupe is making a scale drawing on which each centimeter represents 8 meters. One scale building measures 12 cm by 15 cm. What are the actual dimensions of the building?

    A. 12 m × 15 m
    B. 20 m × 23 m
    (C.) 96 m × 120 m
    D. 12 m × 120 m

22. On a map 1 cm = 25 km. Two cities are 12 cm apart on the map. How far are they actually apart?

    A. 120 km B. 250 km
    (C.) 300 km D. 370 km

For questions 23–24, choose whether the statement is *always, sometimes,* or *never* true.

23. Congruent triangles are similar.

    (A.) always B. sometimes C. never

24. Similar figures have the same shape.

    (A.) always B. sometimes C. never

25. Two rectangles are similar. The first rectangle is 6 in. wide and 10 in. long. The width of the second rectangle is 30 in. Find its length.

    A. 18 in. B. 30 in.
    C. 34 in. **(D.)** 50 in.

26. Two rectangles are similar. The first rectangle has a width of 2 cm and a length of 5 cm. The width of the second rectangle is 0.2 cm. Find the length.

    (A.) 0.5 cm B. 3.8 cm
    C. 10 cm D. 50 cm

Go on to the next page.

Name ______________________

Pretest

CHAPTER 7 PAGE 3

27. A triangle has sides measuring 3 ft, 4 ft, 5 ft. Which of the following could be the lengths of the sides of a similar triangle?

    A. 4 yd, 6 yd, 8 yd
    B. 4 ft, 5 ft, 6 ft
    C. 31 ft, 41 ft, 51 ft
    (D.) 6 ft, 8 ft, 10 ft

28. These two rectangles are similar. Find *x*.

    $\frac{3}{8}$ in / *x* ; $\frac{3}{4}$ in. / 1 in.

    (A.) $\frac{1}{2}$ in. B. $\frac{5}{8}$ in.
    C. $\frac{7}{8}$ in. D. $1\frac{1}{8}$ in.

29. A map scale is 2 cm = 7 km. Mt. Snow is 14 cm from Rayville on the map. How far is that?

    A. 20 km
    B. 30 km
    (C.) 49 km
    D. 100 km

30. The scale on a map is 1 in. = 200 mi. Two cities are $2\frac{1}{2}$ in. apart. If Toni averages 50 mi per hr, how long will it take her to drive from one city to the other?

    A. $8\frac{1}{2}$ hr B. 9 hr
    C. $9\frac{1}{2}$ hr **(D.)** 10 hr

Use the information below to answer questions 31–32.

The formula for finding the circumference of a circle is $C = 2\pi r$, where C = the circumference and r = the radius.

31. Use 3.14 as an approximate value of π. Find the circumference of a circle with a radius of 6 in.

    A. 18.84 in. **(B.)** 37.68 in.
    C. 113.04 in. D. 138.84 in.

32. Use $\frac{22}{7}$ as an approximate value of π. Find the radius of a circle with a circumference of 11 cm.

    A. $1\frac{3}{4}$ cm
    (B.) $3\frac{1}{2}$ cm
    C. $17\frac{1}{2}$ cm
    D. $69\frac{1}{10}$ cm

Stop!

Name ______________________

Posttest

CHAPTER 7 PAGE 1

Choose the letter of the correct answer.

1. At the zoo there are 3 elephants and 8 tigers. Which of the following expresses the ratio of elephants to tigers?

   A. 3 : 11 B. 1 : 5 C. 8 : 3 **(D.)** 3 : 8

2. Which ratio does *not* have the same value as the other three?

   A. $\frac{2}{9}$ **(B.)** $\frac{3}{18}$ C. $\frac{6}{27}$ D. $\frac{10}{45}$

3. Which of the following is *not* a true proportion?

   A. $\frac{3}{4} = \frac{6}{8}$ B. $\frac{1}{5} = \frac{20}{100}$
   C. $\frac{7}{21} = \frac{1}{3}$ **(D.)** $\frac{5}{7} = \frac{55}{70}$

4. Which value of *a* makes the following a true proportion?

   $\frac{a}{12} = \frac{3}{4}$

   A. $a = 3$ B. $a = 4$
   C. $a = 6$ **(D.)** $a = 9$

5. Solve. $\frac{x}{8} = \frac{5}{10}$

   A. $x = 2$ **(B.)** $x = 4$
   C. $x = 16$ D. $x = 40$

6. Solve. $\frac{20}{y} = \frac{25}{2.5}$

   A. $y = 1.0$ B. $y = 1.5$
   (C.) $y = 2.0$ D. $y = 2.5$

7. Solve. $\frac{2a}{3} = \frac{4}{6}$

   (A.) $a = 1$ B. $a = 1\frac{1}{2}$
   C. $a = 2$ D. $a = 2\frac{1}{2}$

8. Solve for *a*. $\frac{4}{a} = \frac{1}{12}$

   A. $a = 3$ B. $a = 16$
   C. $a = 24$ **(D.)** $a = 48$

9. Martha drives 40 miles in 30 minutes. At this rate how far will she drive in 90 minutes?

   A. 68 mi B. 80 mi
   C. 100 mi **(D.)** 120 mi

10. Jim's car runs 32 miles on 1 gallon of gas. At this rate how many gallons of gas will he use to travel 112 miles?

    A. $2\frac{1}{2}$ gal B. 3 gal
    (C.) $3\frac{1}{2}$ gal D. 4 gal

11. Mario is paid $5 for every 2 hours of work. How much does he earn for working 10 hours?

    A. $4 B. $20 **(C.)** $25 D. $50

12. Jon drives $\frac{6}{10}$ mile in one minute. At this rate how far will he drive in 1 hour?

    A. 30 mi **(B.)** 36 mi
    C. 60 mi D. 100 mi

13. A package of 4 light bulbs costs $3.96. What is the unit price?

    A. $0.93 **(B.)** $0.99
    C. $1.02 D. $1.98

14. If 4 light bulbs cost $3.96, how much would 20 light bulbs cost?

    (A.) $19.80 B. $23.76
    C. $39.60 D. $316.80

Go on to the next page.

Name ______________________

Posttest

CHAPTER 7
PAGE 2

15. A package of crackers weighing 10 ounces (oz) costs $1.90; a 16-oz package costs $2.40. Which package has the lower unit price, and by how much per ounce?
    - A. 10-oz, by $0.04
    - B. 16-oz, by $0.15
    - C. 10-oz, by $0.19
    - D. 16-oz, by $0.04

16. If 6 oranges cost $1.98, how many oranges can be bought for $5.94?
    A. 12 oranges B. 14 oranges C. 16 oranges D. 18 oranges

17. On a scale drawing, 1 inch is used to represent 2 feet. Which ratio expresses that relationship?
    A. 1 : 4 B. 1 : 12 C. 1 : 24 D. 2 : 24

18. On Map A, 1 inch = 5 miles. On Map B, 1 inch = 10 miles. A road is 3 inches long on Map A. How long will the road be on Map B?
    A. $1\frac{1}{2}$ in. B. 3 in. C. $4\frac{1}{2}$ in. D. 6 in.

19. Use the scale 3 cm = 8 m. What is the actual length of an object when the scale length is 27 cm?
    A. 9 m B. 32 m C. 72 m D. 216 m

20. Use the scale 1 cm = 25 cm. What is the actual length of an object when the scale length is 0.5 cm?
    A. 5 cm B. 10 cm C. 12.5 cm D. 37.5 cm

21. John is making a scale drawing on which each centimeter represents 5 meters. A field drawn to scale measures 15 cm by 20 cm. What are the actual dimensions of the field?
    - A. 3 m × 4 m
    - B. 20 m × 25 m
    - C. 75 m × 100 m
    - D. 20 m × 100 m

22. On a map 1 inch = 10 miles. Two cities are 10 inches apart on the map. How many miles are they actually apart?
    A. 10 mi B. 100 mi C. 200 mi D. 1,000 mi

For questions 23–24, choose whether the statement is *always*, *sometimes*, or *never* true.

23. Similar triangles are congruent.
    A. always B. sometimes C. never

24. Similar figures have different shapes.
    A. always B. sometimes C. never

25. Two rectangles are similar. The first rectangle has a width of 8 m and a length of 12 m. The width of the second rectangle is 24 m. Find its length.
    A. 20 m B. 24 m C. 28 m D. 36 m

26. Two rectangles are similar. The first rectangle has a width of 3 cm and a length of 6 cm. The width of the second rectangle is 0.5 cm. Find its length.
    A. 0.25 cm B. 1 cm C. 1.5 cm D. 3.5 cm

Go on to the next page.

Name ______________________

Posttest

CHAPTER 7
PAGE 3

27. A triangle has sides measuring 6 cm, 8 cm, and 10 cm. Which of the following could be the lengths of the sides of a similar triangle?
    - A. 6 km, 12 km, 15 km
    - B. 7 cm, 9 cm, 11 cm
    - C. 9 cm, 12 cm, 20 cm
    - D. 9 m, 12 m, 15 m

28. These two triangles are similar. Find a.

    4 mm, 5 mm, 3 mm — 6 mm, a, 4.5 mm

    A. 5.5 mm B. 6.5 mm C. 7.5 mm D. not here

29. A map scale is 2 cm = 9 km. The map distance from Redville to Leetown is 18 cm. The map distance from Leetown to Spring is 3 cm. How far is it from Redville to Spring through Leetown?
    A. 21 km B. 42 km C. 94.5 km D. 189 km

30. The scale on a map is 1 in. = 150 mi. Two cities are 3 in. apart. If Tom averages 50 mi per hr and makes no stops, how long will it take him to drive from one city to the other?
    A. 3 hr B. $7\frac{1}{2}$ hr C. $8\frac{1}{2}$ hr D. 9 hr

Use the information below to answer questions 31–32.

The formula for finding the circumference of a circle is $C = \pi \cdot d$, where C = the circumference and d = the diameter.

31. Use $\frac{22}{7}$ as an approximate value of π. Find the circumference of a circle with a diameter of 28 in.
    A. 9 in. B. $7\frac{1}{7}$ in. C. 88 in. D. $\frac{6}{16}$ in.

32. Use 3.14 as an approximate value of π. Find the diameter of a tree with a circumference of 8.32 m.
    A. 0.38 m B. 2.65 m C. 5.18 m D. 26.12 m

Stop!

Name ______________________

Pretest

CHAPTER 8
PAGE 1

Choose the letter of the correct answer.

1. Which of the following is not equivalent to the others?
   A. 18% B. 1.8 C. $\frac{9}{5}$ D. 180%

2. Which fraction can be most easily expressed as a percent?
   A. $\frac{7}{22}$ B. $\frac{3}{7}$ C. $\frac{2}{10}$ D. $\frac{1}{16}$

3. What is $\frac{90}{1,000}$ as a percent?
   A. 0.9% B. 9% C. 90% D. 99%

4. What is 0.7 as a percent?
   A. 0.7% B. 7% C. 70% D. 700%

5. What is 96% as a fraction in simplest form?
   A. $\frac{3}{20}$ B. $\frac{24}{25}$ C. $\frac{49}{50}$ D. not here

6. What is $\frac{4}{10,000}$ as a decimal?
   A. 0.04 B. 0.004 C. 0.0004 D. not here

7. A bookstore's sales increased by 25% over last year's sales. By what fraction did the store's sales increase?
   A. $\frac{1}{5}$ B. $\frac{1}{4}$ C. $\frac{2}{5}$ D. $\frac{3}{4}$

8. In 1992 attendance at a movie theater increased by $\frac{3}{20}$ in comparison to the attendance in 1991. By what percent did attendance increase?
   A. 3% B. 15% C. 30% D. 32%

9. Estimate. 19% of 397 = __?__
   A. 50 B. 80 C. 90 D. 100

10. Which percent is the best estimate of 6 out of 17?
    A. 16% B. 20% C. 25% D. 33%

11. In a class of 30 students, 8 were absent. Which equation shows the percent who were absent?
    - A. $n\% \times 30 = 8$
    - B. $n\% \times 8 = 30$
    - C. $n\% = 8 \times 30$
    - D. $n\% \times 30 \div 8$

12. John saves $1.50 per week. That is equal to 20% of his allowance. Which is the correct equation to find John's allowance?
    - A. $n = 20\% \times 1.50$
    - B. $n = 20\% \div 1.50$
    - C. $n \times 20\% = 1.50$
    - D. $n \times 1.50 = 20\%$

13. 160% of 200 = __?__
    A. 32 B. 120 C. 160 D. 320

14. $\frac{1}{2}$% of 400 = __?__
    A. 2 B. 4 C. 20 D. 200

15. 25 is what percent of 125?
    A. 20% B. 25% C. 30% D. 50%

16. 20% of what number is 6?
    A. 12 B. 30 C. 40 D. 60

Go on to the next page.

Name ______________________

Pretest

CHAPTER 8
PAGE 2

17. The Jacksons make a down payment of $15,000 on a house. The down payment is 20% of the total price. What is the price of the house?
    A. $30,000 B. $60,000 C. $70,000 D. $75,000

18. Sarah bought a used car for $3,500 plus sales tax. The tax was $210. What was the percent of sales tax?
    A. 3% B. 6% C. 9% D. 21%

19. Mara's Hardware Store usually sells a toolbox for $19.00. Today the store has a sale in which all prices are reduced by 20%. What is the sale price of the toolbox?
    A. $15.20 B. $15.80 C. $16.20 D. $18.80

20. Justin's meal at the restaurant costs $15.00. He wants to leave a 15% tip. How much should he leave for the tip?
    A. $1.50 B. $1.65 C. $1.75 D. $2.25

21. What is the sum of the degrees in the sections of a circle graph?
    A. 180° B. 270° C. 300% D. 360°

22. Rob made a circle graph to show how he spends his leisure time. A central angle of 72° represents what percent of his time?
    A. 10% B. 20% C. 40% D. 72%

Use the circle graph below to answer questions 23–24.

McDonald Family Budget

Child care 5%, Transportation 5%, Clothes 5%, Other 5%, Savings 10%, Food 30%, Rent 40%

23. The McDonald family income is $1,500 per month. How much do they spend per month on food?
    A. $50 B. $100 C. $450 D. $1,000

24. How much do the McDonalds spend on clothes and child care combined?
    A. $25 B. $150 C. $175 D. $225

25. Sales tax is 6%. What is the tax on an item that costs $32?
    A. $0.65 B. $1.92 C. $2.80 D. not here

26. An item that costs $80 is reduced by 25%. What is the sale price?
    A. $20.00 B. $28.00 C. $54.00 D. $60.00

27. Juan buys a sweater that was discounted 25%. He pays $15.00. What was the original price of the sweater?
    A. $18.00 B. $18.75 C. $20.00 D. $35.00

Go on to the next page.

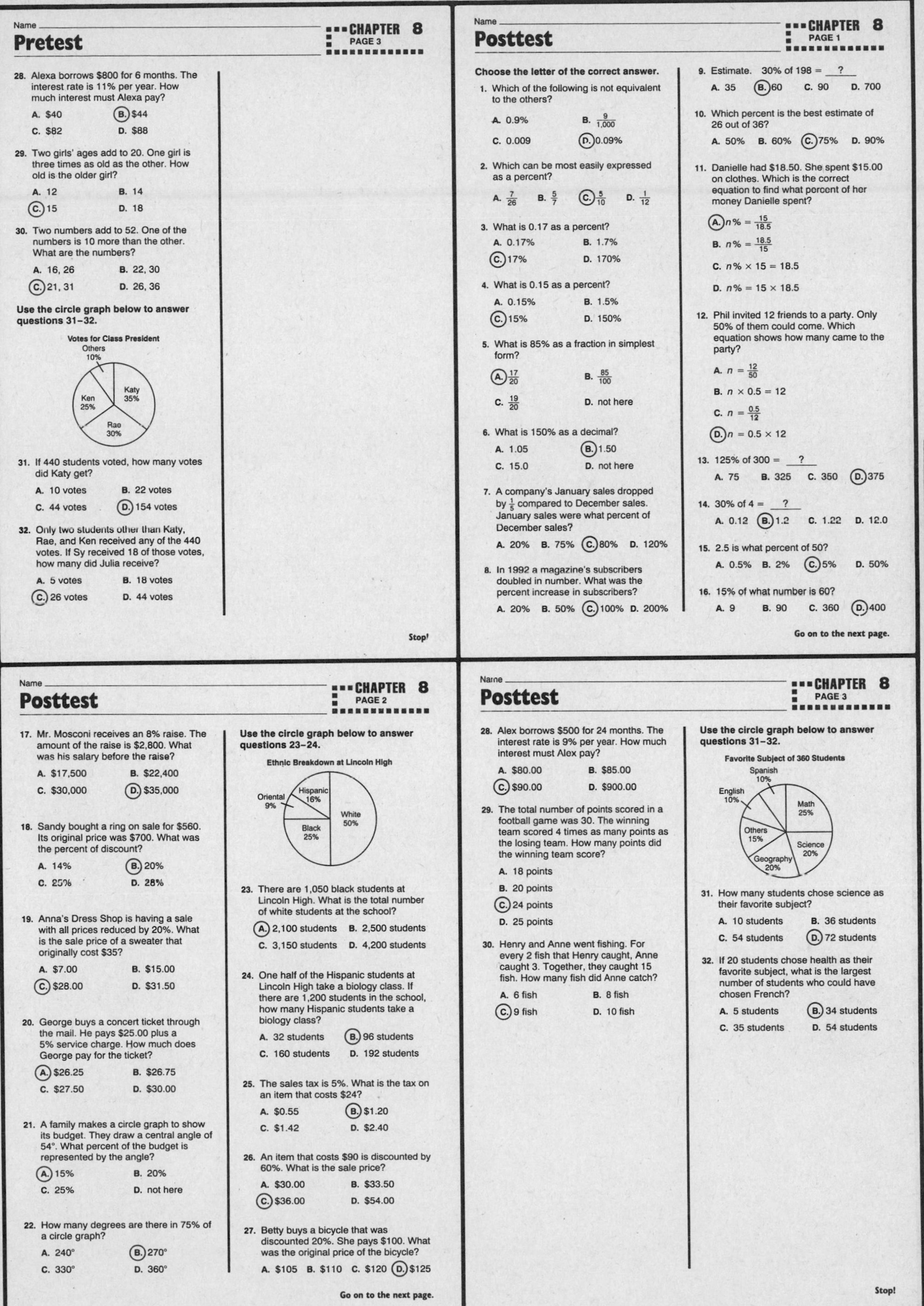

Name

Pretest

CHAPTER 8
PAGE 3

28. Alexa borrows $800 for 6 months. The interest rate is 11% per year. How much interest must Alexa pay?
 A. $40 (B.) $44
 C. $82 D. $88

29. Two girls' ages add to 20. One girl is three times as old as the other. How old is the older girl?
 A. 12 B. 14
 (C.) 15 D. 18

30. Two numbers add to 52. One of the numbers is 10 more than the other. What are the numbers?
 A. 16, 26 B. 22, 30
 (C.) 21, 31 D. 26, 36

Use the circle graph below to answer questions 31–32.

31. If 440 students voted, how many votes did Katy get?
 A. 10 votes B. 22 votes
 C. 44 votes (D.) 154 votes

32. Only two students other than Katy, Rae, and Ken received any of the 440 votes. If Sy received 18 of those votes, how many did Julia receive?
 A. 5 votes B. 18 votes
 (C.) 26 votes D. 44 votes

Stop!

Name

Posttest

CHAPTER 8
PAGE 1

Choose the letter of the correct answer.

1. Which of the following is not equivalent to the others?
 A. 0.9% B. $\frac{9}{1,000}$
 C. 0.009 (D.) 0.09%

2. Which can be most easily expressed as a percent?
 A. $\frac{7}{26}$ B. $\frac{5}{7}$ (C.) $\frac{5}{10}$ D. $\frac{1}{12}$

3. What is 0.17 as a percent?
 A. 0.17% B. 1.7%
 (C.) 17% D. 170%

4. What is 0.15 as a percent?
 A. 0.15% B. 1.5%
 (C.) 15% D. 150%

5. What is 85% as a fraction in simplest form?
 (A.) $\frac{17}{20}$ B. $\frac{85}{100}$
 C. $\frac{19}{20}$ D. not here

6. What is 150% as a decimal?
 A. 1.05 (B.) 1.50
 C. 15.0 D. not here

7. A company's January sales dropped by $\frac{1}{5}$ compared to December sales. January sales were what percent of December sales?
 A. 20% B. 75% (C.) 80% D. 120%

8. In 1992 a magazine's subscribers doubled in number. What was the percent increase in subscribers?
 A. 20% B. 50% (C.) 100% D. 200%

9. Estimate. 30% of 198 = ___?___
 A. 35 (B.) 60 C. 90 D. 700

10. Which percent is the best estimate of 26 out of 36?
 A. 50% B. 60% (C.) 75% D. 90%

11. Danielle had $18.50. She spent $15.00 on clothes. Which is the correct equation to find what percent of her money Danielle spent?
 (A.) $n\% = \frac{15}{18.5}$
 B. $n\% = \frac{18.5}{15}$
 C. $n\% \times 15 = 18.5$
 D. $n\% = 15 \times 18.5$

12. Phil invited 12 friends to a party. Only 50% of them could come. Which equation shows how many came to the party?
 A. $n = \frac{12}{50}$
 B. $n \times 0.5 = 12$
 C. $n = \frac{0.5}{12}$
 (D.) $n = 0.5 \times 12$

13. 125% of 300 = ___?___
 A. 75 B. 325 C. 350 (D.) 375

14. 30% of 4 = ___?___
 A. 0.12 (B.) 1.2 C. 1.22 D. 12.0

15. 2.5 is what percent of 50?
 A. 0.5% B. 2% (C.) 5% D. 50%

16. 15% of what number is 60?
 A. 9 B. 90 C. 360 (D.) 400

Go on to the next page.

Name

Posttest

CHAPTER 8
PAGE 2

17. Mr. Mosconi receives an 8% raise. The amount of the raise is $2,800. What was his salary before the raise?
 A. $17,500 B. $22,400
 C. $30,000 (D.) $35,000

18. Sandy bought a ring on sale for $560. Its original price was $700. What was the percent of discount?
 A. 14% (B.) 20%
 C. 25% D. 28%

19. Anna's Dress Shop is having a sale with all prices reduced by 20%. What is the sale price of a sweater that originally cost $35?
 A. $7.00 B. $15.00
 (C.) $28.00 D. $31.50

20. George buys a concert ticket through the mail. He pays $25.00 plus a 5% service charge. How much does George pay for the ticket?
 (A.) $26.25 B. $26.75
 C. $27.50 D. $30.00

21. A family makes a circle graph to show its budget. They draw a central angle of 54°. What percent of the budget is represented by the angle?
 (A.) 15% B. 20%
 C. 25% D. not here

22. How many degrees are there in 75% of a circle graph?
 A. 240° (B.) 270°
 C. 330° D. 360°

Use the circle graph below to answer questions 23–24.

23. There are 1,050 black students at Lincoln High. What is the total number of white students at the school?
 (A.) 2,100 students B. 2,500 students
 C. 3,150 students D. 4,200 students

24. One half of the Hispanic students at Lincoln High take a biology class. If there are 1,200 students in the school, how many Hispanic students take a biology class?
 A. 32 students (B.) 96 students
 C. 160 students D. 192 students

25. The sales tax is 5%. What is the tax on an item that costs $24?
 A. $0.55 (B.) $1.20
 C. $1.42 D. $2.40

26. An item that costs $90 is discounted by 60%. What is the sale price?
 A. $30.00 B. $33.50
 (C.) $36.00 D. $54.00

27. Betty buys a bicycle that was discounted 20%. She pays $100. What was the original price of the bicycle?
 A. $105 B. $110 C. $120 (D.) $125

Go on to the next page.

Name

Posttest

CHAPTER 8
PAGE 3

28. Alex borrows $500 for 24 months. The interest rate is 9% per year. How much interest must Alex pay?
 A. $80.00 B. $85.00
 (C.) $90.00 D. $900.00

29. The total number of points scored in a football game was 30. The winning team scored 4 times as many points as the losing team. How many points did the winning team score?
 A. 18 points
 B. 20 points
 (C.) 24 points
 D. 25 points

30. Henry and Anne went fishing. For every 2 fish that Henry caught, Anne caught 3. Together, they caught 15 fish. How many fish did Anne catch?
 A. 6 fish B. 8 fish
 (C.) 9 fish D. 10 fish

Use the circle graph below to answer questions 31–32.

31. How many students chose science as their favorite subject?
 A. 10 students B. 36 students
 C. 54 students (D.) 72 students

32. If 20 students chose health as their favorite subject, what is the largest number of students who could have chosen French?
 A. 5 students (B.) 34 students
 C. 35 students D. 54 students

Stop!

Name

Cumulative Test

CHAPTERS 1–8
PAGE 1

Choose the letter of the correct answer.

1. Juan bought 2 books for \$1.95 each, 2 pads at \$1.05 each, and a pen for \$1.95. *Estimate* how many dollars Juan spent.
   A. \$5.00 B. \$6.00 C. \$7.00 **D. \$8.00**

2. Donna had \$31.00. She spent \$7.50 for dinner, \$4.00 for a movie, and \$1.50 for a drink at the movie. How much money did Donna have left?
   A. \$13.00 B. \$17.50 **C. \$18.00** D. \$19.00

3. What is the value of 5^1?
   A. 0 B. 1 **C. 5** D. 50

4. Express 7,302 in scientific notation.
   A. 7.302×10^2 B. 7.32×10^2 **C. 7.302×10^3** D. 7.32×10^3

Use the stem-and-leaf plot below to answer questions 5–6.

Number of Points Scored by a Football Team in 20 Games

Stem	Leaves
1	0 4 4 4 7 7 9
2	0 0 1 1 1 1 4 4 7
3	1 5
4	2 4

5. What was the least number of points scored by the team in any game?
   A. 0 points B. 4 points **C. 10 points** D. 14 points

6. In what percentage of its games did the team score 25 or more points?
   A. 5% **B. 25%** C. 30% D. 40%

Use the double-bar graph to answer questions 7–8.

Voting Results for School Mascot
Mascot
Alligator
Bear
Eagle
Seventh Grade
Eighth Grade
0 10 20 30 40 50 60 70
Number of Students

7. Which mascot was chosen by the greatest number of seventh graders?
   A. Alligator **B. Bear** C. Eagle

8. Which mascot was chosen by more seventh graders than eighth graders?
   A. Alligator B. Bear C. Eagle

Use the information below to answer questions 9–10.

A college basketball player scored the following numbers of points in the first 10 games of the season: 19, 15, 15, 22, 17, 20, 17, 19, 20, 20.

9. What is the mode?
   A. 15 points B. 17 points C. 19 points **D. 20 points**

10. What is the mean number of points scored?
    A. 18.0 points B. 18.2 points **C. 18.4 points** D. 19.0 points

11. Which fraction is *not* expressed in simplest form?
    A. $\frac{1}{13}$ B. $\frac{2}{9}$ **C. $\frac{6}{15}$** D. $\frac{10}{11}$

12. Which group is ordered from least to greatest?
    A. $\frac{1}{6}, \frac{1}{2}, \frac{5}{11}$ **B. $\frac{1}{7}, \frac{1}{2}, \frac{3}{5}$** C. $\frac{3}{8}, \frac{3}{9}, \frac{3}{10}$ D. $\frac{5}{6}, \frac{4}{5}, \frac{3}{4}$

Go on to the next page.

Name

Cumulative Test

CHAPTERS 1–8
PAGE 2

13. Which fraction is midway between $\frac{1}{50}$ and $\frac{2}{50}$?
    A. $\frac{2}{100}$ B. $\frac{5}{200}$ **C. $\frac{3}{100}$** D. $\frac{7}{200}$

14. $\frac{5}{12} + \frac{5}{8} =$ ___?___
    A. $1\frac{1}{24}$ B. $1\frac{1}{12}$ C. $1\frac{1}{8}$ D. not here

15. $7\frac{1}{5} - 6\frac{3}{5} =$ ___?___
    A. $\frac{2}{5}$ **B. $\frac{3}{5}$** C. $1\frac{1}{5}$ D. not here

16. $1\frac{1}{3} \times 1\frac{1}{3} =$ ___?___
    A. 1 B. $1\frac{1}{3}$ C. $1\frac{2}{3}$ **D. $1\frac{7}{9}$**

17. $\frac{1}{5} \div \frac{1}{10} =$ ___?___
    A. $\frac{1}{50}$ B. $\frac{1}{2}$ **C. 2** D. not here

18. Evaluate $\frac{2(a-6)}{3}$ for $a = 15$
    A. 3 **B. 6** C. 28 D. not here

19. Find the value of $(5 + 4) \times (4 - 1)$.
    A. 17 B. 20 **C. 27** D. 35

20. Find the value of $3 + 6 + (4 - 1)^2$.
    A. 12 B. 15 C. 144 **D. not here**

21. Which inequality shows that a number, b, increased by 9 is less than 200?
    A. $b - 9 < 200$ B. $b - 9 > 200$ C. $9b < 200$ **D. $b + 9 < 200$**

22. Solve the inequality. $x + 10 < 10$
    A. $x < 0$ B. $x < 1$ C. $x < 10$ D. $x < 20$

23. Solve. $a - 20 = 0$
    A. $a = 0$ B. $a = 10$ **C. $a = 20$** D. $a = 40$

24. Solve. $\frac{x}{12} = 6$
    A. $x = \frac{1}{2}$ B. $x = 2$ C. $x = 18$ **D. $x = 72$**

25. Are right triangles isosceles?
    A. always **B. sometimes** C. never

26. What is the complement of a 75° angle?
    A. 15° B. 25° C. 105° D. not here

27. Which ratio does *not* have the same meaning as the other three?
    A. $\frac{2}{15}$ **B. $\frac{5}{30}$** C. $\frac{6}{45}$ D. $\frac{8}{60}$

28. Which values of a and b make the following a true proportion?
    $\frac{a}{20} = \frac{6}{b}$
    A. $a = 4, b = 30$ B. $a = 4, b = 25$ C. $a = 3, b = 30$ D. $a = 12, b = 12$

29. A package of 6 cassette tapes costs \$7.80. What is the unit price?
    A. \$1.10 B. \$1.20 **C. \$1.30** D. \$1.80

30. If 8 cakes cost \$3.20 and 12 cakes cost \$4.68, which offer has the lower unit price?
    A. 8 cakes for \$3.20 **B. 12 cakes for \$4.68**

Go on to the next page.

Name

Cumulative Test

CHAPTERS 1–8
PAGE 3

31. On a scale drawing, 1 inch is used to represent 6 feet. Which ratio expresses that relationship?
    A. 1:12 B. 5:12 **C. 1:6** D. 6:1

32. On Map A, 1 inch = 10 miles. On Map B, 1 inch = 30 miles. A road between two cities is 3 inches long on Map A. How long is the same road on Map B?
    A. 1 in. B. $1\frac{1}{2}$ in. C. 2 in. D. 9 in.

33. Do similar figures have the same shape and same dimensions?
    A. always **B. sometimes** C. never

34. A triangle has sides of 7.5 cm, 10 cm, and 12.5 cm. To which of the following triangles is it similar?
    A. 2 cm, 3 cm, 4 cm
    B. 6 cm, 8 cm, 12 cm
    C. 1 cm, 3.5 cm, 6 cm
    D. 9 cm, 12 cm, 15 cm

35. Express $\frac{88}{1,000}$ as a decimal.
    A. 0.0088 **B. 0.088** C. 0.880 D. 0.888

36. Express 0.78 as a fraction in simplest form.
    A. $\frac{39}{500}$ B. $\frac{39}{100}$ C. $\frac{34}{50}$ **D. $\frac{39}{50}$**

37. Estimate. 15% of 498 = ___?___
    A. 55 B. 60 **C. 75** D. 90

38. 40% of what number is 8?
    A. 3.2 **B. 20** C. 32 D. 50

39. Bob made a circle graph to show how he spends his leisure time. A central angle of 90° represents what percent of his time?
    A. $12\frac{1}{2}$% B. 15% **C. 25%** D. not here

Use the circle graph below to answer question 40.

Club Membership at Lincoln School
1 Club (50%)
2 Clubs (30%)
3 Clubs (15%)
0 Clubs (5%)

40. A total of 450 students at Lincoln School are members of two clubs. What is the total number of students at the school?
    A. 750 students B. 1,350 students **C. 1,500 students** D. 3,000 students

41. An item that regularly costs \$200 is reduced by 35%. What is the sale price?
    A. \$35.00 B. \$70.00 **C. \$130.00** D. \$135.00

Go on to the next page.

Name

Cumulative Test

CHAPTERS 1–8
PAGE 4

42. Lee borrows \$1,200. The interest rate is 10.5% per year. The payoff period is 6 months. How much interest will Lee pay? Use the formula $I = prt$.
    A. \$63.00 B. \$66.00 C. \$120.00 D. \$126.00

Use the table below to answer question 43.

Students' Music Preferences		
Type of Music	Boys	Girls
Classical	30	70
Country	40	40
Jazz	60	35
Rock	110	135
	240	280

43. What percentage of boys prefer jazz?
    A. $12\frac{1}{2}$% **B. 25%** C. 35% D. 60%

44. Tony and Jen began a running program. Tony ran 2 miles on day 1 and increased the distance by $\frac{1}{4}$ mile each day. Jen ran 3 miles on day 1 and increased the distance by $\frac{1}{8}$ mile each day. On what day will Tony and Jen run the same distance?
    A. day 5 B. day 8 **C. day 9** D. day 13

45. Justin can bicycle one mile in $\frac{1}{4}$ hour. Don can bicycle one mile in $\frac{1}{2}$ hour. If both boys ride one hour, how much farther will Justin ride?
    A. $\frac{1}{2}$ mi B. 1 mi **C. 2 mi** D. $2\frac{1}{2}$ mi

Use the information below to answer question 46.

Taxi fare is given by the formula \$1.25 + \$0.90 m + \$1.00 b, where m = the distance traveled in miles and b = the number of pieces of luggage carried.

46. Hal took a taxi for a distance of 4 miles. He had 2 pieces of luggage. If he tipped the driver 15%, how much did Hal pay in all?
    A. about \$5.58 B. about \$6.44 C. about \$6.73 **D. about \$7.88**

47. Sid withdrew money from his bank account. He spent \$50 for shoes and \$10 for a tie. He gave $\frac{2}{3}$ of what was left to his son. He then had \$45. How much money did Sid withdraw?
    A. \$120 B. \$150 **C. \$195** D. \$200

48. The scale on a map is 1 inch = 100 miles. Two cities are 2 inches apart. If Tom averages 50 miles per hour and makes no stops, how long will it take him to drive from one city to the other?
    A. 2 hr B. $2\frac{1}{2}$ hr C. 3 hr **D. 4 hr**

49. Two numbers add up to 72. One of the numbers is 10 more than the other. What are the numbers?
    A. 26 and 36 B. 32 and 40 **C. 31 and 41** D. 36 and 46

Go on to the next page.

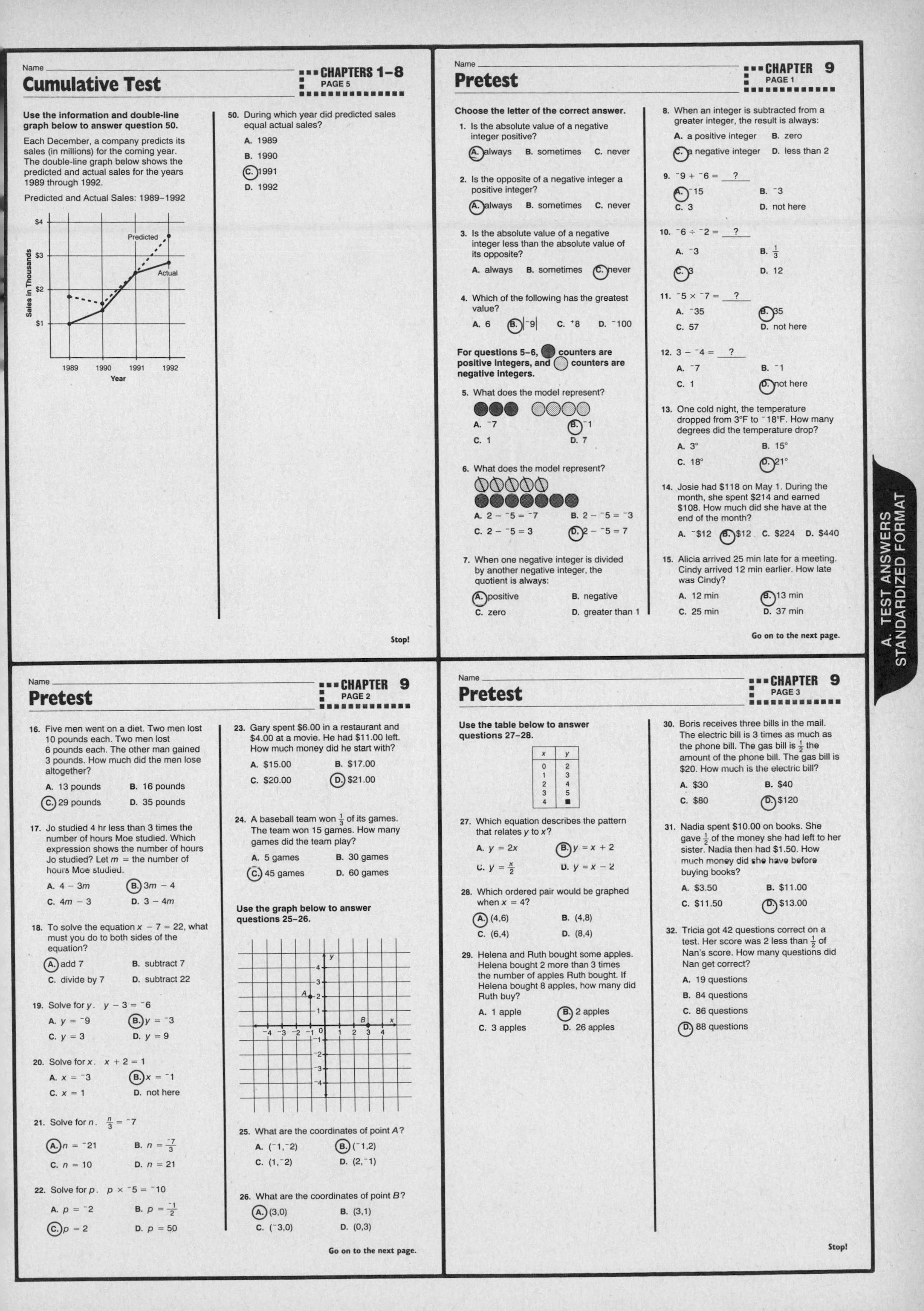

Name

Cumulative Test

CHAPTERS 1–8
PAGE 5

Use the information and double-line graph below to answer question 50.

Each December, a company predicts its sales (in millions) for the coming year. The double-line graph below shows the predicted and actual sales for the years 1989 through 1992.

Predicted and Actual Sales: 1989–1992

50. During which year did predicted sales equal actual sales?
    - A. 1989
    - B. 1990
    - **(C.)** 1991
    - D. 1992

Stop!

Name

Pretest

CHAPTER 9
PAGE 1

Choose the letter of the correct answer.

1. Is the absolute value of a negative integer positive?
   (A.) always B. sometimes C. never
2. Is the opposite of a negative integer a positive integer?
   (A.) always B. sometimes C. never
3. Is the absolute value of a negative integer less than the absolute value of its opposite?
   A. always B. sometimes **(C.)** never
4. Which of the following has the greatest value?
   A. 6 **(B.)** $|{}^{-}9|$ C. $^{+}8$ D. $^{-}100$

For questions 5–6, (dark) counters are positive integers, and (light) counters are negative integers.

5. What does the model represent?
   A. $^{-}7$ **(B.)** $^{-}1$
   C. 1 D. 7
6. What does the model represent?
   A. $2 - {}^{-}5 = {}^{-}7$ B. $2 - {}^{-}5 = {}^{-}3$
   C. $2 - {}^{-}5 = 3$ **(D.)** $2 - {}^{-}5 = 7$
7. When one negative integer is divided by another negative integer, the quotient is always:
   (A.) positive B. negative
   C. zero D. greater than 1
8. When an integer is subtracted from a greater integer, the result is always:
   A. a positive integer B. zero
   (C.) a negative integer D. less than 2
9. $^{-}9 + {}^{-}6 =$ ___?___
   (A.) $^{-}15$ B. $^{-}3$
   C. 3 D. not here
10. $^{-}6 \div {}^{-}2 =$ ___?___
    A. $^{-}3$ B. $\frac{1}{3}$
    (C.) 3 D. 12
11. $^{-}5 \times {}^{-}7 =$ ___?___
    A. $^{-}35$ **(B.)** 35
    C. 57 D. not here
12. $3 - {}^{-}4 =$ ___?___
    A. $^{-}7$ B. $^{-}1$
    C. 1 **(D.)** not here
13. One cold night, the temperature dropped from 3°F to $^{-}$18°F. How many degrees did the temperature drop?
    A. 3° B. 15°
    C. 18° **(D.)** 21°
14. Josie had $118 on May 1. During the month, she spent $214 and earned $108. How much did she have at the end of the month?
    A. $^{-}$$12 **(B.)** $12 C. $224 D. $440
15. Alicia arrived 25 min late for a meeting. Cindy arrived 12 min earlier. How late was Cindy?
    A. 12 min **(B.)** 13 min
    C. 25 min D. 37 min

Go on to the next page.

Name

Pretest

CHAPTER 9
PAGE 2

16. Five men went on a diet. Two men lost 10 pounds each. Two men lost 6 pounds each. The other man gained 3 pounds. How much did the men lose altogether?
    A. 13 pounds B. 16 pounds
    (C.) 29 pounds D. 35 pounds
17. Jo studied 4 hr less than 3 times the number of hours Moe studied. Which expression shows the number of hours Jo studied? Let m = the number of hours Moe studied.
    A. $4 - 3m$ **(B.)** $3m - 4$
    C. $4m - 3$ D. $3 - 4m$
18. To solve the equation $x - 7 = 22$, what must you do to both sides of the equation?
    (A.) add 7 B. subtract 7
    C. divide by 7 D. subtract 22
19. Solve for y. $y - 3 = {}^{-}6$
    A. $y = {}^{-}9$ **(B.)** $y = {}^{-}3$
    C. $y = 3$ D. $y = 9$
20. Solve for x. $x + 2 = 1$
    A. $x = {}^{-}3$ **(B.)** $x = {}^{-}1$
    C. $x = 1$ D. not here
21. Solve for n. $\frac{n}{3} = {}^{-}7$
    (A.) $n = {}^{-}21$ B. $n = \frac{{}^{-}7}{3}$
    C. $n = 10$ D. $n = 21$
22. Solve for p. $p \times {}^{-}5 = {}^{-}10$
    A. $p = {}^{-}2$ B. $p = \frac{{}^{-}1}{2}$
    (C.) $p = 2$ D. $p = 50$
23. Gary spent $6.00 in a restaurant and $4.00 at a movie. He had $11.00 left. How much money did he start with?
    A. $15.00 B. $17.00
    C. $20.00 **(D.)** $21.00
24. A baseball team won $\frac{1}{3}$ of its games. The team won 15 games. How many games did the team play?
    A. 5 games B. 30 games
    (C.) 45 games D. 60 games

Use the graph below to answer questions 25–26.

25. What are the coordinates of point A?
    A. $({}^{-}1, {}^{-}2)$ **(B.)** $({}^{-}1, 2)$
    C. $(1, {}^{-}2)$ D. $(2, {}^{-}1)$
26. What are the coordinates of point B?
    (A.) (3,0) B. (3,1)
    C. $({}^{-}3, 0)$ D. (0,3)

Go on to the next page.

Name

Pretest

CHAPTER 9
PAGE 3

Use the table below to answer questions 27–28.

x	y
0	2
1	3
2	4
3	5
4	■

27. Which equation describes the pattern that relates y to x?
    A. $y = 2x$ **(B.)** $y = x + 2$
    C. $y = \frac{x}{2}$ D. $y = x - 2$
28. Which ordered pair would be graphed when $x = 4$?
    (A.) (4,6) B. (4,8)
    C. (6,4) D. (8,4)
29. Helena and Ruth bought some apples. Helena bought 2 more than 3 times the number of apples Ruth bought. If Helena bought 8 apples, how many did Ruth buy?
    A. 1 apple **(B.)** 2 apples
    C. 3 apples D. 26 apples
30. Boris receives three bills in the mail. The electric bill is 3 times as much as the phone bill. The gas bill is $\frac{1}{2}$ the amount of the phone bill. The gas bill is $20. How much is the electric bill?
    A. $30 B. $40
    C. $80 **(D.)** $120
31. Nadia spent $10.00 on books. She gave $\frac{1}{2}$ of the money she had left to her sister. Nadia then had $1.50. How much money did she have before buying books?
    A. $3.50 B. $11.00
    C. $11.50 **(D.)** $13.00
32. Tricia got 42 questions correct on a test. Her score was 2 less than $\frac{1}{2}$ of Nan's score. How many questions did Nan get correct?
    A. 19 questions
    B. 84 questions
    C. 86 questions
    (D.) 88 questions

Stop!

Name ____________________

Posttest

CHAPTER 9
PAGE 1

Choose the letter of the correct answer.

1. Are the absolute values of an integer and its opposite equal?
   (A.) always B. sometimes C. never

2. Is the opposite of an integer an integer?
   (A.) always B. sometimes C. never

3. Is the absolute value of a negative integer negative?
   A. always B. sometimes (C.) never

4. Which of the following shows the integers in order from least to greatest?
   A. $^{-}2, ^{-}4, 6$ B. $^{-}2, ^{-}4, ^{-}6$
   (C.) $^{-}6, ^{-}4, ^{-}2$ D. $^{-}6, ^{-}2, ^{-}4$

For questions 5–6, ● counters are positive integers, and ○ counters are negative integers.

5. What does the model represent?
   ●●○○○
   A. $^{-}5$ (B.) $^{-}1$
   C. 1 D. 5

6. What does the model represent?
   ●●●●●●●
   ⊘⊘⊘⊘
   A. $3 - {^{-}4} = {^{-}7}$ B. $3 - {^{-}4} = {^{-}1}$
   C. $3 - {^{-}4} = 1$ (D.) $3 - {^{-}4} = 7$

7. The product of two integers with unlike signs is always:
   A. a positive integer
   (B.) a negative integer
   C. a number greater than 1
   D. a number less than 1

8. When an integer is subtracted from a greater integer, the result is always:
   (A.) a positive integer
   B. a negative integer
   C. greater than 1
   D. zero

9. $^{-}15 + {^{-}20} =$ __?__
   A. $^{-}5$ B. 5 C. 35 (D.) not here

10. $^{-}5 \div {^{-}5} =$ __?__
    A. $^{-}5$ B. $^{-}1$ (C.) 1 D. 5

11. $^{-}50 \times {^{-}2} =$ __?__
    A. $^{-}100$ B. 25 (C.) 100 D. not here

12. $100 - {^{-}99} =$ __?__
    A. $^{-}1$ B. 1 C. 99 (D.) 199

13. The temperature increased from $^{-}9$°F to $^{-}7$°F. How many degrees did the temperature change?
    A. $^{-}2$° (B.) 2°
    C. 9° D. 16°

14. A club had $140 on March 1. During the month, the members spent $270 and earned $181. How much did they have at the end of the month?
    A. $^{-}$$311 (B.) $51
    C. $229 D. $591

15. Roni's watch runs 7 sec slow each day. How slow will it be after one week?
    A. 14 sec B. 35 sec
    (C.) 49 sec D. 70 sec

Go on to the next page.

Name ____________________

Posttest

CHAPTER 9
PAGE 2

16. The 10 members of the Garden Club went on diets. One half of the members lost 8 pounds each. Three members lost 5 pounds each. Two members gained 2 pounds each. What was the total amount lost?
    A. 11 pounds (B.) 51 pounds
    C. 91 pounds D. 109 pounds

17. Pat works 3 hr less per week than $\frac{1}{2}$ the time Lyn works. Which expression shows the number of hours per week Pat works? Let h = the number of hours Lyn works.
    A. $3 - \frac{1}{2}h$ (B.) $\frac{1}{2}h - 3$
    C. $3h - \frac{1}{2}$ D. $\frac{1}{2} - 3h$

18. To solve the equation $4x = 32$, what must you do to both sides of the equation?
    (A.) divide by 4
    B. subtract 4
    C. multiply by 4
    D. multiply by 32

19. Solve for y. $y - 11 = {^{-}12}$
    A. $y = {^{-}23}$ (B.) $y = {^{-}1}$
    C. $y = 1$ D. $y = 23$

20. Solve for x. $x + 12 = 2$
    A. $x = {^{-}14}$ (B.) $x = {^{-}10}$
    C. $x = 10$ D. $x = 14$

21. Solve for n. $\frac{n}{10} = {^{-}3}$
    (A.) $n = {^{-}30}$ B. $n = \frac{^{-}10}{3}$
    C. $n = 30$ D. not here

22. Solve for p. $p \times {^{-}1} = {^{-}99}$
    A. $p = {^{-}1}$ B. $p = \frac{^{-}1}{99}$
    C. $p = \frac{1}{99}$ (D.) $p = 99$

23. Hal gave $15 to his brother and $10 to each of his 2 sisters. He was left with $44. How much money did he start with?
    A. $9 B. $35 C. $69 (D.) $79

24. A soccer team lost $\frac{3}{4}$ of its games. The team won 8 games. How many games did the team play?
    A. 6 games B. 12 games
    C. 24 games (D.) 32 games

Use the graph below to answer questions 25–26.

25. What are the coordinates of point A?
    A. $(^{-}1, ^{-}1)$ (B.) $(^{-}1, 1)$
    C. $(1, ^{-}1)$ D. not here

Go on to the next page.

Name ____________________

Posttest

CHAPTER 9
PAGE 3

26. What are the coordinates of point B?
    A. $(0, ^{-}3)$ B. $(1, 3)$
    C. $(0, 3)$ (D.) $(^{-}3, 0)$

Use the table below to answer questions 27–28.

x	y
0	0
1	2
2	4
3	6
4	■

27. Which equation describes the pattern that relates y to x?
    (A.) $y = 2x$ B. $y = x + 2$
    C. $y = \frac{x}{2}$ D. $y = x - 2$

28. Which ordered pair would be graphed when $x = 4$?
    A. (4,7) (B.) (4,8)
    C. (7,4) D. (8,4)

29. Helen played a game three times. Her second score was twice as high as her first score. Her third score was 5 points higher than her second score. Helen's third score was 225. What was her first score?
    A. 100 points (B.) 110 points
    C. 115 points D. 120 points

30. Mike got 5 words wrong on a spelling test. His sister got 3 more than twice as many words wrong. How many words did his sister get wrong?
    A. 7 words B. 10 words
    (C.) 13 words D. 18 words

31. Emile spent $50.00 on groceries. He spent $\frac{1}{5}$ of the money on fruit, $\frac{1}{2}$ on dairy items, and the rest on meat. How much did he spend on meat?
    A. $3.00 (B.) $15.00
    C. $30.00 D. $35.00

32. Bill weighs twice as much as Phil. John weighs $\frac{2}{3}$ as much as Phil. Bill weighs 240 pounds. How much does John weigh?
    A. 60 pounds (B.) 80 pounds
    C. 120 pounds D. 160 pounds

Stop!

Name ____________________

Pretest

CHAPTER 10
PAGE 1

Choose the letter of the correct answer.

1. Are mixed numbers rational numbers?
   (A.) always B. sometimes C. never

2. Are the square roots of numbers that are not perfect squares irrational numbers?
   (A.) always B. sometimes C. never

3. Does a fraction whose denominator has only 2 and 5 as prime factors result in a repeating decimal?
   A. always B. sometimes (C.) never

4. Are irrational numbers real numbers?
   (A.) always B. sometimes C. never

5. Express $\frac{39}{50}$ as a decimal.
   A. 0.075 B. 0.078
   C. 0.395 (D.) 0.78

6. Express 0.825 as a percent.
   A. 0.825% B. 8.25%
   (C.) 82.5% D. not here

7. Express 1.75 as a percent.
   A. 1.75% B. 17.5%
   (C.) 175% D. not here

8. Which of the following fractions can be expressed as a repeating decimal?
   A. $\frac{1}{16}$ (B.) $\frac{6}{9}$ C. $\frac{3}{20}$ D. $\frac{12}{40}$

9. Which of the following fractions can be expressed as a terminating decimal?
   A. $\frac{3}{7}$ B. $\frac{6}{18}$ C. $\frac{1}{12}$ (D.) $\frac{3}{25}$

10. Which square root is an irrational number?
    A. $\sqrt{4}$ (B.) $\sqrt{8}$ C. $\sqrt{16}$ D. $\sqrt{81}$

11. Compare. $^{-}1{,}000$ ○ 0.0001
    (A.) < B. = C. >

12. Compare. $\frac{^{-}1}{50}$ ○ $^{-}0.5$
    A. < B. = (C.) >

13. Which is a rational number between 0.870 and $\frac{7}{8}$?
    A. 0.869 (B.) 0.871
    C. $\frac{13}{16}$ D. 0.876

14. Which is a rational number between $^{-}0.50$ and $^{-}0.51$?
    A. $^{-}0.499$ B. $^{-}0.501$
    C. $^{-}0.511$ (D.) not here

15. Order from least to greatest.
    $\frac{6}{7}, \frac{7}{8}, \frac{^{-}5}{6}, \frac{^{-}9}{8}$
    A. $\frac{^{-}5}{6}, \frac{6}{7}, \frac{7}{8}, \frac{^{-}9}{8}$ B. $\frac{^{-}5}{6}, \frac{^{-}9}{8}, \frac{6}{7}, \frac{7}{8}$
    (C.) $\frac{^{-}9}{8}, \frac{^{-}5}{6}, \frac{6}{7}, \frac{7}{8}$ D. $\frac{^{-}9}{8}, \frac{^{-}5}{6}, \frac{7}{8}, \frac{6}{7}$

16. Order from least to greatest.
    $\frac{1}{6}, \frac{1}{7}, 0.17, 0.14$
    (A.) $0.14, \frac{1}{7}, \frac{1}{6}, 0.17$
    B. $\frac{1}{7}, 0.14, 0.17, \frac{1}{6}$
    C. $\frac{1}{7}, \frac{1}{6}, 0.14, 0.17$
    D. $0.14, 0.17, \frac{1}{7}, \frac{1}{6}$

Go on to the next page.

Name

Pretest

CHAPTER 10 PAGE 2

17. Compare. $\frac{1}{4^3}$ ○ $\frac{1}{2^5}$
    A. > B. = **(C.)** <

18. Compare. $\frac{1}{3^7}$ ○ $\frac{1}{4^8}$
    A. < B. = **(C.)** >

19. 3^4 = ___?___
    A. 34 B. 64 **(C.)** 81 D. 243

20. What is 3^{-4} expressed with a positive exponent?
    (A.) $\frac{1}{3^4}$ B. $\frac{1}{81^2}$
    C. $\frac{3}{3^3}$ D. $\frac{3^3}{33}$

21. What is 4.2×10^3 in standard form?
    A. 420 **(B.)** 4,200
    C. 42,000 D. 420,000

22. What is 502,000,000 in scientific notation?
    A. 5.02×10^6 **(B.)** 5.02×10^8
    C. 5.2×10^7 D. 5.20×10^8

23. Which is a perfect square?
    A. 24 B. 44 C. 99 **(D.)** 144

24. Which is the best estimate of $\sqrt{83}$?
    A. 8.9 **(B.)** 9.1 C. 10.1 D. 11.2

25. $\sqrt{121}$ = ___?___
    (A.) 11 B. 12
    C. 21 D. not here

26. $\sqrt{\frac{9}{16}}$ = ___?___
    A. $\frac{3}{16}$ **(B.)** $\frac{3}{4}$
    C. $\frac{9}{4}$ D. not here

27. $\sqrt{0.0049}$ = ___?___
    A. 0.0007 B. 0.007
    (C.) 0.07 D. 0.7

28. $\sqrt{10,000}$ = ___?___
    A. 10 **(B.)** 100 C. 500 D. 1,000

Use the table below to answer questions 29–32.

Students Attending Grant High

Year	Number
1970	1,050
1975	1,100
1980	1,200
1985	1,350
1990	1,550

29. How many students attended Grant High in 1980?
    A. 1,100 students B. 1,150 students
    (C.) 1,200 students D. 1,550 students

30. If the trend continues, how many students will attend Grant High in 1995?
    A. 1,700 students B. 1,750 students
    (C.) 1,800 students D. 1,850 students

31. If the trend continues, how many students will attend Grant High in 2005?
    A. 2,100 students B. 2,400 students
    (C.) 2,450 students D. 2,500 students

32. How much did the student population increase from 1970 to 1990?
    A. 50 students B. 300 students
    C. 450 students **(D.)** 500 students

Stop!

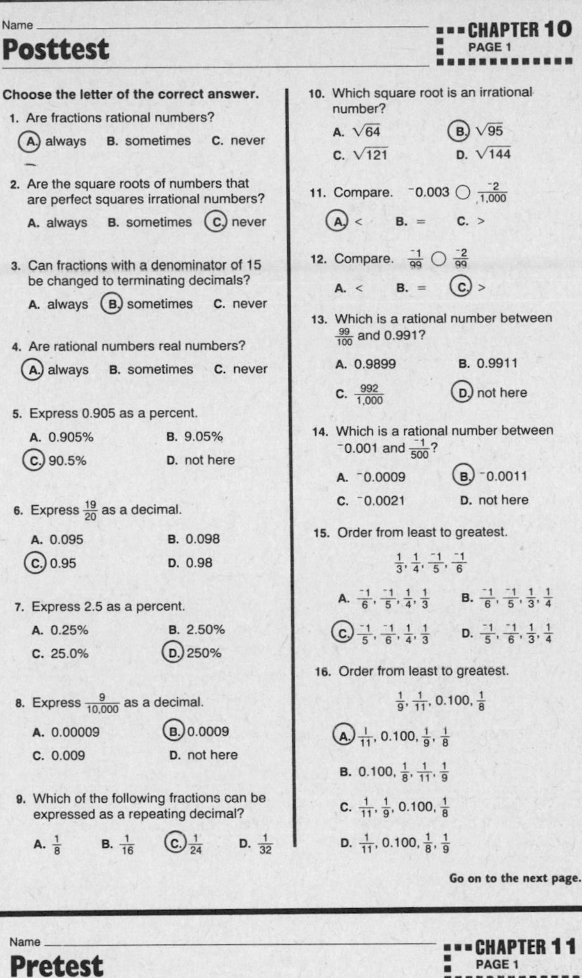

Name

Posttest

CHAPTER 10 PAGE 1

Choose the letter of the correct answer.

1. Are fractions rational numbers?
   (A.) always B. sometimes C. never

2. Are the square roots of numbers that are perfect squares irrational numbers?
   A. always B. sometimes **(C.)** never

3. Can fractions with a denominator of 15 be changed to terminating decimals?
   A. always **(B.)** sometimes C. never

4. Are rational numbers real numbers?
   (A.) always B. sometimes C. never

5. Express 0.905 as a percent.
   A. 0.905% B. 9.05%
   (C.) 90.5% D. not here

6. Express $\frac{19}{20}$ as a decimal.
   A. 0.095 B. 0.098
   (C.) 0.95 D. 0.98

7. Express 2.5 as a percent.
   A. 0.25% B. 2.50%
   C. 25.0% **(D.)** 250%

8. Express $\frac{9}{10,000}$ as a decimal.
   A. 0.00009 **(B.)** 0.0009
   C. 0.009 D. not here

9. Which of the following fractions can be expressed as a repeating decimal?
   A. $\frac{1}{8}$ B. $\frac{1}{16}$ **(C.)** $\frac{1}{24}$ D. $\frac{1}{32}$

10. Which square root is an irrational number?
    A. $\sqrt{64}$ **(B.)** $\sqrt{95}$
    C. $\sqrt{121}$ D. $\sqrt{144}$

11. Compare. $^-0.003$ ○ $\frac{^-2}{1,000}$
    (A.) < B. = C. >

12. Compare. $\frac{^-1}{99}$ ○ $\frac{^-2}{99}$
    A. < B. = **(C.)** >

13. Which is a rational number between $\frac{99}{100}$ and 0.991?
    A. 0.9899 B. 0.9911
    C. $\frac{992}{1,000}$ **(D.)** not here

14. Which is a rational number between $^-0.001$ and $\frac{^-1}{500}$?
    A. $^-0.0009$ **(B.)** $^-0.0011$
    C. $^-0.0021$ D. not here

15. Order from least to greatest.
    $\frac{1}{3}, \frac{1}{4}, \frac{^-1}{5}, \frac{^-1}{6}$
    A. $\frac{^-1}{6}, \frac{^-1}{5}, \frac{1}{4}, \frac{1}{3}$ B. $\frac{^-1}{6}, \frac{^-1}{5}, \frac{1}{3}, \frac{1}{4}$
    (C.) $\frac{^-1}{5}, \frac{^-1}{6}, \frac{1}{4}, \frac{1}{3}$ D. $\frac{^-1}{5}, \frac{^-1}{6}, \frac{1}{3}, \frac{1}{4}$

16. Order from least to greatest.
    $\frac{1}{9}, \frac{1}{11}, 0.100, \frac{1}{8}$
    (A.) $\frac{1}{11}$, 0.100, $\frac{1}{9}, \frac{1}{8}$
    B. 0.100, $\frac{1}{8}, \frac{1}{11}, \frac{1}{9}$
    C. $\frac{1}{11}, \frac{1}{9}$, 0.100, $\frac{1}{8}$
    D. $\frac{1}{11}$, 0.100, $\frac{1}{8}, \frac{1}{9}$

Go on to the next page.

Name

Posttest

CHAPTER 10 PAGE 2

17. What is another way to write $7 \times 7 \times 7 \times 7$?
    A. $4 \cdot 7^2$ B. 4×7
    C. 4^7 **(D.)** 7^4

18. Compare. $\frac{1}{2^3}$ ○ $\frac{1}{3^2}$
    A. < B. = **(C.)** >

19. What is 5^{-3} expressed with a positive exponent?
    A. 3^5 B. $\frac{5^3}{25}$
    C. $\frac{1}{53}$ **(D.)** $\frac{1}{5^3}$

20. $\frac{1,000}{10^4}$ = ___?___
    (A.) $\frac{1}{100}$ B. $\frac{1}{10}$
    C. $\frac{3}{4}$ D. 10

21. What is 1.2×10^{-3} in standard form?
    A. 0.00012 **(B.)** 0.0012
    C. 0.012 D. 0.12

22. What is 32,000 in scientific notation?
    A. 3.20×10^3 B. 3.02×10^5
    (C.) 3.2×10^4 D. 3.20×10^5

23. Which is a perfect square?
    A. 200 B. 300 **(C.)** 400 D. 500

24. Which is the best estimate of $\sqrt{124}$?
    A. 10.9 **(B.)** 11.1
    C. 12.4 D. 13.1

25. $\sqrt{144}$ = ___?___
    (A.) 12 B. 12.5
    C. 13 D. not here

26. $\sqrt{\frac{4}{25}}$ = ___?___
    A. $\frac{2}{25}$ **(B.)** $\frac{2}{5}$
    C. $2\frac{1}{2}$ D. not here

27. $\sqrt{0.09}$ = ___?___
    A. 0.003 B. 0.03
    (C.) 0.3 D. 0.81

28. $\sqrt{1,000,000}$ = ___?___
    A. 100 B. 500
    (C.) 1,000 D. 10,000

Use the table below to answer questions 29–32.

Town Population

Year	Population
1986	2,000
1987	2,100
1988	2,300
1989	2,600
1990	3,000
1991	3,500

29. What was the town population in 1988?
    A. 230 B. 2,100
    (C.) 2,300 D. 23,000

30. If the trend continues, what will be the town population in 1992?
    A. 4,000 **(B.)** 4,100
    C. 4,200 D. 41,000

31. If the trend continues, what will be the town population in 1995?
    A. 5,600 B. 6,400
    (C.) 6,500 D. 65,000

32. By how many people did the population increase from 1986 to 1991?
    A. 1,000 **(B.)** 1,500 C. 2,000 D. 2,500

Stop!

Name

Pretest

CHAPTER 11 PAGE 1

Choose the letter of the correct answer.

1. Jan picks one month at random from a calendar. What is the number of possible outcomes?
   A. 6 outcomes B. 9 outcomes
   (C.) 12 outcomes D. 24 outcomes

2. Al has a spinner with five equal sections, labeled 3, 5, 7, 9, and 10. What is the number of possible outcomes of one spin?
   A. 3 outcomes **(B.)** 5 outcomes
   C. 9 outcomes D. 19 outcomes

3. A restaurant has 3 choices for soup, 6 possible main courses, and 5 desserts. If Sandra orders soup, a main course, and dessert, how many possible selections can she make?
   A. 14 selections B. 18 selections
   C. 30 selections **(D.)** 90 selections

4. What is the number of possible three-digit area codes? Assume that zero *cannot* be used as the first digit.
   A. 29 codes B. 90 codes
   (C.) 900 codes D. 1,000 codes

5. Can a greater number of 2-member committees be selected from a group of 6 people or from a group of 8 people?
   A. from a group of 6 people
   (B.) from a group of 8 people
   C. The same number of committees can be selected from either group.

6. Which of these words provides the *least* number of possible letter arrangements?
   (A.) *dog* B. *bird*
   C. *horse* D. *flower*

7. How many different combinations of 2 items can be selected from a set of 8 items?
   A. 16 combinations
   (B.) 28 combinations
   C. 56 combinations
   D. not here

8. How many different arrangements of the letters *A, B, C, D* are possible?
   A. 12 arrangements
   B. 16 arrangements
   (C.) 24 arrangements
   D. 120 arrangements

9. A tennis team has 6 members. How many possible teams of 2 can be formed?
   A. 12 teams **(B.)** 15 teams
   C. 30 teams D. 120 teams

10. A basketball league has 4 teams. In how many different ways can the teams be arranged in the final standings?
    A. 12 ways B. 16 ways
    C. 20 ways **(D.)** 24 ways

11. A coin is flipped 3 times. What is the probability of obtaining a head on any *one* of the flips?
    A. $\frac{1}{8}$ B. $\frac{1}{4}$ **(C.)** $\frac{1}{2}$ D. $\frac{3}{4}$

12. A coin is flipped 4 times. Which of the following is the most likely outcome?
    A. 4 heads, 0 tails
    B. 3 heads, 1 tail
    (C.) 2 heads, 2 tails
    D. 1 head, 3 tails

Go on to the next page.

A. TEST ANSWERS STANDARDIZED FORMAT

Name

Pretest

CHAPTER 11 PAGE 2

13. Hector randomly picks an integer between 0 and 9. Which of the following is the most likely outcome?
    A. The number is 5.
    B. The number is 9.
    C. The number is odd.
    (D.) The number is not 1.

14. On one roll of a number cube marked 1–6, which of these has the highest probability?
    A. a number greater than 4
    B. a 4
    (C.) a number less than 4
    D. a 1 or a 2

15. A coin is flipped 4 times. What is the probability of obtaining 4 heads?
    A. $\frac{1}{64}$ B. $\frac{1}{32}$ C. $\frac{1}{8}$ (D.) not here

Use the information below to answer questions 16–18.

A spinner has 10 equal sections, numbered from 1 to 10.

16. What is the probability of spinning an odd number?
    A. $\frac{1}{5}$ B. $\frac{4}{10}$ (C.) $\frac{5}{10}$ D. $\frac{6}{10}$

17. What is the probability of spinning a number greater than 6?
    (A.) $\frac{4}{10}$ B. $\frac{5}{10}$ C. $\frac{6}{10}$ D. not here

18. What is the best prediction of how many times the number 10 will be obtained in 900 spins?
    A. 45 times (B.) 90 times
    C. 100 times D. 180 times

Use the information below to answer questions 19–20.

A large jar is filled with red beans and blue beans. It holds 100 beans in all. Bea shakes the jar and removes a handful. Her handful contains 12 red beans and 18 blue beans.

19. What ratio should Bea use to predict the number of blue beans in the jar?
    (A.) $\frac{18}{30}$ B. $\frac{12}{18}$
    C. $\frac{12}{30}$ D. $\frac{12}{100}$

20. What is the best prediction of how many red beans are in the jar?
    A. 36 beans
    (B.) 40 beans
    C. 50 beans
    D. 60 beans

21. How should you compute the probability of two independent events both happening?
    A. add the two probabilities
    B. subtract the lower probability from the higher
    (C.) multiply the two probabilities
    D. divide the higher probability by the lower

22. Which of the following does *not* describe independent events?
    A. flipping a coin 2 times
    B. rolling a number cube 2 times
    C. spinning a spinner 2 times
    (D.) drawing 2 balls from a box without replacing the first ball

Go on to the next page.

Name

Pretest

CHAPTER 11 PAGE 3

To answer questions 23–24, assume that a jar contains 3 red balls and 3 black balls. The first ball drawn is *not* replaced in the jar.

23. Helen takes 2 balls from the jar. What is the probability that both balls are black?
    A. $\frac{1}{10}$ (B.) $\frac{1}{5}$ C. $\frac{1}{2}$ D. $\frac{2}{3}$

24. Mike takes 2 balls from the jar. What is the probability that they are the same color?
    A. $\frac{1}{10}$ B. $\frac{2}{10}$ (C.) $\frac{2}{5}$ D. $\frac{1}{2}$

25. Juan flips a coin and rolls a number cube. What is the probability that the coin lands on heads and the number cube shows a 1?
    A. $\frac{1}{24}$ (B.) $\frac{1}{12}$ C. $\frac{1}{6}$ D. $\frac{1}{2}$

26. Olga flips a coin 2 times. What is the probability that both flips come up heads?
    A. $\frac{1}{16}$ B. $\frac{1}{8}$ (C.) $\frac{1}{4}$ D. $\frac{1}{2}$

27. Two baseball teams play 2 games. Since Team A is the better team, its chance of winning each game is $\frac{2}{3}$. What is the probability that Team A will win both games?
    A. $\frac{1}{3}$ (B.) $\frac{4}{9}$ C. $\frac{2}{3}$ D. $\frac{4}{3}$

28. A jar contains 5 red and 5 white marbles. Jo takes 1 marble from the jar, does not replace it, and then removes another marble. What is the probability that both marbles are white?
    (A.) $\frac{4}{18}$ B. $\frac{1}{5}$ C. $\frac{1}{4}$ D. $\frac{17}{18}$

29. Sam, Tina, and Ray each have a collection. Ray and the stamp collector are best friends. The coin collector lives next to Tina, who enjoys trading bells from her collection. Who has the coin collection?
    (A.) Ray B. Sam C. Tina

Use the information below to answer questions 30–32.

Joe and Edna must take music, health, and art courses during a two-year period. Each student is randomly assigned one of the courses for the current semester. Assume that the probability of assignment to each course is $\frac{1}{3}$.

30. What is the probability that both students will be assigned the art course?
    A. $\frac{3}{9}$ B. $\frac{2}{9}$ C. $\frac{1}{6}$ (D.) $\frac{1}{9}$

31. What is the probability that both students will be assigned any of the three courses together?
    A. $\frac{2}{3}$ (B.) $\frac{3}{9}$ C. $\frac{2}{9}$ D. $\frac{1}{9}$

32. What is the probability that neither student will be assigned the music course?
    A. $\frac{2}{3}$ (B.) $\frac{4}{9}$ C. $\frac{1}{3}$ D. $\frac{4}{81}$

Stop!

Name

Posttest

CHAPTER 11 PAGE 1

Choose the letter of the correct answer.

1. Lisa picks one day of the week from the calendar. What is the number of possible outcomes?
   A. 1 outcome B. 3 outcomes
   (C.) 7 outcomes D. 12 outcomes

2. John has a spinner with 8 equal sections, labeled 1, 2, 3, 4, 5, 6, 7, and 8. What is the number of possible outcomes?
   A. 2 outcomes B. 6 outcomes
   (C.) 8 outcomes D. 12 outcomes

3. Len is at a fruit stand. He has 6 kinds of fruit and 5 kinds of vegetables to choose from. If he buys 1 fruit and 1 vegetable, how many outcomes are possible?
   A. 6 outcomes
   B. 11 outcomes
   C. 25 outcomes
   (D.) 30 outcomes

4. What is the number of possible four-digit numbers? Do *not* use zero in the first place.
   A. 4,000 numbers
   (B.) 9,000 numbers
   C. 10,000 numbers
   D. 100,000 numbers

5. Can a greater number of 2-member teams be selected from a group of 4 girls or from a group of 5 girls?
   A. from a group of 4 girls
   (B.) from a group of 5 girls
   C. The same number of teams can be selected from either group.

6. Which of these words provides the *greatest* number of possible letter arrangements?
   A. *hat* B. *noon*
   C. *seven* (D.) *jacket*

7. How many different combinations of 2 items can be selected from a set of 6 items?
   A. 12 combinations
   (B.) 15 combinations
   C. 36 combinations
   D. not here

8. How many different arrangements of the letters *A, B, C, D, E* are possible?
   A. 12 arrangements
   B. 24 arrangements
   (C.) 120 arrangements
   D. 720 arrangements

9. A baseball team has 12 members. The team needs 2 co-captains. How many different pairs of cocaptains can be selected from the 12 members?
   (A.) 66 pairs
   B. 120 pairs
   C. 121 pairs
   D. 132 pairs

10. A basketball league has 7 teams. In how many different ways can the teams be arranged in the final standings?
    A. 49 ways
    B. 210 ways
    C. 720 ways
    (D.) 5,040 ways

Go on to the next page.

Name

Posttest

CHAPTER 11 PAGE 2

11. What is the probability of an event that is impossible?
    A. a negative number
    (B.) 0
    C. $\frac{1}{2}$
    D. 1

12. A coin is flipped 2 times. Which of the following is the most likely outcome?
    A. 2 heads, 0 tails
    (B.) 1 head, 1 tail
    C. 0 heads, 2 tails
    D. All of these are equally likely.

13. Lupe randomly picks one letter of the alphabet. Which of the following is the most likely outcome?
    A. The letter is *A*.
    B. The letter is *Z*.
    C. The letter is a vowel.
    (D.) The letter is a consonant.

14. On one roll of a number cube marked 1–6, which of these has the highest probability?
    A. an odd number
    B. an even number
    (C.) a number less than 5
    D. a number greater than 5

15. A coin is flipped 3 times. What is the probability of obtaining 3 heads?
    A. $\frac{1}{9}$ (B.) $\frac{1}{8}$
    C. $\frac{1}{3}$ D. not here

Use the information below to answer questions 16–18.

A spinner has 10 equal sections, numbered from 1 to 10.

16. What is the probability of spinning 1, 4, or 9?
    A. $\frac{1}{10}$ B. $\frac{2}{10}$ (C.) $\frac{3}{10}$ D. $\frac{1}{3}$

17. What is the probability of spinning a number greater than 2?
    A. $\frac{2}{10}$ B. $\frac{6}{10}$ C. $\frac{7}{10}$ (D.) not here

18. What is the best prediction of how many times the number 2 will be obtained in 1,000 spins?
    A. 25 times B. 50 times
    (C.) 100 times D. 200 times

Use the information below to answer questions 19–20.

A large jar is filled with black candies and white candies. It holds 200 candies in all. Hal shakes the jar and removes a handful. His handful contains 10 white candies and 15 black candies.

19. What ratio should Hal use to predict the number of white candies in the jar?
    (A.) $\frac{10}{25}$ B. $\frac{10}{15}$ C. $\frac{15}{200}$ D. $\frac{10}{200}$

20. What is the best prediction of the total number of black candies in the jar?
    A. 40 candies B. 80 candies
    (C.) 120 candies D. 160 candies

21. When the outcome of an event is *not* affected by the outcome of an earlier event, what are the events called?
    A. dependent (B.) independent
    C. equally likely D. impossible

Go on to the next page.

Name

Posttest

CHAPTER 11
PAGE 3

22. Which of the following pairs of events are most likely to be independent?
   - A. It rains heavily. You carry an umbrella.
   - **(B)** You are tall. You get an A in math.
   - C. You are sick. You do not go to school.
   - D. You study hard in science. You get an A in science.

To answer questions 23–24, assume that a jar contains 3 red balls and 2 black balls. The first ball drawn is *not* replaced in the jar.

23. Anne takes 2 balls from the jar. What is the probability that both balls are black?
   (A) $\frac{2}{20}$ B. $\frac{2}{10}$ C. $\frac{4}{25}$ D. $\frac{2}{5}$

24. Jake takes 2 balls from the jar. What is the probability that he picks a red ball and then a black ball?
   A. $\frac{3}{20}$ B. $\frac{3}{15}$ **(C)** $\frac{3}{10}$ D. $\frac{3}{30}$

25. Donna flips a coin 3 times. What is the probability of obtaining 3 heads?
   A. $\frac{1}{3}$ B. $\frac{1}{4}$ **(C)** $\frac{1}{8}$ D. $\frac{1}{16}$

26. Lupe rolls a number cube, marked 1–6, twice. What is the probability of rolling a 6 both times?
   (A) $\frac{1}{36}$ B. $\frac{1}{18}$ C. $\frac{1}{12}$ D. $\frac{2}{6}$

27. Tom and Mae play chess against each other. Tom has a $\frac{2}{3}$ probability of beating Mae in each game. What is the probability of Tom winning 2 games in a row?
   A. $\frac{1}{9}$ **(B)** $\frac{4}{9}$ C. $\frac{5}{9}$ D. $\frac{6}{9}$

28. A jar contains 7 blue marbles and 3 black marbles. Rae removes 1 marble, does not replace it, and then removes another marble. What is the probability that both marbles are blue?
   A. $\frac{1}{49}$ B. $\frac{2}{7}$ **(C)** $\frac{42}{90}$ D. $\frac{49}{100}$

29. Bo, Cara, and Don are neighbors. They each have one pet: a dog, a hamster, or a turtle. Bo and the turtle owner are the same age. Bo walks his pet every morning. Don is afraid of turtles. Who owns the hamster?
   A. Bo B. Cara **(C)** Don

Use the information below to answer questions 30–32.

Bob and Cari must take health and music during a two-year period. Each student is randomly assigned one of the courses for the current semester. Assume that the probability of assignment to each course is $\frac{1}{2}$.

30. What is the probability that both students will be assigned the health course?
   A. $\frac{3}{4}$ B. $\frac{1}{2}$ **(C)** $\frac{1}{4}$ D. $\frac{1}{8}$

31. What is the probability that neither student will be assigned the music course?
   A. $\frac{3}{4}$ B. $\frac{1}{2}$ **(C)** $\frac{1}{4}$ D. $\frac{1}{8}$

32. What is the probability that Bob and Cari will be assigned different classes?
   A. $\frac{3}{4}$ **(B)** $\frac{1}{2}$ C. $\frac{1}{4}$ D. $\frac{1}{8}$

Stop!

Name

Cumulative Test

CHAPTERS 1–11
PAGE 1

Choose the letter of the correct answer.

1. Bill wants to buy 2 suits that cost \$199.95 each and 3 shirts that cost \$14.95 each. *Estimate* how much money Bill needs.
   A. \$245.00 B. \$430.00 C. \$435.00 **(D)** \$445.00

2. What is the value of 3^4?
   A. 12 B. 64 **(C)** 81 D. 243

3. Express 506,000 in scientific notation.
   (A) 5.06×10^5 B. 5.6×10^5 C. 5.06×10^6 D. 5.60×10^6

Use the bar graph below to answer questions 4–5.

Unemployment Percents by Month

Percent: 1 2 3 4 5 6 7 8
Month: Sept. Oct. Nov. Dec.

4. Which month had the greatest unemployment percent?
   A. September **(B)** October C. November D. December

5. The unemployment percent in December was what fraction of the unemployment percent in November?
   A. $\frac{1}{5}$ B. $\frac{2}{3}$ C. $\frac{5}{7}$ **(D)** $\frac{5}{6}$

Use the information and box-and-whisker graph below to answer questions 6–7.

A clothing-store owner records the number of suits sold in his store each day for 30 days. He summarizes the information in the box-and-whisker graph below.

Suits Sold

0 5 10 15

6. What was the median number of suits sold in his store each day?
   A. 7 suits **(B)** 11 suits C. 12 suits D. 14 suits

7. What was the greatest number of suits sold on any of the 30 days?
   A. 12 suits **(B)** 14 suits C. 15 suits D. 16 suits

Use the information below to answer questions 8–9.

Li's bowling scores for her last 10 games were 85, 90, 90, 90, 92, 92, 95, 97, 99, and 105.

8. What is the range of these scores?
   (A) 20 B. 85 C. 105 D. 190

9. What is the median of these scores?
   A. 90 **(B)** 92 C. 93 D. 93.5

10. Which fractions are ordered from least to greatest?
   A. $\frac{5}{6}, \frac{4}{5}, \frac{3}{4}$ **(B)** $\frac{1}{4}, \frac{1}{3}, \frac{1}{2}$ C. $\frac{3}{9}, \frac{2}{9}, \frac{1}{9}$ D. $\frac{2}{3}, \frac{2}{4}, \frac{2}{5}$

Go on to the next page.

Name

Cumulative Test

CHAPTERS 1–11
PAGE 2

11. Which fraction falls between $\frac{8}{10}$ and $\frac{9}{10}$?
   A. $\frac{15}{20}$ B. $\frac{31}{40}$ **(C)** $\frac{17}{20}$ D. $\frac{7}{10}$

12. $13\frac{1}{4} - 10\frac{1}{6} =$ ___?___
   A. $2\frac{1}{12}$ B. $2\frac{1}{6}$ **(C)** $3\frac{1}{12}$ D. not here

13. $3\frac{1}{2} \times 1\frac{1}{3} =$ ___?___
   A. $3\frac{1}{6}$ B. $3\frac{2}{3}$ **(C)** $4\frac{2}{3}$ D. not here

14. $\frac{7}{8} \div \frac{8}{7} =$ ___?___
   A. 1 B. $1\frac{1}{8}$ C. $1\frac{1}{7}$ **(D)** not here

15. Find the value of $(2 + 4) \times (7 - 2)$.
   A. 22 B. 28 **(C)** 30 D. 40

16. Solve the inequality. $x + 9 < 11$
   A. $x < 1$ B. $x < \frac{11}{9}$ **(C)** $x < 2$ D. $x < 20$

17. Rectangles *A* and *B* are congruent. Rectangle *A* is 10 cm long by 20 cm wide. Which of the following are possible dimensions of Rectangle *B*?
   A. 1 cm × 2 cm
   B. 5 m × 15 cm
   C. 20 cm × 40 cm
   (D) 10 cm × 20 cm

18. A box of 5 cakes costs \$4.20. What is the unit price?
   A. \$0.21 B. \$0.82 **(C)** \$0.84 D. \$0.94

19. Use the scale 2 cm = 100 cm. What is the actual length of an object when the scale length is 5 cm?
   A. 20 cm B. 40 cm **(C)** 250 cm D. 500 cm

20. Two rectangles are similar. The first rectangle has a width of 8 ft and a length of 20 ft. The width of the second rectangle is 10 ft. Find its length.
   A. 4 ft B. 16 ft C. 22 ft **(D)** 25 ft

21. Express the ratio 9:12 as a percent.
   A. 0.75% B. 7.5% **(C)** 75% D. 750%

22. Express 85% as a fraction.
   A. $\frac{5}{8}$ **(B)** $\frac{17}{20}$ C. $\frac{8}{5}$ D. not here

23. Phil invited 28 friends to a party. Only 25% of them could come. Which is the correct equation to find how many came to the party?
   A. $n = \frac{28}{0.25}$ B. $n \times 0.25 = 28$ C. $n = \frac{0.25}{28}$ **(D)** $n = 0.25 \times 28$

24. 3.5 is what percent of 70?
   A. 0.5% **(B)** 5% C. 20% D. 50%

25. 15% of what number is 120?
   A. 18 B. 80 C. 720 **(D)** 800

Go on to the next page.

Name

Cumulative Test

CHAPTERS 1–11
PAGE 3

26. Jan is buying a bicycle that is marked down 20%. What will be the sale price of a \$200 bicycle?
   (A) \$160 B. \$220 C. \$225 D. \$250

27. Dick borrows \$600. The interest rate is 9.5% per year. The payoff period is 24 months. How much interest will Dick pay? Use the formula $I = prt$.
   A. \$54 B. \$57 C. \$108 **(D)** \$114

28. $^-4 + ^-3 =$ ___?___
   (A) $^-7$ B. $^-1$ C. 1 D. not here

29. $\frac{^-8}{^-4} =$ ___?___
   A. $^-2$ B. $\frac{1}{2}$ **(C)** 2 D. not here

30. To solve the equation $x - 9 = 22$, what must you do to both sides of the equation?
   (A) add 9 B. subtract 9 C. add 22 D. subtract 22

31. Solve for *y*. $y + 6 = ^-12$
   (A) $y = ^-18$ B. $y = ^-6$ C. $y = 6$ D. $y = 18$

32. Are mixed numbers rational?
   (A) always B. sometimes C. never

33. Which square root is an irrational number?
   A. $\sqrt{4}$ **(B)** $\sqrt{6}$ C. $\sqrt{9}$ D. $\sqrt{16}$

34. Compare. $^-500$ ◯ 0.0006
   A. > B. = **(C)** <

35. Compare. 2.30×10^3 ◯ 2.03×10^4
   A. > B. = **(C)** <

36. $\frac{10^4}{10^2} =$ ___?___
   A. 2 B. 10 **(C)** 100 D. 1,000

37. Compare. $\sqrt{144}$ ◯ 13
   A. > B. = **(C)** <

38. How many different combinations of 2 can be selected from a set of 6 items?
   A. 12 combinations
   B. 20 combinations
   C. 30 combinations
   (D) not here

39. On one roll of a number cube numbered from 1 to 6, which has the highest probability?
   A. a 1
   B. a 5
   C. a number less than 3
   (D) a number greater than 3

40. A spinner has 8 equal-sized sections numbered from 1 to 8. What is the probability of spinning an even number?
   A. $\frac{1}{8}$ B. $\frac{2}{8}$ **(C)** $\frac{1}{2}$ D. $\frac{6}{8}$

To answer questions 41–42, assume that there is a vase containing 3 red balls and 3 black balls.

41. Helen takes 2 balls from the vase without replacement. What is the probability that both balls are black?
   (A) $\frac{1}{5}$ B. $\frac{1}{4}$ C. $\frac{3}{5}$ D. $\frac{2}{3}$

Go on to the next page.

A. TEST ANSWERS STANDARDIZED FORMAT

Name

Cumulative Test

CHAPTERS 1–11
PAGE 4

42. Helen takes 1 ball from the vase, replaces it, and then takes a second ball. What is the probability that both balls are red?

A. $\frac{1}{5}$ **(B.) $\frac{1}{4}$** C. $\frac{1}{2}$ D. $\frac{3}{5}$

43. The number of points scored by two football teams totals 50. One team scored 4 times as many points as the other. How many points did the winning team score?

A. 24 points B. 32 points
(C.) 40 points D. 44 points

44. Nadia spent $20 buying books. She gave $\frac{1}{4}$ of the remaining money to her sister. Nadia then had $15. How much money did she have before buying her books?

A. $35 **(B.) $40**
C. $60 D. $80

Use the table below to answer question 45.

Students Graduating from Grant High	
Year	Number
1986	1,250
1987	1,300
1988	1,400
1989	1,550
1990	1,750

45. Assuming the trend continued, how many students graduated from Grant High in 1991?

A. 1,900 students
B. 1,950 students
(C.) 2,000 students
D. 2,050 students

46. Tricia had 32 questions correct on a math test. Her score was 2 items more than $\frac{3}{4}$ of her twin sister's score. How many questions did Tricia's sister have correct?

A. 23 questions B. 26 questions
(C.) 40 questions D. 42 questions

47. Joel and Gary begin a swimming program. Joel swims 4 laps on day 1 and increases the distance by 1 lap each day. Gary swims 4 laps on day 1 and increases the distance by $1\frac{1}{2}$ laps each day. On day 7, how much farther will Gary swim than Joel?

(A.) 3 laps B. $4\frac{1}{2}$ laps
C. 6 laps D. 13 laps

48. There are 8 teams in a softball league. If each team plays each of the other teams twice, how many total games will be played?

A. 28 games **(B.) 56 games**
C. 112 games D. 128 games

49. Heidi watches a television program that lasts $\frac{1}{2}$ hour. Commercials make up $\frac{1}{6}$ of the show. What part of an hour does she spend watching the actual program?

A. $\frac{1}{12}$ hr B. $\frac{1}{3}$ hr
(C.) $\frac{5}{12}$ hr D. $\frac{5}{6}$ hr

50. Mrs. Chan drove at an average speed of 48 miles per hour. She drove a distance of 120 miles. How long did the drive take? (distance = rate × time)

A. $2\frac{1}{4}$ hr **(B.) $2\frac{1}{2}$ hr**
C. $2\frac{3}{4}$ hr D. 3 hr

Stop!

Name

Pretest

CHAPTER 12
PAGE 1

Choose the letter of the correct answer.

1. To measure the width of a closet, which is the most appropriate unit of measure?

A. inch **(B.) foot** C. yard D. mile

2. In which of the following situations would precise measurements be necessary?

A. determining your car's gas mileage
B. determining how much paint is needed to paint your house
C. planning how far you will drive on a vacation
(D.) measuring the time to run a 50-yd dash at a track meet

3. Which might be the height of a man?

A. 3,000 mm B. 2,000 cm
(C.) 1.7 m D. 0.5 km

4. Which might be the weight of a banana?

(A.) 4 oz B. 4 lb C. 4 g D. 4 kg

5. Which is the best estimate of the amount of water it takes to fill an eyedropper?

(A.) 1 mL B. 0.5 L C. 1 qt D. $\frac{1}{4}$ pint

6. Which is the best estimate of the width of a desk?

A. 15 in. **(B.) 3 ft** C. 4 yd D. 4 m

7. Which measurement is most precise?

A. 36 in. B. 3 ft
C. 1 yd **(D.) $1\frac{1}{4}$ yd**

8. Which measurement is most precise?

A. 98 cm B. 1 m
C. 1,000 mL **(D.) 1,000.0 mL**

9. Which measurement is most precise?

A. 4 qt **(B.) $4\frac{1}{4}$ qt**
C. $1\frac{1}{2}$ gal D. $1\frac{3}{4}$ gal

10. Which measurement is most precise?

(A.) 3,600 sec B. 60 min
C. 1 hr D. 1.0 hr

11. What is the perimeter of a regular pentagon when $s = 4\frac{1}{2}$ cm?

A. 18 cm B. $20\frac{1}{4}$ cm
(C.) $22\frac{1}{2}$ cm D. not here

12. What is the circumference of a circle with a radius of 3 in.? Use $\pi = 3.14$.

A. 9.14 in. B. 9.42 in.
(C.) 18.84 in. D. 28.26 in.

13. Jim is making a basketball hoop from a metal bar. He wants the diameter of the hoop to be 50 cm. How long does the metal bar need to be? Use $\pi = 3.14$.

A. 15.7 cm **(B.) 157 cm**
C. 31.4 cm D. 314 cm

14. Doreen wants to put wooden trim along the top of the walls in her room. The room is 9 ft wide by 12 ft long. How many feet of trim does she need?

A. 21 ft **(B.) 42 ft**
C. 108 ft D. 225 ft

Go on to the next page.

Name

Pretest

CHAPTER 12
PAGE 2

15. Which figure has the greater area?

h b h b

A. the triangle
(B.) the parallelogram
C. The areas are equal.

16. Do two trapezoids with the same height have the same area?

A. always **(B.) sometimes** C. never

17. What is the area of a parallelogram with a base of 10 cm and a height of 5 cm?

5 cm 10 cm

A. 15 cm² B. 30 cm²
(C.) 50 cm² D. not here

18. What is the area of a triangle with a base of 9 ft and a height of 7 ft?

7 ft 9 ft

A. 16 ft² **(B.) $31\frac{1}{2}$ ft²**
C. 63 ft² D. 130 ft²

19. What is the area of a circle with a diameter of 7 m? Use $\pi = \frac{22}{7}$.

7 m

A. 22 m² B. 44 m²
C. 154 m² **(D.) not here**

20. What is the area of a trapezoid with bases of 4 cm and 6 cm and a height of 10 cm?

4 cm 10 cm 6 cm

A. 20 cm² B. 30 cm²
C. 100 cm² **(D.) not here**

21. Jack's garden is in the shape of a square. Its perimeter is 26 ft. What is the area of the garden?

A. 26 ft² **(B.) 42.25 ft²**
C. 169 ft² D. 676 ft²

22. Erica arranged a play space for her dog. She put a pole into the ground and attached a 10-ft leash. The dog can go exactly 10 ft in any direction from the pole. What is the area of the play space?

A. 31.4 ft² B. 62.8 ft²
C. 100 ft² **(D.) 314 ft²**

23. Does a figure with line symmetry also have turn symmetry?

A. always **(B.) sometimes** C. never

24. Does translation change a figure's size but not its shape?

A. always B. sometimes **(C.) never**

25. Which of the following letters has line symmetry?

A. P B. R C. F **(D.) E**

26. Which of the following letters has turn symmetry?

A. L B. E **(C.) H** D. K

Go on to the next page.

Name

Pretest

CHAPTER 12
PAGE 3

27. How many lines of symmetry, does a square have?

A. 1 line B. 2 lines
C. 3 lines **(D.) 4 lines**

28. For turn symmetry, what is the angle measure of each turn for a regular six-sided polygon?

A. 30° B. 60°
C. 90° **(D.) 120°**

Use the formulas below to answer questions 29–31.

$°F = \frac{9}{5}°C + 32$ $°C = \frac{5}{9}(°F - 32)$

29. A mixture freezes at 10°C. At what Fahrenheit temperature does it freeze?

A. 22°F B. 42°F
(C.) 50°F D. 75.6°F

30. One day the temperature dropped from 60°F to 42°F in one hour. What was the temperature drop in degrees Celsius?

A. 2° **(B.) 10°** C. 18° D. 42°

31. Alice heated wax and observed its temperature with a Celsius thermometer. If the reading on her thermometer was 45°C, what was the temperature in degrees Fahrenheit?

A. 13°F B. 25°F
C. 81°F **(D.) 113°F**

32. Harry wants to make a wallpaper design from congruent regular polygons. Which of the following regular polygons do *not* tessellate a plane?

A. triangles **(B.) squares**
C. hexagons D. pentagons

Stop!

Name

Posttest

CHAPTER 12 PAGE 1

Choose the letter of the correct answer.

1. To measure the distance between two cities, which is the most appropriate unit of measure?
   A. millimeter B. centimeter C. meter (D.) kilometer

2. In which situation would estimation be most appropriate?
   A. pouring a dose of medicine
   B. measuring your ring size for a jeweler
   (C.) measuring the temperature of meat you are cooking
   D. measuring a table leg that you must replace

3. Which might be the height of a room in a home?
   A. 1,000 mm B. 5,000 cm (C.) 3.5 m D. 0.1 km

4. Which might be the weight of an egg?
   (A.) 2 oz B. 1 lb C. 2 g D. 1 kg

5. Which is the best estimate of the amount of water it takes to fill a cup?
   A. 10 mL B. 1 L (C.) 8 oz D. 2 pt

6. Which is the best estimate of the width of a refrigerator?
   A. 15 in. (B.) 3 ft C. 4 yd D. 5 m

7. Which measurement is most precise?
   A. 35 in. (B.) $36\frac{1}{2}$ in. C. 3 ft D. $3\frac{1}{2}$ ft

8. Which measurement is most precise?
   A. 98 ml (B.) 1 ml C. 1 m D. 1.13 m

9. Which measurement is most precise?
   (A.) 128 fl oz B. 8 pt C. 2 qt D. $\frac{1}{2}$ gal

10. Which measurement is most precise?
    A. 2 sec B. 1.1 sec C. $1\frac{1}{2}$ sec (D.) 1.001 sec

11. What is the perimeter of a regular hexagon when $s = 2.5$ cm?
    A. 6.25 cm B. 12.5 cm (C.) 15 cm D. 17.5 cm

12. What is the circumference of a circle with a radius of 2 ft? Use $\pi = 3.14$.
    A. 6.28 ft (B.) 12.56 ft C. 18.84 ft D. not here

13. Harold wants to build a hoop with a circumference of 16 ft. What will be the diameter? Use $\pi = 3.14$.
    A. 2.55 ft (B.) 5.10 ft C. 25.12 ft D. 50.24 ft

14. The perimeter of a rectangle is 10 cm. The shorter sides are each 2 cm. What is the length of each of the longer sides?
    A. $2\frac{1}{2}$ cm (B.) 3 cm C. 4 cm D. 5 cm

Go on to the next page.

Name

Posttest

CHAPTER 12 PAGE 2

15. Which triangle has the greater area?
    A. the equilateral triangle
    B. the right triangle
    (C.) The areas are equal.

16. Do two circles with the same circumference have the same area?
    (A.) always B. sometimes C. never

17. What is the area of a parallelogram with a base of 20 cm and a height of 15 cm?
    A. 70 cm² B. 150 cm² (C.) 300 cm² D. not here

18. What is the area of a triangle with a base of 3 ft and a height of 4 ft?
    A. $3\frac{1}{2}$ ft² (B.) 6 ft² C. 12 ft² D. 49 ft²

19. What is the area of a circle with a diameter of 10 m? Use $\pi = 3.14$.
    A. 15.7 m² B. 31.4 m² C. 314 m² (D.) not here

20. What is the area of a trapezoid with bases of 10 cm and 20 cm and a height of 10 cm?
    A. 50 cm² (B.) 150 cm² C. 200 cm² D. 300 cm²

21. Jack's garden is in the shape of a square. Its area is 36 ft². What is the perimeter of the garden?
    A. 6 ft B. 12 ft C. 18 ft (D.) 24 ft

22. Hilda's garden is circular. The area is 3.14 ft². What is the diameter?
    A. $\frac{1}{2}$ ft B. 1 ft C. $1\frac{1}{2}$ ft (D.) 2 ft

23. Does a circle have more than 4 lines of symmetry?
    (A.) always B. sometimes C. never

24. Does reflection change both the size and the shape of a figure?
    A. always B. sometimes (C.) never

25. Which of the following letters has line symmetry?
    A. J B. Q C. G (D.) H

26. Which of the following letters has turn symmetry?
    A. A B. P C. W (D.) X

27. How many lines of symmetry does an isosceles triangle have?
    (A.) 1 line B. 2 lines C. 3 lines D. 4 lines

Go on to the next page.

Name

Posttest

CHAPTER 12 PAGE 3

28. For turn symmetry, what is the angle measure of each turn for a square?
    A. 30° B. 60° (C.) 90° D. 120°

Use the formulas below to answer questions 29–31.

$F = \frac{9}{5}C + 32$ $C = \frac{5}{9}(F - 32)$

29. Water boils at 100°C. At what Fahrenheit temperature does it boil?
    A. 112°F B. 132°F C. 100°F (D.) 212°F

30. One day the temperature increased from 70°F to 79°F in one hour. What was the temperature increase in degrees Celsius?
    A. 1° (B.) 5° C. 9° D. 18°

31. Berto chilled water and observed its temperature with a Fahrenheit thermometer. If the reading on his thermometer was 45°F, what was the temperature in degrees Celsius?
    A. 13°C B. 25°C C. 77°C (D.) not here

32. John wants to make a quilt design from congruent regular polygons. He can use any regular polygon for which the measures of the angles meeting at a vertex have a sum of which of the following?
    A. 90° B. 180° (C.) 360° D. 720°

Stop!

Name

Pretest

CHAPTER 13 PAGE 1

Choose the letter of the correct answer.

1. Paco is drawing a pattern for a solid figure. The base of the figure is a polygon. Each side of the base is the base of an isosceles triangle. Which figure is Paco drawing?
   A. cone B. cylinder (C.) pyramid D. triangular prism

2. Helen built a solid figure with 2 flat surfaces and 1 curved surface. Which figure did she build?
   (A.) cylinder B. sphere C. cone D. polyhedron

3. Which of these figures has no flat surfaces?
   A. cone B. cylinder C. pyramid (D.) sphere

4. What do prisms and pyramids have in common?
   A. Their lateral faces are all triangles.
   (B.) They are polyhedrons.
   C. They have congruent, parallel bases.
   D. None of their surfaces are polygons.

5. How many faces does a triangular prism have?
   A. 3 faces B. 4 faces (C.) 5 faces D. 6 faces

6. A polyhedron has 6 faces and 8 vertices. How many edges does it have?
   (A.) 12 edges B. 14 edges C. 16 edges D. 18 edges

7. How many faces does a rectangular pyramid have?
   A. 4 faces (B.) 5 faces C. 6 faces D. 8 faces

8. A solid figure has 6 faces: 5 triangles and 1 pentagon. What is the figure?
   (A.) pentagonal pyramid
   B. triangular prism
   C. pentagonal prism
   D. not here

9. What is the surface area of this cylinder? Use $\pi = 3.14$.
   A. 207.24 cm² (B.) 244.92 cm² C. 282.6 cm² D. not here

10. The dimensions of a rectangular prism are 5 ft × 4 ft × 4 ft. What is its surface area?
    A. 80 ft² B. 104 ft² (C.) 112 ft² D. 400 ft²

11. A cube has a surface area of 600 cm². What is the length of each face?
    A. 50 cm B. 60 cm C. 100 cm (D.) not here

12. What is the surface area of this square pyramid?
    A. 120 in.² (B.) 145 in.² C. 240 in.² D. 300 in.²

Go on to the next page.

A. TEST ANSWERS STANDARDIZED FORMAT

Name

Pretest

13. What is the volume of a cylinder whose radius is 2 m and whose height is 0.25 m? Use $\pi = 3.14$.
    A. 3.14 m³ B. 6.28 m³
    C. 9.42 m³ D. 12.56 m³

14. What is the volume of this rectangular pyramid?
    A. 8 ft³
    B. 12 ft³
    C. 16 ft³
    D. 24 ft³

    4 ft
    2 ft 3 ft

15. What is the volume of a rectangular prism with the dimensions 4 in. × 6 in. × 8 in.?
    A. 18 in.³ B. 72 in.³
    C. 144 in.³ D. 192 in.³

16. What volume of water is held by a spoon with a capacity of 10 mL?
    A. 1 cm³ B. 10 cm³
    C. 100 cm³ D. not here

17. A swimming pool is 25 m long, 10 m wide, and 2 m deep. What is the capacity of the pool?
    A. 50 kL B. 250 kL
    C. 500 kL D. 5,000 kL

18. The volume of a pitcher is 1,500 cm³. What mass of water will fill it?
    A. 1.5 kg B. 15 kg
    C. 150 kg D. 1,500 kg

19. Lena's aquarium is in the shape of a cylinder. Its radius is 5 cm and its height is 10 cm. What is the volume? Use $\pi = 3.14$.
    A. 157 cm³ B. 785 cm³
    C. 1,000 cm D. 1,570 cm³

20. Harry packs yogurt into tubs that are in the shape of a cylinder 8 in. high and 8 in. in diameter. What are the dimensions of a tub that will hold 4 times as much yogurt?
    A. h = 16 in.; d = 8 in.
    B. h = 8 in.; d = 16 in.
    C. h = 12 in.; d = 12 in.
    D. h = 16 in.; d = 16 in.

21. *Estimate* how much money you need in order to buy 3 pairs of socks at $1.98 a pair and 4 cans of tennis balls at $2.99 a can.
    A. $17.00 B. $18.00
    C. $18.50 D. $19.00

22. *About* how many square feet of carpet are needed for a rectangular room 9 ft 11 in. by 17 ft 2 in.?
    A. 150 ft² B. 170 ft²
    C. 180 ft² D. 190 ft²

23. Joe is wrapping a box. *Estimate* how many square feet of wrapping paper he needs if the box measures $11\frac{1}{4}$ in. by $12\frac{1}{2}$ in. by 23 in.
    A. 2 ft² B. 6 ft²
    C. 8 ft² D. 10 ft²

24. Aquarium A and aquarium B are both cylinders. They have the same radius, but A is twice as high as B. What is the ratio of the volume of water held by A to the volume of water held by B?
    A. $\frac{1}{2}$ to 1 B. 2 to 1
    C. 4 to 1 D. 8 to 1

Stop!

Name

Posttest

Choose the letter of the correct answer.

1. Lorna is building a solid figure. The figure has 2 parallel bases that are congruent triangles. Its other faces are rectangles. What is the figure?
    A. rectangular pyramid
    B. rectangular prism
    C. triangular pyramid
    D. triangular prism

2. Sean built a solid figure with 1 flat surface and 1 curved surface. Which figure did he build?
    A. cylinder B. sphere
    C. cone D. polyhedron

3. Which of these figures is *not* a polyhedron?
    A. cylinder B. prism
    C. pyramid D. cube

4. What do prisms and cylinders have in common?
    A. Their bases are circles.
    B. They have parallel, congruent bases.
    C. Their bases are polygons.
    D. They have curved surfaces.

5. How many faces does a triangular pyramid have?
    A. 3 faces B. 4 faces
    C. 5 faces D. 6 faces

6. A polyhedron has 8 faces and 12 vertices. How many edges does it have?
    A. 16 edges B. 18 edges
    C. 20 edges D. 22 edges

7. How many faces does a rectangular prism have?
    A. 4 faces B. 5 faces
    C. 6 faces D. 8 faces

8. A solid figure has 6 rectangular surfaces. What is the figure?
    A. rectangular prism
    B. hexagonal prism
    C. rectangular pyramid
    D. not here

9. What is the surface area of this cylinder? Use $\pi = 3.14$.

    10 cm
    25 cm

    A. 628 cm² B. 1,570 cm²
    C. 2,198 cm² D. not here

10. What is the surface area of this rectangular prism?
    A. 32 ft²
    B. 48 ft²
    C. 72 ft²
    D. 256 ft²

    2 ft 8 ft
    2 ft

11. A cube has a surface area of 96 cm². What is the length of each face?
    A. 8 cm B. 16 cm
    C. 24 cm D. not here

12. The base of a square pyramid is 4 in. × 4 in. The height of each triangular face is 6 in. What is the surface area of the pyramid?
    A. 48 in.² B. 64 in.²
    C. 96 in.² D. 112 in.²

Go on to the next page.

Name

Posttest

13. What is the volume of a cylinder with a radius of 1 m and a height of 0.5 m? Use $\pi = 3.14$.
    A. 0.785 m³ B. 1.57 m³
    C. 3.14 m³ D. not here

14. What is the volume of a rectangular pyramid that has a base 3 ft × 4 ft and a height of 5 ft?
    A. 20 ft³ B. 40 ft³
    C. 48 ft³ D. 60 ft³

15. What is the volume of this rectangular prism?
    A. 60 in.³
    B. 120 in.³
    C. 540 in.³
    D. 600 in.³

    5 in.
    12 in. 10 in.

16. A cup with a volume of 50 cm³ has what liquid capacity?
    A. 5 mL B. 50 mL
    C. 500 mL D. not here

17. A swimming pool is 30 m long, 15 m wide, and 1.5 m deep. What is the capacity of the pool?
    A. 67.5 kL B. 450 kL
    C. 675 kL D. 6,750 kL

18. The volume of a bowl is 3,500 cm³. What mass of water will fill it?
    A. 3.5 kg B. 35 kg
    C. 350 kg D. 3,500 kg

19. Eli's aquarium is in the shape of a cylinder. It has a radius of 10 cm and height of 20 cm. What is the volume? Use $\pi = 3.14$.
    A. 314 cm³ B. 628 cm³
    C. 3,140 cm³ D. 6,280 cm³

20. Joel packs food into containers that are in the shape of a cone 6 cm high and 6 cm in diameter. What are the dimensions of a cone that will hold 2 times as much food?
    A. h = 12 cm; d = 6 cm
    B. h = 6 cm; d = 12 cm
    C. h = 9 cm; d = 9 cm
    D. h = 12 cm; d = 12 cm

21. *Estimate* how much money Lee needs in order to buy 2 shirts at $14.95 each, 2 pairs of shoes at $49.50 a pair, and 3 ties at $19.95 each.
    A. $160.00 B. $175.00
    C. $185.00 D. $190.00

22. *Estimate* the area of a rectangular garden that has the dimensions 15 ft 11 in. by 20 ft 2 in.
    A. 300 ft² B. 315 ft²
    C. 320 ft² D. 350 ft²

23. Emily is wrapping a box. *About* how many square feet of wrapping paper will she need if the dimensions of the box are $11\frac{1}{4}$ in. by $24\frac{1}{2}$ in. by 23 in.?
    A. 4 ft² B. 10 ft²
    C. 12 ft² D. 16 ft²

24. Aquarium A and aquarium B are both rectangular prisms. Their bases have the same dimensions, but A is twice as high as B. What is the ratio of the volume of water held by A to that held by B?
    A. $\frac{1}{2}$ to 1 B. 2 to 1
    C. 4 to 1 D. 8 to 1

Stop!

Name

Pretest

Choose the letter of the correct answer.

1. What are the first four square numbers?
    A. 1, 3, 6, 12 B. 1, 4, 9, 16
    C. 1, 5, 10, 15 D. 1, 2, 4, 8

2. How many dots would come next in this pattern?
    A. 9 B. 10 C. 12 D. not here

3. What is the next number in the pattern? 1, 3, 6, 10, 15
    A. 20 B. 21
    C. 22 D. not here

4. What is the next number in the pattern? 243, 81, 27
    A. 3 B. 9 C. 18 D. not here

5. What are the first four numbers in this pattern? The first number is 7. Add 12 each time.
    A. 7, 12, 24, 36 B. 7, 19, 26, 38
    C. 7, 19, 31, 43 D. 12, 24, 36, 43

6. What are the first four numbers in this pattern? The first number is 2. Multiply by 3 and subtract 1 each time.
    A. 2, 3, 8, 23 B. 2, 5, 14, 41
    C. 2, 5, 15, 44 D. 2, 6, 5, 15

To answer questions 7–8, use the relation {(0,0), (1,2), (2,4), (3,6)}.

7. What is the domain of the relation?
    A. 0, 1 B. 2, 3
    C. 0, 1, 2, 3 D. 0, 2, 4, 6

8. What is the range of the relation?
    A. 0, 2 B. 4, 6
    C. 0, 1, 2, 3 D. 0, 2, 4, 6

9. Which ordered pairs are graphed?
    A. (0,⁻2), (2,0)
    B. (⁻2,0), (2,0)
    C. (0,2), (2,0)
    D. (⁻2,0), (0,2)

10. If the length and width of a rectangle both triple, what happens to the area of the rectangle?
    A. It doubles.
    B. It triples.
    C. It increases by 6 times.
    D. It increases by 9 times.

11. Which equation represents the relation {(0,0), (1,3), (2,6), (3,9)}?
    A. $y = x + 6$ B. $y = \frac{1}{3}x$
    C. $y = x + 3$ D. $y = 3x$

Go on to the next page.

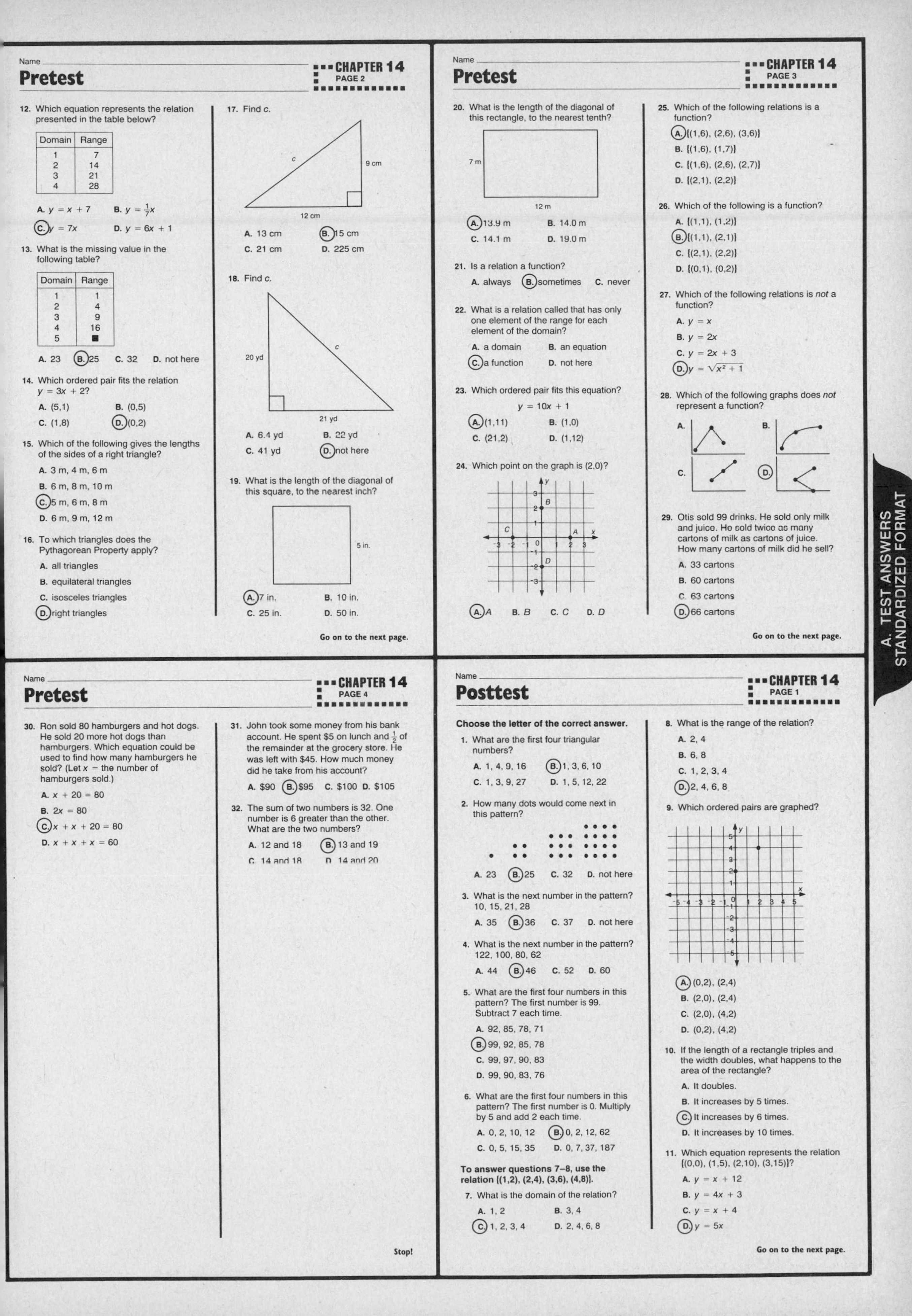

Name

Pretest

CHAPTER 14
PAGE 2

12. Which equation represents the relation presented in the table below?

Domain	Range
1	7
2	14
3	21
4	28

A. $y = x + 7$ B. $y = \frac{1}{7}x$
(C.) $y = 7x$ D. $y = 6x + 1$

13. What is the missing value in the following table?

Domain	Range
1	1
2	4
3	9
4	16
5	■

A. 23 **(B.)** 25 C. 32 D. not here

14. Which ordered pair fits the relation $y = 3x + 2$?

A. (5,1) B. (0,5)
C. (1,8) **(D.)** (0,2)

15. Which of the following gives the lengths of the sides of a right triangle?

A. 3 m, 4 m, 6 m
B. 6 m, 8 m, 10 m
(C.) 5 m, 6 m, 8 m
D. 6 m, 9 m, 12 m

16. To which triangles does the Pythagorean Property apply?

A. all triangles
B. equilateral triangles
C. isosceles triangles
(D.) right triangles

17. Find *c*.

A. 13 cm **(B.)** 15 cm
C. 21 cm D. 225 cm

18. Find *c*.

A. 6.4 yd B. 22 yd
C. 41 yd **(D.)** not here

19. What is the length of the diagonal of this square, to the nearest inch?

(A.) 7 in. B. 10 in.
C. 25 in. D. 50 in.

Go on to the next page.

Name

Pretest

CHAPTER 14
PAGE 3

20. What is the length of the diagonal of this rectangle, to the nearest tenth?

(A.) 13.9 m B. 14.0 m
C. 14.1 m D. 19.0 m

21. Is a relation a function?

A. always **(B.)** sometimes C. never

22. What is a relation called that has only one element of the range for each element of the domain?

A. a domain B. an equation
(C.) a function D. not here

23. Which ordered pair fits this equation?

$y = 10x + 1$

(A.) (1,11) B. (1,0)
C. (21,2) D. (1,12)

24. Which point on the graph is (2,0)?

(A.) A B. B C. C D. D

25. Which of the following relations is a function?

(A.) {(1,6), (2,6), (3,6)}
B. {(1,6), (1,7)}
C. {(1,6), (2,6), (2,7)}
D. {(2,1), (2,2)}

26. Which of the following is a function?

A. {(1,1), (1,2)}
(B.) {(1,1), (2,1)}
C. {(2,1), (2,2)}
D. {(0,1), (0,2)}

27. Which of the following relations is *not* a function?

A. $y = x$
B. $y = 2x$
C. $y = 2x + 3$
(D.) $y = \sqrt{x^2 + 1}$

28. Which of the following graphs does *not* represent a function?

A. B. C. **(D.)**

29. Otis sold 99 drinks. He sold only milk and juice. He sold twice as many cartons of milk as cartons of juice. How many cartons of milk did he sell?

A. 33 cartons
B. 60 cartons
C. 63 cartons
(D.) 66 cartons

Go on to the next page.

Name

Pretest

CHAPTER 14
PAGE 4

30. Ron sold 80 hamburgers and hot dogs. He sold 20 more hot dogs than hamburgers. Which equation could be used to find how many hamburgers he sold? (Let x = the number of hamburgers sold.)

A. $x + 20 = 80$
B. $2x = 80$
(C.) $x + x + 20 = 80$
D. $x + x + x = 60$

31. John took some money from his bank account. He spent $5 on lunch and $\frac{1}{2}$ of the remainder at the grocery store. He was left with $45. How much money did he take from his account?

A. $90 **(B.)** $95 C. $100 D. $105

32. The sum of two numbers is 32. One number is 6 greater than the other. What are the two numbers?

A. 12 and 18 **(B.)** 13 and 19
C. 14 and 18 D. 14 and 20

Stop!

Name

Posttest

CHAPTER 14
PAGE 1

Choose the letter of the correct answer.

1. What are the first four triangular numbers?

A. 1, 4, 9, 16 **(B.)** 1, 3, 6, 10
C. 1, 3, 9, 27 D. 1, 5, 12, 22

2. How many dots would come next in this pattern?

A. 23 **(B.)** 25 C. 32 D. not here

3. What is the next number in the pattern? 10, 15, 21, 28

A. 35 **(B.)** 36 C. 37 D. not here

4. What is the next number in the pattern? 122, 100, 80, 62

A. 44 **(B.)** 46 C. 52 D. 60

5. What are the first four numbers in this pattern? The first number is 99. Subtract 7 each time.

A. 92, 85, 78, 71
(B.) 99, 92, 85, 78
C. 99, 97, 90, 83
D. 99, 90, 83, 76

6. What are the first four numbers in this pattern? The first number is 0. Multiply by 5 and add 2 each time.

A. 0, 2, 10, 12 **(B.)** 0, 2, 12, 62
C. 0, 5, 15, 35 D. 0, 7, 37, 187

To answer questions 7–8, use the relation {(1,2), (2,4), (3,6), (4,8)}.

7. What is the domain of the relation?

A. 1, 2 B. 3, 4
(C.) 1, 2, 3, 4 D. 2, 4, 6, 8

8. What is the range of the relation?

A. 2, 4
B. 6, 8
C. 1, 2, 3, 4
(D.) 2, 4, 6, 8

9. Which ordered pairs are graphed?

(A.) (0,2), (2,4)
B. (2,0), (2,4)
C. (2,0), (4,2)
D. (0,2), (4,2)

10. If the length of a rectangle triples and the width doubles, what happens to the area of the rectangle?

A. It doubles.
B. It increases by 5 times.
(C.) It increases by 6 times.
D. It increases by 10 times.

11. Which equation represents the relation {(0,0), (1,5), (2,10), (3,15)}?

A. $y = x + 12$
B. $y = 4x + 3$
C. $y = x + 4$
(D.) $y = 5x$

Go on to the next page.

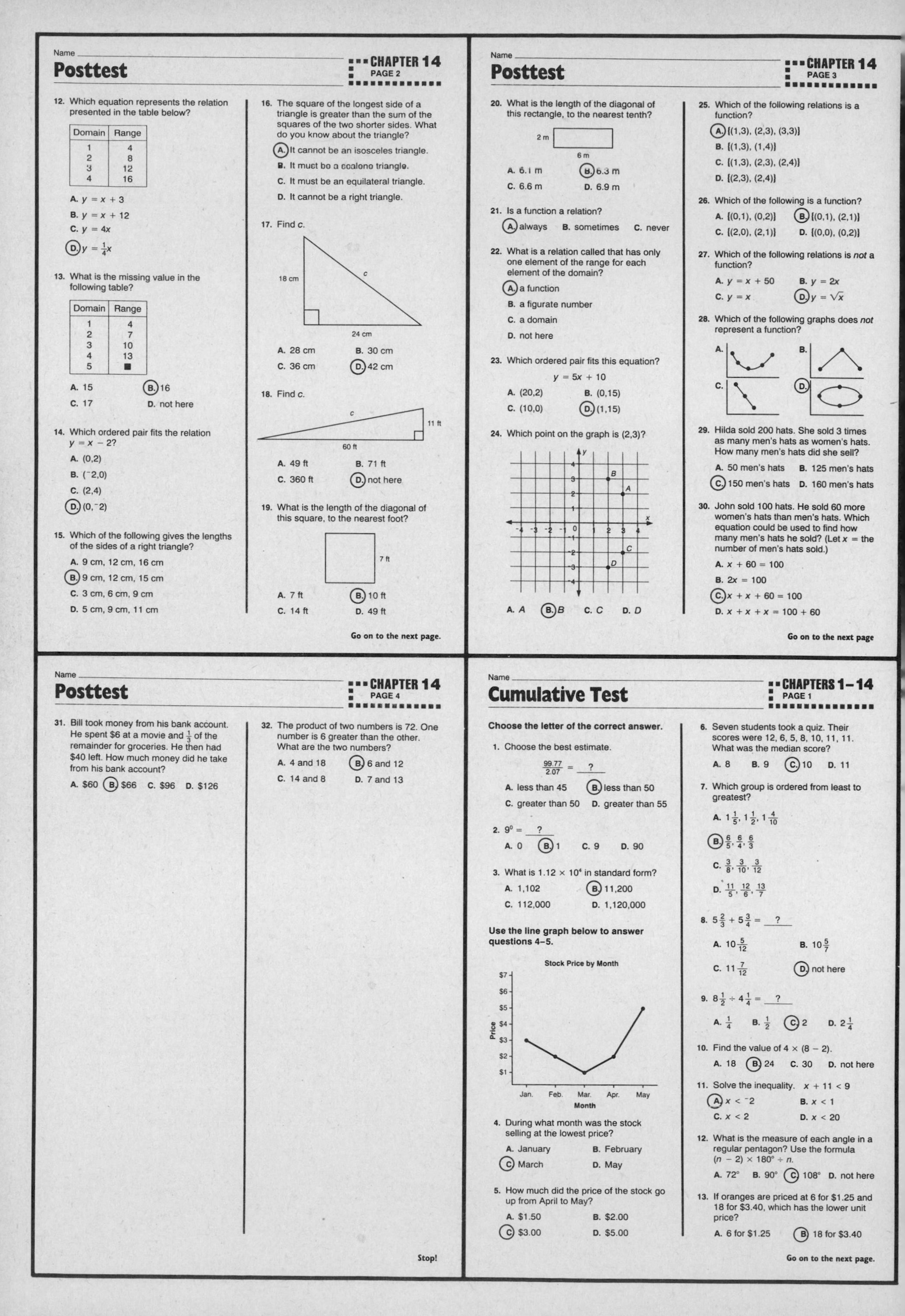

Name

Posttest

CHAPTER 14 PAGE 2

12. Which equation represents the relation presented in the table below?

Domain	Range
1	4
2	8
3	12
4	16

A. $y = x + 3$
B. $y = x + 12$
C. $y = 4x$
(D.) $y = \frac{1}{4}x$

13. What is the missing value in the following table?

Domain	Range
1	4
2	7
3	10
4	13
5	■

A. 15 (B.) 16
C. 17 D. not here

14. Which ordered pair fits the relation $y = x - 2$?
A. (0,2)
B. (⁻2,0)
C. (2,4)
(D.) (0,⁻2)

15. Which of the following gives the lengths of the sides of a right triangle?
A. 9 cm, 12 cm, 16 cm
(B.) 9 cm, 12 cm, 15 cm
C. 3 cm, 6 cm, 9 cm
D. 5 cm, 9 cm, 11 cm

16. The square of the longest side of a triangle is greater than the sum of the squares of the two shorter sides. What do you know about the triangle?
(A.) It cannot be an isosceles triangle.
B. It must be a scalene triangle.
C. It must be an equilateral triangle.
D. It cannot be a right triangle.

17. Find *c*.

18 cm, *c*, 24 cm

A. 28 cm B. 30 cm
C. 36 cm (D.) 42 cm

18. Find *c*.

c, 11 ft, 60 ft

A. 49 ft B. 71 ft
C. 360 ft (D.) not here

19. What is the length of the diagonal of this square, to the nearest foot?

7 ft

A. 7 ft (B.) 10 ft
C. 14 ft D. 49 ft

Go on to the next page.

Name

Posttest

CHAPTER 14 PAGE 3

20. What is the length of the diagonal of this rectangle, to the nearest tenth?

2 m, 6 m

A. 6.1 m (B.) 6.3 m
C. 6.6 m D. 6.9 m

21. Is a function a relation?
(A.) always B. sometimes C. never

22. What is a relation called that has only one element of the range for each element of the domain?
(A.) a function
B. a figurate number
C. a domain
D. not here

23. Which ordered pair fits this equation?

$y = 5x + 10$

A. (20,2) B. (0,15)
C. (10,0) (D.) (1,15)

24. Which point on the graph is (2,3)?

A. *A* (B.) *B* C. *C* D. *D*

25. Which of the following relations is a function?
(A.) {(1,3), (2,3), (3,3)}
B. {(1,3), (1,4)}
C. {(1,3), (2,3), (2,4)}
D. {(2,3), (2,4)}

26. Which of the following is a function?
A. {(0,1), (0,2)} (B.) {(0,1), (2,1)}
C. {(2,0), (2,1)} D. {(0,0), (0,2)}

27. Which of the following relations is *not* a function?
A. $y = x + 50$ B. $y = 2x$
C. $y = x$ (D.) $y = \sqrt{x}$

28. Which of the following graphs does *not* represent a function?
A. B.
C. (D.)

29. Hilda sold 200 hats. She sold 3 times as many men's hats as women's hats. How many men's hats did she sell?
A. 50 men's hats B. 125 men's hats
(C.) 150 men's hats D. 160 men's hats

30. John sold 100 hats. He sold 60 more women's hats than men's hats. Which equation could be used to find how many men's hats he sold? (Let x = the number of men's hats sold.)
A. $x + 60 = 100$
B. $2x = 100$
(C.) $x + x + 60 = 100$
D. $x + x + x = 100 + 60$

Go on to the next page

Name

Posttest

CHAPTER 14 PAGE 4

31. Bill took money from his bank account. He spent \$6 at a movie and $\frac{1}{3}$ of the remainder for groceries. He then had \$40 left. How much money did he take from his bank account?
A. \$60 (B.) \$66 C. \$96 D. \$126

32. The product of two numbers is 72. One number is 6 greater than the other. What are the two numbers?
A. 4 and 18 (B.) 6 and 12
C. 14 and 8 D. 7 and 13

Stop!

Name

Cumulative Test

CHAPTERS 1–14 PAGE 1

Choose the letter of the correct answer.

1. Choose the best estimate.

$\frac{99.77}{2.07} = $?

A. less than 45 (B.) less than 50
C. greater than 50 D. greater than 55

2. $9^0 = $?
A. 0 (B.) 1 C. 9 D. 90

3. What is 1.12×10^4 in standard form?
A. 1,102 (B.) 11,200
C. 112,000 D. 1,120,000

Use the line graph below to answer questions 4–5.

Stock Price by Month

Price: \$1, \$2, \$3, \$4, \$5, \$6, \$7
Month: Jan., Feb., Mar., Apr., May

4. During what month was the stock selling at the lowest price?
A. January B. February
(C) March D. May

5. How much did the price of the stock go up from April to May?
A. \$1.50 B. \$2.00
(C) \$3.00 D. \$5.00

6. Seven students took a quiz. Their scores were 12, 6, 5, 8, 10, 11, 11. What was the median score?
A. 8 B. 9 (C.) 10 D. 11

7. Which group is ordered from least to greatest?
A. $1\frac{1}{5}, 1\frac{1}{2}, 1\frac{4}{10}$
(B.) $\frac{6}{5}, \frac{6}{4}, \frac{6}{3}$
C. $\frac{3}{8}, \frac{3}{10}, \frac{3}{12}$
D. $\frac{11}{5}, \frac{12}{6}, \frac{13}{7}$

8. $5\frac{2}{3} + 5\frac{3}{4} = $?
A. $10\frac{5}{12}$ B. $10\frac{5}{7}$
C. $11\frac{7}{12}$ (D.) not here

9. $8\frac{1}{2} \div 4\frac{1}{4} = $?
A. $\frac{1}{4}$ B. $\frac{1}{2}$ (C.) 2 D. $2\frac{1}{4}$

10. Find the value of $4 \times (8 - 2)$.
A. 18 (B.) 24 C. 30 D. not here

11. Solve the inequality. $x + 11 < 9$
(A.) $x < ^-2$ B. $x < 1$
C. $x < 2$ D. $x < 20$

12. What is the measure of each angle in a regular pentagon? Use the formula $(n - 2) \times 180° \div n$.
A. 72° B. 90° (C.) 108° D. not here

13. If oranges are priced at 6 for \$1.25 and 18 for \$3.40, which has the lower unit price?
A. 6 for \$1.25 (B.) 18 for \$3.40

Go on to the next page.

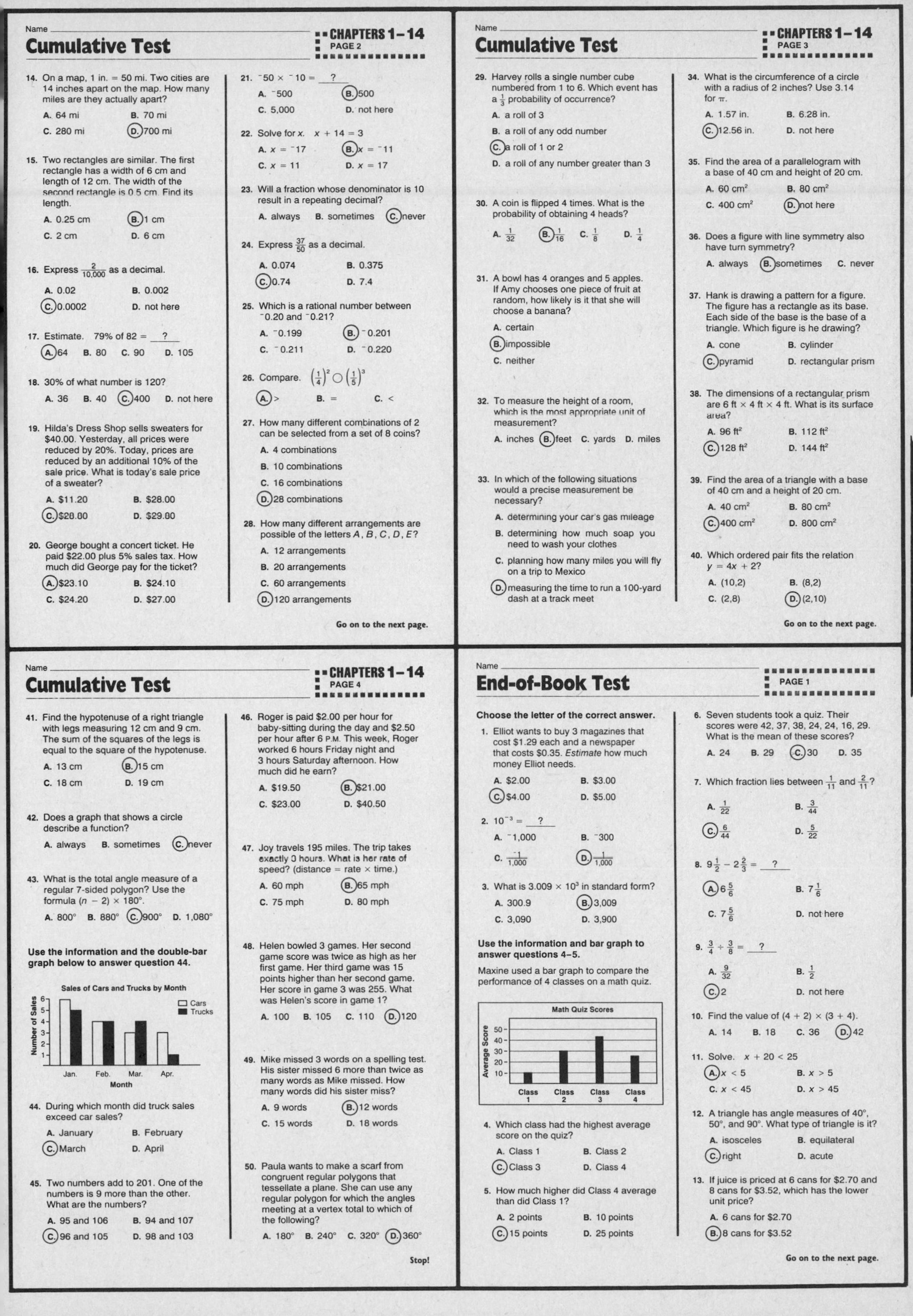

Name

Cumulative Test

CHAPTERS 1–14
PAGE 2

14. On a map, 1 in. = 50 mi. Two cities are 14 inches apart on the map. How many miles are they actually apart?
A. 64 mi B. 70 mi C. 280 mi D. 700 mi

15. Two rectangles are similar. The first rectangle has a width of 6 cm and length of 12 cm. The width of the second rectangle is 0.5 cm. Find its length.
A. 0.25 cm B. 1 cm C. 2 cm D. 6 cm

16. Express $\frac{2}{10,000}$ as a decimal.
A. 0.02 B. 0.002 C. 0.0002 D. not here

17. Estimate. 79% of 82 = ___?___
A. 64 B. 80 C. 90 D. 105

18. 30% of what number is 120?
A. 36 B. 40 C. 400 D. not here

19. Hilda's Dress Shop sells sweaters for $40.00. Yesterday, all prices were reduced by 20%. Today, prices are reduced by an additional 10% of the sale price. What is today's sale price of a sweater?
A. $11.20 B. $28.00 C. $28.80 D. $29.00

20. George bought a concert ticket. He paid $22.00 plus 5% sales tax. How much did George pay for the ticket?
A. $23.10 B. $24.10 C. $24.20 D. $27.00

21. $^-50 \times {}^-10$ = ___?___
A. $^-500$ B. 500 C. 5,000 D. not here

22. Solve for x. $x + 14 = 3$
A. $x = {}^-17$ B. $x = {}^-11$ C. $x = 11$ D. $x = 17$

23. Will a fraction whose denominator is 10 result in a repeating decimal?
A. always B. sometimes C. never

24. Express $\frac{37}{50}$ as a decimal.
A. 0.074 B. 0.375 C. 0.74 D. 7.4

25. Which is a rational number between $^-0.20$ and $^-0.21$?
A. $^-0.199$ B. $^-0.201$ C. $^-0.211$ D. $^-0.220$

26. Compare. $\left(\frac{1}{4}\right)^2 \bigcirc \left(\frac{1}{5}\right)^3$
A. > B. = C. <

27. How many different combinations of 2 can be selected from a set of 8 coins?
A. 4 combinations
B. 10 combinations
C. 16 combinations
D. 28 combinations

28. How many different arrangements are possible of the letters *A, B, C, D, E*?
A. 12 arrangements
B. 20 arrangements
C. 60 arrangements
D. 120 arrangements

Go on to the next page.

Name

Cumulative Test

CHAPTERS 1–14
PAGE 3

29. Harvey rolls a single number cube numbered from 1 to 6. Which event has a $\frac{1}{3}$ probability of occurrence?
A. a roll of 3
B. a roll of any odd number
C. a roll of 1 or 2
D. a roll of any number greater than 3

30. A coin is flipped 4 times. What is the probability of obtaining 4 heads?
A. $\frac{1}{32}$ B. $\frac{1}{16}$ C. $\frac{1}{8}$ D. $\frac{1}{4}$

31. A bowl has 4 oranges and 5 apples. If Amy chooses one piece of fruit at random, how likely is it that she will choose a banana?
A. certain
B. impossible
C. neither

32. To measure the height of a room, which is the most appropriate unit of measurement?
A. inches B. feet C. yards D. miles

33. In which of the following situations would a precise measurement be necessary?
A. determining your car's gas mileage
B. determining how much soap you need to wash your clothes
C. planning how many miles you will fly on a trip to Mexico
D. measuring the time to run a 100-yard dash at a track meet

34. What is the circumference of a circle with a radius of 2 inches? Use 3.14 for π.
A. 1.57 in. B. 6.28 in. C. 12.56 in. D. not here

35. Find the area of a parallelogram with a base of 40 cm and height of 20 cm.
A. 60 cm² B. 80 cm² C. 400 cm² D. not here

36. Does a figure with line symmetry also have turn symmetry?
A. always B. sometimes C. never

37. Hank is drawing a pattern for a figure. The figure has a rectangle as its base. Each side of the base is the base of a triangle. Which figure is he drawing?
A. cone B. cylinder C. pyramid D. rectangular prism

38. The dimensions of a rectangular prism are 6 ft × 4 ft × 4 ft. What is its surface area?
A. 96 ft² B. 112 ft² C. 128 ft² D. 144 ft²

39. Find the area of a triangle with a base of 40 cm and a height of 20 cm.
A. 40 cm² B. 80 cm² C. 400 cm² D. 800 cm²

40. Which ordered pair fits the relation $y = 4x + 2$?
A. (10,2) B. (8,2) C. (2,8) D. (2,10)

Go on to the next page.

A. TEST ANSWERS
STANDARDIZED FORMAT

Name

Cumulative Test

CHAPTERS 1–14
PAGE 4

41. Find the hypotenuse of a right triangle with legs measuring 12 cm and 9 cm. The sum of the squares of the legs is equal to the square of the hypotenuse.
A. 13 cm B. 15 cm C. 18 cm D. 19 cm

42. Does a graph that shows a circle describe a function?
A. always B. sometimes C. never

43. What is the total angle measure of a regular 7-sided polygon? Use the formula $(n - 2) \times 180°$.
A. 800° B. 880° C. 900° D. 1,080°

Use the information and the double-bar graph below to answer question 44.

Sales of Cars and Trucks by Month
Cars
Trucks
Number of Sales
Jan. Feb. Mar. Apr.
Month

44. During which month did truck sales exceed car sales?
A. January B. February C. March D. April

45. Two numbers add to 201. One of the numbers is 9 more than the other. What are the numbers?
A. 95 and 106 B. 94 and 107 C. 96 and 105 D. 98 and 103

46. Roger is paid $2.00 per hour for baby-sitting during the day and $2.50 per hour after 6 P.M. This week, Roger worked 6 hours Friday night and 3 hours Saturday afternoon. How much did he earn?
A. $19.50 B. $21.00 C. $23.00 D. $40.50

47. Joy travels 195 miles. The trip takes exactly 3 hours. What is her rate of speed? (distance = rate × time.)
A. 60 mph B. 65 mph C. 75 mph D. 80 mph

48. Helen bowled 3 games. Her second game score was twice as high as her first game. Her third game was 15 points higher than her second game. Her score in game 3 was 255. What was Helen's score in game 1?
A. 100 B. 105 C. 110 D. 120

49. Mike missed 3 words on a spelling test. His sister missed 6 more than twice as many words as Mike missed. How many words did his sister miss?
A. 9 words B. 12 words C. 15 words D. 18 words

50. Paula wants to make a scarf from congruent regular polygons that tessellate a plane. She can use any regular polygon for which the angles meeting at a vertex total to which of the following?
A. 180° B. 240° C. 320° D. 360°

Stop!

Name

End-of-Book Test

PAGE 1

Choose the letter of the correct answer.

1. Elliot wants to buy 3 magazines that cost $1.29 each and a newspaper that costs $0.35. *Estimate* how much money Elliot needs.
A. $2.00 B. $3.00 C. $4.00 D. $5.00

2. 10^{-3} = ___?___
A. $^-1,000$ B. $^-300$ C. $\frac{^-1}{1,000}$ D. $\frac{1}{1,000}$

3. What is 3.009×10^3 in standard form?
A. 300.9 B. 3,009 C. 3,090 D. 3,900

Use the information and bar graph to answer questions 4–5.

Maxine used a bar graph to compare the performance of 4 classes on a math quiz.

Math Quiz Scores
Average Score
Class 1 Class 2 Class 3 Class 4

4. Which class had the highest average score on the quiz?
A. Class 1 B. Class 2 C. Class 3 D. Class 4

5. How much higher did Class 4 average than did Class 1?
A. 2 points B. 10 points C. 15 points D. 25 points

6. Seven students took a quiz. Their scores were 42, 37, 38, 24, 24, 16, 29. What is the mean of these scores?
A. 24 B. 29 C. 30 D. 35

7. Which fraction lies between $\frac{1}{11}$ and $\frac{2}{11}$?
A. $\frac{1}{22}$ B. $\frac{3}{44}$ C. $\frac{6}{44}$ D. $\frac{5}{22}$

8. $9\frac{1}{2} - 2\frac{2}{3}$ = ___?___
A. $6\frac{5}{6}$ B. $7\frac{1}{6}$ C. $7\frac{5}{6}$ D. not here

9. $\frac{3}{4} \div \frac{3}{8}$ = ___?___
A. $\frac{9}{32}$ B. $\frac{1}{2}$ C. 2 D. not here

10. Find the value of $(4 + 2) \times (3 + 4)$.
A. 14 B. 18 C. 36 D. 42

11. Solve. $x + 20 < 25$
A. $x < 5$ B. $x > 5$ C. $x < 45$ D. $x > 45$

12. A triangle has angle measures of 40°, 50°, and 90°. What type of triangle is it?
A. isosceles B. equilateral C. right D. acute

13. If juice is priced at 6 cans for $2.70 and 8 cans for $3.52, which has the lower unit price?
A. 6 cans for $2.70
B. 8 cans for $3.52

Go on to the next page.

Name ______

End-of-Book Test

PAGE 2

14. On Map A, 1 cm = 50 km. On Map B, 1 cm = 75 km. Two cities are 10 cm apart on Map B. How far apart are they on Map A?

A. 15 cm B. 25 cm
C. 250 cm D. 375 cm

15. A triangle has sides measuring 6 cm, 8 cm, and 10 cm. Which of the following could be the lengths of the sides of a similar triangle?

A. 7 cm, 9 cm, 11 cm
B. 16 km, 18 km, 20 km
C. 9 yd, 12 yd, 20 yd
D. 9 m, 12 m, 15 m

16. Express 0.3% as a decimal.

A. 0.0003 **B.** 0.003
C. 0.03 D. not here

17. 27 is what percent of 108?

A. 19% **B.** 25%
C. 27% D. 400%

18. 20% of what number is 200?

A. 40 B. 400 **C.** 1,000 D. 10,000

19. Nikki bought a shirt for $22.00 plus 7% sales tax. How much did Nikki pay for the shirt?

A. $22.70 B. $23.44
C. $23.54 D. $24.54

20. Andy is buying a suit that is marked down 25%. What is the sale price of a $240 suit?

A. $175 **B.** $180 C. $215 D. $300

21. $^-25 \times 20 =$ ___?___

A. $^-250$ B. $^-50$
C. 500 **D.** not here

22. Solve for x. $x + 20 = 18$

A. $x = {}^-2$ B. $x = {}^-1$
C. $x = 2$ D. $x = 38$

23. Are mixed numbers rational numbers?

A. always B. sometimes C. never

24. Which of these fractions can be expressed as a terminating decimal?

A. $\frac{2}{18}$ B. $\frac{1}{7}$ **C.** $\frac{1}{5}$ D. $\frac{1}{3}$

25. Which is a rational number between 0.61 and 0.62?

A. 0.601 B. 0.609
C. 0.6110 D. 0.6201

26. Compare. $10^{-5} \bigcirc 10^{3}$

A. > B. = **C.** <

27. How many different combinations of 2 can be selected from a set of 20 items?

A. 90 combinations
B. 190 combinations
C. 380 combinations
D. not here

28. In how many different orders can 5 people arrange themselves in a line?

A. 24 orders B. 60 orders
C. 120 orders D. 360 orders

29. Harvey rolls a single number cube numbered from 1 to 6. What is the probability that he will roll an even number?

A. $\frac{1}{6}$ B. $\frac{1}{3}$ **C.** $\frac{1}{2}$ D. $\frac{3}{5}$

Go on to the next page.

Name ______

End-of-Book Test

PAGE 3

30. A coin is flipped 3 times. What is the probability of obtaining 3 heads?

A. $\frac{1}{16}$ B. $\frac{1}{9}$ **C.** $\frac{1}{8}$ D. $\frac{1}{6}$

31. Jon and Ruth play 2 chess matches. Ruth is a better player than Jon. Her probability of winning each game is $\frac{3}{4}$. What is the probability that Ruth will win both games?

A. $\frac{3}{8}$ **B.** $\frac{9}{16}$ C. $\frac{2}{3}$ D. $\frac{3}{4}$

32. Which unit of measurement is the most appropriate for measuring the length of a book?

A. millimeter **B.** centimeter
C. meter D. kilometer

33. Which might be the weight of a peach?

A. 4 oz B. 4 lb
C. 4 g D. 4 kg

34. What is the perimeter of a regular pentagon with a side measuring 3.5 cm?

A. 14 cm B. 15.5 cm
C. 17.5 cm D. not here

35. Find the area of a triangle with a base of 20 in. and height of 35 in.

A. 55 in.^{2} B. 110 in.^{2}
C. 700 in.^{2} **D.** not here

36. Which of these letters has line symmetry?

A. the letter "J" B. the letter "R"
C. the letter "L" **D.** the letter "H"

37. Which of these figures is a polyhedron?

A. sphere B. cone
C. cylinder **D.** prism

38. The dimensions of a rectangular prism are 3 ft × 1 ft × 1 ft. What is its surface area?

A. 3 ft^2 B. 10 ft^2
C. 14 ft^2 D. 25 ft^2

39. Find the volume of a rectangular prism with the dimensions 2 in. × 4 in. × 8 in.

A. 14 in.^{3} **B.** 64 in.^{3}
C. 112 in.^{3} D. not here

40. Which ordered pair fits the relation $y = 3x + 2$?

A. (5,0) B. (0,5)
C. (2,0) **D.** (0,2)

41. Find the length of the hypotenuse of this right triangle. The sum of the squares of the legs is equal to the square of the hypotenuse.

21 in.
?
28 in.

A. 30 in. **B.** 35 in.
C. 42 in. D. 49 in.

42. A relation that has only one element of the range for each element of the domain is called

A. a domain. **B.** a function.
C. an equation. D. a figurate number.

43. The scale of a road map is 1 inch = 100 miles. On the map, the highway distance from Budd Lake to Mount Tom is $\frac{3}{4}$ in. What is the distance in miles between the two points?

A. 3 miles B. 25 miles
C. 75 miles D. 125 miles

Go on to the next page.

Name ______

End-of-Book Test

PAGE 4

Use the information and the double-bar graph to answer question 44.

Joan used a double-bar graph to compare the monthly sales of hatchbacks and sedans over a period of 4 months.

Sales in Millions
2
1.5
1
0.5
Jan. Feb. March Apr.
Month
hatchbacks
sedans

44. During which month was the difference in sales of hatchbacks and sedans the greatest?

A. January **B.** February
C. March D. April

45. Two numbers add to 360. One of the numbers is 3 times the other. What is the smaller number?

A. 60 B. 80 **C.** 90 D. 120

46. Nancy walks 5 blocks to school. It takes her $2\frac{1}{2}$ minutes to walk each block. Pat walks 6 blocks to school. It takes her only $1\frac{1}{2}$ minutes to walk each block. How much longer does it take Nancy to walk to school than it takes Pat?

A. $2\frac{1}{4}$ min B. $2\frac{1}{2}$ min
C. $3\frac{1}{2}$ min D. $3\frac{3}{4}$ min

47. Art drove a distance of 275 miles in his car. His average speed was 50 miles per hour. How long did the drive take? (distance = rate × time)

A. $5\frac{1}{4}$ hr **B.** $5\frac{1}{2}$ hr
C. $5\frac{3}{4}$ hr D. $6\frac{1}{2}$ hr

48. Mimi spent $\frac{1}{2}$ of her savings on Monday. On Tuesday, after she spent $50.00, she had only $2.00 left. How much money did Mimi start with?

A. $98.00 B. $102.00
C. $104.00 D. $106.00

49. Helena and Mark went shopping. Mark spent $2.00 more than 3 times the amount Helena spent. Helena spent $5.50. How much did Mark spend?

A. $16.50 B. $17.50
C. $18.50 D. $19.50

Use the table to answer question 50.

Music Preferred by Students

Music	Frequency	Cumulative Frequency
Rock	50	50
Jazz	20	70
Country	20	90
Rap	10	100

50. How many more students chose rock music than jazz music?

A. 20 students **B.** 30 students
C. 50 students D. 100 students

Stop!

Name ___

Inventory Test

Read each question. Find the answer.

1. What is the value of the 7 in 147,523,968?

___

2. What is two ten-thousandths in standard form?

___

3. $18 - 4 \times 3 + 2 =$ ___

4. $\begin{array}{r} 2.471 \\ -0.69 \\ \hline \end{array}$

5. $\begin{array}{r} 3,219 \\ \times \quad 42 \\ \hline \end{array}$

6. $26.163 \div 5.7 =$ ___

7. There are 9 chairs in a row. How many rows are needed to seat 387 people?

___

8. Lisa wants to compare the numbers of cars sold by two sales representatives during a six-month period. Ring the kind of graph that she should use.

circle graph

line graph

double-bar graph

stem-and-leaf plot

Use the stem-and-leaf plot below to answer question 9.

Stem	Leaves
4	0 1 2 3 3 6
5	1 2 3 4 5 6 6 6
6	0 1 2 2 3 5

9. What is the mode?

___

10. What is the range of 1, 3, 5, 7, 9, 11, 13?

___

11. How many choices are possible for 9 pairs of socks and 5 pairs of shoes?

___

12. What is the prime factorization of 14?

___

13. What is the greatest common factor (GCF) of 8 and 16?

___

14. $\frac{1}{4} + \frac{1}{5} =$ ___

Go on to the next page.

Name ___

Inventory Test

15. $1\frac{1}{6} + 4\frac{3}{6} =$ ___

16. $\frac{3}{8} \times \frac{4}{5} =$ ___

17. Jack had $1\frac{1}{4}$ quarts of milk. He used $\frac{1}{3}$ of it. How much milk did he use?

___

18. Write <, >, or =. $\frac{2}{5} \div \frac{1}{7}$ ◯ 1

19. $\frac{3}{4} \div \frac{1}{6} =$ ___

20. What is $\frac{25}{100}$ written as a decimal?

___

21.

$$\begin{array}{r} 2 \text{ hr } 10 \text{ min} \\ -\ 1 \text{ hr } 55 \text{ min} \\ \hline \end{array}$$

22. Joel has to compute the total weight of 3 objects. The first one weighs 2.41 g, the second 1.94 g, and the third 0.83 g. What is the total weight of the objects?

___

23. Write 4 out of 5 as a ratio.

___

24. Ring the term that makes these ratios equivalent.

$\frac{1}{6}, \frac{6}{■}$

1   11   12   36

25. Ring the cross products.

$\frac{4}{5} = \frac{y}{35}$

$4 \times y = 5 \times 35$

$5 \times y = 4 \times 35$

$4 \times 5 = y \times 35$

26. A scale model of a new theater has a parking lot 15 in. long and a lawn 10 in. long. If the actual parking lot is 240 ft long, how long is the lawn?

___

27. If 40% of the fruit in a jar is apples, what fraction of the fruit is apples?

___

Go on to the next page.

Name ______________________________

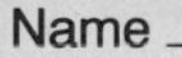

Inventory Test

28. Write 8% as a fraction in simplest form.

____

29. 30% of 90 = __________

30. A store advertises sneakers for 20% off the regular price of $29.95. About how much is the sale price?

____

31. What is an angle called that measures less than 90°?

____

32. Angles *A* and *B* are complementary. If Angle *A* measures 10°, what is the measure of Angle *B*?

____

33. Find the perimeter of this triangle.

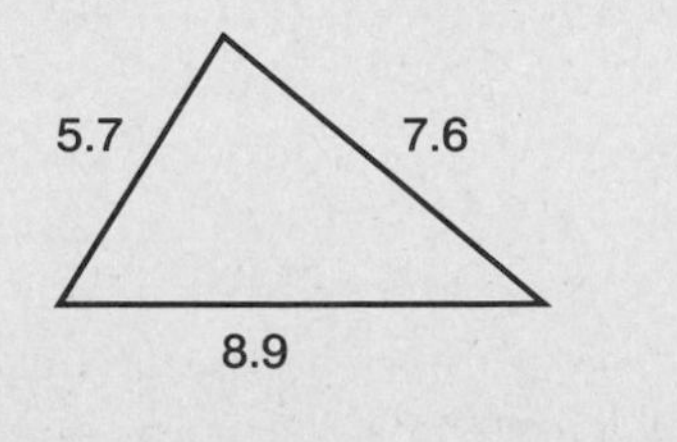

____

34. Ring the formula that will find the area of this figure.

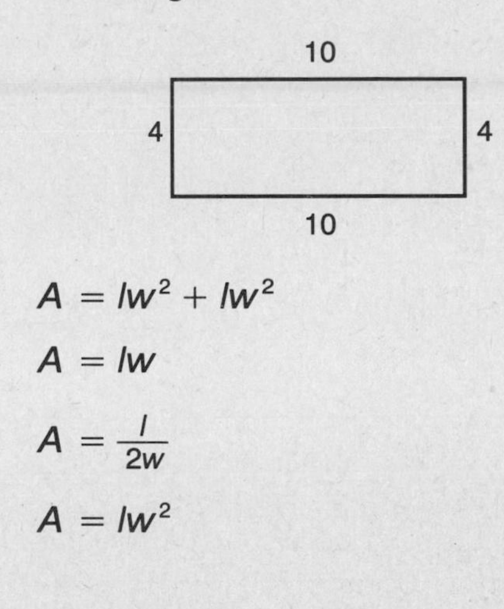

$A = lw^2 + lw^2$

$A = lw$

$A = \frac{l}{2w}$

$A = lw^2$

35. Find the area of a triangle with a base of 12 ft and a height of 5 ft.

____

36. Find the circumference of a circle with a diameter of 4 in. Use $\pi = 3.14$.

____

37. Ring the number that has the greatest value.

$^-12$ 0   3 $^-180$

38. Write <, >, or =.   8 ◯ $^-8$

39. $12 - {}^-15 =$ __________

40. Ring the expression that is the same as 12 more than *y*.

$y - 12$ $y + 12$

$12y$ $\frac{12}{y}$

41. Evaluate $7z$ for $z = 0.5$.

____

Go on to the next page.

Name ______________________________

Inventory Test

42. Ring the number line that shows the solution for $p > {}^{-}2$.

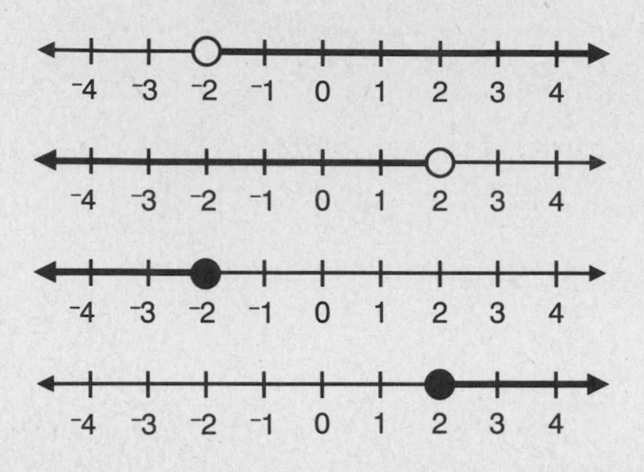

43. Peter spent $\frac{1}{2}$ of his money to buy a tennis racket and $\frac{1}{4}$ for shorts. He had $10 left to buy tennis balls. How much money did he have to start with?

____

44. Mike saved $10.50 in January, $8.75 in February, and $12.60 in March. On April 1, he had $50.00 in his bank. How much was in his bank before January?

____

45. Lee is twice as old as Marty. Bridget is twice as old as Lee, and Heidi is twice as old as Bridget. If Marty is 3, how old is Heidi?

____

Use the table below to answer questions 46–47.

Dry Cleaning Rates		
Day	**First Pound**	**Each Additional Pound**
Mon–Fri	$2.50	$1.95
Sat–Sun	2.75	2.25
Holiday	3.00	2.50

46. What is the cost of dry cleaning 5 pounds on a Wednesday?

____

47. What is the cost of dry cleaning 3 pounds on a holiday?

____

48. The difference between two numbers is 15. Their sum is 125. What is the larger number?

____

49. A contractor wants to put a 3-foot-wide door in the middle of a 14-foot-wide wall. If the door is centered, how far will it be from each end of the wall?

____

50. Jed is making a bird feeder. He bought 3 boards for $2.19 each, 35 nails for $0.03 each, and 6 sheets of sandpaper for $0.21 each. What was the total cost of the supplies?

____

Stop!

Name ________________________________

Pretest

Read each question. Find the answer.

1. Ring the *best* estimate.

   3,198 + 2,025

   less than 4,000

   less than 5,000

   greater than 5,000

2. Ring the *best* estimate. 982 − 386

   less than 600

   greater than 600

   greater than 800

3. Ring the *best* estimate. 3.1 × 15.02

   less than 40

   greater than 45

   greater than 50

4. Ring the *best* estimate.

   49.97 ÷ 2.11

   less than 20

   less than 25

   greater than 25

5. Ring the *best* estimate.

   355 + 1,650 + 422

   2,100 2,200

   2,500 2,600

6. Ring the *best* estimate.

   1,153.2 − 166.5

   700 800

   1,000 1,100

7. Ring the *best* estimate. 59.95 × 1,000

   6,000 58,000

   60,000 599,000

8. Ring the *best* estimate. 2,005 ÷ 9.57

   200 950

   2,000 18,000

9. Jane wants to buy 3 pens that cost $1.19 each and 2 pads that cost $1.15 each. Ring the *best* estimate for the amount of money that Jane needs.

   $3.00 $4.00

   $4.50 $6.00

10. Bill had $18.99 to spend. He spent $3.09 on Friday, $2.10 on Saturday, and $2.95 on Sunday. Ring the *best* estimate for the amount of money he had left.

    $5.00 $8.00

    $11.00 $14.00

11. Ring the quotient with the greatest value.

    99.2 ÷ 3 99.2 ÷ 5

    99.2 ÷ 4 99.2 ÷ 10

12. Ring the difference with the greatest value.

    142.5 − 12 14.25 − 12

    1,425.0 − 12 1.1425 − 12

13. Ring the product with the greatest value.

    4 × 1,001 5 × 101

    7 × 10.1 9 × 1.01

Go on to the next page.

Pretest

14. Ring the quotient with the greatest value.

$50 \div 0.1$ $50 \div 0.01$

$50 \div 0.001$ $50 \div 0.0001$

15. Find the sum. $7.86 + 2.99 + 3.07$

____

16. Find the difference. $8.07 - 2.09$

____

17. Find the product. 9.90×11.20

____

18. Find the quotient. $16.4 \div 4.1$

____

19. Julio bought 5.5 pounds of cheese for $13.75. How much did he pay per pound of cheese?

____

20. Myra runs 2.5 miles every weekday, 3.0 miles each Saturday, and 3.5 miles each Sunday. How many miles does she run in a 2-week period?

____

21. Find the value of 5^3.

____

22. Find the value of 9^0.

____

23. What is the value of 4 to the third power?

____

24. What is the value of 10^1?

____

25. Express 605,000,000 in scientific notation.

____

26. Express 4.5×10^3 in standard form.

____

27. Express 8,203 in scientific notation.

____

28. Express 3.5×10^1 in standard form.

____

Go on to the next page.

Name

Pretest

Use this information to answer questions 29–30.

The menu at Joe's Restaurant lists three main dishes: hamburger, hot dog, and pizza. It lists four salads: tossed, macaroni, potato, and coleslaw.

29. In how many combinations could Natasha order one main dish and one salad?

____

30. If Natasha orders a hamburger and two salads, how many combinations are possible for her to order?

____

31. A total of 56 boys and girls are in the school chorus. There are 20 more girls than boys. How many boys are in the chorus?

____

32. The Andersons went to a concert. They spent $39 for tickets. The tickets cost $12 for each adult and $5 for each child. How many children went to the concert?

____

Stop!

Name ______________________________

Posttest

Read each question. Find the answer.

1. Ring the *best* estimate.

   $4,125 + 3,002$

   less than 6,000

   less than 7,000

   greater than 7,000

2. Ring the *best* estimate.

   $655 - 462$

   less than 100

   less than 200

   less than 300

3. Ring the *best* estimate.

   4.1×25.05

   less than 90

   less than 100

   greater than 100

4. Ring the *best* estimate.

   $99.98 \div 2.05$

   less than 33

   less than 50

   greater than 60

5. Ring the *best* estimate.

   $544 + 2,499 + 351$

   2,800   3,100

   3,400   3,500

6. Ring the *best* estimate.

   $2,742.1 - 653.5$

   1,900   1,950

   2,100   2,300

7. Ring the *best* estimate.

   $69.97 \times 10,000$

   70,000   650,000

   700,000   750,000

8. Ring the *best* estimate.

   $5,005 \div 9.67$

   500   700

   2,500   45,000

9. Mario wants to buy 2 suits that cost \$149.95 each and 4 shirts that cost \$19.95 each. Ring the *best* estimate for the amount of money that Mario needs.

   \$320.00   \$340.00

   \$360.00   \$380.00

10. Doris had \$25.92 to spend. She spent \$9.65 on dinner, \$5.00 on a movie, and \$2.95 to park her car. Ring the *best* estimate for the amount of money she had left.

    \$5.00   \$6.00

    \$8.00   \$11.00

11. Ring the quotient with the greatest value.

    $33.3 \div 6$ $33.3 \div 8$

    $33.3 \div 7$ $33.3 \div 10$

12. Ring the difference with the greatest value.

    $162 - 11.92$ $162 - 119.2$

    $162 - 1.192$ $162 - 0.1192$

13. Ring the product with the greatest value.

    $4 \times 2,001$ 6×201

    8×20.1 10×2.01

Go on to the next page.

Posttest

14. Ring the quotient with the greatest value.

800 ÷ 0.3 800 ÷ 0.03

800 ÷ 0.003 800 ÷ 0.0003

15. Find the sum. 3.92 + 6.14 + 11.90

____

16. Find the difference. 8.68 − 4.79

____

17. Find the product. 8.80 × 10.2

____

18. Solve. $5.1\overline{)25.5}$ __________

19. Juan bought 7.25 ounces of steak for $5.80. How much did he pay per ounce of steak?

____

20. Betty works 8 hours per day from Monday through Friday. She also works 2 hours per evening on Monday and Wednesday. How many hours does she work per week?

____

21. Find the value of 2^4.

____

22. Find the value of 7^0.

____

23. What is the value of 5 to the third power?

____

24. What is the value of 11^1?

____

25. Express 105,000 in scientific notation.

____

26. Express 5.6×10^4 in standard form.

____

27. Express 7,015 in scientific notation.

____

28. Express 0.25×10^1 in standard form.

____

Go on to the next page.

Name ______________________

Posttest

Use this information to answer questions 29–30.

Sonya decides to spend New Year's Day watching football games. There are 3 games on television in the early afternoon, 3 in the late afternoon, and 2 at night.

29. Sonya decides to watch one game in the early afternoon, one in the late afternoon, and one at night. In how many different ways can she choose games to watch?

____

30. If Sonya decides to watch one game in the early afternoon, one in the late afternoon, but none at night, in how many different ways can she choose games to watch?

____

31. A total of 108 parents attended the seventh-grade open house. There were 20 more mothers present than fathers. How many parents that attended were fathers?

____

32. The Masons went to a play. They spent $34 for tickets. The tickets cost $9 for adults and $4 for children. How many children went to the play?

____

Stop!

Name ______________________________

Pretest

Read each question. Find the answer.

Use the bar graph below to answer questions 1–2.

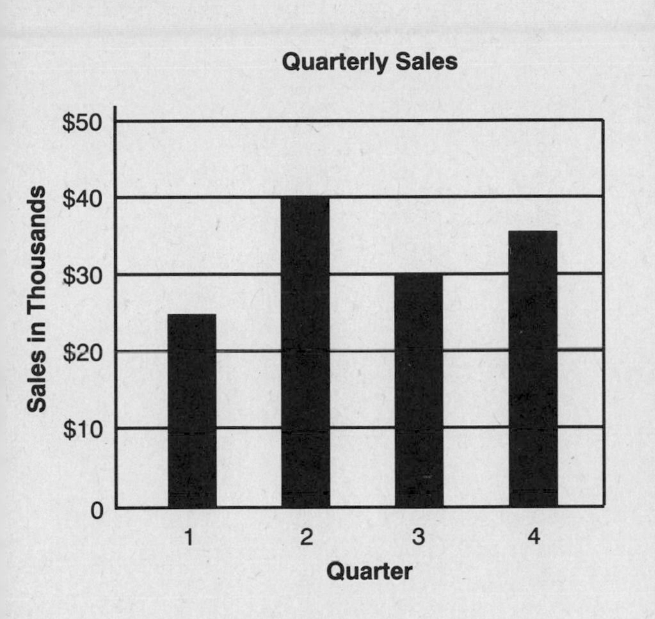

1. Which quarter had the greatest sales?

____

2. How much did sales drop between Quarters 2 and 3?

____

Use the circle graph to answer questions 3–4.

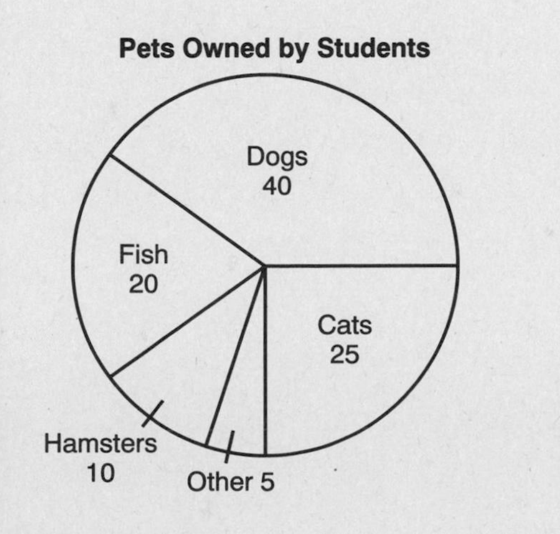

3. How many of the students own a cat or a hamster?

____

4. How many of the students own a dog, a cat, or fish?

____

Use the stem-and-leaf plot to answer questions 5–6.

Number of Points Scored by a Football Team in 16 Games

Stem	Leaves
1	4 4 7 7
2	0 1 4 4 7 7 8 8
3	1 5
4	1 2

5. What was the least number of points scored by the team?

____

6. In how many of its games did the team score 30 or more points?

____

Go on to the next page.

Name ______________________________

Pretest

Use the information and the double-line graph to answer questions 7–10.

Each December, a store predicts its sales (in millions) for the coming year. The double-line graph shows the predicted and actual sales for a five-year period.

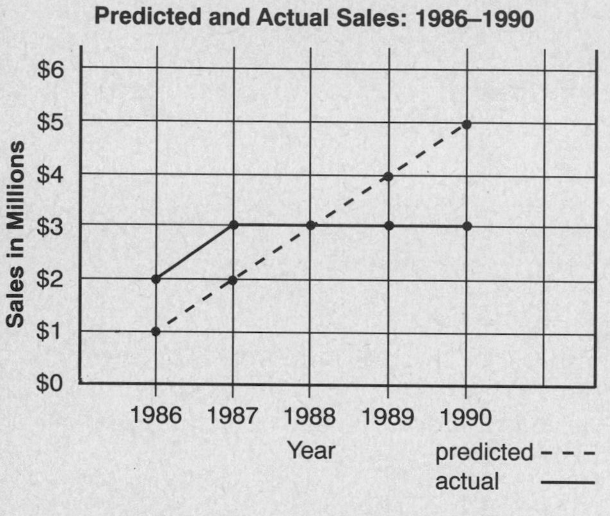

7. For which year did predicted sales equal actual sales?

____

8. For which year were actual sales *most* behind predicted sales?

____

9. How much did actual sales increase between 1986 and 1989?

____

10. For how many years were actual sales at $3,000,000?

____

Use the table below to answer questions 11–14.

Grades Earned by English Students		
Grade	Frequency	Cumulative Frequency
A	6	6
B	10	16
C	8	24
D	7	31
F	1	32

11. Which grade was received by the most students?

____

12. How many students received a C?

____

13. How many students received a grade of B or better?

____

14. How many students received a grade of D?

____

Use the information below to answer questions 15–18.

These are the scores of 10 students on a spelling test.

10, 13, 13, 13, 15, 15, 17, 18, 19, 20

15. What is the mode for the scores?

____

16. What is the range of the scores?

____

Go on to the next page.

Name ______________________

Pretest

17. What is the mean of the scores?

____

18. What is the median score?

____

Use the information and the box-and-whisker graph below to answer questions 19–20.

Liz recorded the number of customers in her store each day for 20 days. The box-and-whisker graph below summarizes the information.

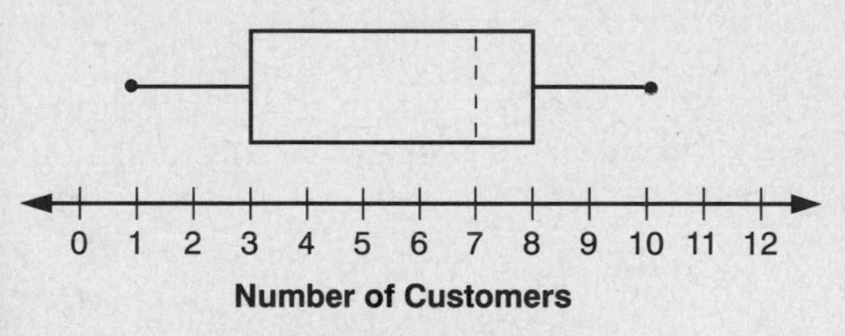

19. What was the median number of customers in the store?

____

20. What was the greatest number of customers in the store on any of the 20 days?

____

Use the table below to answer questions 21–23.

Senior Class Elections

Candidate	Girls' Votes	Boys' Votes
John	85	70
Thomas	40	85
Darlene	100	50
Vera	95	75
Total	320	280

21. Which candidate received the most votes?

____

22. How many of the boys' votes did John receive?

____

23. If half of the boys who voted for Darlene had voted for John, who would have received the most votes?

____

24. A manager wants to draw a graph comparing the monthly sales of male and female sales staff. Ring the type of graph that she should draw.

double-bar graph

circle graph

box-and-whisker graph

stem-and-leaf plot

Stop!

Name ______________________________

Read each question. Find the answer.

Use the bar graph below to answer questions 1–2.

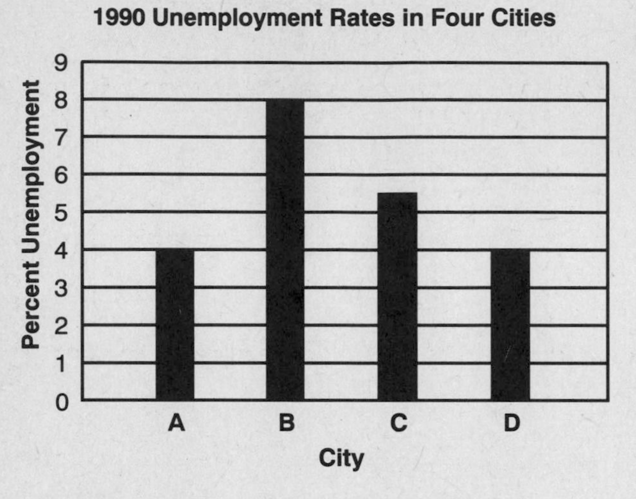

1. Which city had the highest rate of unemployment?

____

2. The percent of unemployment in City A was how much less than the percent of unemployment in City B?

____

Use the circle graph below to answer questions 3–4.

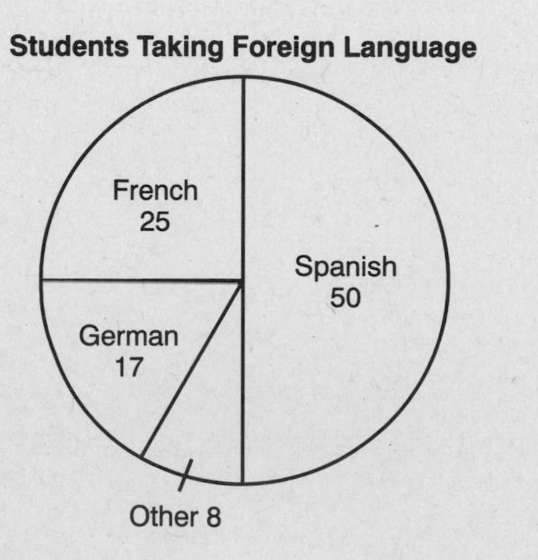

3. How many students are taking French?

____

4. How many students are taking Spanish or German?

____

Use the stem-and-leaf plot below to answer questions 5–6.

Number of Points Scored by a Basketball Team in 25 Games

Stem	Leaves
6	0 1 3 5 7
7	2 4 5 6 7 8 9 9
8	1 3 3 3 5 6 7
9	1 4 5 8 9

5. What was the greatest number of points scored by the team?

____

6. In how many of its games did the team score 95 or more points?

____

Go on to the next page.

Name ____________________

Posttest

Use the information and the double-line graph below to answer questions 7–10.

Each month, a real-estate agent predicts how many homes his office will sell that month. The double-line graph below shows the predicted and actual sales for a five-month period.

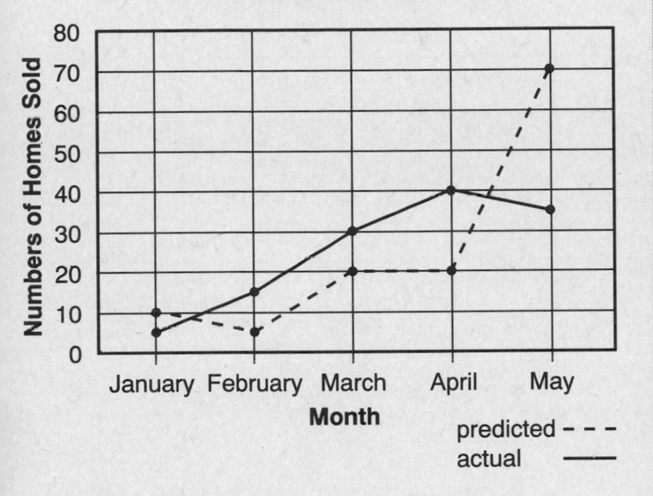

7. For which month were actual sales greatest?

____

8. For which month were actual sales *most* behind predicted sales?

____

9. By how many homes did actual sales increase between March and April?

____

10. During April, how many more homes were sold than predicted?

____

Use the table below to answer questions 11–14.

Weight Loss in Pounds During First Week in Diet Class

Weight Loss	Frequency	Cumulative Frequency
0	2	2
1	4	6
2	6	12
3	8	20
4	12	32
5	5	37
6	3	40

11. Which weight loss was most frequent?

____

12. How many of the dieters lost 2 pounds?

____

13. How many of the dieters lost 4 or fewer pounds?

____

14. How many dieters lost no weight?

____

Go on to the next page.

Name ______________________

Posttest

Use the information below to answer questions 15–18.

These are the scores of 9 students on a math test.

15, 16, 16, 19, 21, 23, 24, 25, 30

15. What is the mode for the scores?

____

16. What is the range of the scores?

____

17. What is the mean of the scores?

____

18. What is the median score?

____

Use the information and the box-and-whisker graph below to answer questions 19–20.

Helen recorded the number of suits she sold each day for 30 days. The box-and-whisker graph below summarizes the sales.

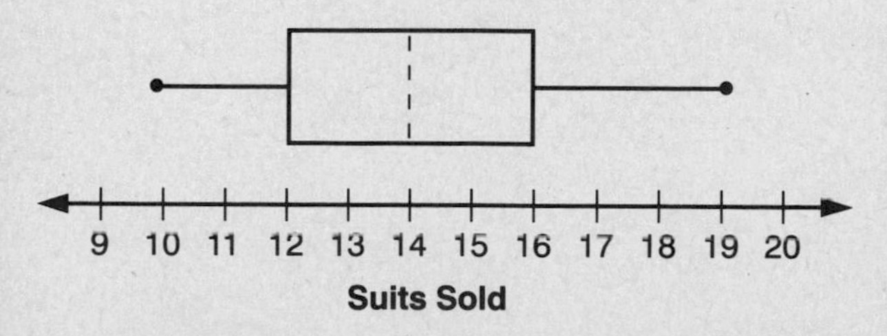

19. What was the median number of suits sold?

____

20. What was the greatest number of suits sold on any of the 30 days?

____

Use the table below to answer questions 21–23.

Music Preference Among Taft High School Students

Type of Music	Boys	Girls
Classical	30	90
Country	20	20
Jazz	40	60
Rock	150	100
Total	240	270

21. Which music received more votes from boys than from girls?

____

22. How many of the girls preferred classical music?

____

23. What music was *least* preferred by both groups?

____

24. A writer wants to draw a graph showing the favorite sports of teenage girls. Ring the type of graph that she should draw.

bar graph

line graph

box-and-whisker graph

stem-and-leaf plot

Stop!

Name ______________________________

Pretest

Read each question. Find the answer.

1. Ring the prime number.

   5   10   15   35

2. Ring the composite number.

   11   17   19   21

3. Ring the prime number.

   26   29   35   51

4. How many factors does 37 have?

   ____

5. What are the prime factors of 30?

   ____

6. Ring the prime factors of 52.

   2, 2, 13   2, 26

   4, 13   50, 2

7. Ring the prime factorization of 24.

   $2^3 \times 3$ $2^2 \times 6$ $4 \times 2 \times 3$

8. Ring the prime factorization of 1,000.

   $2^3 \times 25^2$ $2^3 \times 5^3$

   $2^2 \times 250$ 2×500

9. What is the greatest common factor of 24 and 36?

   ____

10. What is the greatest common factor of 32 and 72?

    ____

11. What is the least common multiple of 5 and 4?

    ____

12. What is the least common multiple of 6 and 9?

    ____

13. Ring the fraction that is expressed in simplest form.

    $\frac{3}{15}$ $\frac{5}{15}$

    $\frac{55}{100}$ $\frac{99}{100}$

14. Ring the fraction that is *not* expressed in simplest form.

    $\frac{1}{2}$ $\frac{2}{3}$ $\frac{4}{8}$ $\frac{11}{12}$

15. Ring the fraction that is greater than $1\frac{3}{4}$.

    $\frac{27}{16}$ $\frac{32}{20}$

    $\frac{173}{100}$ $\frac{352}{200}$

16. Ring the group that is ordered from least to greatest.

    $\frac{2}{5}, \frac{1}{2}, \frac{3}{10}$ $\frac{1}{3}, \frac{1}{2}, \frac{3}{4}$

    $\frac{3}{4}, \frac{3}{5}, \frac{3}{6}$ $\frac{2}{3}, \frac{4}{5}, \frac{7}{12}$

17. Write <, >, or =. $\frac{9}{27}$ ◯ $\frac{1}{3}$

18. Write <, >, or =. $3\frac{1}{7}$ ◯ $3\frac{9}{10}$

19. Write $\frac{75}{90}$ in simplest form.

    ____

Go on to the next page.

Name ______________________________

Pretest

20. Ring the fraction that is equivalent to $\frac{2}{3}$.

$\frac{3}{4}$ $\frac{5}{7}$

$\frac{7}{9}$ $\frac{6}{9}$

Use the information below to answer questions 21–24.

A taxi charges $1.00 for the first mile and $0.20 for each additional $\frac{1}{6}$ mile.

21. How much does it cost to ride $2\frac{1}{2}$ miles?

____

22. How far can you travel for $5.00?

____

23. Jo rode 2 miles and Sam rode 3 miles. How much more did Sam pay?

____

24. Rashawn and a friend take a taxi for $1\frac{1}{2}$ miles. The two riders split the fare evenly. How much does each pay?

____

25. Ring the fraction that falls between $\frac{4}{10}$ and $\frac{5}{10}$.

$\frac{9}{20}$ $\frac{11}{20}$ $\frac{16}{40}$ $\frac{9}{10}$

26. Ring the fraction that falls between $\frac{3}{8}$ and $\frac{4}{8}$.

$\frac{6}{16}$ $\frac{8}{16}$ $\frac{14}{32}$ $\frac{17}{32}$

27. Ring the fraction that falls between $\frac{1}{5}$ and $\frac{2}{5}$.

$\frac{3}{10}$ $\frac{4}{10}$

$\frac{1}{20}$ $\frac{9}{20}$

28. Ring the fraction that falls between $\frac{1}{3}$ and $\frac{1}{2}$.

$\frac{3}{6}$ $\frac{2}{6}$

$\frac{5}{12}$ $\frac{7}{12}$

Use the information below to answer questions 29–32.

At Jones Pier the lake is 20 feet deep. Every $\frac{1}{4}$ mile from the pier the lake becomes 1 foot deeper until it reaches a maximum depth of 30 feet.

29. How deep is the water $1\frac{1}{2}$ miles from the pier?

____

30. How many miles from the pier does the lake reach its maximum depth?

____

31. How much deeper is the water 2 miles from the pier than it is 1 mile out?

____

32. How far from the pier does the water reach 22 feet in depth?

____

Stop!

Name ______________________________

Posttest

Read each question. Find the answer.

1. Ring the prime number.

   3 6 10 12

2. Ring the composite number.

   13 23 33 43

3. Ring the prime number.

   42 59 65 81

4. How many factors does 41 have?

   ____

5. What are the prime factors of 42?

   ____

6. Ring the prime factors of 68.

   2, 2, 17 2, 34

   4, 17 1, 68

7. Ring the prime factorization of 48.

   $2^4 \times 3$ $2^3 \times 6$

   $4 \times 4 \times 4$ 8×6

8. Ring the prime factorization of 100.

   $2^2 \times 25^2$ $2^2 \times 5^2$

   $2^2 \times 25$ $2 \times 5 \times 10$

9. What is the greatest common factor of 32 and 40?

   ____

10. What is the greatest common factor of 16 and 24?

    ____

11. What is the least common multiple of 3 and 4?

    ____

12. What is the least common multiple of 9 and 12?

    ____

13. Ring the fraction that is expressed in simplest form.

    $\frac{4}{18}$ $\frac{5}{20}$

    $\frac{55}{200}$ $\frac{97}{102}$

14. Ring the fraction that is *not* expressed in simplest form.

    $\frac{1}{3}$ $\frac{2}{5}$ $\frac{4}{10}$ $\frac{5}{11}$

15. Ring the fraction that is greater than $1\frac{1}{4}$.

    $\frac{24}{20}$ $\frac{32}{30}$

    $\frac{123}{100}$ $\frac{257}{200}$

16. Ring the group that is ordered from least to greatest.

    $\frac{1}{5}, \frac{2}{3}, \frac{4}{10}$ $\frac{1}{4}, \frac{1}{2}, \frac{3}{4}$

    $\frac{3}{6}, \frac{3}{7}, \frac{3}{8}$ $\frac{3}{4}, \frac{7}{8}, \frac{6}{7}$

17. Write <, >, or =. $\frac{13}{18}$ ◯ $\frac{3}{8}$

18. Write <, >, or =. $2\frac{7}{8}$ ◯ $2\frac{21}{24}$

19. Write $\frac{7}{28}$ in simplest form.

    ____

Go on to the next page.

Name ______________________________

Posttest

20. Ring the fraction that is equivalent to $\frac{3}{4}$.

$\frac{4}{5}$ $\frac{5}{7}$ $\frac{6}{9}$ $\frac{6}{8}$

Use the information below to answer questions 21–24.

A parking lot charges $3.00 for the first hour and $1.00 for each additional $\frac{1}{2}$ hour.

21. How much does it cost to park for $2\frac{1}{2}$ hours?

____

22. For how long can you park for $4.00?

____

23. Alan parks for 4 hours and Alexa parks for 6 hours. How much more does Alexa pay than Alan?

____

24. Justin and Julie drive to work together. They split the cost of parking. If they park for $7\frac{1}{2}$ hours, how much will each pay?

____

25. Ring the fraction that falls between $\frac{4}{12}$ and $\frac{5}{12}$.

$\frac{9}{24}$ $\frac{10}{24}$

$\frac{16}{48}$ $\frac{20}{48}$

26. Ring the fraction that falls between $\frac{3}{10}$ and $\frac{4}{10}$.

$\frac{6}{20}$ $\frac{8}{20}$ $\frac{13}{40}$ $\frac{16}{40}$

27. Ring the fraction that falls between $\frac{1}{8}$ and $\frac{2}{8}$.

$\frac{3}{16}$ $\frac{4}{16}$

$\frac{15}{32}$ $\frac{17}{32}$

28. Ring the fraction that falls between $\frac{1}{4}$ and $\frac{1}{3}$.

$\frac{3}{12}$ $\frac{4}{12}$

$\frac{7}{24}$ $\frac{6}{24}$

Use the information below to answer questions 29–32.

Bob and Peg began a running program. Bob ran 1 mile on day 1 and increased the distance by $\frac{1}{10}$ mile each day. Peg ran 2 miles on day 1 and increased the distance by $\frac{1}{20}$ mile each day.

29. How far did Bob run on day 5?

____

30. How far did Peg run on day 6?

____

31. How far did Bob run on day 11?

____

32. How far did Peg run on day 15?

____

Stop!

Name ______________________________

Pretest

Read each question. Find the answer.

1. To estimate the sum of $6\frac{4}{9} + 3\frac{1}{8}$, what should $6\frac{4}{9}$ be rounded to? Ring the answer.

   5   $6\frac{1}{4}$ $6\frac{1}{2}$ 8

2. To estimate the quotient of $5\frac{7}{8} \div 2\frac{1}{8}$, what should $2\frac{1}{8}$ be rounded to? Ring the answer.

   2   $2\frac{3}{4}$ 3   4

3. Ring the best estimate. $13\frac{5}{8} + 9\frac{2}{5}$

   20   21   23   24

4. Ring the best estimate. $10\frac{7}{8} - 1\frac{3}{5}$

   8   $8\frac{1}{2}$ $9\frac{1}{2}$ 11

5. Ring the best estimate. $\frac{7}{6} \times \frac{4}{9}$

   $\frac{1}{8}$ $\frac{1}{2}$ 1 $1\frac{1}{2}$

6. Ring the best estimate. $12\frac{1}{9} \div 3\frac{9}{10}$

   3   $4\frac{1}{2}$ 5   9

7. Emilio bought two melons weighing $2\frac{7}{8}$ pounds (lb) and $3\frac{9}{10}$ pounds. Ring the best estimate for the weight of the two melons combined.

   4 lb   6 lb
   7 lb   8 lb

8. Three farmers split $14\frac{8}{10}$ gallons (gal) of gas equally. Ring the best estimate for the number of gallons each received.

   3 gal   4 gal
   5 gal   7 gal

9. To find the sum of $\frac{7}{8} + \frac{1}{3}$, $\frac{7}{8}$ should be renamed to which equivalent fraction?

   ____

10. To find the difference of $\frac{3}{4} - \frac{2}{5}$, $\frac{2}{5}$ should be renamed to which equivalent fraction?

    ____

11. Find the sum. $\frac{5}{8} + \frac{7}{8}$

    ____

12. Find the sum. $3\frac{1}{3} + 4\frac{3}{4}$

    ____

13. Find the difference. $\frac{1}{2} - \frac{1}{6}$

    ____

14. Find the difference. $12\frac{2}{5} - 11\frac{3}{5}$

    ____

Go on to the next page.

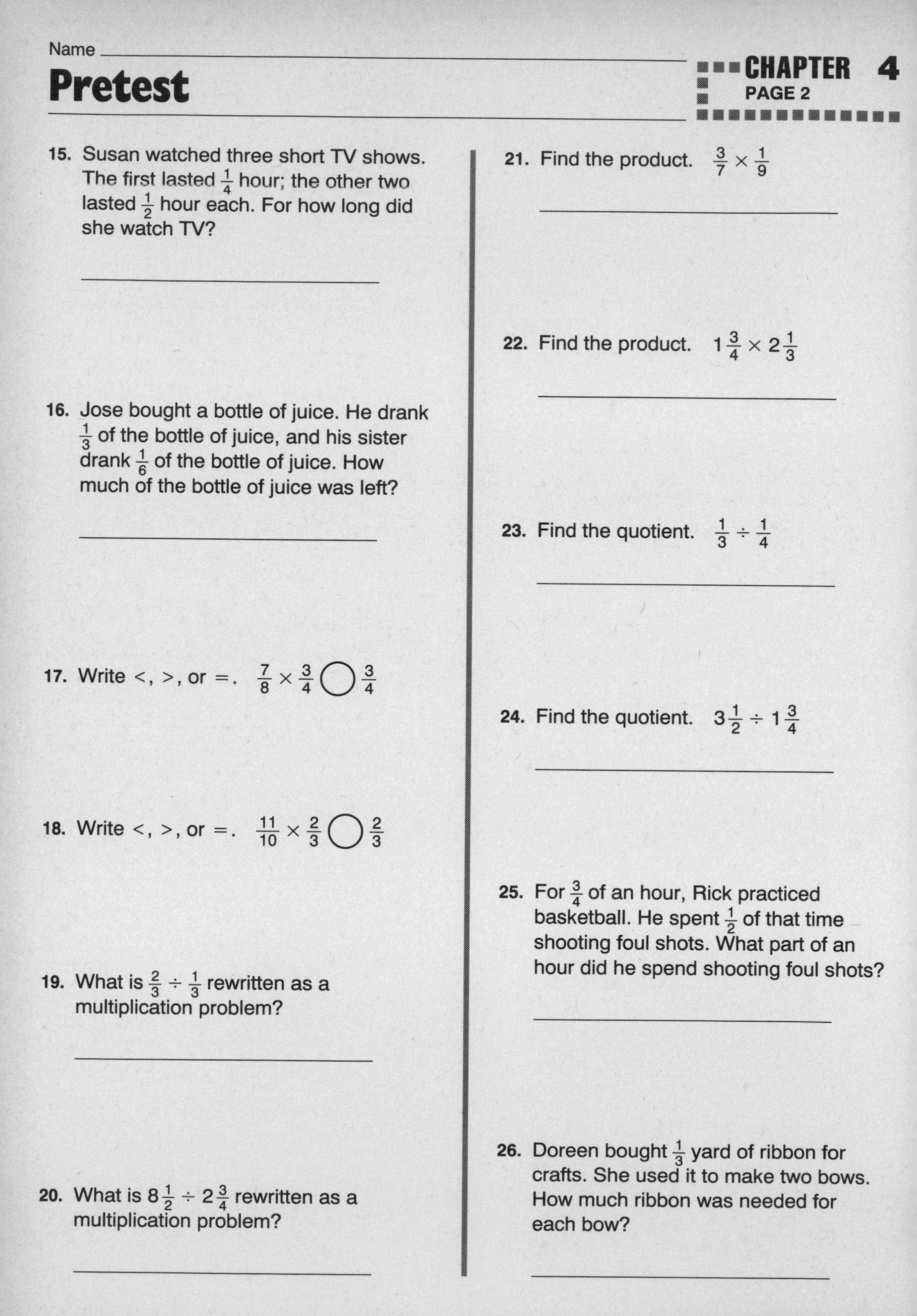

Name ______________________

Pretest

15. Susan watched three short TV shows. The first lasted $\frac{1}{4}$ hour; the other two lasted $\frac{1}{2}$ hour each. For how long did she watch TV?

____

16. Jose bought a bottle of juice. He drank $\frac{1}{3}$ of the bottle of juice, and his sister drank $\frac{1}{6}$ of the bottle of juice. How much of the bottle of juice was left?

____

17. Write $<$, $>$, or $=$. $\frac{7}{8} \times \frac{3}{4} \bigcirc \frac{3}{4}$

18. Write $<$, $>$, or $=$. $\frac{11}{10} \times \frac{2}{3} \bigcirc \frac{2}{3}$

19. What is $\frac{2}{3} \div \frac{1}{3}$ rewritten as a multiplication problem?

____

20. What is $8\frac{1}{2} \div 2\frac{3}{4}$ rewritten as a multiplication problem?

____

21. Find the product. $\frac{3}{7} \times \frac{1}{9}$

____

22. Find the product. $1\frac{3}{4} \times 2\frac{1}{3}$

____

23. Find the quotient. $\frac{1}{3} \div \frac{1}{4}$

____

24. Find the quotient. $3\frac{1}{2} \div 1\frac{3}{4}$

____

25. For $\frac{3}{4}$ of an hour, Rick practiced basketball. He spent $\frac{1}{2}$ of that time shooting foul shots. What part of an hour did he spend shooting foul shots?

____

26. Doreen bought $\frac{1}{3}$ yard of ribbon for crafts. She used it to make two bows. How much ribbon was needed for each bow?

____

Go on to the next page.

Name ______________________________

Pretest

27. A carton containing 6 identical boxes weighs $12\frac{1}{2}$ pounds. How many pounds does each box weigh?

____

28. Each floor tile is $\frac{10}{12}$ foot long. How many tiles are along one wall of a room that is 25 feet long?

____

29. Liza mowed lawns for $2\frac{1}{3}$ hours each day on Monday, Tuesday, and Wednesday. She raked leaves for $4\frac{1}{2}$ hours on Friday and $3\frac{2}{3}$ hours on Saturday. For how many hours did she work this week?

____

30. There are 12 teams in a baseball league. If each team plays each of the others once, how many games will be played?

____

Use the information below to answer questions 31–32.

Nancy must walk 5 blocks to school. It takes her $2\frac{1}{2}$ minutes to walk each block. Pat walks 7 blocks to school. It takes her only $1\frac{1}{2}$ minutes to walk each block.

31. How much longer does it take Nancy to walk to school than it takes Pat?

____

32. How long does it take Nancy to walk $2\frac{1}{2}$ blocks?

____

Stop!

Name ______________________

Posttest

Read each question. Find the answer.

1. To estimate the sum of $7\frac{5}{9} + 4\frac{1}{8}$, what should $7\frac{5}{9}$ be rounded to? Ring the answer.

   7 $\quad$ $7\frac{1}{8}$ $\quad$ $7\frac{1}{2}$ $\quad$ 9

2. To estimate the product of $5\frac{1}{6} \times 1\frac{1}{10}$, what should $1\frac{1}{10}$ be rounded to? Ring the answer.

   1 $\quad$ $1\frac{1}{6}$ $\quad$ $1\frac{1}{2}$ $\quad$ 2

3. Ring the best estimate. $4\frac{4}{7} + 9\frac{4}{9}$

   11 $\quad$ $12\frac{1}{2}$ $\quad$ $13\frac{1}{2}$ $\quad$ 14

4. Ring the best estimate. $10\frac{6}{7} - 2\frac{2}{5}$

   $7\frac{1}{2}$ $\quad$ $8\frac{1}{2}$ $\quad$ $9\frac{1}{2}$ $\quad$ 10

5. Ring the best estimate. $\frac{10}{9} \times \frac{4}{9}$

   $\frac{1}{2}$ $\quad$ $\frac{3}{4}$ $\quad$ $1\frac{1}{2}$ $\quad$ 2

6. Ring the best estimate. $15\frac{1}{8} \div 4\frac{9}{11}$

   2 $\quad$ 3 $\quad$ 4 $\quad$ 5

7. Lucy bought two bags of peanuts. The first weighed $1\frac{7}{8}$ pounds (lb), and the second weighed $1\frac{1}{16}$ pounds. Ring the best estimate for the weight of the two bags combined.

   $1\frac{1}{2}$ lb $\quad$ 3 lb $\quad$ $3\frac{1}{2}$ lb $\quad$ 4 lb

8. Three friends divided $3\frac{1}{4}$ pints (pt) of yogurt equally. Ring the best estimate for the amount of yogurt that each person had.

   $\frac{1}{2}$ pt $\quad$ 1 pt

   $1\frac{1}{2}$ pt $\quad$ 2 pt

9. To find the sum of $\frac{7}{8} + \frac{1}{5}$, $\frac{7}{8}$ should be renamed to which equivalent fraction?

   ____

10. To find the difference of $\frac{2}{3} - \frac{2}{5}$, $\frac{2}{5}$ should be renamed to which equivalent fraction?

    ____

11. Find the sum. $\frac{5}{12} + \frac{5}{6}$

    ____

12. Find the sum. $4\frac{2}{3} + 4\frac{3}{4}$

    ____

13. Find the difference. $\frac{3}{4} - \frac{1}{2}$

    ____

Go on to the next page.

14. Find the difference. $11\frac{1}{6} - 10\frac{1}{3}$

______

15. Julie played three video games. The first lasted $\frac{1}{2}$ hour; the other two lasted $\frac{1}{4}$ hour each. For how long did she play video games?

______

16. Beth and her two sisters each ate $\frac{1}{4}$ of a pizza. What fraction of the whole pizza was left?

______

17. Write <, >, or =. $\frac{9}{10} \times \frac{2}{3} \bigcirc \frac{2}{3}$

18. Write <, >, or =. $\frac{21}{20} \times \frac{4}{5} \bigcirc \frac{4}{5}$

19. What is $\frac{3}{5} \div \frac{1}{5}$ rewritten as a multiplication problem?

______

20. What is $9\frac{1}{3} \div 2\frac{2}{3}$ rewritten as a multiplication problem?

______

21. Find the product. $\frac{3}{11} \times \frac{1}{3}$

______

22. Find the product. $2\frac{1}{2} \times 1\frac{1}{4}$

______

23. Find the quotient. $\frac{1}{2} \div \frac{1}{3}$

______

24. Find the quotient. $6\frac{1}{2} \div 3\frac{1}{4}$

______

25. Paul spends $1\frac{1}{2}$ hours on his music every day. He spends $\frac{1}{3}$ of that time composing music. How much time does he spend composing daily?

______

Go on to the next page.

Posttest

26. Heidi watches a television show that lasts $\frac{1}{2}$ hour. Commercials make up $\frac{1}{6}$ of the show. What part of an hour does she spend watching commercials?

____

27. A carton containing 15 identical books weighs $22\frac{1}{2}$ pounds. How many pounds does each book weigh?

____

28. Ramon works for $3\frac{1}{4}$ hours on Saturday, $5\frac{1}{4}$ hours on Sunday, and $2\frac{1}{2}$ hours each weekday. For how many hours does he work in 2 weeks?

____

29. There are 12 girls in the state tennis tournament. Each girl will play each of the other girls once. How many matches will be played?

____

30. Jeri worked for 14 hours at $4.50 per hour. Alyce worked for $20\frac{1}{2}$ hours at $3.50 per hour. Who earned more, and by how much?

____

Use the information below to answer questions 31–32.

Justin can bicycle one mile in $\frac{1}{4}$ hour. Don can bicycle one mile in $\frac{1}{3}$ hour.

31. If both boys ride for two hours, how much farther will Justin have traveled?

____

32. How long will it take Justin to ride $2\frac{1}{2}$ miles?

____

Stop!

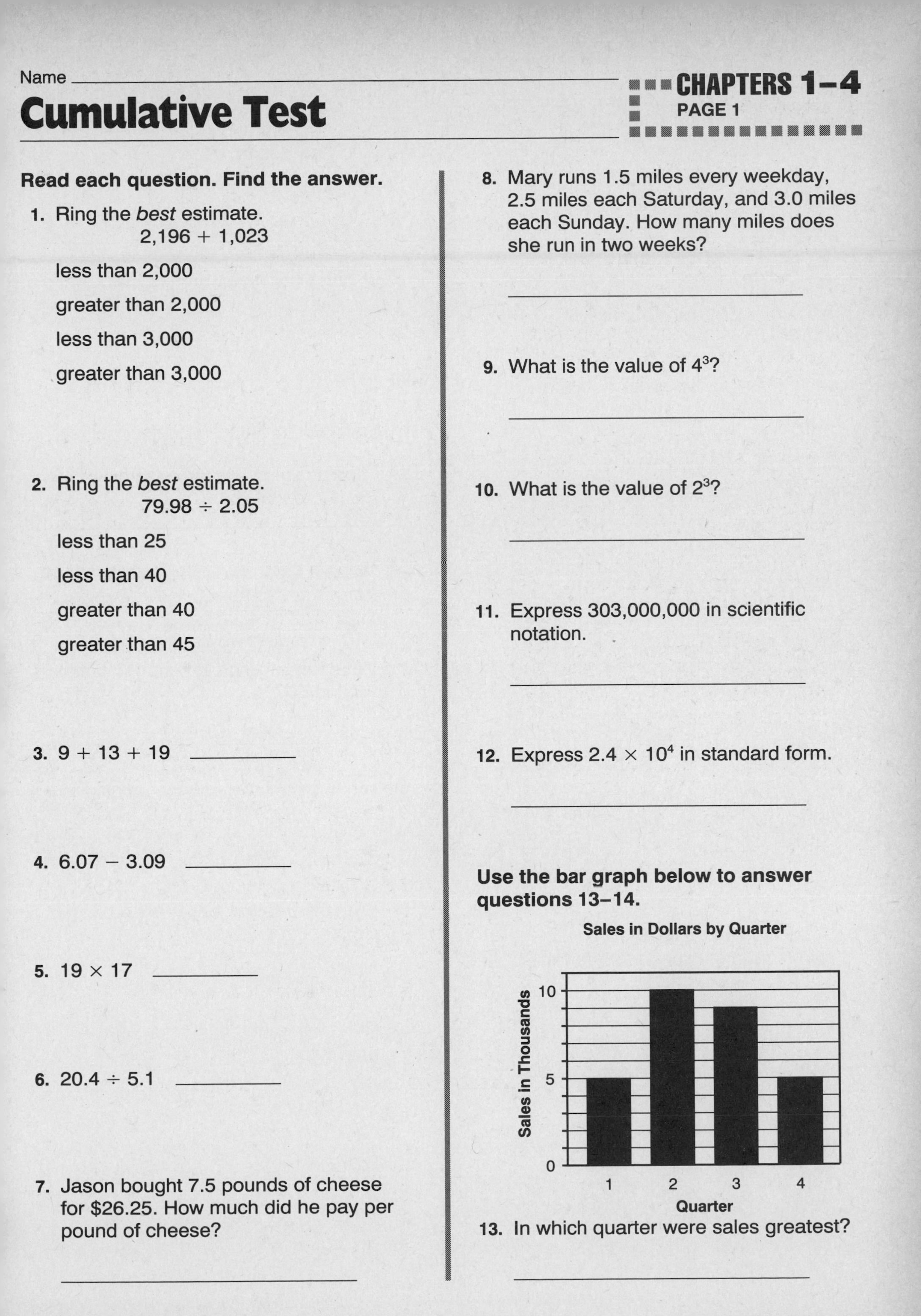

Name ____________________

Cumulative Test

CHAPTERS 1–4
PAGE 1

Read each question. Find the answer.

1. Ring the *best* estimate.
   2,196 + 1,023

   less than 2,000

   greater than 2,000

   less than 3,000

   greater than 3,000

2. Ring the *best* estimate.
   79.98 ÷ 2.05

   less than 25

   less than 40

   greater than 40

   greater than 45

3. 9 + 13 + 19 __________

4. 6.07 − 3.09 __________

5. 19 × 17 __________

6. 20.4 ÷ 5.1 __________

7. Jason bought 7.5 pounds of cheese for $26.25. How much did he pay per pound of cheese?

   ____

8. Mary runs 1.5 miles every weekday, 2.5 miles each Saturday, and 3.0 miles each Sunday. How many miles does she run in two weeks?

   ____

9. What is the value of 4^3?

   ____

10. What is the value of 2^3?

    ____

11. Express 303,000,000 in scientific notation.

    ____

12. Express 2.4×10^4 in standard form.

    ____

Use the bar graph below to answer questions 13–14.

Sales in Dollars by Quarter

Sales in Thousands
10
5
0
1
2
3
4
Quarter

13. In which quarter were sales greatest?

    ____

Go on to the next page.

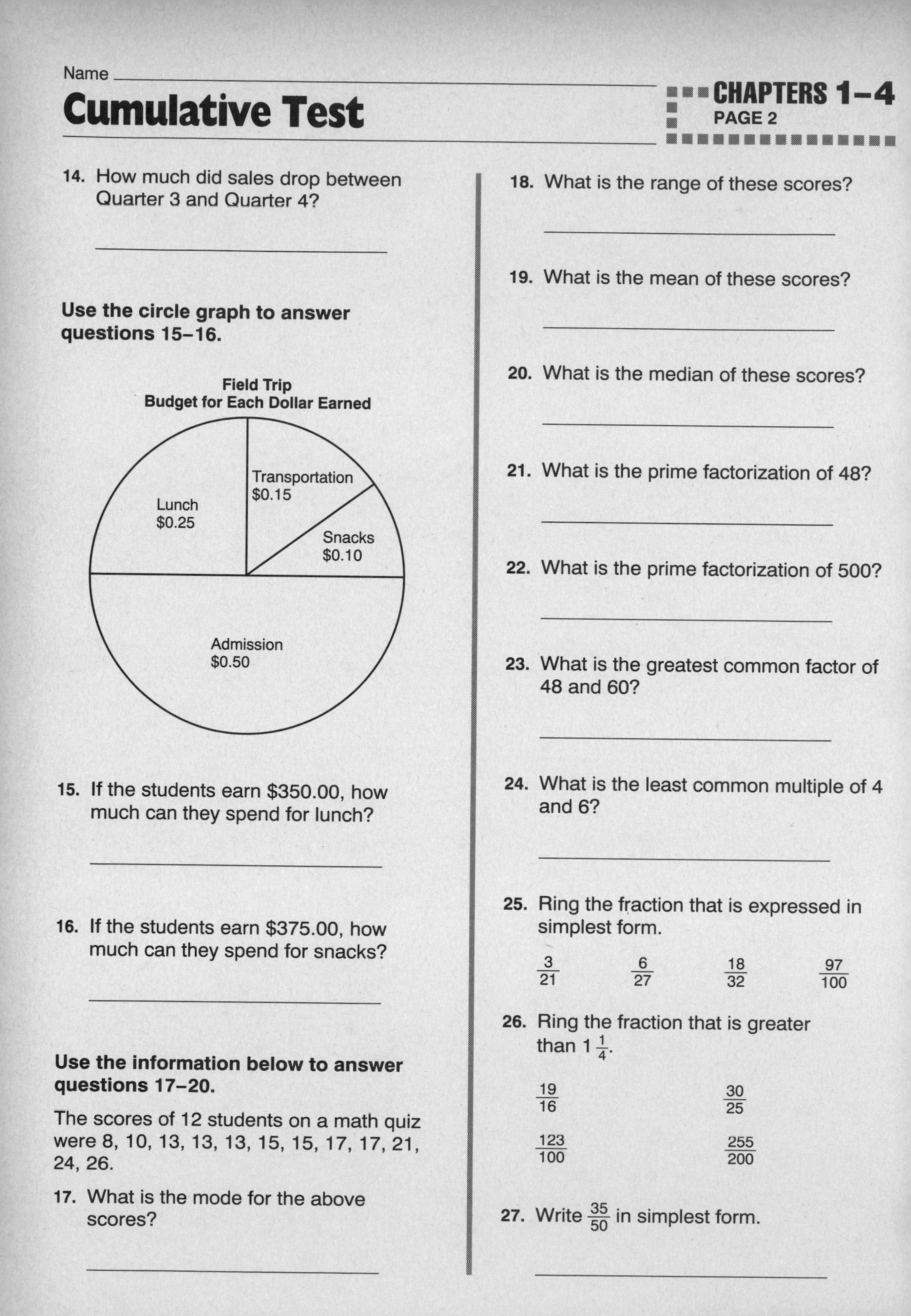

Name ______________________

Cumulative Test

14. How much did sales drop between Quarter 3 and Quarter 4?

____

Use the circle graph to answer questions 15–16.

Field Trip
Budget for Each Dollar Earned

15. If the students earn $350.00, how much can they spend for lunch?

____

16. If the students earn $375.00, how much can they spend for snacks?

____

Use the information below to answer questions 17–20.

The scores of 12 students on a math quiz were 8, 10, 13, 13, 13, 15, 15, 17, 17, 21, 24, 26.

17. What is the mode for the above scores?

____

18. What is the range of these scores?

____

19. What is the mean of these scores?

____

20. What is the median of these scores?

____

21. What is the prime factorization of 48?

____

22. What is the prime factorization of 500?

____

23. What is the greatest common factor of 48 and 60?

____

24. What is the least common multiple of 4 and 6?

____

25. Ring the fraction that is expressed in simplest form.

$\frac{3}{21}$ $\frac{6}{27}$ $\frac{18}{32}$ $\frac{97}{100}$

26. Ring the fraction that is greater than $1\frac{1}{4}$.

$\frac{19}{16}$ $\frac{30}{25}$

$\frac{123}{100}$ $\frac{255}{200}$

27. Write $\frac{35}{50}$ in simplest form.

____

Go on to the next page.

Name ____________________

Cumulative Test

28. Write <, >, or =. $\frac{3}{9} \bigcirc \frac{9}{27}$

29. Ring the fraction that falls between $\frac{4}{20}$ and $\frac{5}{20}$.

$\frac{2}{9}$ $\frac{11}{40}$

$\frac{16}{80}$ $\frac{20}{80}$

30. Ring the fraction that falls between $\frac{1}{3}$ and $\frac{1}{4}$.

$\frac{3}{12}$ $\frac{4}{12}$

$\frac{7}{24}$ $\frac{9}{24}$

31. Ring the fraction that falls between $\frac{1}{10}$ and $\frac{2}{10}$.

$\frac{3}{20}$ $\frac{4}{20}$

$\frac{5}{40}$ $\frac{7}{40}$

32. Ring the fraction that falls between $\frac{1}{5}$ and $\frac{1}{6}$.

$\frac{5}{30}$ $\frac{6}{30}$ $\frac{11}{60}$ $\frac{12}{60}$

33. $\frac{3}{6} + \frac{4}{6}$ ____________

34. $3\frac{2}{3} + 5\frac{1}{4}$ ____________

35. $\frac{1}{4} - \frac{1}{5}$ ____________

36. $11\frac{1}{4} - 10\frac{3}{4}$ ____________

37. Susan watched three movies. The first lasted $\frac{1}{2}$ hour; the other two lasted $\frac{3}{4}$ hour each. How long did she watch movies?

____

38. Tim had a bottle of juice. He drank $\frac{1}{4}$ of the bottle and his sister drank $\frac{1}{2}$ of the bottle. How much was left?

____

39. $\frac{3}{5} \times \frac{1}{7}$ ____________

40. $1\frac{1}{4} \times 1\frac{1}{3}$ ____________

Go on to the next page.

Name ______________________

Cumulative Test

41. $\frac{1}{2} \div \frac{1}{5}$ __________

42. $1\frac{3}{4} \div 1\frac{1}{2}$ __________

43. The lake at Jones Pier is 15 feet deep. Every $\frac{1}{5}$ mile from the pier it becomes 1 foot deeper until it reaches its maximum depth of 30 feet. How many miles from the pier does the lake reach its maximum depth?

____

Use the table below to answer questions 44–45.

Senior Class Elections		
Candidate	Girls' Votes	Boys' Votes
Jack	50	80
Carol	60	84
Darla	100	51
Bill	90	65
	300	280

44. Who received the most votes?

____

45. How many more votes did Bill receive than Carol?

____

46. A principal wants a graph that compares the monthly absences of boys and girls for the year. Ring the type of graph that should be used.

double-bar graph

histogram

stem-and-leaf plot

circle graph

47. A family wants a graph that shows how they spend their monthly income. Ring the type of graph that they should use.

line graph

double-bar graph

box-and-whisker graph

circle graph

48. Tom gave Ben $1.50 to buy stamps. Tom gave Ben only quarters and dimes. He gave Ben 9 coins in all. How many quarters did Tom give Ben?

____

49. There are 10 players in a tennis league. If each player plays one game with each of the other players, how many games will be played?

____

50. Nancy must walk 4 blocks to school. It takes her $2\frac{1}{2}$ minutes to walk each block. Pat walks 5 blocks to school. It takes her only $1\frac{1}{2}$ minutes to walk each block. How much longer does it take Nancy to walk to school than it takes Pat?

____

Stop!

Name ______________________

Pretest

Read each question. Find the answer.

1. Write the algebraic expression for the product of 9 and a number, y.

____

2. Write the algebraic expression for a number, b, decreased by 2.

____

For questions 3–6, let $a = 15$ and $b = 6$.

3. Evaluate. $4a - 6$ __________

4. Evaluate. $3b - 10$ __________

5. Evaluate. $2b - 12$ __________

6. Evaluate. $3a + 2b$ __________

7. Write an expression showing the number of minutes in h hours.

____

8. Mary and June scored a total of 25 points in a basketball game. Let m = the number of points scored by Mary. Write an expression showing how many points June scored.

____

9. Find the value. $(3 + 5) \cdot (3 - 1)$

____

10. Find the value. $(2 + 3) + (5 - 1)^2$

____

11. Find the value. $(2 + 1 + 3) \div (3 - 2)$

____

12. Find the value. $5 \times (9 - 3)$

____

13. Write an inequality showing that a number, b, increased by 5 is less than 100.

____

14. Write an inequality showing that a number, c, decreased by 10 is greater than 10.

____

15. Solve the inequality. $x + 1 > 5$

____

16. Solve the inequality. $x + 5 < 14$

____

17. To solve the equation $y + 3 = 10$, what must be done to both sides of the equation?

____

18. To solve the equation $x - 5 = 7$, what must be done to both sides of the equation?

____

Go on to the next page.

Name ______________________________

Pretest

19. To solve the equation $3x = 15$, what must be done to both sides of the equation?

____

20. To solve the equation $\frac{y}{20} = 2$, what must be done to both sides of the equation?

____

21. Solve. $p + 9 = 12$ ____________

22. Solve. $p - 10 = 0$ ____________

23. Solve. $3x = 30$ ____________

24. Solve. $\frac{x}{5} = 25$ ____________

25. After Joe gave 10 of his baseball cards to Bob, he had 92 cards left. How many cards did Joe have before giving cards to Bob?

____

26. Li earned $11.00 baby-sitting. That brought her savings to $51.50. How much money did Li have before baby-sitting?

____

27. Jon has 3 times as many stamps as Ali. If Jon has 15 stamps, how many does Ali have?

____

28. Ed's salary is $\frac{1}{4}$ of Bill's salary. Ed earns $20,000. How much does Bill earn?

____

29. Mr. Hynes withdrew money from the bank. He spent $30 on shoes and $10 on a tie. He gave $\frac{1}{2}$ of what was left to his wife. He then had $45. How much money did Mr. Hynes withdraw?

____

30. Betty joined a running club. After one month she doubled the distance she ran each day. After two months she doubled the distance again. She then was running 4 miles a day. How many miles a day did she run at the start?

____

31. Over the past 5 weeks, Tom lost 3 pounds per week. He now weighs 200 pounds. How much did Tom weigh before losing the weight?

____

32. Alicia collects stamps. On Monday she added 5 stamps to her collection. On Tuesday she doubled the number of stamps in her collection. She then had 90 stamps. How many stamps did she have on Sunday?

____

Stop!

Name ______________________________

Posttest

Read each question. Find the answer.

1. Write the algebraic expression for the product of 5 and a number, c.

   ____

2. Write the algebraic expression for a number, x, increased by 3.

   ____

For questions 3–6, let $a = 20$ and $b = 5$.

3. Evaluate. $4b + 1$ __________

4. Evaluate. $2a - 27$ __________

5. Evaluate. $3b - 15$ __________

6. Evaluate. $4b - a$ __________

7. Write an expression showing the number of hours in d days.

   ____

8. Hank and Bobby scored a total of 6 goals in a hockey game. Let h = the number of goals scored by Hank. Write an expression showing how many goals Bobby scored.

   ____

9. Find the value. $(2 + 6) \cdot (4 - 2)$

   ____

10. Find the value. $(1 + 2) + (4 - 1)^2$

    ____

11. Find the value. $(3 + 2 + 7) \div (7 - 6)$

    ____

12. Find the value. $10 \times (8 - 4)$

    ____

13. Write an inequality showing that a number, a, increased by 7 is less than 99.

    ____

14. Write an inequality showing that a number, y, increased by 100 is greater than 100.

    ____

15. Solve the inequality. $x + 5 > 12$

    ____

16. Solve the inequality. $x + 4 < 9$

    ____

17. To solve the equation $y + 5 = 15$, what must be done to both sides of the equation?

    ____

18. To solve the equation $x - 3 = 6$, what must be done to both sides of the equation?

    ____

Go on to the next page.

Name ______________________________

Posttest

19. To solve the equation $4x = 20$, what must be done to both sides of the equation?

____

20. To solve the equation $\frac{y}{30} = 3$, what must be done to both sides of the equation?

____

21. Solve. $m + 12 = 14$ __________

22. Solve. $m - 50 = 10$ __________

23. Solve. $5y = 100$ __________

24. Solve. $\frac{x}{7} = 7$ __________

25. Ben sells newspapers. One day he sold 110 papers. He had 45 left. How many papers did he have at the beginning of the day?

____

26. Ann is a waitress. One day she earned $12.50 in tips. She made $35.00 that day, including her salary. What was Ann's salary for the day?

____

27. In a basketball game, Nat scored 3 times as many points as Phil. Nat scored 18 points. How many points did Phil score?

____

28. Ed's weight is $\frac{3}{4}$ of Tony's weight. Ed weighs 120 pounds. How many pounds does Tony weigh?

____

29. Linda withdrew money from the bank. She gave $\frac{1}{2}$ of the money to her husband and then spent $50 on groceries. She then had $50. How much money did she withdraw?

____

30. Betty joined an exercise club. In one month she doubled the number of sit-ups she could do. After two months she doubled the number of sit-ups again. She then could do 60. How many sit-ups could she do at the start?

____

31. Over the past 6 weeks, Tom burned 3 logs per week. Tom now has 22 logs. How many logs did Tom have 6 weeks ago?

____

32. Natasha invested in the stock market. The first week she lost $\frac{1}{2}$ of her money. The second week she lost $\frac{1}{2}$ of the remainder. She was left with $1,200. How much money did she invest?

____

Stop!

Name ______________________________

Pretest

Read each question. Find the answer.

For questions 1–4, write *always*, *sometimes*, or *never*.

1. Intersecting lines are perpendicular.

____

2. A right triangle is isosceles.

____

3. Adjacent angles have a common vertex and a common ray.

____

4. A regular hexagon has 6 congruent angles.

____

5. What is the total angle measure of the interior angles of a quadrilateral?

____

6. What is the complement of a 65° angle?

____

7. Two angles of a triangle measure 40° and 60°. What is the measure of the third angle?

____

8. One acute angle of a right triangle is twice as large as the other. What are the measures of the two acute angles?

____

9. If a triangle has angle measures of 45°, 50°, and 85°, what type of triangle is it?

____

10. Classify the following triangle according to the lengths of its sides: 4 in., 5 in., 6 in.

____

11. Ring the measure of the *reflex* central angle formed by the hands of a clock at 12:40.

60° 100°

120° 240°

12. The measures of two sides of an isosceles triangle are 9 cm and 4 cm. What is the measure of the third side?

____

13. Ring the construction that Figure 2 represents.

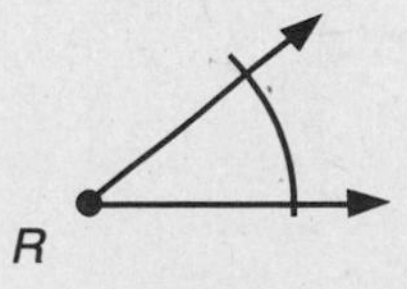

R

Figure 1

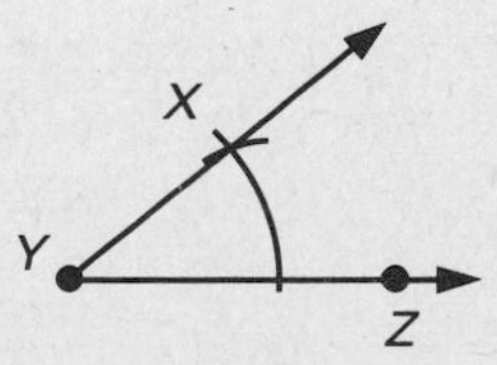

X Y Z

Figure 2

congruent line segment

angle bisector

line bisector

congruent angle

Go on to the next page.

Name ________________________________

Pretest

14. Ring the construction that the figure represents.

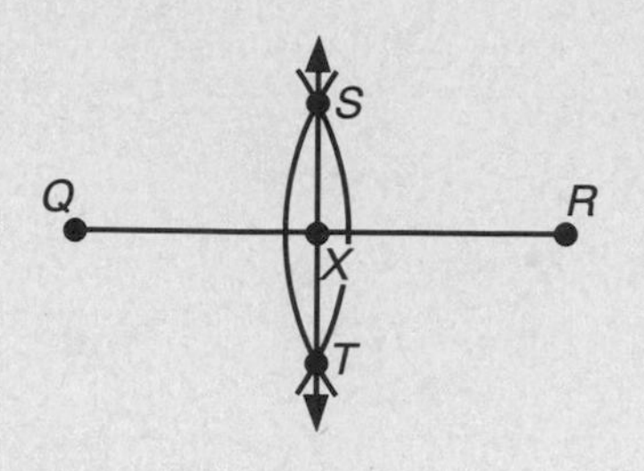

bisected line segment

bisected angle

congruent angle

congruent triangle

15. Ring the construction that Figure 2 represents.

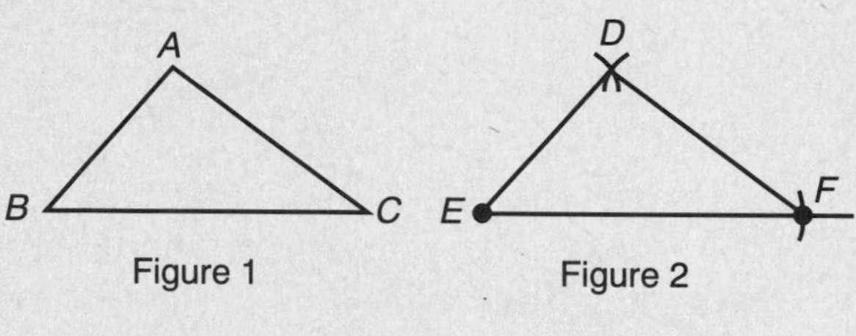

congruent line segment

congruent triangle

congruent angle

angle bisector

16. Ring the construction that the figure represents.

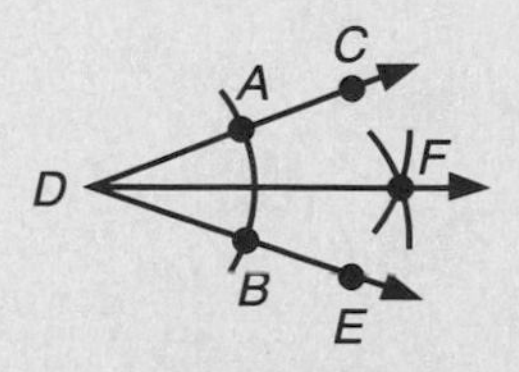

congruent angle

congruent line segment

bisected angle

bisected line segment

Use the Venn diagram below to answer questions 17–20. Write *yes* or *no*.

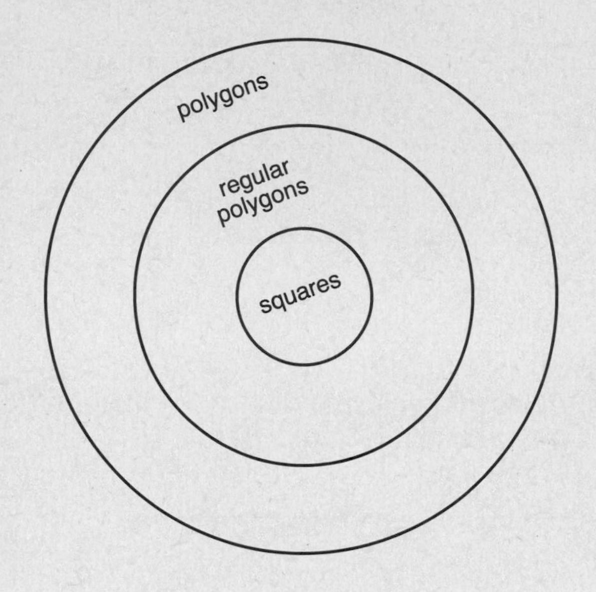

17. Does the diagram show that a polygon is always a square?

____

18. Does the diagram show that all polygons are regular polygons?

____

19. Does the diagram show that all squares are regular polygons?

____

20. Does the diagram show that some polygons are not squares?

____

Go on to the next page.

Name ______________________________

Pretest

The table below shows the total angle measure of the interior angles in regular polygons. Use the table to answer questions 21–22.

Regular Polygons	
Number of Angles	**Total Angle Measure**
3	180°
4	360°
5	540°
6	720°
7	900°
8	?

21. Ring the total interior angle measure of a regular 8-sided polygon.

980° 1,080°

1,000° 1,800°

22. Ring the measure of each interior angle in a regular 10-sided polygon.

90° 144°

126° 180°

Use the information below to answer questions 23–24.

The taxi fare in a city is given by the formula $\$1.50 + 0.75m$, where m = the distance traveled in miles.

23. Jan took a taxi for a distance of 4.2 miles. What was the taxi fare?

____

24. What would be the cost of a 15.5-mile taxi ride?

____

Stop!

Name ____________________

Posttest

CHAPTER 6
PAGE 1

Read each question. Find the answer.

For questions 1–4, write *always*, *sometimes*, or *never*.

1. Parallel lines intersect at one point.

 ____

2. Two line segments with the same length are congruent.

 ____

3. An acute triangle is equilateral.

 ____

4. A square is a rhombus.

 ____

5. Ring the quadrilateral that has exactly one pair of parallel sides.

 rhombus   trapezoid

 rectangle   parallelogram

6. Angle *A* is congruent to angle *X*. If angle *A* measures 50°, what is the measure of angle *X*?

 ____

7. What is the supplement of a 30° angle?

 ____

8. Two angles of a triangle measure 40° and 50°. What is the measure of the third angle?

 ____

9. One acute angle of a right triangle is four times as large as the other. What are the measures of the two acute angles?

 ____

10. Classify the following triangle according to the measures of its sides: 4 cm, 4 cm, and 3 cm.

 ____

11. The measure of $\angle DEF$ is 70°. The measure of adjacent $\angle FEG$ is 10°. Ring the measure of $\angle DEG$.

 10°   60°

 80°   100°

12. Ring the measure of the *acute* central angle formed by the hands of a clock at 12:05.

 30°   60°

 300°   330°

13. Ring the construction that the figure represents.

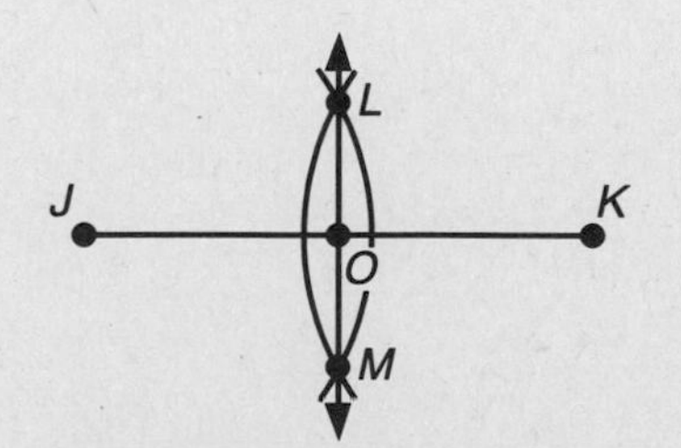

 bisected angle

 bisected line segment

 congruent angle

 congruent line segment

Go on to the next page.

14. Ring the construction that Figure 2 represents.

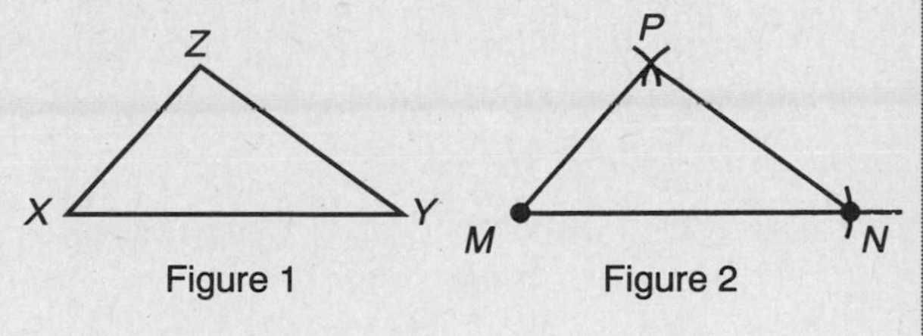

congruent triangle
congruent angle
perpendicular lines
bisected angle

15. Ring the construction that Figure 2 represents.

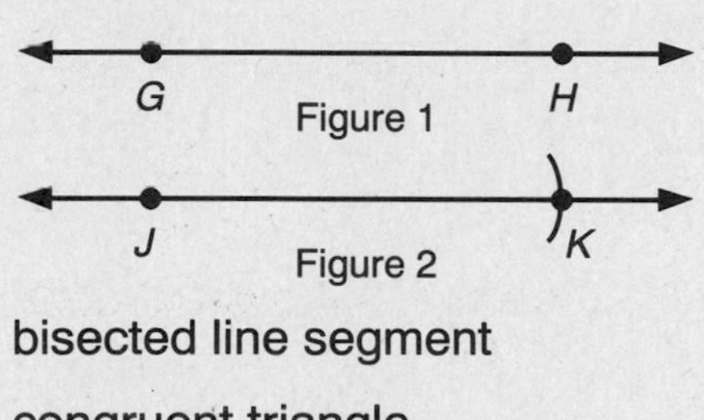

bisected line segment
congruent triangle
congruent line segment
congruent angle

16. Ring the construction that the figure represents.

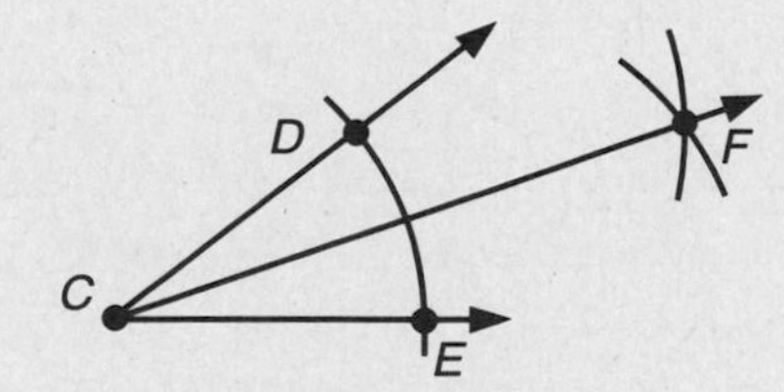

bisected line segment
bisected angle
congruent angle
congruent triangle

Use the Venn diagram below to answer questions 17–20. Write *yes* or *no*.

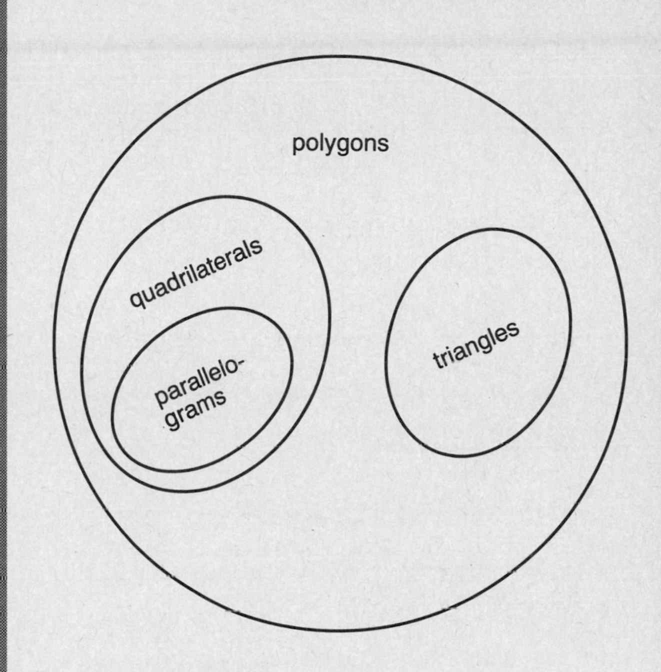

17. Does the diagram show that a parallelogram is always a quadrilateral?

___

18. Does the diagram show that all triangles are polygons?

___

19. Does the diagram show that some parallelograms are triangles?

___

20. Does the diagram show that all polygons are quadrilaterals?

___

Go on to the next page.

Name ____________________

Posttest

The table below shows the number of diagonals in regular polygons. Use it to answer questions 21–22.

Regular Polygons	
Number of Sides	**Number of Diagonals**
3	0
4	2
5	5
6	9
7	14
8	?

21. How many diagonals are in a regular polygon with 8 sides?

____

22. How many diagonals are in a regular 10-sided polygon?

____

Use this information to answer questions 23–24.

The formula for the total angle measure of a regular polygon is $(n - 2) \times 180°$, where n = the number of angles.

23. What is the total angle measure of a 15-sided regular polygon?

____

24. What is the measure of each interior angle of a regular 10-sided polygon?

____

Stop!

Name ____________________

Pretest

Read each question. Find the answer.

1. A box contains 9 red marbles and 3 blue marbles. Ring the answer that expresses the ratio of red to blue marbles.

   9 to 3   9 to 12

   9 to 1   1 to 3

2. Ring the ratio that does *not* have the same value as the other three.

   1:5   2:10   4:20   5:30

3. Ring the proportion that is *not* true.

   $\frac{2}{3} = \frac{4}{6}$ $\frac{4}{5} = \frac{80}{100}$

   $\frac{5}{6} = \frac{155}{186}$ $\frac{7}{8} = \frac{75}{80}$

4. Which value of *a* makes the following a true proportion?

   $$\frac{a}{10} = \frac{4}{20}$$

   ____

5. Solve. $\frac{x}{4} = \frac{3}{6}$ __________

6. Solve. $\frac{10}{y} = \frac{25}{7.5}$ __________

7. Solve. $\frac{3a}{9} = \frac{8}{6}$ __________

8. Solve for *a*. $\frac{5}{a} = \frac{25}{55}$ __________

9. Betty drives 10 miles in 15 minutes. At this rate how far will she drive in 75 minutes?

   ____

10. Jose's car runs 26 miles on 1 gallon of gas. At this rate how many gallons of gas will he use to travel 117 miles?

    ____

11. Mary is paid $7 for every 2 hours of work. How much does she earn for working 14 hours?

    ____

12. Sean drives $\frac{9}{10}$ mile in 1 minute. At this rate how far will he drive in 1 hour?

    ____

13. A package of 8 batteries costs $7.60. What is the unit price?

    ____

14. If 8 batteries cost $7.60, how much would 20 batteries cost?

    ____

Go on to the next page.

15. A box of soap powder weighing 4 pounds costs $3.96; a 6-pound box costs $6.30. Which package has the lower unit price, and by how much per pound?

____

16. If 4 rolls cost $2.16, how many rolls can be bought for $4.32?

____

17. On a scale drawing, 1 inch is used to represent 4 feet. Ring the ratio expressing that relationship.

1:8   1:12   1:48   4:48

18. On Map A, 1 inch = 10 miles. On Map B, 1 inch = 20 miles. A road is 4 inches long on Map A. How long will the road be on Map B?

____

19. Use the scale 6 cm = 25 cm. What is the actual length of an object when the scale length is 24 cm?

____

20. Use the scale 1 m = 50 km. What is the actual length of an object when the scale length is $\frac{1}{4}$ m?

____

21. Lupe is making a scale drawing on which each centimeter represents 8 meters. One scale building measures 12 cm by 15 cm. What are the actual dimensions of the building?

____

22. On a map 1 cm = 25 km. Two cities are 12 cm apart on the map. How far are they actually apart?

____

For questions 23–24, write *always*, *sometimes*, or *never*.

23. Congruent triangles are similar.

____

24. Similar figures have the same shape.

____

25. Two rectangles are similar. The first rectangle is 6 inches wide and 10 inches long. The width of the second rectangle is 30 inches. Find its length.

____

26. Two rectangles are similar. The first rectangle has a width of 2 cm and a length of 5 cm. The width of the second rectangle is 0.2 cm. Find the length.

____

Go on to the next page.

Name ________________________________

Pretest

27. A triangle has sides measuring 3 ft, 4 ft, 5 ft. Ring the answer that could be the lengths of the sides of a similar triangle.

4 yards, 6 yards, 8 yards

4 feet, 5 feet, 6 feet

31 feet, 41 feet, 51 feet

6 feet, 8 feet, 10 feet

28. These two rectangles are similar. Find x.

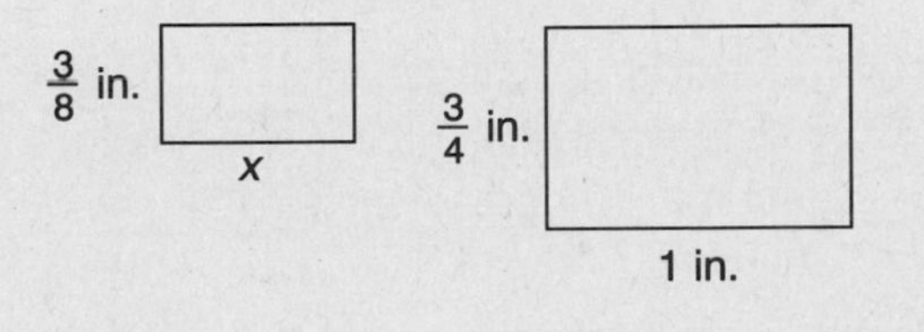

____

29. A map scale is 2 cm = 7 km. Mt. Snow is 14 cm from Rayville on the map. How far is that?

____

30. The scale on a map is 1 inch = 200 miles. Two cities are $2\frac{1}{2}$ inches apart. If Toni averages 50 miles per hour, how long will it take her to drive from one city to the other?

____

Use the information below to answer questions 31–32.

The formula for finding the circumference of a circle is $C = 2\pi r$, where C = the circumference and r = the radius.

31. Use 3.14 as an approximate value of π. Find the circumference of a circle with a radius of 6 inches.

____

32. Use $\frac{22}{7}$ as an approximate value of π. Find the radius of a circle with a circumference of 11 cm.

____

Stop!

Name ____________________

Posttest

Read each question. Find the answer.

1. At the zoo there are 3 elephants and 8 tigers. Ring the answer that expresses the ratio of elephants to tigers.

   3:11 1:5 8:3 3:8

2. Ring the ratio that does *not* have the same value as the other three.

   $\frac{2}{9}$ $\frac{3}{18}$ $\frac{6}{27}$ $\frac{10}{45}$

3. Ring the proportion that is *not* true.

   $\frac{3}{4} = \frac{6}{8}$ $\frac{1}{5} = \frac{20}{100}$

   $\frac{7}{21} = \frac{1}{3}$ $\frac{5}{7} = \frac{55}{70}$

4. Which value of *a* makes the following a true proportion?

   $$\frac{a}{12} = \frac{3}{4}$$

   ____

5. Solve. $\frac{x}{8} = \frac{5}{10}$ ____________

6. Solve. $\frac{20}{y} = \frac{25}{2.5}$ ____________

7. Solve. $\frac{2a}{3} = \frac{4}{6}$ ____________

8. Solve for *a*. $\frac{4}{a} = \frac{1}{12}$ ____________

9. Martha drives 40 miles in 30 minutes. At this rate how far will she drive in 90 minutes?

   ____

10. Jim's car runs 32 miles on 1 gallon of gas. At this rate how many gallons of gas will he use to travel 112 miles?

    ____

11. Mario is paid $5 for every 2 hours of work. How much does he earn for working 10 hours?

    ____

12. Jon drives $\frac{6}{10}$ mile in 1 minute. At this rate how far will he drive in 1 hour?

    ____

13. A package of 4 light bulbs costs $3.96. What is the unit price?

    ____

14. If 4 light bulbs cost $3.96, how much would 20 light bulbs cost?

    ____

Go on to the next page.

Name ______________________________

Posttest

15. A package of crackers weighing 10 ounces costs $1.90; a 16-ounce package costs $2.40. Which package has the lower unit price, and by how much per ounce?

____

16. If 6 oranges cost $1.98, how many oranges can be bought for $5.94?

____

17. On a scale drawing, 1 inch is used to represent 2 feet. Ring the ratio expressing that relationship.

1:4 1:12 1:24 2:24

18. On Map A, 1 inch = 5 miles. On Map B, 1 inch = 10 miles. A road is 3 inches long on Map A. How long will the road be on Map B?

____

19. Use the scale 3 cm = 8 m. What is the actual length of an object when the scale length is 27 cm?

____

20. Use the scale 1 cm = 25 cm. What is the actual length of an object when the scale length is 0.5 cm?

____

21. John is making a scale drawing on which each centimeter represents 5 m. A field drawn to scale measures 15 cm by 20 cm. What are the actual dimensions of the field?

____

22. On a map 1 inch = 10 miles. Two cities are 10 inches apart on the map. How many miles are they actually apart?

____

For questions 23–24, write *always, sometimes,* or *never.*

23. Similar triangles are congruent.

____

24. Similar figures have different shapes.

____

25. Two rectangles are similar. The first rectangle has a width of 8 m and a length of 12 m. The width of the second rectangle is 24 m. Find its length.

____

26. Two rectangles are similar. The first rectangle has a width of 3 cm and a length of 6 cm. The width of the second rectangle is 0.5 cm. Find its length.

____

Go on to the next page.

Name ______________________________

Posttest

27. A triangle has sides measuring 6 cm, 8 cm, and 10 cm. Ring the answer that could be the lengths of the sides of a similar triangle.

6 km, 12 km, 15 km

7 cm, 9 cm, 11 cm

9 cm, 12 cm, 20 cm

9 m, 12 m, 15 m

28. These two triangles are similar. Find *a*.

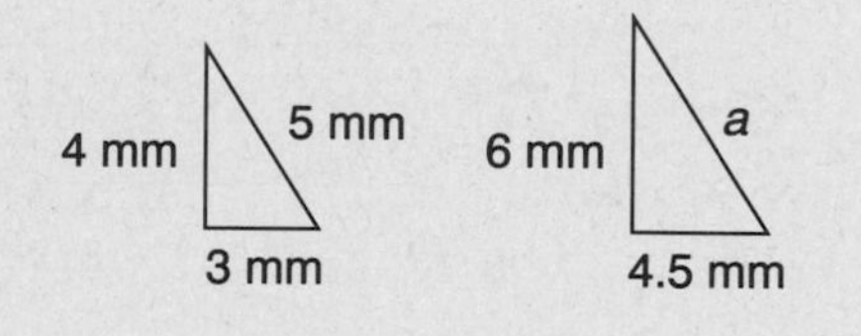

____

29. A map scale is 2 cm = 9 km. The map distance from Redville to Leetown is 18 cm. The map distance from Leetown to Spring is 3 cm. How far is it from Redville to Spring through Leetown?

____

30. The scale on a map is 1 inch = 150 miles. Two cities are 3 inches apart. If Tom averages 50 miles per hour and makes no stops, how long will it take him to drive from one city to the other?

____

Use the information below to answer questions 31–32.

The formula for finding the circumference of a circle is $C = \pi \cdot d$, where C = the circumference and d = the diameter.

31. Use $\frac{22}{7}$ as an approximate value of π. Find the circumference of a circle with a diameter of 28 in.

____

32. Use 3.14 as an approximate value of π. Find the diameter of a tree with a circumference of 8.32 m.

____

Stop!

Name ______________________________

Pretest

Read each question. Find the answer.

1. Ring the answer that is not equivalent to the others.

   18% 1.8 $\frac{9}{5}$ 180%

2. Ring the fraction that can be most easily expressed as a percent.

   $\frac{7}{22}$ $\frac{3}{7}$ $\frac{2}{10}$ $\frac{1}{16}$

3. What is $\frac{90}{1,000}$ as a percent?

   ____

4. What is 0.7 as a percent?

   ____

5. What is 96% as a fraction in simplest form?

   ____

6. What is $\frac{4}{10,000}$ as a decimal?

   ____

7. A bookstore's sales increased by 25% over last year's sales. By what fraction did the store's sales increase?

   ____

8. In 1992 attendance at a movie theater increased by $\frac{3}{20}$ in comparison to the attendance in 1991. By what percent did attendance increase?

   ____

9. Ring the best estimate. 19% of 397

   50 80 90 100

10. Ring the percent that is the best estimate of 6 out of 17.

    16% 20% 25% 33%

11. In a class of 30 students, 8 were absent. Ring the equation that shows what percent were absent.

    $n\% \times 30 = 8$ $n\% \times 8 = 30$

    $n\% = 8 \times 30$ $n\% \times 30 \div 8$

12. John saves $1.50 per week. That is equal to 20% of his allowance. Ring the correct equation for finding John's allowance.

    $n = 20\% \times 1.50$

    $n = 20\% \div 1.50$

    $n \times 20\% = 1.50$

    $n \times 1.50 = 20\%$

13. Solve. 160% of 200

    ____

14. Solve. $\frac{1}{2}$% of 400

    ____

15. 25 is what percent of 125?

    ____

16. 20% of what number is 6?

    ____

Go on to the next page.

Pretest

17. The Jacksons make a down payment of $15,000 on a house. The down payment is 20% of the total price. What is the price of the house?

____

18. Sarah bought a used car for $3,500 plus sales tax. The tax was $210. What was the percent of sales tax?

____

19. Mara's Hardware Store usually sells a toolbox for $19. Today the store has a sale in which all prices are reduced by 20%. What is the sale price of the toolbox?

____

20. Justin's meal at the restaurant costs $15. He wants to leave a 15% tip. How much should he leave for the tip?

____

21. What is the sum of the degrees in the sections of a circle graph?

____

22. Rob made a circle graph to show how he spends his leisure time. A central angle of 72° represents what percent of his time?

____

Use the circle graph below to answer questions 23–24.

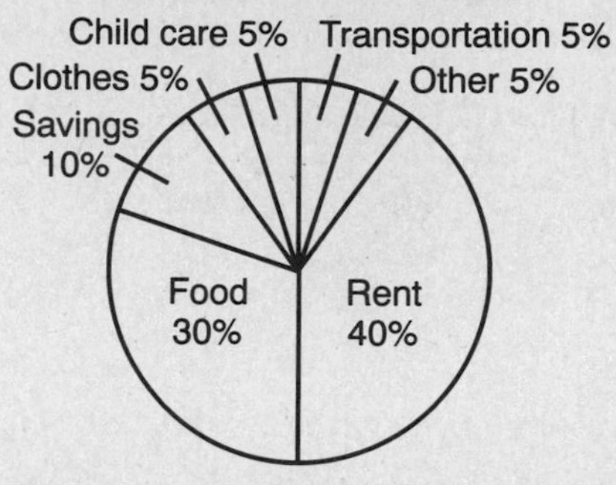

23. The McDonald family income is $1,500 per month. How much do they spend per month on food?

____

24. How much do the McDonalds spend on clothes and child care combined?

____

25. Sales tax is 6%. What is the tax on an item that costs $32?

____

26. An item that costs $80 is reduced by 25%. What is the sale price?

____

27. Juan buys a sweater that was discounted 25%. He pays $15. What was the original price of the sweater?

____

Go on to the next page.

Name ______________________

Pretest

28. Alexa borrows $800 for 6 months. The interest rate is 11% per year. How much interest must Alexa pay?

____

29. Two girls' ages add to 20. One girl is three times as old as the other. How old is the older girl?

____

30. Two numbers add to 52. One of the numbers is 10 more than the other. What are the numbers?

____

Use the circle graph below to answer questions 31–32.

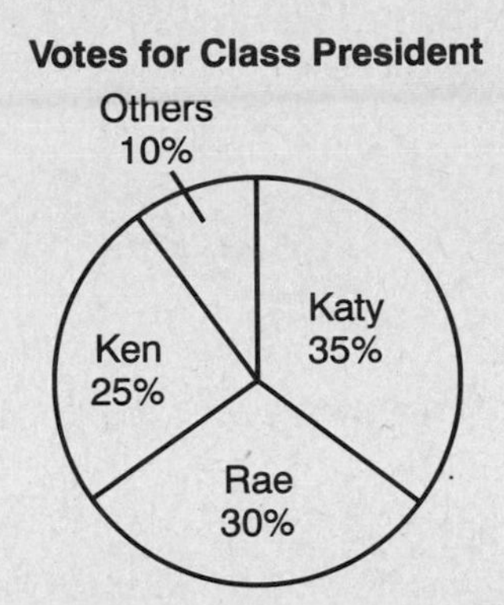

31. If 440 students voted, how many votes did Katy get?

____

32. Only two students other than Katy, Rae, and Ken received any of the 440 votes. If Sy received 18 of those votes, how many did Julia receive?

____

Stop!

Name ____________________

Posttest

Read each question. Find the answer.

1. Ring the answer that is not equivalent to the others.

   0.9% $\frac{9}{1,000}$

   0.009 0.09%

2. Ring the answer with the greatest value.

   $\frac{4}{1,000}$ 0.3% 0.02 0.06%

3. What is 0.17 as a percent?

   ____

4. What is 0.15 as a percent?

   ____

5. What is 85% as a fraction in simplest form?

   ____

6. What is 150% as a decimal?

   ____

7. A company's January sales dropped by $\frac{1}{5}$ compared to December sales. January sales were what percent of December sales?

   ____

8. In 1992 a magazine's subscribers doubled in number. What was the percent increase in subscribers?

   ____

9. Ring the best estimate. 30% of 189

   35 60 90 700

10. Ring the percent that is the best estimate of 26 out of 36.

    50% 60% 75% 90%

11. Danielle had $18.50. She spent $15.00 on clothes. Ring the correct equation for finding what percent of her money Danielle spent.

    $n\% = \frac{15}{18.5}$

    $n\% = \frac{18.5}{15}$

    $n\% \times 15 = 18.5$

    $n\% = 15 \times 18.5$

12. Phil invited 12 friends to a party. Only 50% of them could come. Ring the equation that shows how many came to the party.

    $n = \frac{12}{50}$

    $n \times 0.5 = 12$

    $n = \frac{0.5}{12}$

    $n = 0.5 \times 12$

13. Solve. 125% of 300 __________

14. Solve. 30% of 4 __________

15. 2.5 is what percent of 50?

    ____

16. 15% of what number is 60?

    ____

Go on to the next page.

Posttest

17. Mr. Mosconi receives an 8% raise. The amount of the raise is $2,800. What was his salary before the raise?

____

18. Sandy bought a ring on sale for $560. Its original price was $700. What was the percent of discount?

____

19. Anna's Dress Shop is having a sale with all prices reduced by 20%. What is the sale price of a sweater that originally cost $35?

____

20. George buys a concert ticket through the mail. He pays $25 plus a 5% service charge. How much does George pay for the ticket?

____

21. A family makes a circle graph to show its budget. One section has a central angle of 54°. What percent of the budget is represented by the angle?

____

22. How many degrees are in 75% of a circle graph?

____

Use the circle graph below to answer questions 23–24.

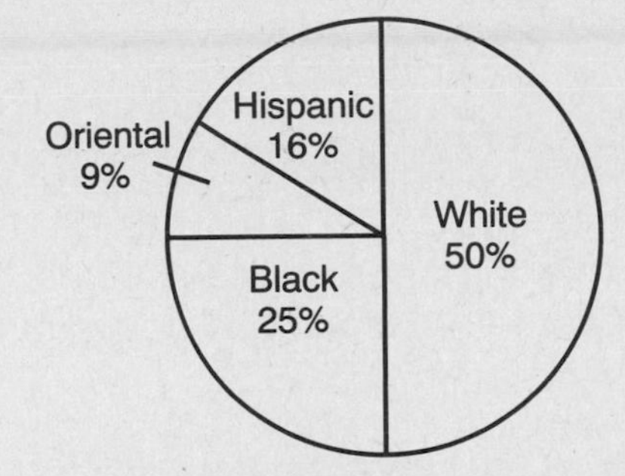

Ethnic Breakdown at Lincoln High

23. There are 1,050 black students at Lincoln High. What is the total number of white students at the school?

____

24. One half of the Hispanic students at Lincoln High take a biology class. If there are 1,200 students in the school, how many Hispanic students take a biology class?

____

25. The sales tax is 5%. What is the tax on an item that costs $24?

____

26. An item that costs $90 is discounted by 60%. What is the sale price?

____

27. Betty buys a bicycle that was discounted 20%. She pays $100. What was the original price of the bicycle?

____

Go on to the next page.

Posttest

28. Alex borrows $500 for 24 months. The interest rate is 9% per year. How much interest must Alex pay?

____

29. The total number of points scored in a football game was 30. The winning team scored 4 times as many points as the losing team. How many points did the winning team score?

____

30. Henry and Anne went fishing. For every 2 fish that Henry caught, Anne caught 3. Together, they caught 15 fish. How many fish did Anne catch?

____

Use the circle graph below to answer questions 31–32.

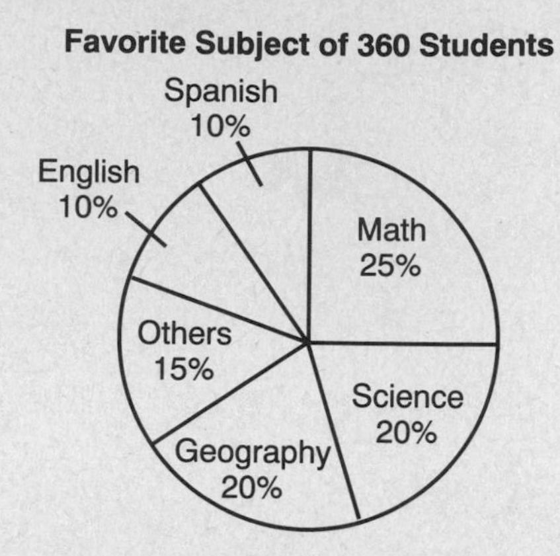

31. How many students chose science as their favorite subject?

____

32. If 20 students chose health as their favorite subject, what is the largest number of students who could have chosen French?

____

Stop!

Name 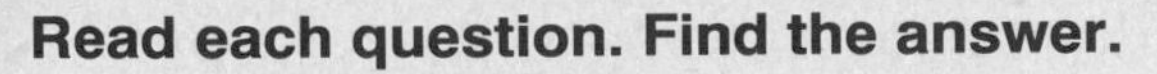

Cumulative Test

Read each question. Find the answer.

1. Juan bought 2 books for $1.95 each, 2 pads at $1.05 each, and a pen for $1.95. Ring the *best estimate* of how many dollars Juan spent.

   $5.00 $6.00

   $7.00 $8.00

2. Donna had $31.00. She spent $7.50 for dinner, $4.00 for a movie, and $1.50 for a drink at the movie. How much money did Donna have left?

   ____

3. What is the value of 5^1?

   ____

4. Express 7,302 in scientific notation.

   ____

Use the stem-and-leaf plot below to answer questions 5–6.

Number of Points Scored by a Football Team in 20 Games

Stem	Leaves
1	0 4 4 4 7 7 9
2	0 0 1 1 1 1 4 4 7
3	1 5
4	2 4

5. What was the least number of points scored by the team in any game?

   ____

6. In what percentage of its games did the team score 25 or more points?

   ____

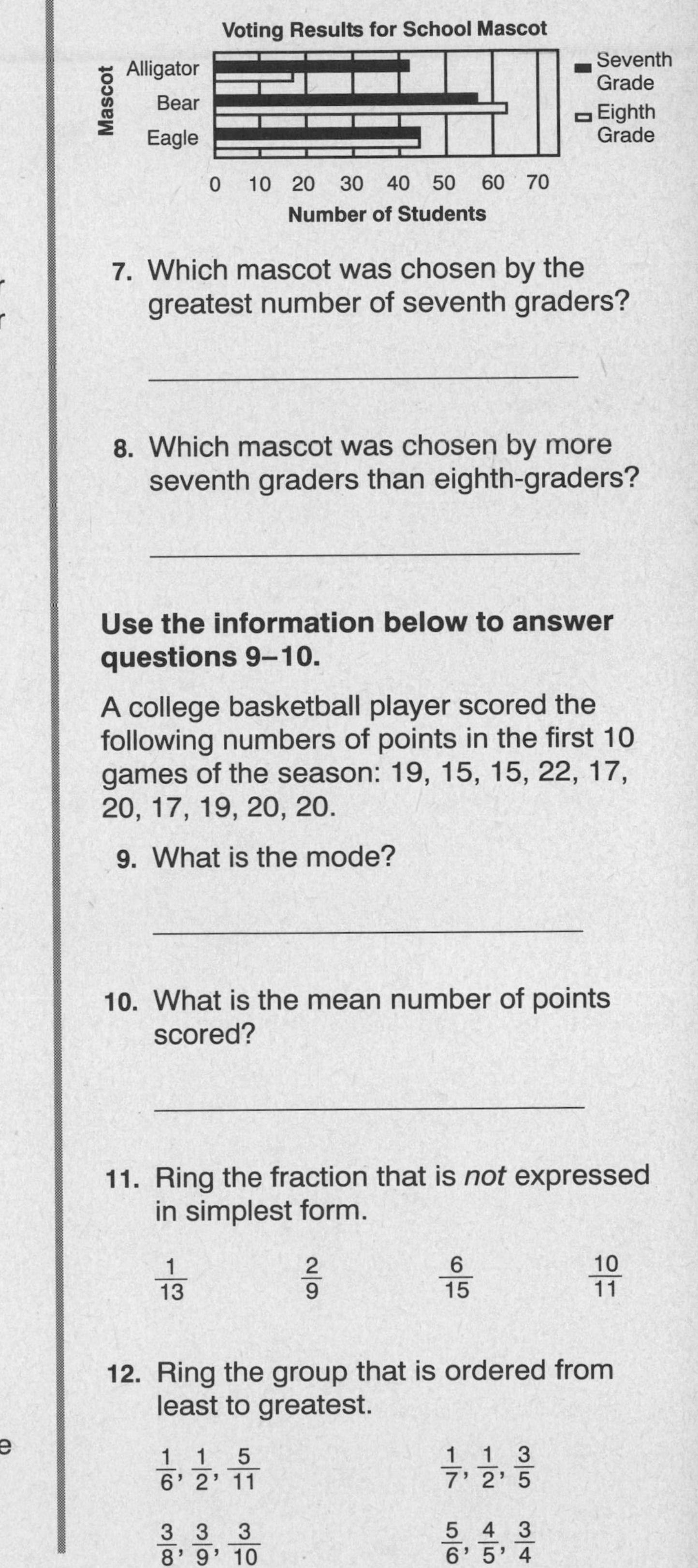

Use the double-bar graph to answer questions 7–8.

7. Which mascot was chosen by the greatest number of seventh graders?

   ____

8. Which mascot was chosen by more seventh graders than eighth-graders?

   ____

Use the information below to answer questions 9–10.

A college basketball player scored the following numbers of points in the first 10 games of the season: 19, 15, 15, 22, 17, 20, 17, 19, 20, 20.

9. What is the mode?

   ____

10. What is the mean number of points scored?

    ____

11. Ring the fraction that is *not* expressed in simplest form.

    $\frac{1}{13}$ $\frac{2}{9}$ $\frac{6}{15}$ $\frac{10}{11}$

12. Ring the group that is ordered from least to greatest.

    $\frac{1}{6}, \frac{1}{2}, \frac{5}{11}$ $\frac{1}{7}, \frac{1}{2}, \frac{3}{5}$

    $\frac{3}{8}, \frac{3}{9}, \frac{3}{10}$ $\frac{5}{6}, \frac{4}{5}, \frac{3}{4}$

Go on to the next page.

Name ____________________

Cumulative Test

13. Ring the fraction that is midway between $\frac{1}{50}$ and $\frac{2}{50}$.

$\frac{2}{100}$ $\frac{5}{200}$

$\frac{3}{100}$ $\frac{7}{200}$

14. $\frac{5}{12} + \frac{5}{8}$ ________

15. $7\frac{1}{5} - 6\frac{3}{5}$ ________

16. $1\frac{1}{3} \times 1\frac{1}{3}$ ________

17. $\frac{1}{5} \div \frac{1}{10}$ ________

18. Evaluate $\frac{2(a-6)}{b}$ for $a = 15$ and $b = 3$.

____

19. Find the value of $(5 + 4) \times (4 - 1)$.

____

20. Find the value of $3 + 6 + (4 - 1)^2$.

____

21. Ring the inequality that shows that a number, b, increased by 9 is less than 200.

$b - 9 < 200$ $b - 9 > 200$

$9b < 200$ $b + 9 < 200$

22. Solve the inequality. $x + 10 < 10$

____

23. Solve. $a - 20 = 0$ ________

24. Solve. $\frac{x}{12} = 6$ ________

25. Are right triangles isosceles? Write *always, sometimes,* or *never.*

____

26. What is the complement of a 75° angle?

____

27. Ring the ratio that does *not* have the same meaning as the other three.

$\frac{2}{15}$ $\frac{5}{30}$ $\frac{6}{45}$ $\frac{8}{60}$

28. Ring the values of a and b that make the following a true proportion.

$$\frac{a}{20} = \frac{6}{b}$$

$a = 4, b = 30$ $a = 4, b = 25$

$a = 3, b = 30$ $a = 12, b = 12$

29. A package of 6 cassette tapes costs \$7.80. What is the unit price?

____

30. If 8 cakes cost \$3.20 and 12 cakes cost \$4.68, which offer has the lower unit price?

____

Go on to the next page.

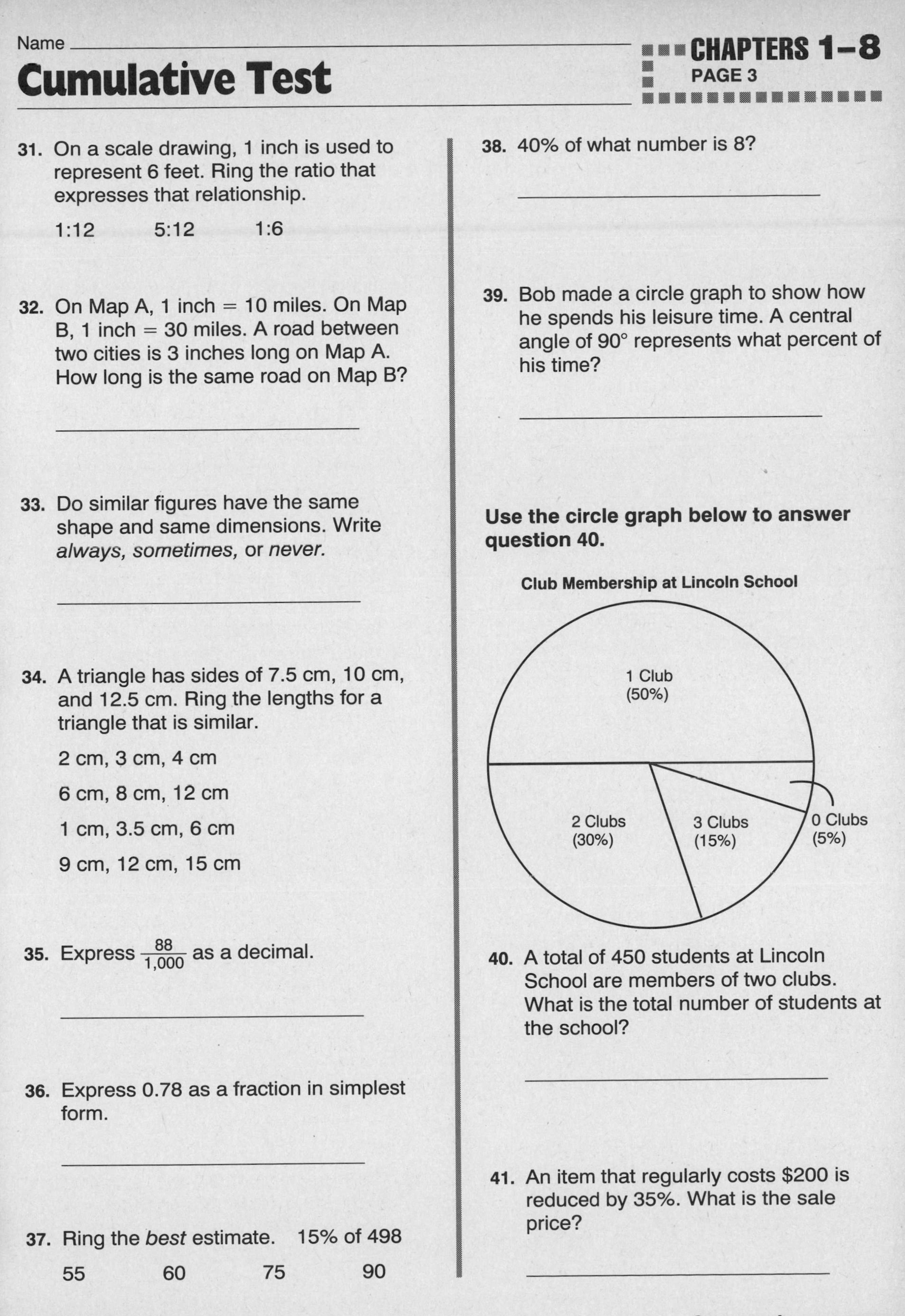

31. On a scale drawing, 1 inch is used to represent 6 feet. Ring the ratio that expresses that relationship.

1:12   5:12   1:6

32. On Map A, 1 inch = 10 miles. On Map B, 1 inch = 30 miles. A road between two cities is 3 inches long on Map A. How long is the same road on Map B?

____

33. Do similar figures have the same shape and same dimensions. Write *always, sometimes,* or *never.*

____

34. A triangle has sides of 7.5 cm, 10 cm, and 12.5 cm. Ring the lengths for a triangle that is similar.

2 cm, 3 cm, 4 cm

6 cm, 8 cm, 12 cm

1 cm, 3.5 cm, 6 cm

9 cm, 12 cm, 15 cm

35. Express $\frac{88}{1,000}$ as a decimal.

____

36. Express 0.78 as a fraction in simplest form.

____

37. Ring the *best* estimate.   15% of 498

55   60   75   90

38. 40% of what number is 8?

____

39. Bob made a circle graph to show how he spends his leisure time. A central angle of 90° represents what percent of his time?

____

Use the circle graph below to answer question 40.

40. A total of 450 students at Lincoln School are members of two clubs. What is the total number of students at the school?

____

41. An item that regularly costs $200 is reduced by 35%. What is the sale price?

____

Go on to the next page.

Name ______________________

Cumulative Test

42. Lee borrows $1,200. The interest rate is 10.5% per year. The payoff period is 6 months. How much interest will Lee pay? Use the formula $I = prt$.

____

Use the table below to answer question 43.

Students' Music Preferences		
Type of Music	**Boys**	**Girls**
Classical	30	70
Country	40	40
Jazz	60	35
Rock	110	135
	240	280

43. What percentage of boys prefer jazz?

____

44. Tony and Jen began a running program. Tony ran 2 miles on day 1 and increased the distance by $\frac{1}{4}$ mile each day. Jen ran 3 miles on day 1 and increased the distance by $\frac{1}{8}$ mile each day. On what day will Tony and Jen run the same distance?

____

45. Justin can bicycle one mile in $\frac{1}{4}$ hour. Don can bicycle one mile in $\frac{1}{2}$ hour. If both boys ride one hour, how much farther will Justin ride?

____

Use the information below to answer question 46.

Taxi fare is given by the formula $\$1.25 + \$0.90\,m$, where m = the distance traveled in miles.

46. Hal took a taxi for a distance of 4 miles. If he tipped the driver 15%, how much did Hal pay in all?

____

47. Sid withdrew money from his bank account. He spent $50 for shoes and $10 for a tie. He gave $\frac{2}{3}$ of what was left to his son. He then had $45. How much money did Sid withdraw?

____

48. The scale on a map is 1 inch = 100 miles. Two cities are 2 inches apart. If Tom averages 50 miles per hour and makes no stops, how long will it take him to drive from one city to the other?

____

49. Two numbers add up to 72. One of the numbers is 10 more than the other. What are the numbers?

____

Go on to the next page.

Name ______________________________

Cumulative Test

Use the information and the double-line graph below to answer question 50.

Each December, a company predicts its sales (in millions) for the coming year. The double-line graph below shows the predicted and actual sales for the years 1989 through 1992.

Predicted and Actual Sales: 1989–1992

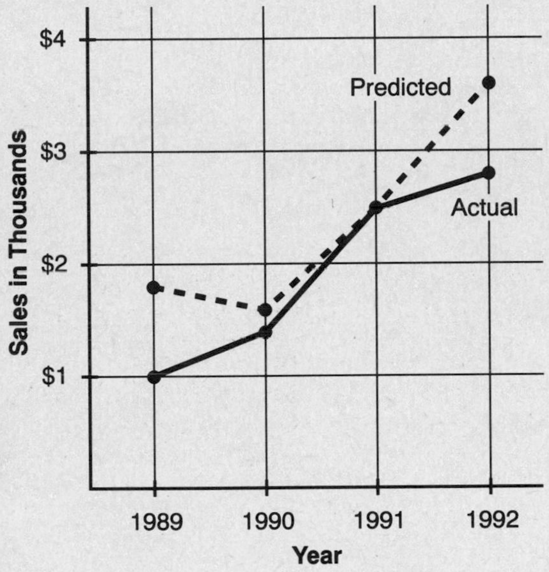

$4
$3
$2
$1
Sales in Thousands
Predicted
Actual
1989 1990 1991 1992
Year

50. During which year did predicted sales equal actual sales?

____

Stop!

Name ______________________________

Pretest

Read each question. Find the answer. Write *always, sometimes,* or *never* for questions 1–3.

1. Is the absolute value of a negative integer positive?

   ____

2. Is the opposite of a negative integer a positive integer?

   ____

3. Is the absolute value of a negative integer less than the absolute value of its opposite?

   ____

4. Ring the answer with the greatest value.

   6　　　$|^-9|$　　　$^+8$　　　$^-100$

For questions 5–6, ● counters are positive integers, and ○ counters are negative integers.

5. What integer does the model represent?

   ____

6. What does the model represent? Ring the answer.

   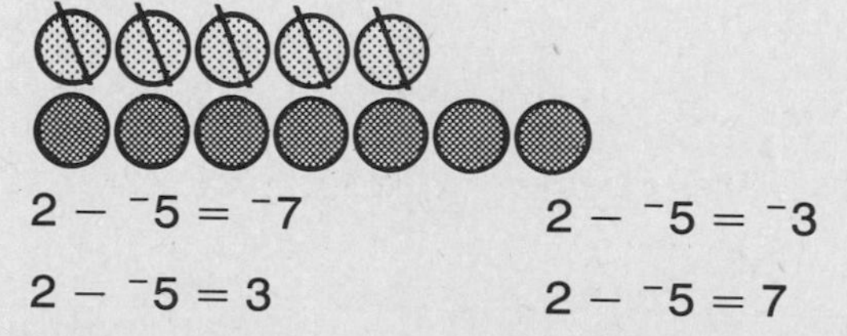

   $2 - {}^-5 = {}^-7$　　　$2 - {}^-5 = {}^-3$

   $2 - {}^-5 = 3$　　　$2 - {}^-5 = 7$

7. When one negative integer is divided by another negative integer, will the result be positive or negative?

   ____

8. When an integer is subtracted from a greater integer, will the result be positive or negative?

   ____

9. Solve. $^-9 + {}^-6$ __________

10. Solve. $^-6 + {}^-2$ __________

11. Solve. $^-5 \times {}^-7$ __________

12. Solve. $3 - {}^-4$ __________

13. One cold night, the temperature dropped from 3°F to $^-18$°F. How many degrees did the temperature drop?

    ____

14. Josie had $118 on May 1. During the month, she spent $214 and earned $108. How much did she have at the end of the month?

    ____

15. Alicia arrived 25 min late for a meeting. Cindy arrived 12 min earlier. How late was Cindy?

    ____

Go on to the next page.

Name ________________________________

Pretest

16. Five men went on a diet. Two men lost 10 pounds each. Two men lost 6 pounds each. The other man gained 3 pounds. How much did the men lose altogether?

____

17. Jo studied 4 hr less than 3 times the number of hours Moe studied. Ring the expression that shows the number of hours Jo studied. Let m = the number of hours Moe studied.

$4 - 3m$ $\quad$ $3m - 4$

$4m - 3$ $\quad$ $3 - 4m$

18. To solve the equation $x - 7 = 22$, what must you do to both sides of the equation?

____

19. Solve for y. $y - 3 = {}^-6$

____

20. Solve for x. $x + 2 = 1$

____

21. Solve for n. $\frac{n}{3} = {}^-7$

____

22. Solve for p. $p \cdot {}^-5 = {}^-10$

____

23. Gary spent \$6 in a restaurant and \$4 at a movie. He had \$11 left. How much money did he start with?

____

24. A baseball team won $\frac{1}{3}$ of its games. The team won 15 games. How many games did the team play?

____

Use the graph below to answer questions 25–26.

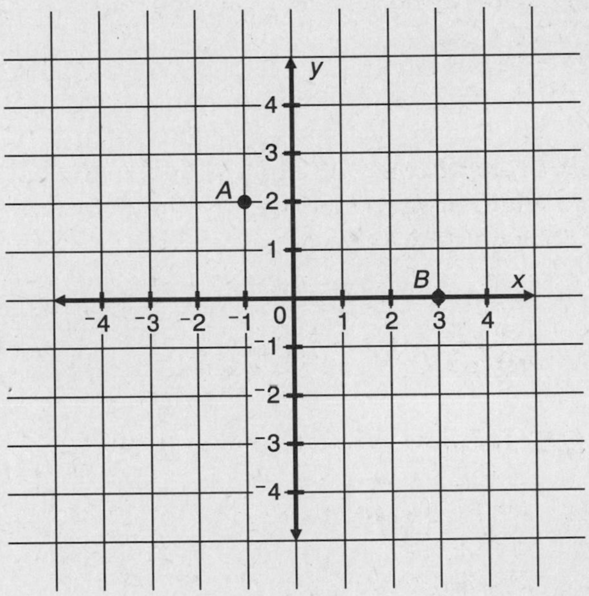

y
4
3
A
2
1
B
x
−4
−3
−2
−1
0
1
2
3
4
−1
−2
−3
−4

25. What are the coordinates of point A?

____

26. What are the coordinates of point B?

____

Go on to the next page.

Name ______________________________

Pretest

Use the table below to answer questions 27–28.

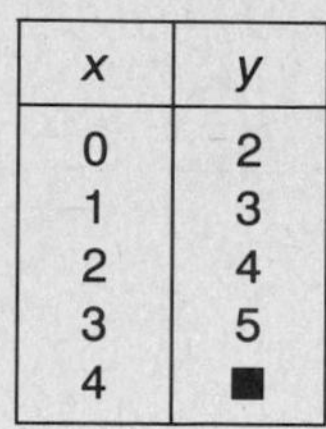

x	y
0	2
1	3
2	4
3	5
4	■

27. Ring the equation that describes the pattern that relates y to x.

$y = 2x$ $\qquad$ $y = x + 2$

$y = \frac{x}{2}$ $\qquad$ $y = x - 2$

28. Ring the ordered pair that would be graphed when $x = 4$.

(4,6) $\qquad$ (4,8)

(6,4) $\qquad$ (8,4)

29. Helena and Ruth bought some apples. Helena bought 2 more than 3 times the number of apples Ruth bought. If Helena bought 8 apples, how many did Ruth buy?

____

30. Boris receives three bills in the mail. The electric bill is 3 times as much as the phone bill. The gas bill is $\frac{1}{2}$ the amount of the phone bill. The gas bill is \$20. How much is the electric bill?

____

31. Nadia spent \$10.00 on books. She gave $\frac{1}{2}$ of the money she had left to her sister. Nadia then had \$1.50. How much money did she have before buying books?

____

32. Tricia got 42 questions correct on a test. Her score was 2 less than $\frac{1}{2}$ of Nan's score. How many questions did Nan get correct?

____

Stop!

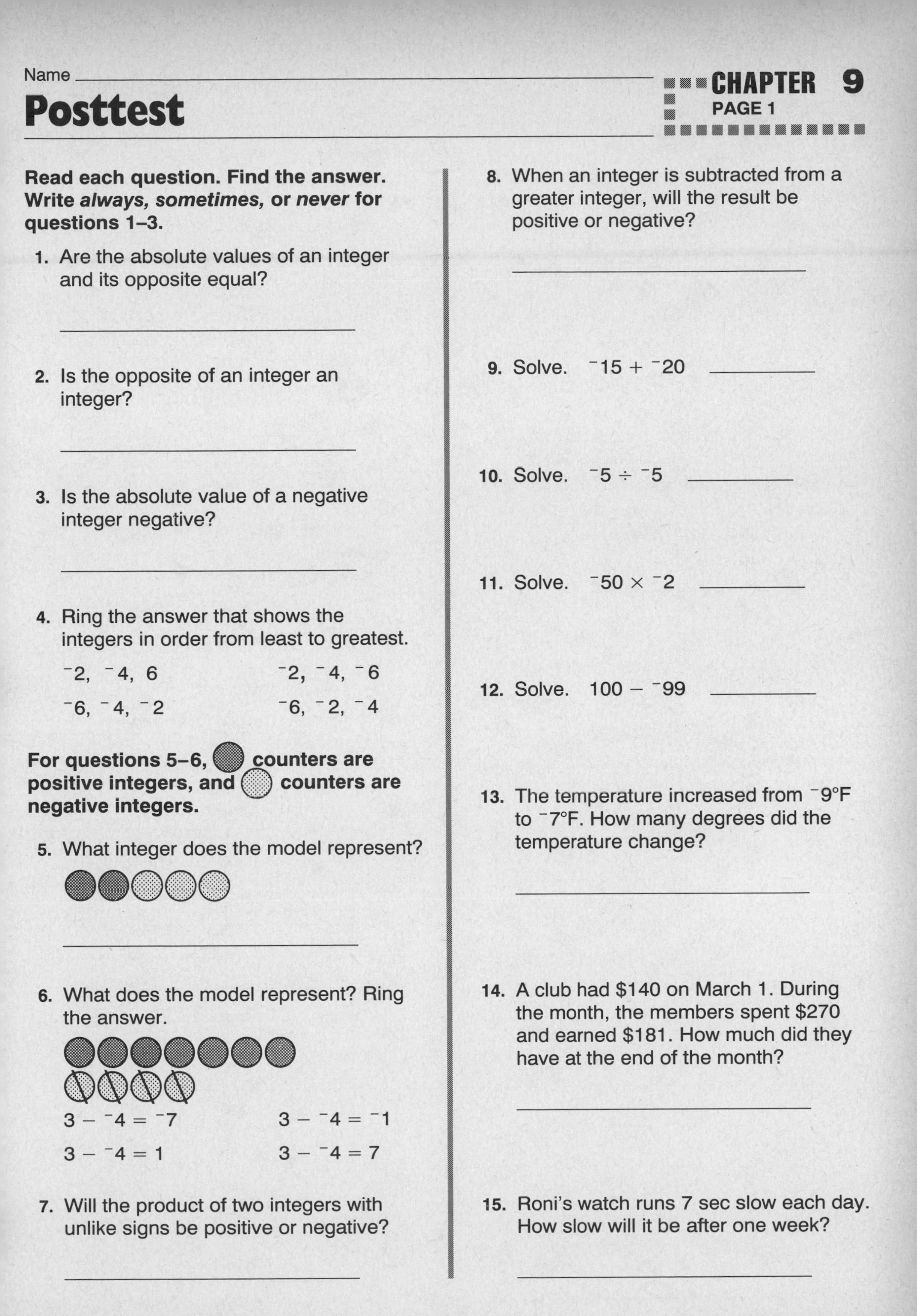

Name ____________________

Posttest

Read each question. Find the answer. Write *always, sometimes,* or *never* for questions 1–3.

1. Are the absolute values of an integer and its opposite equal?

   ____

2. Is the opposite of an integer an integer?

   ____

3. Is the absolute value of a negative integer negative?

   ____

4. Ring the answer that shows the integers in order from least to greatest.

   $^{-}2,\ ^{-}4,\ 6$　　　$^{-}2,\ ^{-}4,\ ^{-}6$

   $^{-}6,\ ^{-}4,\ ^{-}2$　　　$^{-}6,\ ^{-}2,\ ^{-}4$

For questions 5–6, (dark) counters are positive integers, and (light) counters are negative integers.

5. What integer does the model represent?

   ____

6. What does the model represent? Ring the answer.

   $3 - {}^{-}4 = {}^{-}7$　　　$3 - {}^{-}4 = {}^{-}1$

   $3 - {}^{-}4 = 1$　　　$3 - {}^{-}4 = 7$

7. Will the product of two integers with unlike signs be positive or negative?

   ____

8. When an integer is subtracted from a greater integer, will the result be positive or negative?

   ____

9. Solve. $^{-}15 + {}^{-}20$ ____________

10. Solve. $^{-}5 \div {}^{-}5$ ____________

11. Solve. $^{-}50 \times {}^{-}2$ ____________

12. Solve. $100 - {}^{-}99$ ____________

13. The temperature increased from $^{-}9$°F to $^{-}7$°F. How many degrees did the temperature change?

    ____

14. A club had $140 on March 1. During the month, the members spent $270 and earned $181. How much did they have at the end of the month?

    ____

15. Roni's watch runs 7 sec slow each day. How slow will it be after one week?

    ____

Go on to the next page.

16. The 10 members of the Garden Club went on diets. One half of the members lost 8 pounds each. Three members lost 5 pounds each. Two members gained 2 pounds each. What was the total amount lost?

______

17. Pat works 3 hr less per week than $\frac{1}{2}$ the time Lyn works. Ring the expression that shows the number of hours per week Pat works. Let h = the number of hours Lyn works.

$3 - \frac{1}{2}h$ $\frac{1}{2}h - 3$

$3h - \frac{1}{2}$ $\frac{1}{2} - 3h$

18. To solve the equation $4x = 32$, what must you do to both sides of the equation?

______

19. Solve for y. $y - 11 = {}^{-}12$

______

20. Solve for x. $x + 12 = 2$

______

21. Solve for n. $\frac{n}{10} = {}^{-}3$

______

22. Solve for p. $p \cdot {}^{-}1 = {}^{-}99$

______

23. Hal gave \$15 to his brother and \$10 to each of his 2 sisters. He was left with \$44. How much money did he start with?

______

24. A soccer team lost $\frac{3}{4}$ of its games. The team won 8 games. How many games did the team play?

______

Use the graph below to answer questions 25–26.

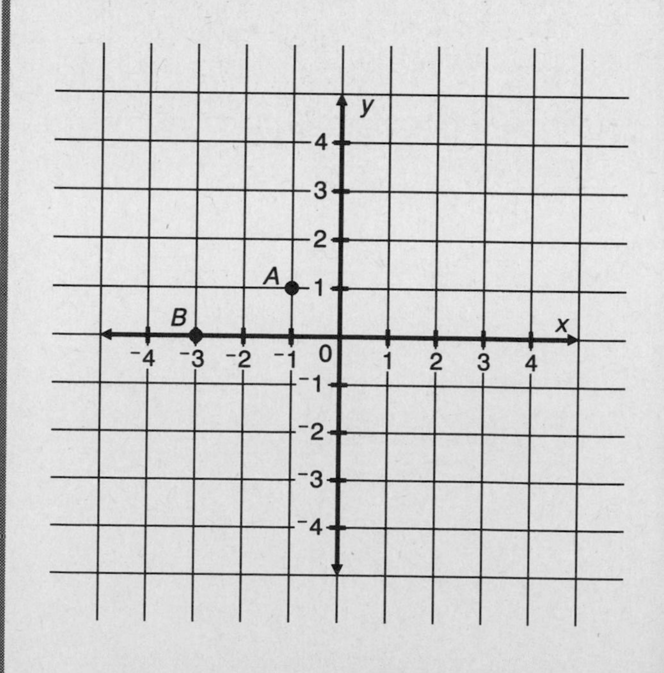

25. What are the coordinates of point A?

______

Go on to the next page.

Posttest

26. What are the coordinates of point B?

____

Use the table below to answer questions 27–28.

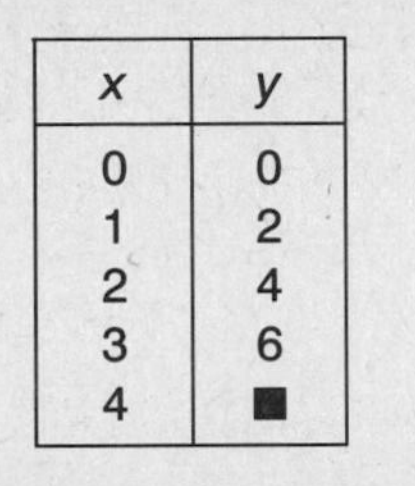

x	y
0	0
1	2
2	4
3	6
4	■

27. Ring the equation that describes the pattern that relates y to x.

$y = 2x$ $\quad$ $y = x + 2$

$y = \frac{x}{2}$ $\quad$ $y = x - 2$

28. Ring the ordered pair that would be graphed when $x = 4$.

(4,7) $\quad$ (4,8)

(7,4) $\quad$ (8,4)

29. Helen played a game three times. Her second score was twice as high as her first score. Her third score was 5 points higher than her second score. Helen's third score was 225. What was her first score?

____

30. Mike got 5 words wrong on a spelling test. His sister got 3 more than twice as many words wrong. How many words did his sister get wrong?

____

31. Emile spent \$50 on groceries. He spent $\frac{1}{5}$ of the money on fruit, $\frac{1}{2}$ on dairy items, and the rest on meat. How much did he spend on meat?

____

32. Bill weighs twice as much as Phil. John weighs $\frac{2}{3}$ as much as Phil. Bill weighs 240 pounds. How much does John weigh?

____

Stop!

Name ______________________________

Pretest

Read each question. Find the answer. Write *always, sometimes,* or *never* for questions 1–4.

1. Are mixed numbers rational numbers?

____

2. Are the square roots of numbers that are not perfect squares irrational numbers?

____

3. Does a fraction whose denominator has only 2 and 5 as prime factors result in a repeating decimal?

____

4. Are irrational numbers real numbers?

____

5. Express $\frac{39}{50}$ as a decimal.

____

6. Express 0.825 as a percent.

____

7. Express 1.75 as a percent.

____

8. Ring the fraction that can be expressed as a repeating decimal.

$\frac{1}{16}$ $\frac{6}{9}$ $\frac{3}{20}$ $\frac{12}{40}$

9. Ring the fraction that can be expressed as a terminating decimal.

$\frac{3}{7}$ $\frac{6}{18}$ $\frac{1}{12}$ $\frac{3}{25}$

10. Ring the square root that is an irrational number.

$\sqrt{4}$ $\sqrt{8}$ $\sqrt{16}$ $\sqrt{81}$

11. Write $<$, $>$, or $=$. $^{-}1{,}000 \bigcirc 0.0001$

12. Write $<$, $>$, or $=$. $\frac{^{-}1}{50} \bigcirc {^{-}0.5}$

13. Ring the rational number between 0.870 and $\frac{7}{8}$.

0.869 0.871

$\frac{13}{16}$ 0.876

14. Ring the rational number between $^{-}0.50$ and $^{-}0.51$.

$^{-}0.499$ $^{-}0.511$

$^{-}0.512$ $^{-}0.501$

15. Order from least to greatest.

$\frac{6}{7}, \frac{7}{8}, \frac{^{-}5}{6}, \frac{^{-}9}{8}$

____

16. Order from least to greatest.

$\frac{1}{6}, \frac{1}{7}, 0.17, 0.14$

____

Go on to the next page.

Name ____________________

Pretest

17. Write <, >, or =. $\frac{1}{4^3} \bigcirc \frac{1}{2^5}$

18. Write <, >, or =. $\frac{1}{3^7} \bigcirc \frac{1}{4^8}$

19. Find the value of 3^4.

____

20. What is 3^{-4} expressed with a positive exponent?

____

21. What is 4.2×10^3 in standard form?

____

22. What is 502,000,000 in scientific notation?

____

23. Ring the perfect square.

24 44 99 144

24. Ring the best estimate of $\sqrt{83}$.

8.9 9.1 10.1 11.2

25. Find the value of $\sqrt{121}$.

____

26. Find the value of $\sqrt{\frac{9}{16}}$.

____

27. Find the value of $\sqrt{0.0049}$.

____

28. Find the value of $\sqrt{10,000}$.

____

Use the table below to answer questions 29–32.

Students Attending Grant High

Year	Number
1970	1,050
1975	1,100
1980	1,200
1985	1,350
1990	1,550

29. How many students attended Grant High in 1980?

____

30. If the trend continues, how many students will attend Grant High in 1995?

____

31. If the trend continues, how many students will attend Grant High in 2005?

____

32. How much did the student population increase from 1970 to 1990?

____

Stop!

Name ______________________________

Posttest

Read each question. Find the answer. Write *always, sometimes,* or *never* for questions 1–4.

1. Are fractions rational numbers?

____

2. Are the square roots of numbers that are perfect squares irrational numbers?

____

3. Can fractions with a denominator of 15 be changed to terminating decimals?

____

4. Are rational numbers real numbers?

____

5. Express 0.905 as a percent.

____

6. Express $\frac{19}{20}$ as a decimal.

____

7. Express 2.5 as a percent.

____

8. Express $\frac{9}{10,000}$ as a decimal.

____

9. Ring the fraction that can be expressed as a repeating decimal.

$\frac{1}{8}$ $\frac{1}{16}$ $\frac{1}{24}$ $\frac{1}{32}$

10. Ring the square root that is an irrational number.

$\sqrt{64}$ $\sqrt{95}$

$\sqrt{121}$ $\sqrt{144}$

11. Write <, >, or =. $^-0.003 \bigcirc \frac{^-2}{1,000}$

12. Write <, >, or =. $\frac{^-1}{99} \bigcirc \frac{^-2}{99}$

13. Ring the rational number between $\frac{99}{100}$ and 0.991.

0.9899 0.9911

$\frac{992}{1,000}$ 0.9901

14. Ring the rational number between $^-0.001$ and $\frac{^-1}{500}$.

$^-0.0009$ $^-0.0011$

$^-0.0021$ $^-0.0091$

15. Order from least to greatest.

$\frac{1}{3}, \frac{1}{4}, \frac{^-1}{5}, \frac{^-1}{6}$

____

16. Order from least to greatest.

$\frac{1}{9}, \frac{1}{11}, 0.100, \frac{1}{8}$

____

Go on to the next page.

Name ______________________________

Posttest

17. Write $7 \times 7 \times 7 \times 7$ in exponent form.

____

18. Write <, >, or =. $\frac{1}{2^3} \bigcirc \frac{1}{3^2}$

19. What is 5^{-3} expressed with a positive exponent?

____

20. Find the value of $\frac{1{,}000}{10^4}$.

____

21. What is 1.2×10^{-3} in standard form?

____

22. What is 32,000 in scientific notation?

____

23. Ring the perfect square.

200 300 400 500

24. Ring the best estimate of $\sqrt{124}$.

10.9 11.1

12.4 13.1

25. Find the value of $\sqrt{144}$.

____

26. Find the value of $\sqrt{\frac{4}{25}}$.

____

27. Find the value of $\sqrt{0.09}$.

____

28. Find the value of $\sqrt{1{,}000{,}000}$.

____

Use the table below to answer questions 29–32.

Town Population

Year	Population
1986	2,000
1987	2,100
1988	2,300
1989	2,600
1990	3,000
1991	3,500

29. What was the town population in 1988?

____

30. If the trend continues, what will be the town population in 1992?

____

31. If the trend continues, what will be the town population in 1995?

____

32. By how many people did the population increase from 1986 to 1991?

____

Stop!

Name ______________________________

Pretest

Read each question. Find the answer.

1. Jan picks one month at random from a calendar. What is the number of possible outcomes?

   ____

2. Al has a spinner with five equal sections, labeled 3, 5, 7, 9, and 10. What is the number of possible outcomes of one spin?

   ____

3. A restaurant has 3 choices for soup, 6 possible main courses, and 5 desserts. If Sandra orders soup, a main course, and dessert, how many possible selections can she make?

   ____

4. What is the number of possible three-digit area codes? Assume that zero *cannot* be used as the first digit.

   ____

5. Can a greater number of 2-member committees be selected from a group of 6 people or from a group of 8 people?

   ____

6. Ring the word that provides the *least* number of possible letter arrangements.

   dog *bird*

   horse *flower*

7. How many different combinations of 2 items can be selected from a set of 8 items?

   ____

8. How many different arrangements of the letters *A, B, C, D* are possible?

   ____

9. A tennis team has 6 members. How many possible teams of 2 can be formed?

   ____

10. A basketball league has 4 teams. In how many different ways can the teams be arranged in the final standings?

    ____

11. A coin is flipped 3 times. What is the probability of obtaining a head on any *one* of the flips?

    ____

12. A coin is flipped 4 times. What is the most likely outcome?

    ____

Go on to the next page.

Name ______________________________

Pretest

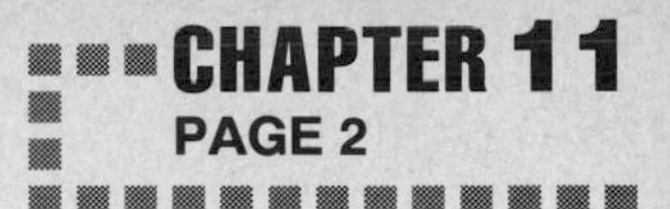

13. Hector randomly picks an integer between 0 and 9. Ring the answer that is the most likely outcome.

The number is 5.

The number is 9.

The number is odd.

The number is not 1.

14. Ring the answer that has the highest probability on one roll of a number cube.

a number greater than 4

a 4

a number less than 4

a 1 or a 2

15. A coin is flipped 4 times. What is the probability of obtaining 4 heads?

____

Use the information below to answer questions 16–18.

A spinner has 10 equal sections, numbered from 1 to 10.

16. What is the probability of spinning an odd number?

____

17. What is the probability of spinning a number greater than 6?

____

18. What is the best prediction of how many times the number 10 will be obtained in 900 spins?

____

Use the information below to answer questions 19–20.

A large jar is filled with red beans and blue beans. It holds 100 beans in all. Bea shakes the jar and removes a handful. Her handful contains 12 red beans and 18 blue beans.

19. What ratio should Bea use to predict the number of blue beans in the jar?

____

20. What is the best prediction of how many red beans are in the jar?

____

21. How should you compute the probability of two independent events both happening?

____

____

22. Ring the answer that does *not* describe independent events.

flipping a coin 2 times

rolling a number cube 2 times

spinning a spinner 2 times

drawing 2 balls from a box without replacing the first ball

Go on to the next page.

Name ____________________

Pretest

To answer questions 23–24, assume that a jar contains 3 red balls and 3 black balls. The first ball drawn is *not* replaced in the jar.

23. Helen takes 2 balls from the jar. What is the probability that both balls are black?

____

24. Mike takes 2 balls from the jar. What is the probability that they are the same color?

____

25. Juan flips a coin and rolls a number cube. What is the probability that the coin lands on heads and the number cube shows a 1?

____

26. Olga flips a coin 2 times. What is the probability that both flips come up heads?

____

27. Two baseball teams play 2 games. Since Team A is the better team, its chance of winning each game is $\frac{2}{3}$. What is the probability that Team A will win both games?

____

28. A jar contains 5 red and 5 white marbles. Jo takes 1 marble from the jar, does not replace it, and then removes another marble. What is the probability that both marbles are white?

____

29. Sam, Tina, and Ray each have a collection. Ray and the stamp collector are best friends. The coin collector lives next to Tina, who enjoys trading bells from her collection. Who has the coin collection?

____

Use the information below to answer questions 30–32.

Joe and Edna must take music, health, and art courses during a two-year period. Each student is randomly assigned one of the courses for the current semester. Assume that the probability of assignment to each course is $\frac{1}{3}$.

30. What is the probability that both students will be assigned the art course?

____

31. What is the probability that both students will be assigned any of the three courses together?

____

32. What is the probability that neither student will be assigned the music course?

____

Stop!

Name ______________________________

Posttest

Read each question. Find the answer.

1. Lisa picks one day of the week from the calendar. What is the number of possible outcomes?

   ____

2. John has a spinner with 8 equal sections, labeled 1, 2, 3, 4, 5, 6, 7, and 8. What is the number of possible outcomes?

   ____

3. Len is at a fruit stand. He has 6 kinds of fruit and 5 kinds of vegetables to choose from. If he buys 1 fruit and 1 vegetable, how many outcomes are possible?

   ____

4. What is the number of possible four-digit numbers? Do *not* use zero in the first place.

   ____

5. Can a greater number of 2-member teams be selected from a group of 4 girls or from a group of 5 girls?

   ____

6. Ring the word that provides the *greatest* number of possible letter arrangements.

   hat *noon*

   seven *jacket*

7. How many different combinations of 2 items can be selected from a set of 6 items?

   ____

8. How many different arrangements of the letters *A, B, C, D, E* are possible?

   ____

9. A baseball team has 12 members. The team needs 2 co-captains. How many different pairs of co-captains can be selected from the 12 members?

   ____

10. A basketball league has 7 teams. In how many different ways can the teams be arranged in the final standings?

    ____

Go on to the next page.

Posttest

11. What is the probability of an event that is impossible?

____

12. A coin is flipped 2 times. What is the most likely outcome?

____

13. Lupe randomly picks one letter of the alphabet. Ring the answer that is the most likely outcome.

The letter is *A.*

The letter is *Z.*

The letter is a vowel.

The letter is a consonant.

14. Ring the answer that has the highest probability on one roll of a number cube.

an odd number

an even number

a number less than 5

a number greater than 5

15. A coin is flipped 3 times. What is the probability of obtaining 3 heads?

____

Use the information below to answer questions 16–18.

A spinner has 10 equal sections, numbered from 1 to 10.

16. What is the probability of spinning 1, 4, or 9?

____

17. What is the probability of spinning a number greater than 2?

____

18. What is the best prediction of how many times the number 2 will be obtained in 1,000 spins?

____

Use the information below to answer questions 19–20.

A large jar is filled with black candies and white candies. It holds 200 candies in all. Hal shakes the jar and removes a handful. His handful contains 10 white candies and 15 black candies.

19. What ratio should Hal use to predict the number of white candies in the jar?

____

20. What is the best prediction of the total number of black candies in the jar?

____

21. When the outcome of an event is *not* affected by the outcome of an earlier event, what are the events called?

____

Go on to the next page.

Name ______________________________

Posttest

22. Ring the pair of events that are most likely to be independent.

It rains heavily.
You carry an umbrella.

You are tall.
You get an A in math.

You are sick.
You do not go to school.

You study hard in science.
You get an A in science.

To answer questions 23–24, assume that a jar contains 3 red balls and 2 black balls. The first ball drawn is *not* replaced in the jar.

23. Anne takes 2 balls from the jar. What is the probability that both balls are black?

____

24. Jake takes 2 balls from the jar. What is the probability that he picks a red ball and then a black ball?

____

25. Donna flips a coin 3 times. What is the probability of obtaining 3 heads?

____

26. Lupe rolls a number cube, marked 1–6, twice. What is the probability of rolling a 6 both times?

____

27. Tom and Mae play chess against each other. Tom has a $\frac{2}{3}$ probability of beating Mae in each game. What is the probability of Tom winning 2 games in a row?

____

28. A jar contains 7 blue marbles and 3 black marbles. Rae removes 1 marble, does not replace it, and then removes another marble. What is the probability that both marbles are blue?

____

29. Bo, Cara, and Don are neighbors. They each have one pet: a dog, a hamster, or a turtle. Bo and the turtle owner are the same age. Bo walks his pet every morning. Don is afraid of turtles. Who owns the hamster?

____

Use the information below to answer questions 30–32.

Bob and Cari must take health and music during a two-year period. Each student is randomly assigned one of the courses for the current semester. Assume that the probability of assignment to each course is $\frac{1}{2}$.

30. What is the probability that both students will be assigned the health course?

____

31. What is the probability that neither student will be assigned the music course?

____

32. What is the probability that Bob and Cari will be assigned different classes?

____

Stop!

Name ____________________

Cumulative Test

Read each question. Find the answer.

1. Bill wants to buy 2 suits that cost $199.95 each and 3 shirts that cost $14.95 each. Ring the *best estimate* of how much money Bill needs.

   $245.00 $430.00

   $435.00 $445.00

2. What is the value of 3^4?

   ____

3. Express 506,000 in scientific notation.

   ____

Use the bar graph below to answer questions 4–5.

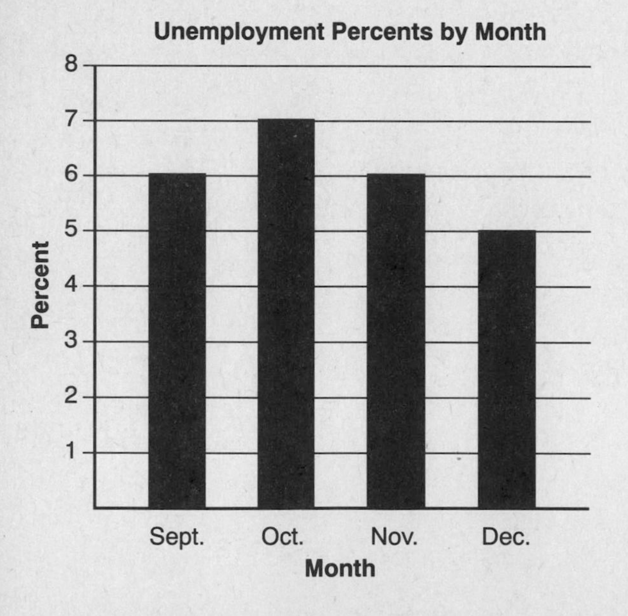

4. Which month had the greatest unemployment percent?

   ____

5. The unemployment percent in December was what fraction of the unemployment percent in November?

   ____

Use the information and box-and-whisker graph below to answer questions 6–7.

A clothing-store owner records the number of suits sold in his store each day for 30 days. He summarizes the information in the box-and-whisker graph below.

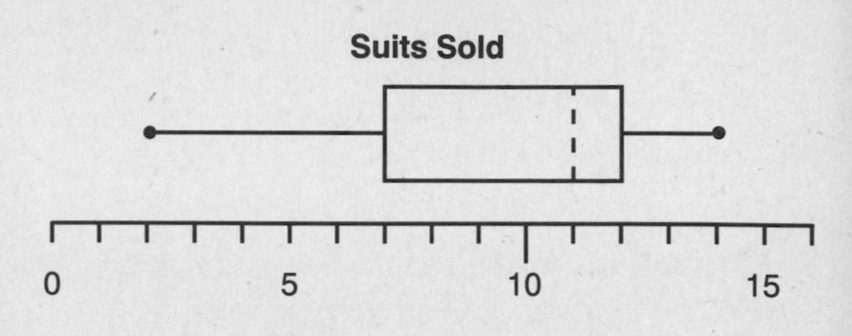

6. What was the median number of suits sold in his store each day?

   ____

7. What was the greatest number of suits sold on any of the 30 days?

   ____

Use the information below to answer questions 8–9.

Li's bowling scores for her last 10 games were 85, 90, 90, 90, 92, 92, 95, 97, 99, and 105.

8. What is the range of these scores?

   ____

9. What is the median of these scores?

   ____

10. Ring the fractions that are ordered from least to greatest.

    $\frac{5}{6}, \frac{4}{5}, \frac{3}{4}$ $\frac{1}{4}, \frac{1}{3}, \frac{1}{2}$

    $\frac{3}{9}, \frac{2}{9}, \frac{1}{9}$ $\frac{2}{3}, \frac{2}{4}, \frac{2}{5}$

Go on to the next page.

Name ____________________

Cumulative Test

11. Ring the fraction that falls between $\frac{8}{10}$ and $\frac{9}{10}$.

$\frac{15}{20}$ $\frac{31}{40}$

$\frac{17}{20}$ $\frac{7}{10}$

12. Solve. $13\frac{1}{4} - 10\frac{1}{6}$ ________

13. Solve. $3\frac{1}{2} \times 1\frac{1}{3}$ ________

14. Solve. $\frac{7}{8} \div \frac{8}{7}$ ________

15. Find the value of $(2 + 4) \times (7 - 2)$.

____

16. Solve the inequality. $x + 9 < 11$

____

17. Rectangles *A* and *B* are congruent. Rectangle *A* is 10 cm long by 20 cm wide. What are the dimensions of Rectangle *B*?

____

18. A box of 5 cakes costs $4.20. What is the unit price?

____

19. Use the scale 2 cm = 100 cm. What is the actual length of an object when the scale length is 5 cm?

____

20. Two rectangles are similar. The first rectangle has a width of 8 ft and a length of 20 ft. The width of the second rectangle is 10 ft. Find its length.

____

21. Express the ratio 9:12 as a percent.

____

22. Express 85% as a fraction in simplest form.

____

23. Phil invited 28 friends to a party. Only 25% of them could come. Ring the correct equation to find how many came to the party.

$n = \frac{28}{0.25}$ $n \times 0.25 = 28$

$n = \frac{0.25}{28}$ $n = 0.25 \times 28$

24. 3.5 is what percent of 70?

____

25. 15% of what number is 120?

____

Go on to the next page.

Name ____________________

Cumulative Test

26. Jan is buying a bicycle that is marked down 20%. What will be the sale price of a $200 bicycle?

____

27. Dick borrows $600. The interest rate is 9.5% per year. The payoff period is 24 months. How much interest will Dick pay? Use the formula $I = prt$.

____

28. Solve. $^{-}4 + {}^{-}3$ __________

29. Solve. $\frac{^{-}8}{^{-}4}$ __________

30. To solve the equation $x - 9 = 22$, what must you do to both sides of the equation?

____

31. Solve for y. $y + 6 = {}^{-}12$

____

32. Are mixed numbers rational? Write *always, sometimes,* or *never.*

____

33. Ring the square root that is an irrational number.

$\sqrt{4}$ $\sqrt{6}$ $\sqrt{9}$ $\sqrt{16}$

34. Write <, >, or =. $^{-}500$ ◯ 0.0006

35. Write <, >, or =. 2.30×10^3 ◯ 2.03×10^4

36. Solve. $\frac{10^4}{10^2}$ __________

37. Write <, >, or =. $\sqrt{144}$ ◯ 13

38. How many different combinations of 2 can be selected from a set of 6 items?

____

39. Ring the answer that has the highest probability on one roll of a number cube numbered from 1 to 6.

a 1

a 5

a number less than 3

a number greater than 3

40. A spinner has 8 equal-sized sections numbered from 1 to 8. What is the probability of spinning an even number?

____

To answer questions 41–42, assume that there is a vase containing 3 red balls and 3 black balls.

41. Helen takes 2 balls from the vase without replacement. What is the probability that both balls are black?

____

Go on to the next page.

Cumulative Test

42. Helen takes 1 ball from the vase, replaces it, and then takes a second ball. What is the probability that both balls are red?

____

43. The number of points scored by two football teams totals 50. One team scored 4 times as many points as the other. How many points did the winning team score?

____

44. Nadia spent $20 buying books. She gave $\frac{1}{4}$ of the remaining money to her sister. Nadia then had $15. How much money did she have before buying her books?

____

Use the table below to answer question 45.

Students Graduating from Grant High	
Year	**Number**
1986	1,250
1987	1,300
1988	1,400
1989	1,550
1990	1,750

45. Assuming the trend continued, how many students graduated from Grant High in 1991?

____

46. Tricia had 32 questions correct on a math test. Her score was 2 items more than $\frac{3}{4}$ of her twin sister's score. How many questions did Tricia's sister have correct?

____

47. Joel and Gary begin a swimming program. Joel swims 4 laps on day 1 and increases the distance by 1 lap each day. Gary swims 4 laps on day 1 and increases the distance by $1\frac{1}{2}$ laps each day. On day 7, how much farther will Gary swim than Joel?

____

48. There are 8 teams in a softball league. If each team plays each of the other teams twice, how many total games will be played?

____

49. Heidi watches a television program that lasts $\frac{1}{2}$ hour. Commercials make up $\frac{1}{6}$ of the show. What part of an hour does she spend watching the actual program?

____

50. Mrs. Chan drove at an average speed of 48 miles per hour. She drove a distance of 120 miles. How long did the drive take? (distance = rate × time)

____

Stop!

Name ______________________________

Pretest

Read each question. Find the answer.

1. Ring the most appropriate unit of measure for measuring the width of a closet.

   inch foot yard mile

2. Ring the situation in which a precise measurement would be necessary.

   determining your car's gas mileage

   determining how much paint is needed to paint your house

   planning how far you will drive on a vacation

   measuring the time to run a 50-yd dash at a track meet

3. Ring the best estimate for the height of a man.

   3,000 mm 2,000 cm

   1.7 m 0.5 km

4. Ring the best estimate for the weight of a banana.

   4 oz 4 lb 4 g 4 kg

5. Ring the best estimate of the amount of water it takes to fill an eyedropper.

   1 mL 0.5 L 1 qt $\frac{1}{4}$ pt

6. Ring the best estimate of the width of a desk.

   15 in. 3 ft 4 yd 4 m

7. Ring the most precise measurement.

   36 in. 3 ft

   1 yd $1\frac{1}{4}$ yd

8. Ring the most precise measurement.

   98 cm 1 m

   1,000 mL 1,000.0 mL

9. Ring the most precise measurement.

   4 qt $4\frac{1}{4}$ qt

   $1\frac{1}{2}$ gal $1\frac{3}{4}$ gal

10. Ring the most precise measurement.

    3,600 sec 60 min

    1 hr 1.0 hr

11. What is the perimeter of a regular pentagon when $s = 4\frac{1}{2}$ cm?

    ____

12. What is the circumference of a circle with a radius of 3 in.? Use $\pi = 3.14$.

    ____

13. Jim is making a basketball hoop from a metal bar. He wants the diameter of the hoop to be 50 cm. How long does the metal bar need to be? Use $\pi = 3.14$.

    ____

14. Doreen wants to put wooden trim along the top of the walls in her room. The room is 9 ft wide by 12 ft long. How many feet of trim does she need?

    ____

Go on to the next page.

Name ______________________________

Pretest

15. Write the name of the figure with the greater area.

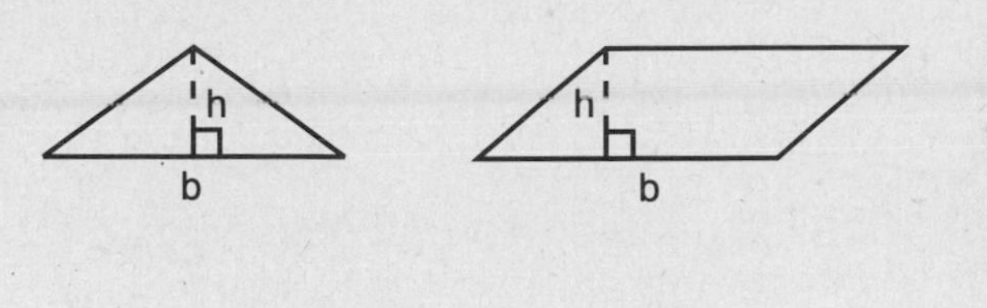

____

16. Do two trapezoids with the same height have the same area? Write *always, sometimes,* or *never.*

____

17. What is the area of a parallelogram with a base of 10 cm and a height of 5 cm?

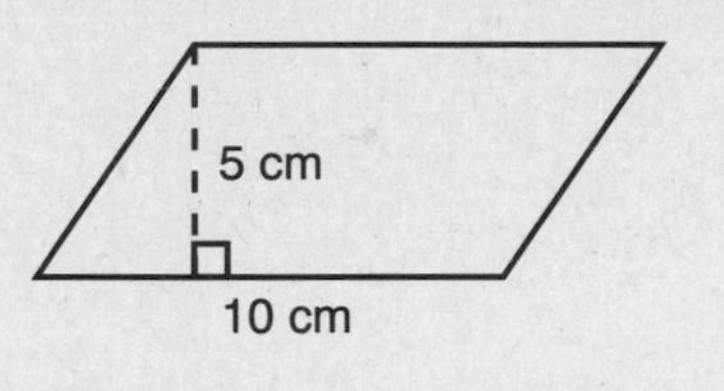

____

18. What is the area of a triangle with a base of 9 ft and a height of 7 ft?

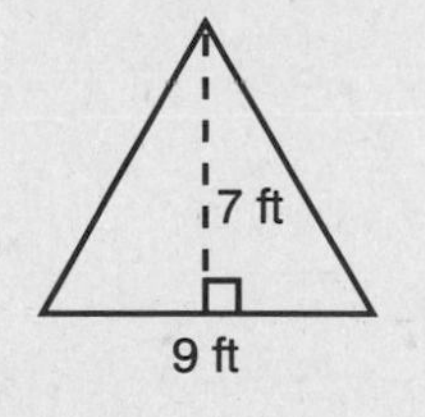

____

19. What is the area of a circle with a diameter of 7 m? Use $\pi = \frac{22}{7}$.

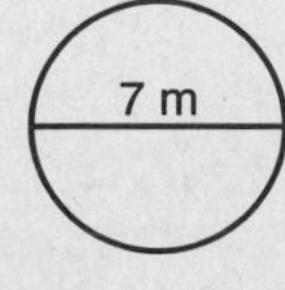

____

20. What is the area of a trapezoid with bases of 4 cm and 6 cm and a height of 10 cm?

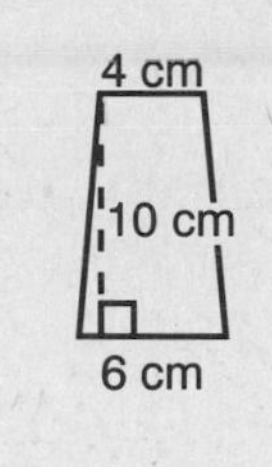

____

21. Jack's garden is in the shape of a square. Its perimeter is 26 ft. What is the area of the garden?

____

22. Erica arranged a play space for her dog. She put a pole into the ground and attached a 10-ft leash. The dog can go exactly 10 ft in any direction from the pole. What is the area of the play space?

____

For questions 23–24, write *always, sometimes,* or *never.*

23. Does a figure with line symmetry also have turn symmetry?

____

24. Does translation change a figure's size but not its shape?

____

25. Ring the letter that has line symmetry.

P R F E

26. Ring the letter that has turn symmetry.

L E H K

Go on to the next page.

Name ______________________________

Pretest

27. How many lines of symmetry does a square have?

____

28. For turn symmetry, what is the angle measure of each turn for a regular six-sided polygon?

____

Use the formulas below to answer questions 29–31.

$F = \frac{9}{5}C + 32$ $C = \frac{5}{9}(F - 32)$

29. A mixture freezes at 10°C. At what Fahrenheit temperature does it freeze?

____

30. One day the temperature dropped from 60°F to 42°F in one hour. What was the temperature drop in degrees Celsius?

____

31. Alice heated wax and observed its temperature with a Celsius thermometer. If the reading on her thermometer was 45°C, what was the temperature in degrees Fahrenheit?

____

32. Harry wants to make a wallpaper design from congruent regular polygons. Ring the name of the regular polygons that do *not* tessellate a plane.

triangles squares

hexagons pentagons

Stop!

Name ______________________________

Posttest

Read each question. Find the answer.

1. Ring the most appropriate unit of measure for measuring the distance between two cities.

   millimeter   centimeter

   meter   kilometer

2. Ring the situation in which an estimation would be most appropriate.

   pouring a dose of medicine

   measuring your ring size for a jeweler

   measuring the temperature of meat you are cooking

   measuring a table leg that you must replace

3. Ring the best estimate for the height of a room in a home.

   1,000 mm   5,000 cm

   3.5 m   0.1 km

4. Ring the best estimate for the weight of an egg.

   2 oz   1 lb   2 g   1 kg

5. Ring the best estimate of the amount of water it takes to fill a cup.

   10 mL   1 L   8 oz   2 pt

6. Ring the best estimate of the width of a refrigerator.

   15 in.   3 ft   4 yd   5 m

7. Ring the most precise measurement.

   35 in. $36\frac{1}{2}$ in.

   3 ft $3\frac{1}{2}$ ft

8. Ring the most precise measurement.

   98 cm   98.1 cm

   1 m   1.13 m

9. Ring the most precise measurement.

   128 oz   8 pt

   2 qt $\frac{1}{2}$ gal

10. Ring the most precise measurement.

    2 sec   1.1 sec

    $1\frac{1}{2}$ sec   1.001 sec

11. What is the perimeter of a regular hexagon when $s = 2.5$ cm?

    ____

12. What is the circumference of a circle with a radius of 2 ft? Use $\pi = 3.14$.

    ____

13. Harold wants to make a hoop with a circumference of 16 ft. What will be the diameter? Use $\pi = 3.14$.

    ____

14. The perimeter of a rectangle is 10 cm. The shorter sides are each 2 cm. What is the length of each of the longer sides?

    ____

Go on to the next page.

Name ______________________________

Posttest

15. Which triangle has the greater area? Ring the answer.

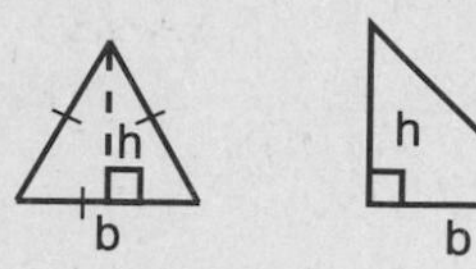

the equilateral triangle

the right triangle

The areas are equal.

16. Do two circles with the same circumference have the same area? Write *always, sometimes,* or *never.*

____

17. What is the area of a parallelogram with a base of 20 cm and a height of 15 cm?

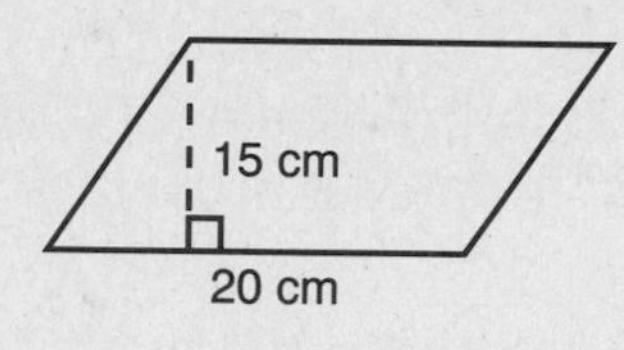

____

18. What is the area of a triangle with a base of 3 ft and a height of 4 ft?

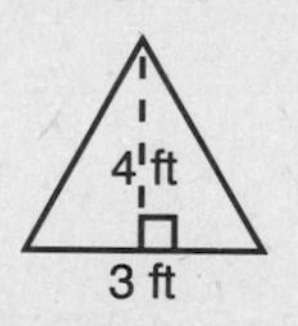

____

19. What is the area of a circle with a diameter of 10 m? Use $\pi = 3.14$.

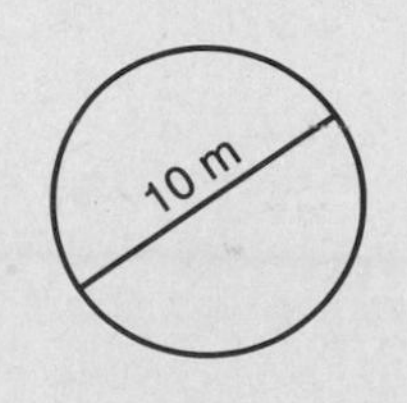

____

20. What is the area of a trapezoid with bases of 10 cm and 20 cm and a height of 10 cm?

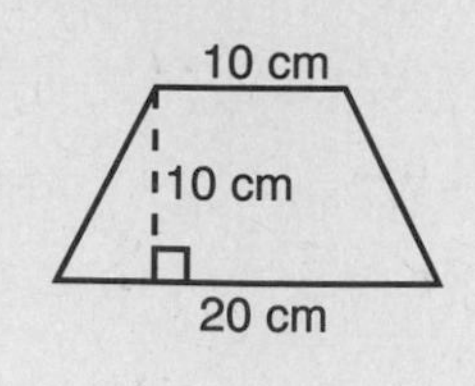

____

21. Jack's garden is in the shape of a square. Its area is 36 ft^2. What is the perimeter of the garden?

____

22. Hilda's garden is circular. The area is 3.14 ft^2. What is the diameter?

____

For questions 23–24, write *always, sometimes,* or *never*.

23. Does a circle have more than 4 lines of symmetry?

____

24. Does reflection change both the size and the shape of a figure?

____

25. Ring the letter that has line symmetry.

J Q G H

26. Ring the letter that has turn symmetry.

A P W X

27. How many lines of symmetry does an isosceles triangle have?

____

Go on to the next page.

Posttest

28. For turn symmetry, what is the angle measure of each turn for a square?

____

Use the formulas below to answer questions 29–31.

$F = \frac{9}{5}C + 32$ $C = \frac{5}{9}(F - 32)$

29. Water boils at 100°C. At what Fahrenheit temperature does it boil?

____

30. One day the temperature increased from 70°F to 79°F in one hour. What was the temperature increase in degrees Celsius?

____

31. Berto chilled water and observed its temperature with a Fahrenheit thermometer. If the reading on his thermometer was 50°F, what was the temperature in degrees Celsius?

____

32. John wants to make a quilt design from congruent regular polygons. He can use any regular polygon for which the measures of the angles meeting at a vertex have a sum of how many degrees?

____

Stop!

Name ______________________________

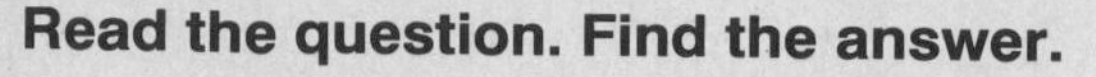

Pretest

Read the question. Find the answer.

1. Paco is drawing a pattern for a solid figure. The base of the figure is a polygon. Each side of the base is the base of an isosceles triangle. Ring the figure that Paco is drawing.

   cone　　　cylinder

   pyramid　　　triangular prism

2. Helen built a solid figure with 2 flat surfaces and 1 curved surface. Ring the figure that she built.

   cylinder　　　sphere

   cone　　　polyhedron

3. Ring the figure that has no flat surfaces.

   cone　　　cylinder

   pyramid　　　sphere

4. What do prisms and pyramids have in common? Ring the answer.

   Their lateral faces are all triangles.

   They are polyhedrons.

   They have congruent, parallel bases.

   None of their surfaces are polygons.

5. How many faces does a triangular prism have?

   ____

6. A polyhedron has 6 faces and 8 vertices. How many edges does it have?

   ____

7. How many faces does a rectangular pyramid have?

   ____

8. A solid figure has 6 faces: 5 triangles and 1 pentagon. Ring the figure.

   pentagonal pyramid

   triangular prism

   pentagonal prism

   not here

9. What is the surface area of this cylinder? Use $\pi = 3.14$.

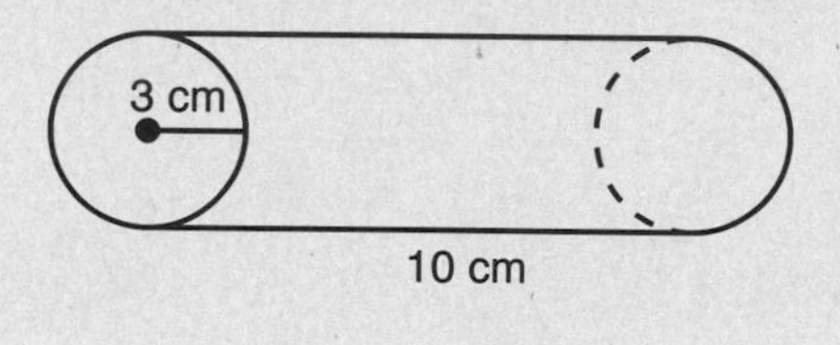

   ____

10. The dimensions of a rectangular prism are 5 ft $\times$ 4 ft $\times$ 4 ft. What is its surface area?

   ____

11. A cube has a surface area of 600 cm^2. What is the length of each face?

   ____

12. What is the surface area of this square pyramid?

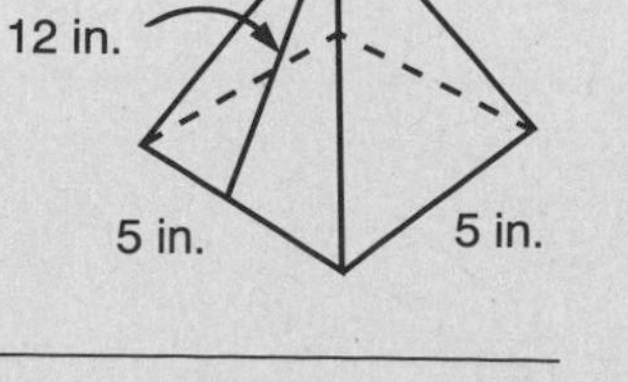

   ____

Go on to the next page.

13. What is the volume of a cylinder whose radius is 2 m and whose height is 0.25 m? Use $\pi = 3.14$.

___

14. What is the volume of this rectangular pyramid?

4 ft
2 ft
3 ft

___

15. What is the volume of a rectangular prism with the dimensions 4 in. $\times$ 6 in. $\times$ 8 in.?

___

16. What volume of water is held by a spoon with a capacity of 10 mL?

___

17. A swimming pool is 25 m long, 10 m wide, and 2 m deep. What is the capacity of the pool in kiloliters?

___

18. The volume of a pitcher is 1,500 cm^3. What mass of water will fill it?

___

19. Lena's aquarium is in the shape of a cylinder. Its radius is 5 cm and its height is 10 cm. What is the volume? Use $\pi = 3.14$.

___

20. Harry packs yogurt into tubs that are in the shape of a cylinder 8 in. high and 8 in. in diameter. Ring the dimensions of a tub that will hold 4 times as much yogurt.

$h = 16$ in.; $d = 8$ in.

$h = 8$ in.; $d = 16$ in.

$h = 12$ in.; $d = 12$ in.

$h = 16$ in.; $d = 16$ in.

21. Ring the best *estimate* of how much money you need in order to buy 3 pairs of socks at $1.98 a pair and 4 cans of tennis balls at $2.99 a can.

$17.00 $18.00

$18.50 $19.00

22. *About* how many square feet of carpet are needed for a rectangular room 9 ft 11 in. by 17 ft 2 in.? Ring the best estimate.

about 150 ft^2 about 170 ft^2

about 180 ft^2 about 190 ft^2

23. Joe is wrapping a box. Ring the best *estimate* of how many square feet of wrapping paper he needs if the box measures $11\frac{1}{4}$ in. by $12\frac{1}{2}$ in. by 23 in.

2 ft^2 6 ft^2

8 ft^2 10 ft^2

24. Aquarium A and aquarium B are both cylinders. They have the same radius, but A is twice as high as B. Ring the ratio of the volume of water held by A to the volume of water held by B.

$\frac{1}{2}$ to 1 2 to 1

4 to 1 8 to 1

Stop!

Posttest

Read the question. Find the answer.

1. Lorna is building a solid figure. The figure has 2 parallel bases that are congruent triangles. Its other faces are rectangles. Ring the figure.

   rectangular pyramid

   rectangular prism

   triangular pyramid

   triangular prism

2. Sean built a solid figure with 1 flat surface and 1 curved surface. Ring the figure that he built.

   cylinder   sphere

   cone   polyhedron

3. Ring the figure that is *not* a polyhedron.

   cylinder   prism

   pyramid   cube

4. What do prisms and cylinders have in common? Ring the answer.

   Their bases are circles.

   They have parallel, congruent bases.

   Their bases are polygons.

   They have curved surfaces.

5. How many faces does a triangular pyramid have?

   ____

6. A polyhedron has 8 faces and 12 vertices. How many edges does it have?

   ____

7. How many faces does a rectangular prism have?

   ____

8. A solid figure has 6 rectangular surfaces. Ring the figure.

   rectangular prism

   hexagonal prism

   rectangular pyramid

   not here

9. What is the surface area of this cylinder? Use $\pi = 3.14$.

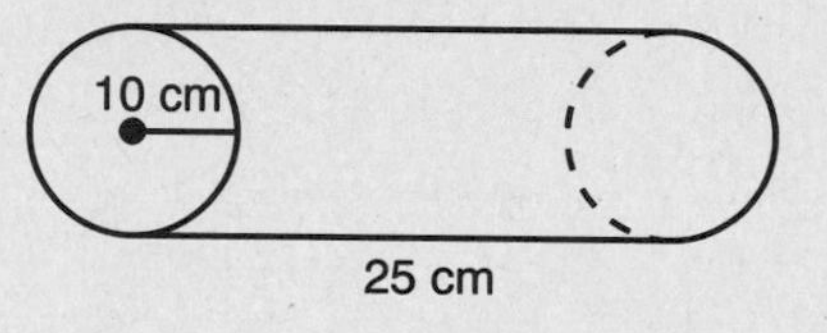

   ____

10. What is the surface area of this rectangular prism?

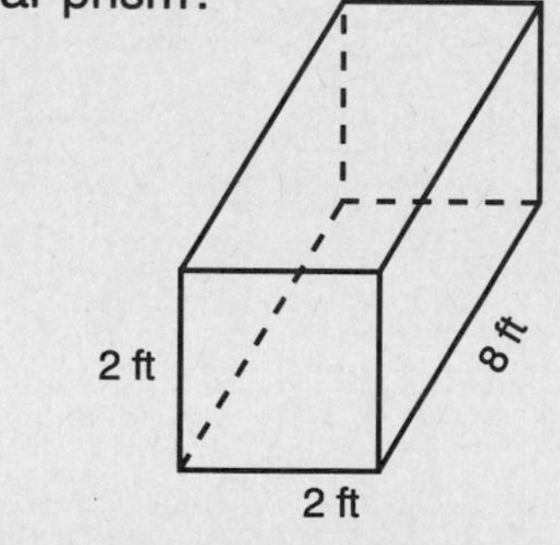

   ____

11. A cube has a surface area of 96 cm^2. What is the length of each face?

   ____

12. The base of a square pyramid is 4 in. × 4 in. The height of each triangular face is 6 in. What is the surface area of the pyramid?

   ____

Go on to the next page.

Name ______________________________

Posttest

13. What is the volume of a cylinder with a radius of 1 m and a height of 0.5 m? Use $\pi = 3.14$.

____

14. What is the volume of a rectangular pyramid that has a base 3 ft × 4 ft and a height of 5 ft?

____

15. What is the volume of this rectangular prism?

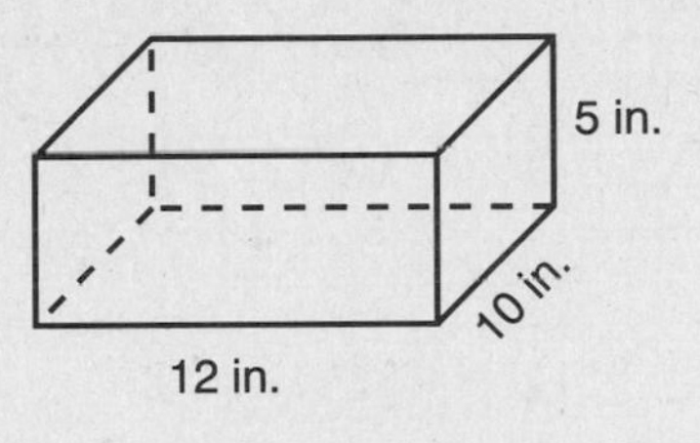

____

16. A cup with a volume of 50 cm^3 has what liquid capacity?

____

17. A swimming pool is 30 m long, 15 m wide, and 1.5 m deep. What is the capacity of the pool in kiloliters?

____

18. The volume of a bowl is 3,500 cm^3. What mass of water will fill it?

____

19. Eli's aquarium is in the shape of a cylinder. It has a radius of 10 cm and height of 20 cm. What is the volume? Use $\pi = 3.14$.

____

20. Joel packs food into containers that are in the shape of a cone 6 cm high and 6 cm in diameter. Ring the dimensions of a cone that will hold 2 times as much food.

$h = 12$ cm; $d = 6$ cm

$h = 6$ cm; $d = 12$ cm

$h = 9$ cm; $d = 9$ cm

$h = 12$ cm; $d = 12$ cm

21. Ring the best *estimate* of how much money Lee needs in order to buy 2 shirts at $14.95 each, 2 pairs of shoes at $49.50 a pair, and 3 ties at $19.95 each.

$160.00 $175.00

$185.00 $190.00

22. Ring the best *estimate* of the area of a rectangular garden that has the dimensions 15 ft 11 in. by 20 ft 2 in.

300 ft^2 315 ft^2

320 ft^2 350 ft^2

23. Emily is wrapping a box. *About* how many square feet of wrapping paper will she need if the dimensions of the box are $11\frac{1}{4}$ in. by $24\frac{1}{2}$ in. by 23 in.? Ring the best estimate.

about 4 ft^2 about 10 ft^2

about 12 ft^2 about 16 ft^2

24. Aquarium A and aquarium B are both rectangular prisms. Their bases have the same dimensions, but A is twice as high as B. Ring the ratio of the volume of water held by A to that held by B.

$\frac{1}{2}$ to 1 2 to 1

4 to 1 8 to 1

Stop!

Name ________________________________

Pretest

Read each question. Find the answer.

1. Ring the first four square numbers.

   1, 3, 6, 12   1, 4, 9, 16

   1, 5, 10, 15   1, 2, 4, 8

2. How many dots would come next in this pattern?

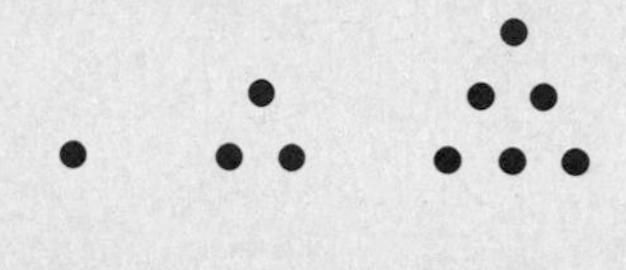

   ____

3. What is the next number in the pattern?
   1, 3, 6, 10, 15

   ____

4. What is the next number in the pattern?
   243, 81, 27

   ____

5. What are the first four numbers in this pattern? The first number is 7. Add 12 each time.

   ____

6. What are the first four numbers in this pattern? The first number is 2. Multiply by 3 and subtract 1 each time.

   ____

To answer questions 7–8, use the relation {(0,0), (1,2), (2,4), (3,6)}.

7. Ring the domain of the relation.

   0, 1   2, 3

   0, 1, 2, 3   0, 2, 4, 6

8. Ring the range of the relation.

   0, 2   4, 6

   0, 1, 2, 3   0, 2, 4, 6

9. Which ordered pairs are graphed?

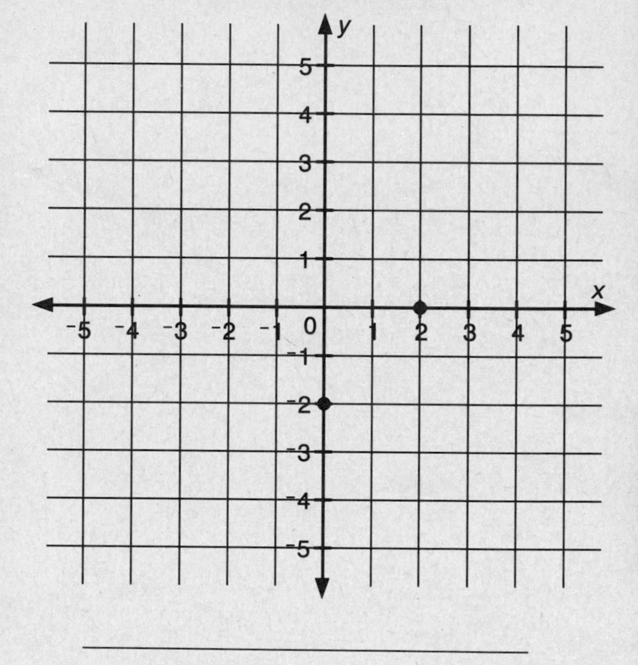

   ____

10. If the length and width of a rectangle both triple, what happens to the area of the rectangle? Ring the answer.

    It doubles.

    It triples.

    It increases by 6 times.

    It increases by 9 times.

11. Ring the equation that represents the relation {(0,0), (1,3), (2,6), (3,9)}.

    $y = x + 6$ $y = \frac{1}{3}x$

    $y = x + 3$ $y = 3x$

Go on to the next page.

Name ___

Pretest

12. Ring the equation that represents the relation presented in the table below.

Domain	Range
1	7
2	14
3	21
4	28

$y = x + 7$ $\qquad$ $y = \frac{1}{7}x$

$y = 7x$ $\qquad$ $y = 6x + 1$

13. What is the missing value in the following table?

Domain	Range
1	1
2	4
3	9
4	16
5	■

___

14. Ring the ordered pair that fits the relation $y = 3x + 2$.

(5,1) $\qquad$ (0,5)

(1,8) $\qquad$ (0,2)

15. Ring the answer that gives the lengths of the sides of a right triangle.

3 m, 4 m, 6 m

6 m, 8 m, 10 m

5 m, 6 m, 8 m

6 m, 9 m, 12 m

16. To which triangles does the Pythagorean Property apply?

___

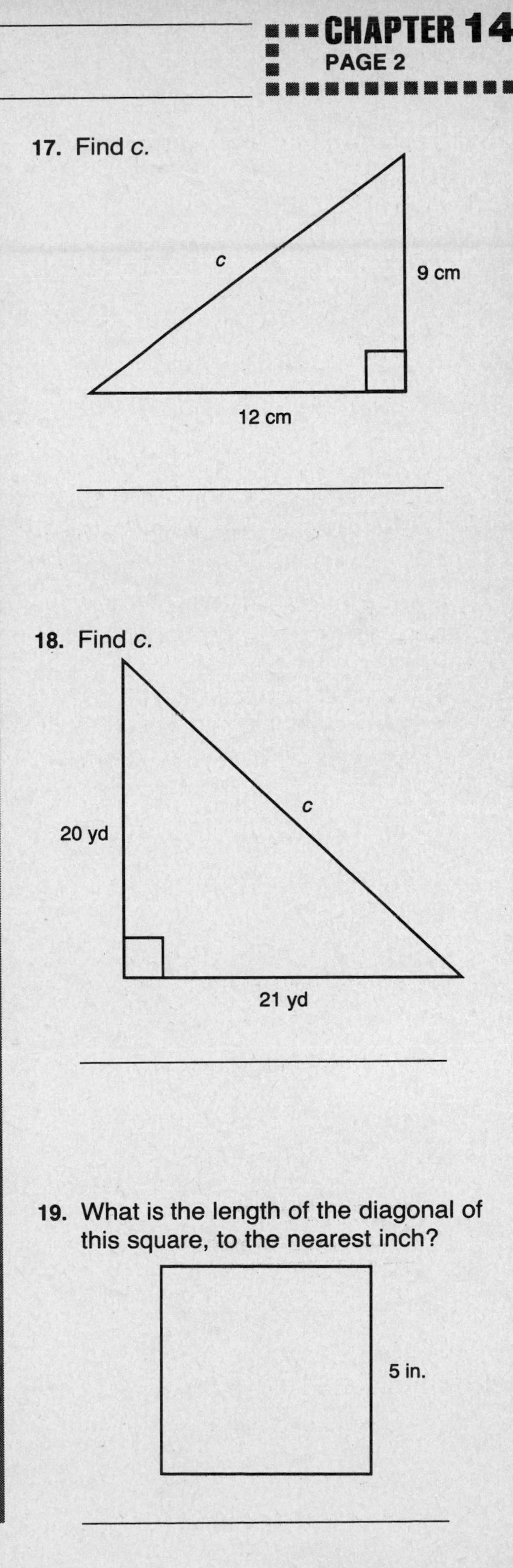

17. Find c.

___

18. Find c.

___

19. What is the length of the diagonal of this square, to the nearest inch?

___

Go on to the next page.

Name ______________________________

Pretest

20. What is the length of the diagonal of this rectangle, to the nearest tenth?

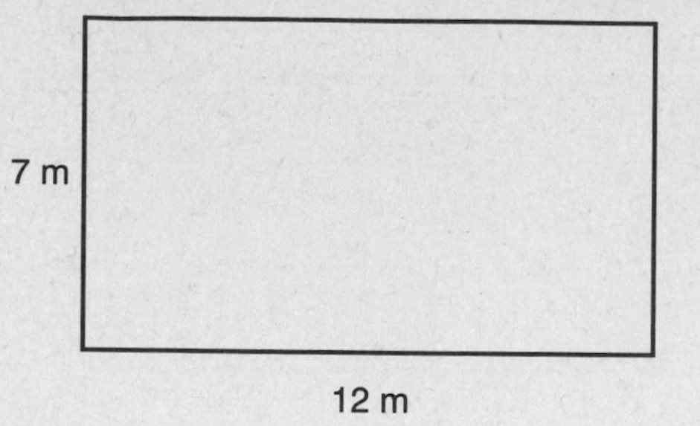

____

21. Is a relation a function? Write *always, sometimes,* or *never.*

____

22. What is a relation called that has only one element of the range for each element of the domain?

____

23. Ring the ordered pair that fits this equation.

$y = 10x + 1$

(1,11) (1,0)

(21,2) (1,12)

24. Which point on the graph is (2,0)?

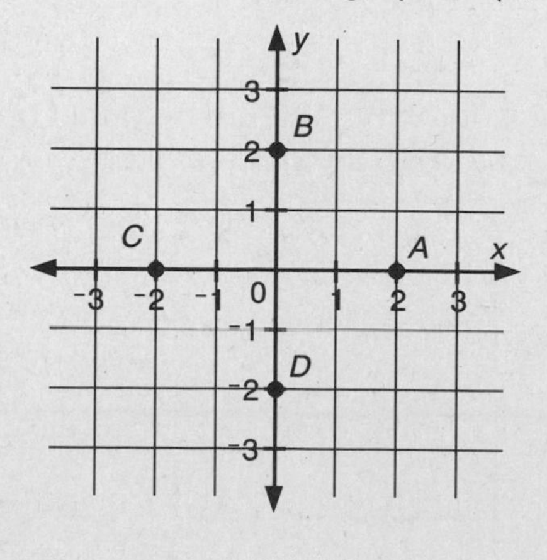

____

25. Ring the relation that is a function.

{(1,6), (2,6), (3,6)}

{(1,6), (1,7)}

{(1,6), (2,6), (2,7)}

{(2,1), (2,2)}

26. Ring the function.

{(1,1), (1,2)}

{(1,1), (2,1)}

{(2,1), (2,2)}

{(0,1), (0,2)}

27. Ring the relation that is *not* a function.

$y = x$

$y = 2x$

$y = 2x + 3$

$y = \sqrt{x^2 + 1}$

28. Ring the graph that does *not* represent a function.

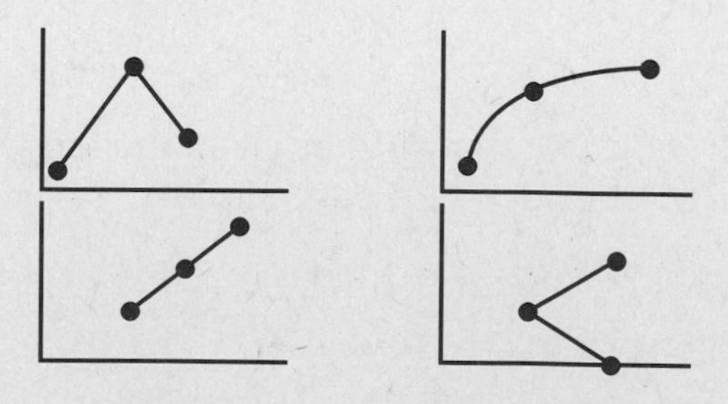

29. Otis sold 99 drinks. He sold only milk and juice. He sold twice as many cartons of milk as cartons of juice. How many cartons of milk did he sell?

____

Go on to the next page.

Name ______________________________

Pretest

30. Ron sold 80 hamburgers and hot dogs. He sold 20 more hot dogs than hamburgers. Ring the equation that could be used to find how many hamburgers he sold. (Let x = the number of hamburgers sold.)

$x + 20 = 80$

$2x = 80$

$x + x + 20 = 80$

$x + x + x = 60$

31. John took some money from his bank account. He spent $5 on lunch and $\frac{1}{2}$ of the remainder at the grocery store. He was left with $45. How much money did he take from his account?

____

32. The sum of two numbers is 32. One number is 6 greater than the other. What are the two numbers?

____

Stop!

Name ______________________

Posttest

Read each question. Find the answer.

1. What are the first four triangular numbers?

   1, 4, 9, 16   1, 3, 6, 10

   1, 3, 9, 27   1, 5, 12, 22

2. How many dots would come next in this pattern?

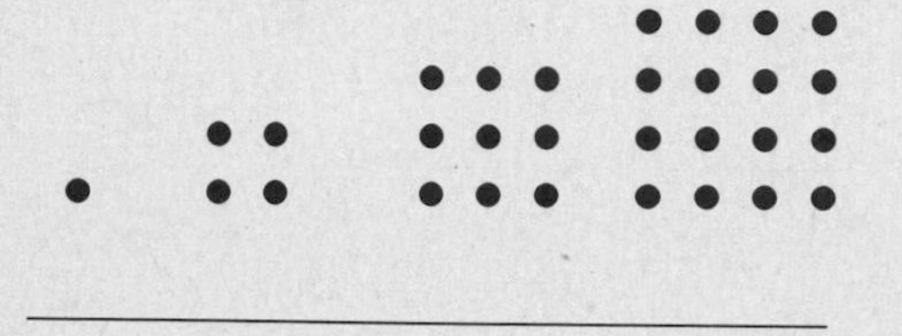

   ____

3. What is the next number in the pattern? 10, 15, 21, 28

   ____

4. What is the next number in the pattern? 122, 100, 80, 62

   ____

5. What are the first four numbers in this pattern? The first number is 99. Subtract 7 each time.

   ____

6. What are the first four numbers in this pattern? The first number is 0. Multiply by 5 and add 2 each time.

   ____

To answer questions 7–8, use the relation {(1,2), (2,4), (3,6), (4,8)}.

7. Ring the domain of the relation.

   1, 2   3, 4

   1, 2, 3, 4   2, 4, 6, 8

8. Ring the range of the relation.

   2, 4

   6, 8

   1, 2, 3, 4

   2, 4, 6, 8

9. Which ordered pairs are graphed?

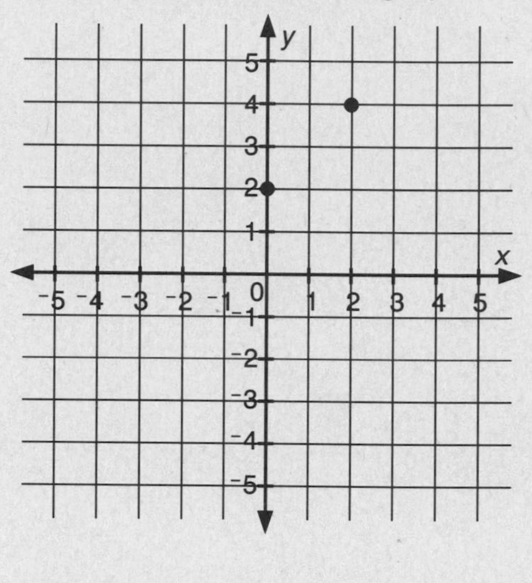

   ____

10. If the length of a rectangle triples and the width doubles, what happens to the area of the rectangle? Ring the answer.

    It doubles.

    It increases by 5 times.

    It increases by 6 times.

    It increases by 10 times.

11. Ring the equation that represents the relation {(0,0), (1,5), (2,10), (3,15)}.

    $y = x + 12$

    $y = 4x + 3$

    $y = x + 4$

    $y = 5x$

Go on to the next page.

Name ___

Posttest

12. Ring the equation that represents the relation presented in the table below.

Domain	Range
1	4
2	8
3	12
4	16

$y = x + 3$

$y = x + 12$

$y = 4x$

$y = \frac{1}{4}x$

13. What is the missing value in the following table?

Domain	Range
1	4
2	7
3	10
4	13
5	■

___

14. Ring the ordered pair that fits the relation $y = x - 2$.

(0,2)

($^-$2,0)

(2,4)

(0,$^-$2)

15. Ring the answer that gives the lengths of the sides of a right triangle.

9 cm, 12 cm, 16 cm

9 cm, 12 cm, 15 cm

3 cm, 6 cm, 9 cm

5 cm, 9 cm, 11 cm

16. The square of the longest side of a triangle is greater than the sum of the squares of the two shorter sides. What do you know about the triangle? Ring the answer.

It cannot be an isosceles triangle.

It must be a scalene triangle.

It must be an equilateral triangle.

It cannot be a right triangle.

17. Find *c*.

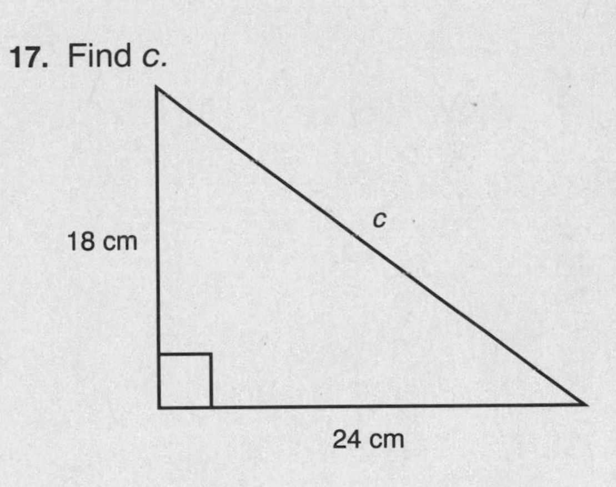

___

18. Find *c*.

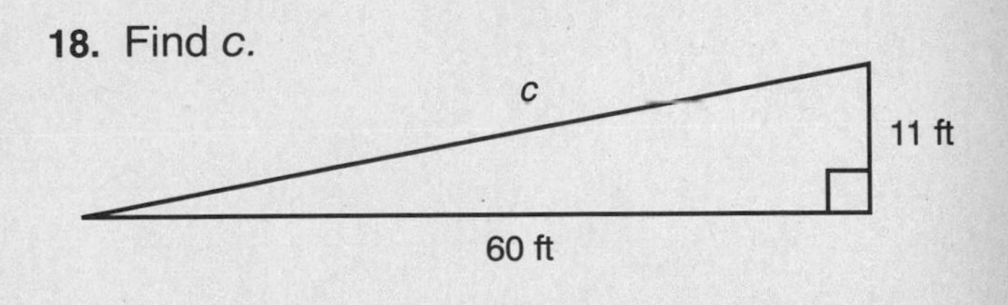

___

19. What is the length of the diagonal of this square, to the nearest foot?

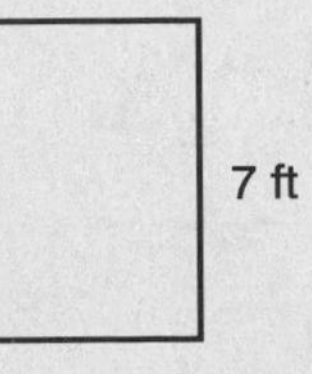

___

Go on to the next page.

Name ______________________________

Posttest

20. What is the length of the diagonal of this rectangle, to the nearest tenth?

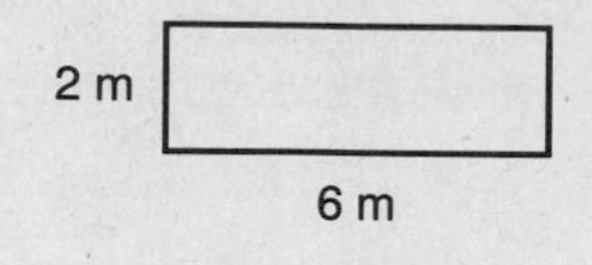

____

21. Is a function a relation? Write *always, sometimes,* or *never.*

____

22. What is a relation called that has only one element of the range for each element of the domain?

____

23. Ring the ordered pair that fits this equation.

$y = 5x + 10$

(20,2) (0,15)

(10,0) (1,15)

24. Which point on the graph is (2,3)?

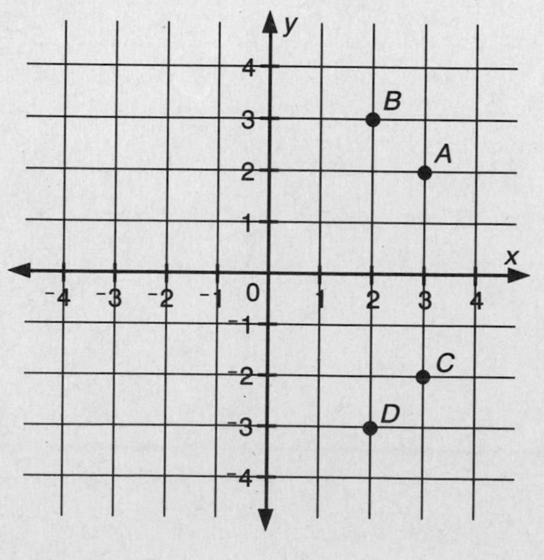

____

25. Ring the relation that is a function.

{(1,3), (2,3), (3,3)}

{(1,3), (1,4)}

{(1,3), (2,3), (2,4)}

{(2,3), (2,4)}

26. Ring the function.

{(0,1), (0,2)} {(0,1), (2,1)}

{(2,0), (2,1)} {(0,0), (0,2)}

27. Ring the relation that is *not* a function.

$y = x + 50$ $y = 2x$

$y = x$ $y = \sqrt{x}$

28. Ring the graph that does *not* represent a function.

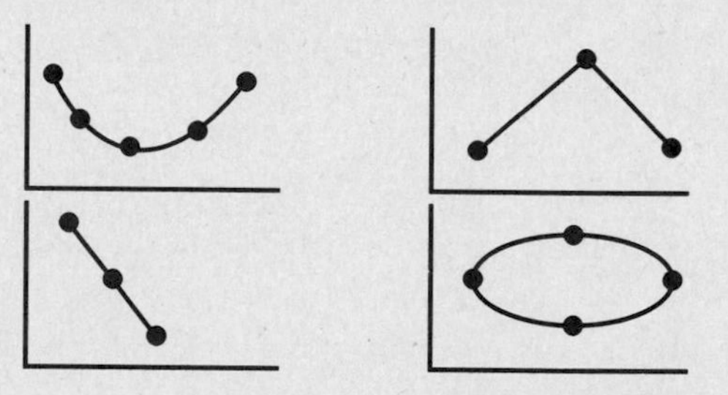

29. Hilda sold 200 hats. She sold 3 times as many men's hats as women's hats. How many men's hats did she sell?

____

30. John sold 100 hats. He sold 60 more women's hats than men's hats. Ring the equation that could be used to find how many men's hats he sold. (Let x = the number of men's hats sold.)

$x + 60 = 100$

$2x = 100$

$x + x + 60 = 100$

$x + x + x = 100 + 60$

Go on to the next page.

Name ______________________________

Posttest

31. Bill took money from his bank account. He spent \$6 at a movie and $\frac{1}{3}$ of the remainder for groceries. He then had \$40 left. How much money did he take from his bank account?

____

32. The product of two numbers is 72. One number is 6 greater than the other. What are the two numbers?

____

Stop!

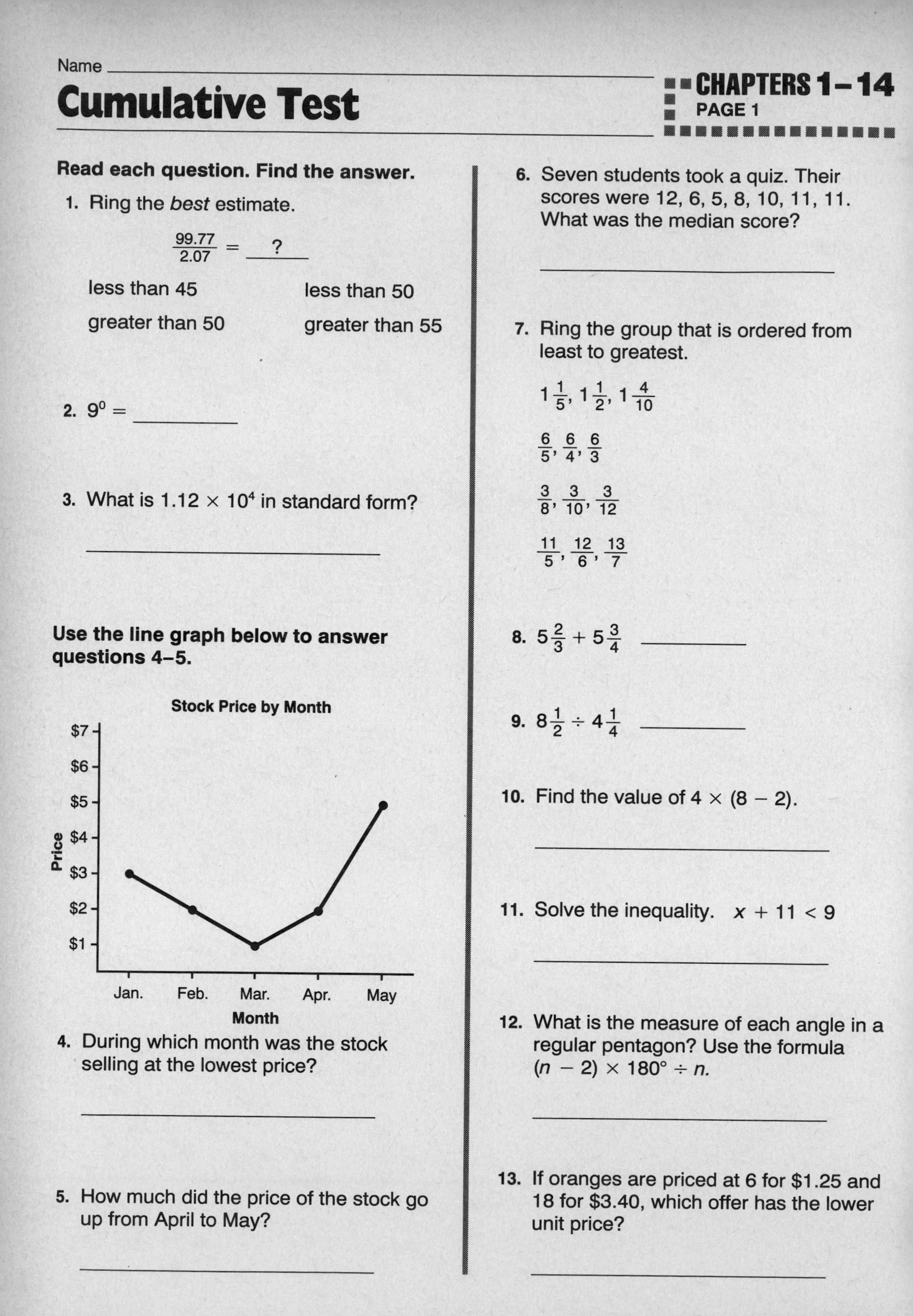

Name ______________________________

Cumulative Test

CHAPTERS 1–14
PAGE 1

Read each question. Find the answer.

1. Ring the *best* estimate.

$$\frac{99.77}{2.07} = \underline{\quad ? \quad}$$

less than 45   less than 50

greater than 50   greater than 55

2. $9^0 =$ __________

3. What is 1.12×10^4 in standard form?

____

Use the line graph below to answer questions 4–5.

Stock Price by Month

Price: $7, $6, $5, $4, $3, $2, $1

Jan. Feb. Mar. Apr. May

Month

4. During which month was the stock selling at the lowest price?

____

5. How much did the price of the stock go up from April to May?

____

6. Seven students took a quiz. Their scores were 12, 6, 5, 8, 10, 11, 11. What was the median score?

____

7. Ring the group that is ordered from least to greatest.

$1\frac{1}{5}, 1\frac{1}{2}, 1\frac{4}{10}$

$\frac{6}{5}, \frac{6}{4}, \frac{6}{3}$

$\frac{3}{8}, \frac{3}{10}, \frac{3}{12}$

$\frac{11}{5}, \frac{12}{6}, \frac{13}{7}$

8. $5\frac{2}{3} + 5\frac{3}{4}$ __________

9. $8\frac{1}{2} \div 4\frac{1}{4}$ __________

10. Find the value of $4 \times (8 - 2)$.

____

11. Solve the inequality. $x + 11 < 9$

____

12. What is the measure of each angle in a regular pentagon? Use the formula $(n - 2) \times 180° \div n$.

____

13. If oranges are priced at 6 for $1.25 and 18 for $3.40, which offer has the lower unit price?

____

Go on to the next page.

Name ______________________________

Cumulative Test

14. On a map, 1 in. = 50 mi. Two cities are 14 inches apart on the map. How many miles are they actually apart?

____

15. Two rectangles are similar. The first rectangle has a width of 6 cm and length of 12 cm. The width of the second rectangle is 0.5 cm. Find its length.

____

16. Express $\frac{2}{10{,}000}$ as a decimal.

____

17. Ring the *best* estimate. 79% of 82

64 80 90 105

18. 30% of what number is 120?

____

19. Hilda's Dress Shop sells sweaters for $40.00. Yesterday, all prices were reduced by 20%. Today, prices are reduced by an additional 10% of the sale price. What is today's sale price of a sweater?

____

20. George bought a concert ticket. He paid $22.00 plus 5% sales tax. How much did George pay for the ticket?

____

21. $^{-}50 \times {}^{-}10$ ______________

22. Solve for *x*. $x + 14 = 3$

____

23. Will a fraction whose denominator is 10 result in a repeating decimal? Write *always, sometimes,* or *never.*

____

24. Express $\frac{37}{50}$ as a decimal.

____

25. Ring the rational number between $^{-}0.20$ and $^{-}0.21$.

$^{-}0.199$ $^{-}0.201$

$^{-}0.211$ $^{-}0.220$

26. Write <, >, or =. $\left(\frac{1}{4}\right)^2 \bigcirc \left(\frac{1}{5}\right)^3$

27. How many different combinations of 2 can be selected from a set of 8 coins?

____

28. How many different arrangements are possible of the letters *A*, *B*, *C*, *D*, *E*?

____

Go on to the next page.

Name ______________________________

Cumulative Test

29. Harvey rolls a single number cube numbered from 1 to 6. Ring the event that has a $\frac{1}{3}$ probability of occurrence.

a roll of 3

a roll of any odd number

a roll of 1 or 2

a roll of any number greater than 3

30. A coin is flipped 4 times. What is the probability of obtaining 4 heads?

____

31. A bowl has 4 oranges and 5 apples. If Amy chooses one piece of fruit at random, how likely is it that she will choose a banana? Write *certain, impossible,* or *neither.*

____

32. Ring the most appropriate unit of measurement for measuring the height of a room.

inches   feet   yards   miles

33. Ring the situation in which a precise measurement would be necessary.

determining your car's gas mileage

determining how much soap you need to wash your clothes

planning how many miles you will fly on a trip to Mexico

measuring the time to run a 100-yard dash at a track meet

34. What is the circumference of a circle with a radius of 2 inches? Use 3.14 for π.

____

35. Find the area of a parallelogram with a base of 40 cm and height of 20 cm.

____

36. Does a figure with line symmetry also have turn symmetry? Write *always, sometimes,* or *never.*

____

37. Hank is drawing a pattern for a figure. The figure has a rectangle as its base. Each side of the base is the base of a triangle. What figure is he drawing?

____

38. The dimensions of a rectangular prism are 6 ft $\times$ 4 ft $\times$ 4 ft. What is its surface area?

____

39. Find the area of a triangle with a base of 40 cm and a height of 20 cm.

____

40. Ring the ordered pair that fits the relation $y = 4x + 2$.

(10,2)   (8,2)

(2,8)   (2,10)

Go on to the next page.

Name ______________________

Cumulative Test

41. Find the hypotenuse of a right triangle with legs measuring 12 cm and 9 cm. (The sum of the squares of the legs is equal to the square of the hypotenuse.)

42. Does a graph that shows a circle describe a function? Write *always, sometimes,* or *never.*

43. What is the total angle measure of a regular 7-sided polygon? Use the formula $(n - 2) \times 180°$.

Use the information and the double-bar graph below to answer question 44.

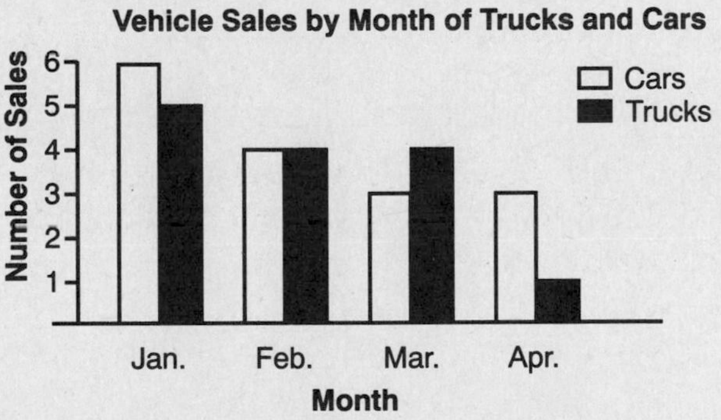

44. During which month did truck sales exceed car sales?

45. Two numbers add to 201. One of the numbers is 9 more than the other. What are the numbers?

46. Roger is paid \$2.00 per hour for baby-sitting during the day and \$2.50 per hour after 6 P.M. This week, Roger worked 6 hours Friday night and 3 hours Saturday afternoon. How much did he earn?

47. Joy travels 195 miles. The trip takes exactly 3 hours. What is her rate of speed? (distance = rate × time)

48. Helen bowled 3 games. Her second game score was twice as high as her first game. Her third game was 15 points higher than her second game. Her score in game 3 was 255. What was Helen's score in game 1?

49. Mike missed 3 words on a spelling test. His sister missed 6 more than twice as many words as Mike missed. How many words did his sister miss?

50. Paula wants to make a scarf from congruent regular polygons that tessellate a plane. She can use any regular polygon for which the angles meeting at a vertex total to which of the following?

Stop!

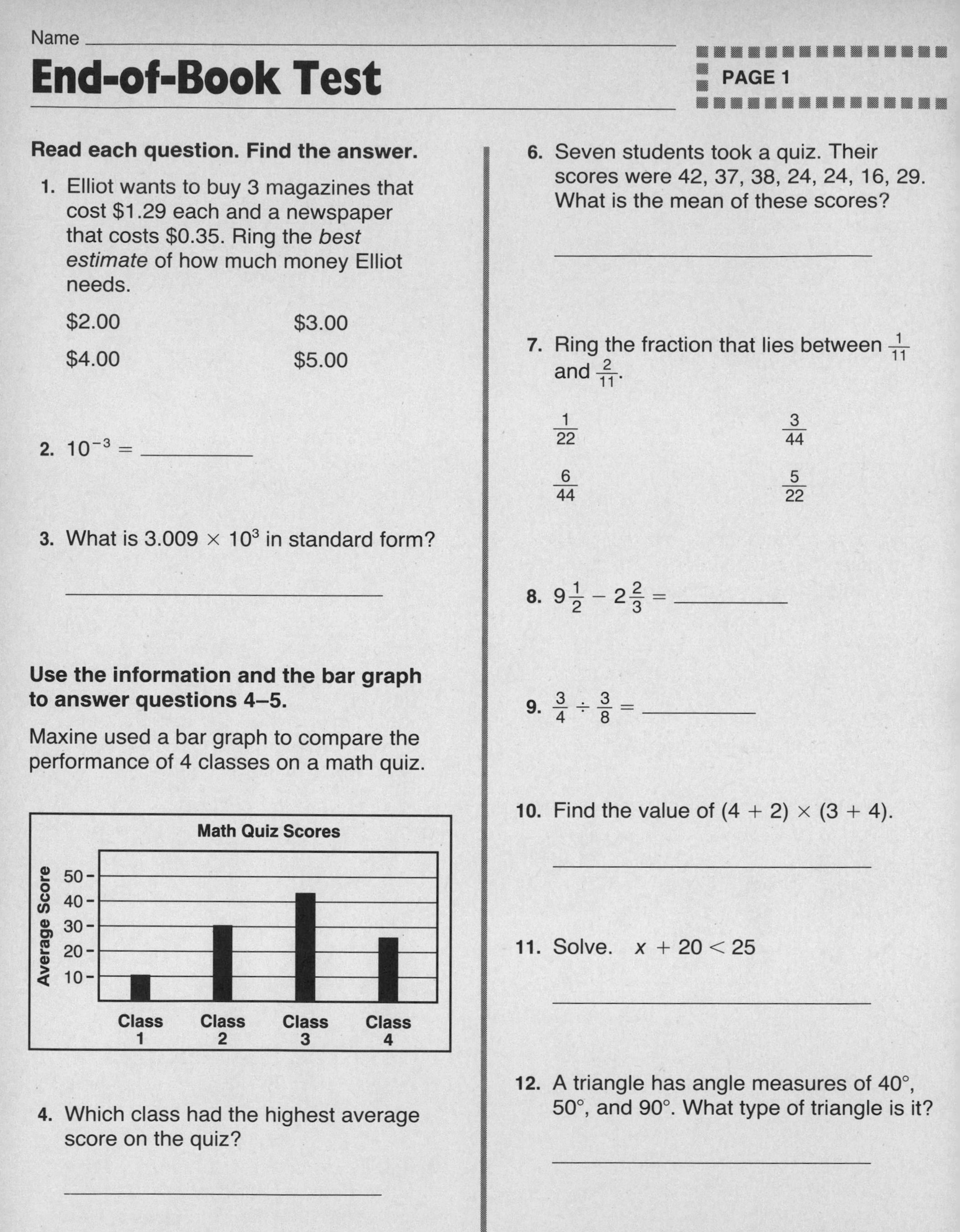

Name ______

End-of-Book Test

Read each question. Find the answer.

1. Elliot wants to buy 3 magazines that cost \$1.29 each and a newspaper that costs \$0.35. Ring the *best estimate* of how much money Elliot needs.

   \$2.00   \$3.00

   \$4.00   \$5.00

2. $10^{-3} =$ ______

3. What is 3.009×10^3 in standard form?

   ______

Use the information and the bar graph to answer questions 4–5.

Maxine used a bar graph to compare the performance of 4 classes on a math quiz.

4. Which class had the highest average score on the quiz?

   ______

5. How much higher did Class 4 average than did Class 1?

   ______

6. Seven students took a quiz. Their scores were 42, 37, 38, 24, 24, 16, 29. What is the mean of these scores?

   ______

7. Ring the fraction that lies between $\frac{1}{11}$ and $\frac{2}{11}$.

   $\frac{1}{22}$ $\frac{3}{44}$

   $\frac{6}{44}$ $\frac{5}{22}$

8. $9\frac{1}{2} - 2\frac{2}{3} =$ ______

9. $\frac{3}{4} \div \frac{3}{8} =$ ______

10. Find the value of $(4 + 2) \times (3 + 4)$.

    ______

11. Solve. $x + 20 < 25$

    ______

12. A triangle has angle measures of 40°, 50°, and 90°. What type of triangle is it?

    ______

13. If juice is priced at 6 cans for \$2.70 and 8 cans for \$3.52, which offer has the lower unit price?

    ______

Go on to the next page.

Name ____________________

End-of-Book Test

14. On Map A, 1 cm = 50 km. On Map B, 1 cm = 75 km. Two cities are 10 cm apart on Map B. How far apart are they on Map A?

____

15. A triangle has sides measuring 6 cm, 8 cm, and 10 cm. Ring the lengths of the sides of a similar triangle.

7 cm, 9 cm, 11 cm

16 km, 18 km, 20 km

9 yd, 12 yd, 20 yd

9 m, 12 m, 15 m

16. Express 0.3% as a decimal.

____

17. 27 is what percent of 108?

____

18. 20% of what number is 200?

____

19. Nikki bought a shirt for $22.00 plus 7% sales tax. How much did Nikki pay for the shirt?

____

20. Andy is buying a suit that is marked down 25%. What is the sale price of the $240 suit?

____

21. $^{-}25 \times 20 =$ __________

22. Solve for x. $x + 20 = 18$

____

23. Are mixed numbers rational numbers? Write *always, sometimes,* or *never.*

____

24. Ring the fraction that can be expressed as a terminating decimal.

$\frac{2}{18}$ $\frac{1}{7}$ $\frac{1}{5}$ $\frac{1}{3}$

25. Ring a rational number between 0.61 and 0.62.

0.601 0.609

0.6110 0.6201

26. Write $<$, $>$, or $=$. 10^{-5} ◯ 10^{3}

27. How many different combinations of 2 can be selected from a set of 20 items?

____

28. In how many different orders can 5 people arrange themselves in a line?

____

29. Harvey rolls a single number cube numbered from 1 to 6. What is the probability that he will roll an even number?

____

Go on to the next page.

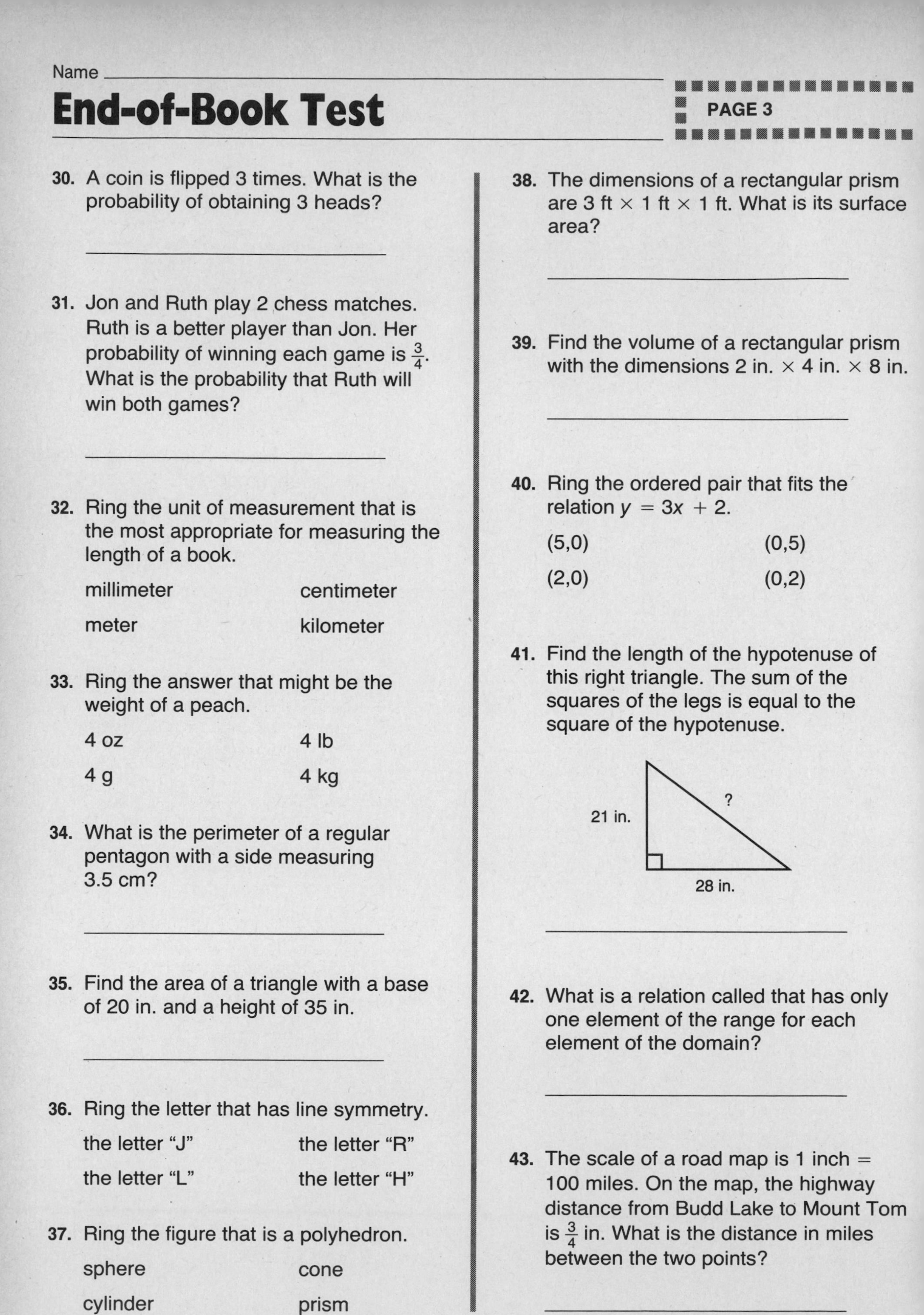

Name ______________________________

End-of-Book Test

30. A coin is flipped 3 times. What is the probability of obtaining 3 heads?

____

31. Jon and Ruth play 2 chess matches. Ruth is a better player than Jon. Her probability of winning each game is $\frac{3}{4}$. What is the probability that Ruth will win both games?

____

32. Ring the unit of measurement that is the most appropriate for measuring the length of a book.

millimeter	centimeter
meter	kilometer

33. Ring the answer that might be the weight of a peach.

4 oz	4 lb
4 g	4 kg

34. What is the perimeter of a regular pentagon with a side measuring 3.5 cm?

____

35. Find the area of a triangle with a base of 20 in. and a height of 35 in.

____

36. Ring the letter that has line symmetry.

the letter "J"	the letter "R"
the letter "L"	the letter "H"

37. Ring the figure that is a polyhedron.

sphere	cone
cylinder	prism

38. The dimensions of a rectangular prism are 3 ft × 1 ft × 1 ft. What is its surface area?

____

39. Find the volume of a rectangular prism with the dimensions 2 in. × 4 in. × 8 in.

____

40. Ring the ordered pair that fits the relation $y = 3x + 2$.

(5,0)	(0,5)
(2,0)	(0,2)

41. Find the length of the hypotenuse of this right triangle. The sum of the squares of the legs is equal to the square of the hypotenuse.

21 in.
?
28 in.

____

42. What is a relation called that has only one element of the range for each element of the domain?

____

43. The scale of a road map is 1 inch = 100 miles. On the map, the highway distance from Budd Lake to Mount Tom is $\frac{3}{4}$ in. What is the distance in miles between the two points?

____

Go on to the next page.

Name ______________________________

End-of-Book Test

Use the information and the double-bar graph to answer question 44.

Joan used a double-bar graph to compare the monthly sales of hatchbacks and sedans over a period of 4 months.

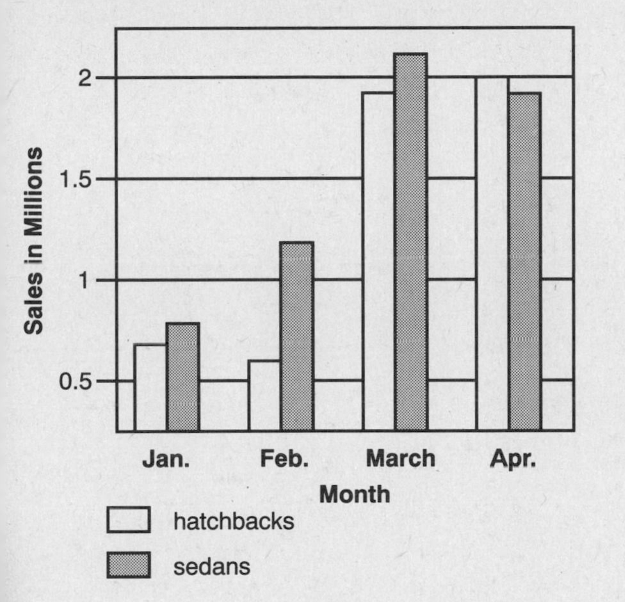

44. During which month was the difference in sales of hatchbacks and sedans the greatest?

____

45. Two numbers add to 360. One of the numbers is 3 times the other. What is the smaller number?

____

46. Nancy walks 5 blocks to school. It takes her $2\frac{1}{2}$ minutes to walk each block. Pat walks 6 blocks to school. It takes her only $1\frac{1}{2}$ minutes to walk each block. How much longer does it take Nancy to walk to school than it takes Pat?

____

47. Art drove a distance of 275 miles in his car. His average speed was 50 miles per hour. How long did the drive take? (distance = rate × time)

____

48. Mimi spent $\frac{1}{2}$ of her savings on Monday. On Tuesday, after she spent \$50.00, she had only \$2.00 left. How much money did Mimi start with?

____

49. Helena and Mark went shopping. Mark spent \$2.00 more than 3 times the amount Helena spent. Helena spent \$5.50. How much did Mark spend?

____

Use the table to answer question 50.

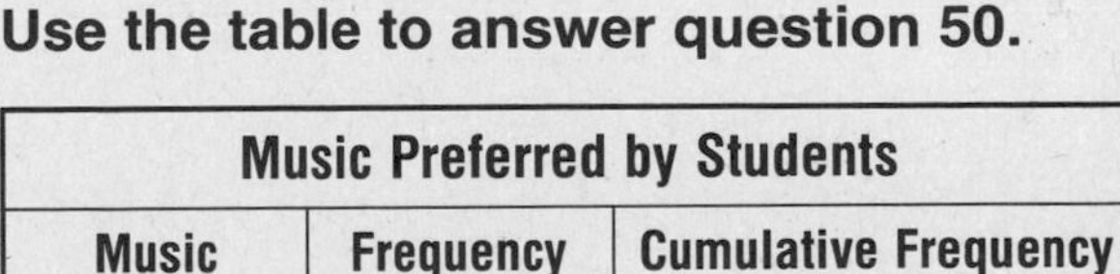
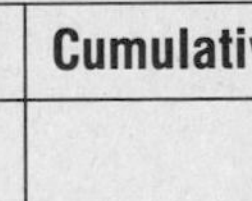

Music Preferred by Students		
Music	**Frequency**	**Cumulative Frequency**
Rock	50	50
Jazz	20	70
Country	20	90
Rap	10	100

50. How many more students chose rock music than jazz music?

____

Stop!

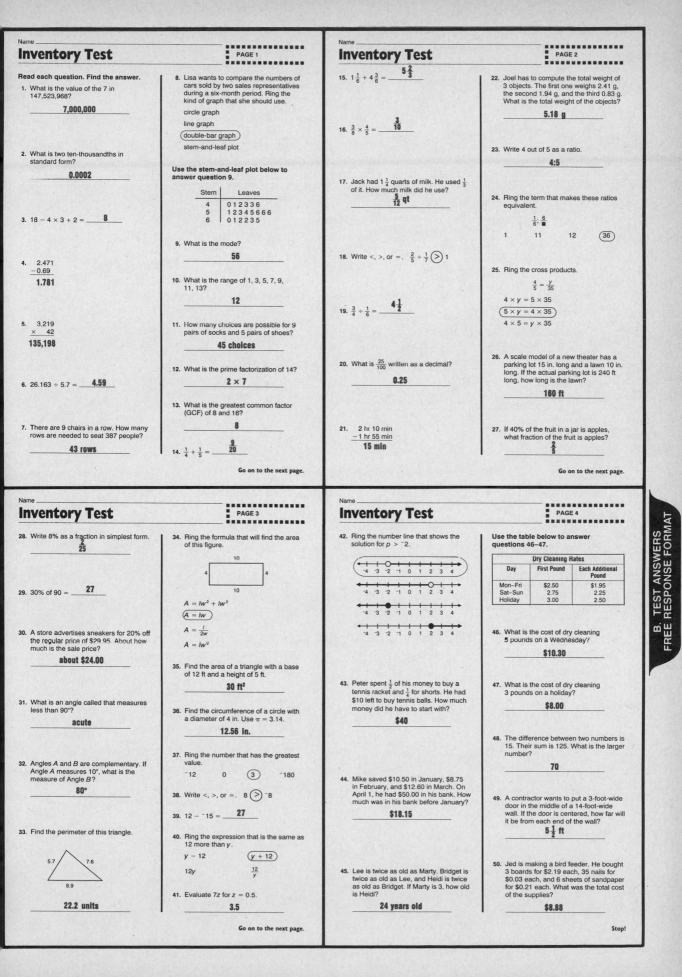

Name

Inventory Test

PAGE 1

Read each question. Find the answer.

1. What is the value of the 7 in 147,523,968?

7,000,000

2. What is two ten-thousandths in standard form?

0.0002

3. $18 - 4 \times 3 + 2 =$ **8**

4. $\begin{array}{r} 2.471 \\ -0.69 \\ \hline \textbf{1.781} \end{array}$

5. $\begin{array}{r} 3{,}219 \\ \times \quad 42 \\ \hline \textbf{135,198} \end{array}$

6. $26.163 \div 5.7 =$ **4.59**

7. There are 9 chairs in a row. How many rows are needed to seat 387 people?

43 rows

8. Lisa wants to compare the numbers of cars sold by two sales representatives during a six-month period. Ring the kind of graph that she should use.

circle graph

line graph

(double-bar graph)

stem-and-leaf plot

Use the stem-and-leaf plot below to answer question 9.

Stem	Leaves
4	0 1 2 3 3 6
5	1 2 3 4 5 6 6 6
6	0 1 2 2 3 5

9. What is the mode?

56

10. What is the range of 1, 3, 5, 7, 9, 11, 13?

12

11. How many choices are possible for 9 pairs of socks and 5 pairs of shoes?

45 choices

12. What is the prime factorization of 14?

2 × 7

13. What is the greatest common factor (GCF) of 8 and 16?

8

14. $\frac{1}{4} + \frac{1}{5} =$ $\mathbf{\frac{9}{20}}$

Go on to the next page.

Name

Inventory Test

PAGE 2

15. $1\frac{1}{6} + 4\frac{3}{6} =$ $\mathbf{5\frac{2}{3}}$

16. $\frac{3}{8} \times \frac{4}{5} =$ $\mathbf{\frac{3}{10}}$

17. Jack had $1\frac{1}{4}$ quarts of milk. He used $\frac{1}{3}$ of it. How much milk did he use?

$\mathbf{\frac{5}{12}}$ **qt**

18. Write <, >, or =. $\frac{2}{5} \div \frac{1}{7}$ (>) 1

19. $\frac{3}{4} \div \frac{1}{6} =$ $\mathbf{4\frac{1}{2}}$

20. What is $\frac{25}{100}$ written as a decimal?

0.25

21. $\begin{array}{r} 2 \text{ hr } 10 \text{ min} \\ -1 \text{ hr } 55 \text{ min} \\ \hline \textbf{15 min} \end{array}$

22. Joel has to compute the total weight of 3 objects. The first one weighs 2.41 g, the second 1.94 g, and the third 0.83 g. What is the total weight of the objects?

5.18 g

23. Write 4 out of 5 as a ratio.

4:5

24. Ring the term that makes these ratios equivalent.

$\frac{1}{6}, \frac{6}{■}$

1   11   12   (36)

25. Ring the cross products.

$\frac{4}{5} = \frac{y}{35}$

$4 \times y = 5 \times 35$

($5 \times y = 4 \times 35$)

$4 \times 5 = y \times 35$

26. A scale model of a new theater has a parking lot 15 in. long and a lawn 10 in. long. If the actual parking lot is 240 ft long, how long is the lawn?

160 ft

27. If 40% of the fruit in a jar is apples, what fraction of the fruit is apples?

$\mathbf{\frac{2}{5}}$

Go on to the next page.

Name

Inventory Test

PAGE 3

28. Write 8% as a fraction in simplest form.

$\mathbf{\frac{2}{25}}$

29. 30% of 90 = **27**

30. A store advertises sneakers for 20% off the regular price of $29.95. About how much is the sale price?

about $24.00

31. What is an angle called that measures less than 90°?

acute

32. Angles *A* and *B* are complementary. If Angle *A* measures 10°, what is the measure of Angle *B*?

80°

33. Find the perimeter of this triangle.

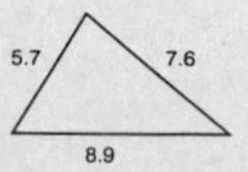

22.2 units

34. Ring the formula that will find the area of this figure.

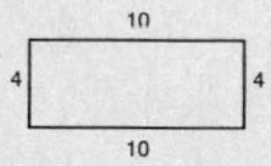

$A = lw^2 + lw^2$

($A = lw$)

$A = \frac{l}{2w}$

$A = lw^2$

35. Find the area of a triangle with a base of 12 ft and a height of 5 ft.

30 ft²

36. Find the circumference of a circle with a diameter of 4 in. Use $\pi = 3.14$.

12.56 in.

37. Ring the number that has the greatest value.

$^-12$ 0   (3) $^-180$

38. Write <, >, or =. 8 (>) $^-8$

39. $12 - {}^-15 =$ **27**

40. Ring the expression that is the same as 12 more than *y*.

$y - 12$ ($y + 12$)

$12y$ $\frac{12}{y}$

41. Evaluate $7z$ for $z = 0.5$.

3.5

Go on to the next page.

Name

Inventory Test

PAGE 4

42. Ring the number line that shows the solution for $p > {}^-2$.

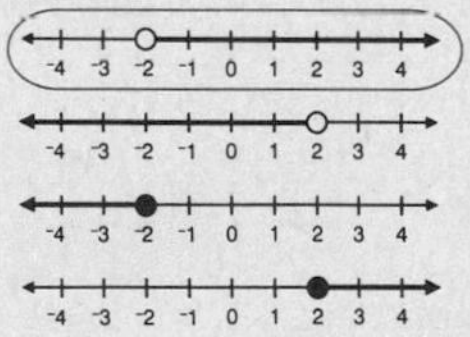

43. Peter spent $\frac{1}{2}$ of his money to buy a tennis racket and $\frac{1}{4}$ for shorts. He had $10 left to buy tennis balls. How much money did he have to start with?

$40

44. Mike saved $10.50 in January, $8.75 in February, and $12.60 in March. On April 1, he had $50.00 in his bank. How much was in his bank before January?

$18.15

45. Lee is twice as old as Marty. Bridget is twice as old as Lee, and Heidi is twice as old as Bridget. If Marty is 3, how old is Heidi?

24 years old

Use the table below to answer questions 46–47.

Dry Cleaning Rates		
Day	First Pound	Each Additional Pound
Mon–Fri	$2.50	$1.95
Sat–Sun	2.75	2.25
Holiday	3.00	2.50

46. What is the cost of dry cleaning 5 pounds on a Wednesday?

$10.30

47. What is the cost of dry cleaning 3 pounds on a holiday?

$8.00

48. The difference between two numbers is 15. Their sum is 125. What is the larger number?

70

49. A contractor wants to put a 3-foot-wide door in the middle of a 14-foot-wide wall. If the door is centered, how far will it be from each end of the wall?

$\mathbf{5\frac{1}{2}}$ **ft**

50. Jed is making a bird feeder. He bought 3 boards for $2.19 each, 35 nails for $0.03 each, and 6 sheets of sandpaper for $0.21 each. What was the total cost of the supplies?

$8.88

Stop!

Name

Pretest

CHAPTER 1
PAGE 1

Read each question. Find the answer.

1. Ring the *best* estimate.

   3,198 + 2,025

   less than 4,000

   less than 5,000

   greater than 5,000

2. Ring the *best* estimate. 982 − 386

   less than 600

   greater than 600

   greater than 800

3. Ring the *best* estimate. 3.1 × 15.02

   less than 40

   greater than 45

   greater than 50

4. Ring the *best* estimate.

   49.97 ÷ 2.11

   less than 20

   less than 25

   greater than 25

5. Ring the *best* estimate.

   355 + 1,650 + 422

2,100	2,200
2,500	2,600

6. Ring the *best* estimate.

   1,153.2 − 166.5

700	800
1,000	1,100

7. Ring the *best* estimate. 59.95 × 1,000

6,000	58,000
60,000	599,000

8. Ring the *best* estimate. 2,005 ÷ 9.57

200	950
2,000	18,000

9. Jane wants to buy 3 pens that cost $1.19 each and 2 pads that cost $1.15 each. Ring the *best* estimate for the amount of money that Jane needs.

$3.00	$4.00
$4.50	**$6.00**

10. Bill had $18.99 to spend. He spent $3.09 on Friday, $2.10 on Saturday, and $2.95 on Sunday. Ring the *best* estimate for the amount of money he had left.

$5.00	$8.00
$11.00	$14.00

11. Ring the quotient with the greatest value.

99.2 ÷ 3	99.2 ÷ 5
99.2 ÷ 4	99.2 ÷ 10

12. Ring the difference with the greatest value.

142.5 − 12	14.25 − 12
1,425.0 − 12	1.1425 − 12

13. Ring the product with the greatest value.

4 × 1,001	5 × 101
7 × 10.1	9 × 1.01

Go on to the next page.

Name

Pretest

CHAPTER 1
PAGE 2

14. Ring the quotient with the greatest value.

50 ÷ 0.1	50 ÷ 0.01
50 ÷ 0.001	**50 ÷ 0.0001**

15. Find the sum. 7.86 + 2.99 + 3.07

    13.92

16. Find the difference. 8.07 − 2.09

    5.98

17. Find the product. 9.90 × 11.20

    110.88

18. Find the quotient. 16.4 ÷ 4.1

    4

19. Julio bought 5.5 pounds of cheese for $13.75. How much did he pay per pound of cheese?

    $2.50 per pound

20. Myra runs 2.5 miles every weekday, 3.0 miles each Saturday, and 3.5 miles each Sunday. How many miles does she run in a 2-week period?

    38 miles

21. Find the value of 5^3.

    125

22. Find the value of 9^0.

    1

23. What is the value of 4 to the third power?

    64

24. What is the value of 10^1?

    10

25. Express 605,000,000 in scientific notation.

    6.05×10^8

26. Express 4.5×10^3 in standard form.

    4,500

27. Express 8,203 in scientific notation.

    8.203×10^3

28. Express 3.5×10^1 in standard form.

    35

Go on to the next page.

Name

Pretest

CHAPTER 1
PAGE 3

Use this information to answer questions 29–30.

The menu at Joe's Restaurant lists three main dishes: hamburger, hot dog, and pizza. It lists four salads: tossed, macaroni, potato, and coleslaw.

29. In how many combinations could Natasha order one main dish and one salad?

    12 combinations

30. If Natasha orders a hamburger and two salads, how many combinations are possible for her to order?

    6 combinations

31. A total of 56 boys and girls are in the school chorus. There are 20 more girls than boys. How many boys are in the chorus?

    18 boys

32. The Andersons went to a concert. They spent $39 for tickets. The tickets cost $12 for each adult and $5 for each child. How many children went to the concert?

    3 children

Stop!

Name

Posttest

CHAPTER 1
PAGE 1

Read each question. Find the answer.

1. Ring the *best* estimate.

   4,125 + 3,002

   less than 6,000

   less than 7,000

   greater than 7,000

2. Ring the *best* estimate.

   655 − 462

   less than 100

   less than 200

   less than 300

3. Ring the *best* estimate.

   4.1 × 25.05

   less than 90

   less than 100

   greater than 100

4. Ring the *best* estimate.

   99.98 ÷ 2.05

   less than 33

   less than 50

   greater than 60

5. Ring the *best* estimate.

   544 + 2,499 + 351

2,800	3,100
3,400	3,500

6. Ring the *best* estimate.

   2,742.1 − 653.5

1,900	1,950
2,100	2,300

7. Ring the *best* estimate.

   69.97 × 10,000

70,000	650,000
700,000	750,000

8. Ring the *best* estimate.

   5,005 ÷ 9.67

500	700
2,500	45,000

9. Mario wants to buy 2 suits that cost $149.95 each and 4 shirts that cost $19.95 each. Ring the *best* estimate for the amount of money that Mario needs.

$320.00	**$340.00**
$360.00	$380.00

10. Doris had $25.92 to spend. She spent $9.65 on dinner, $5.00 on a movie, and $2.95 to park her car. Ring the *best* estimate for the amount of money she had left.

$5.00	$6.00
$8.00	$11.00

11. Ring the quotient with the greatest value.

33.3 ÷ 6	33.3 ÷ 8
33.3 ÷ 7	33.3 ÷ 10

12. Ring the difference with the greatest value.

162 − 11.92	**162 − 119.2**
162 − 1.192	162 − 0.1192

13. Ring the product with the greatest value.

4 × 2,001	6 × 201
8 × 20.1	10 × 2.01

Go on to the next page.

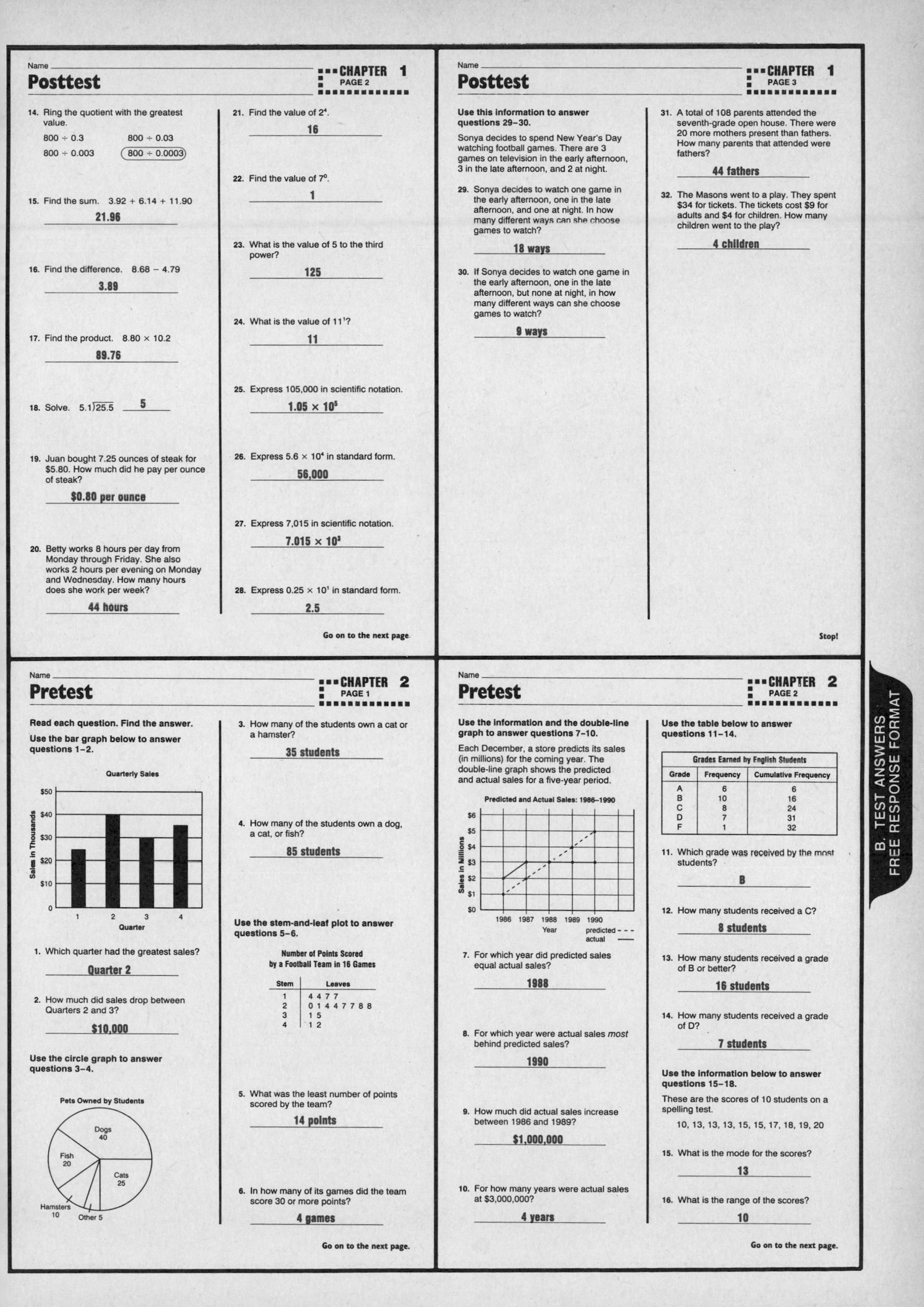

Name

Posttest

CHAPTER 1 PAGE 2

14. Ring the quotient with the greatest value.

800 ÷ 0.3   800 ÷ 0.03

800 ÷ 0.003   (800 ÷ 0.0003)

15. Find the sum. 3.92 + 6.14 + 11.90
21.96

16. Find the difference. 8.68 − 4.79
3.89

17. Find the product. 8.80 × 10.2
89.76

18. Solve. $5.1\overline{)25.5}$ **5**

19. Juan bought 7.25 ounces of steak for $5.80. How much did he pay per ounce of steak?
$0.80 per ounce

20. Betty works 8 hours per day from Monday through Friday. She also works 2 hours per evening on Monday and Wednesday. How many hours does she work per week?
44 hours

21. Find the value of 2^4.
16

22. Find the value of 7^0.
1

23. What is the value of 5 to the third power?
125

24. What is the value of 11^1?
11

25. Express 105,000 in scientific notation.
1.05×10^5

26. Express 5.6×10^4 in standard form.
56,000

27. Express 7,015 in scientific notation.
7.015×10^3

28. Express 0.25×10^1 in standard form.
2.5

Go on to the next page.

Name

Posttest

CHAPTER 1 PAGE 3

Use this information to answer questions 29–30.

Sonya decides to spend New Year's Day watching football games. There are 3 games on television in the early afternoon, 3 in the late afternoon, and 2 at night.

29. Sonya decides to watch one game in the early afternoon, one in the late afternoon, and one at night. In how many different ways can she choose games to watch?
18 ways

30. If Sonya decides to watch one game in the early afternoon, one in the late afternoon, but none at night, in how many different ways can she choose games to watch?
9 ways

31. A total of 108 parents attended the seventh-grade open house. There were 20 more mothers present than fathers. How many parents that attended were fathers?
44 fathers

32. The Masons went to a play. They spent $34 for tickets. The tickets cost $9 for adults and $4 for children. How many children went to the play?
4 children

Stop!

Name

Pretest

CHAPTER 2 PAGE 1

Read each question. Find the answer.

Use the bar graph below to answer questions 1–2.

Quarterly Sales

Sales in Thousands: $0, $10, $20, $30, $40, $50

Quarter: 1, 2, 3, 4

1. Which quarter had the greatest sales?
Quarter 2

2. How much did sales drop between Quarters 2 and 3?
$10,000

Use the circle graph to answer questions 3–4.

Pets Owned by Students

Dogs 40; Fish 20; Cats 25; Hamsters 10; Other 5

3. How many of the students own a cat or a hamster?
35 students

4. How many of the students own a dog, a cat, or fish?
85 students

Use the stem-and-leaf plot to answer questions 5–6.

Number of Points Scored by a Football Team in 16 Games

Stem	Leaves
1	4 4 7 7
2	0 1 4 4 7 7 8 8
3	1 5
4	1 2

5. What was the least number of points scored by the team?
14 points

6. In how many of its games did the team score 30 or more points?
4 games

Go on to the next page.

Name

Pretest

CHAPTER 2 PAGE 2

Use the information and the double-line graph to answer questions 7–10.

Each December, a store predicts its sales (in millions) for the coming year. The double-line graph shows the predicted and actual sales for a five-year period.

Predicted and Actual Sales: 1986–1990

Sales in Millions: $0, $1, $2, $3, $4, $5, $6

Year: 1986, 1987, 1988, 1989, 1990

predicted - - -
actual ——

7. For which year did predicted sales equal actual sales?
1988

8. For which year were actual sales *most* behind predicted sales?
1990

9. How much did actual sales increase between 1986 and 1989?
$1,000,000

10. For how many years were actual sales at $3,000,000?
4 years

Use the table below to answer questions 11–14.

Grades Earned by English Students

Grade	Frequency	Cumulative Frequency
A	6	6
B	10	16
C	8	24
D	7	31
F	1	32

11. Which grade was received by the most students?
B

12. How many students received a C?
8 students

13. How many students received a grade of B or better?
16 students

14. How many students received a grade of D?
7 students

Use the information below to answer questions 15–18.

These are the scores of 10 students on a spelling test.

10, 13, 13, 13, 15, 15, 17, 18, 19, 20

15. What is the mode for the scores?
13

16. What is the range of the scores?
10

Go on to the next page.

B. TEST ANSWERS
FREE RESPONSE FORMAT

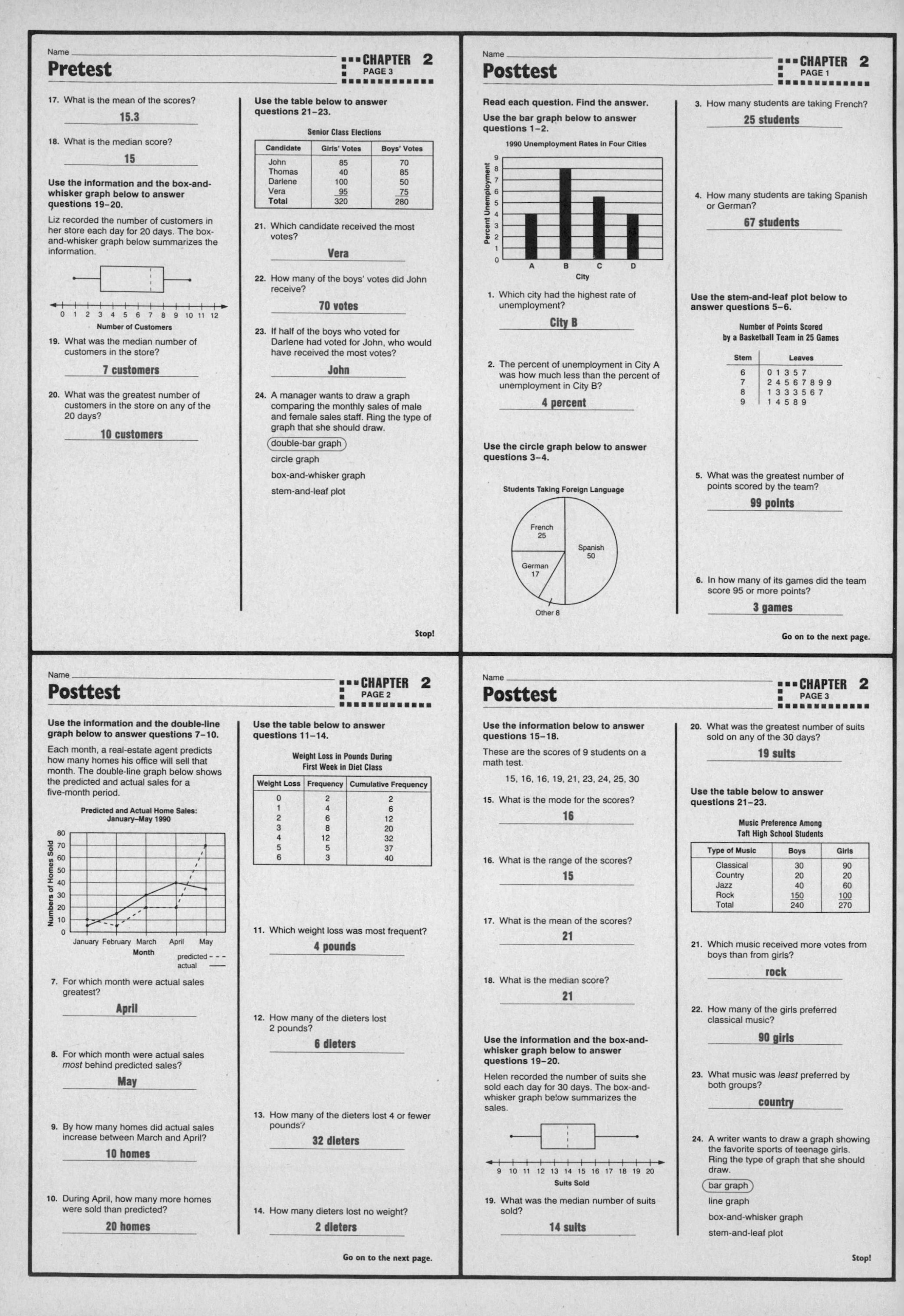

Name

Pretest

CHAPTER 2
PAGE 3

17. What is the mean of the scores?
15.3

18. What is the median score?
15

Use the information and the box-and-whisker graph below to answer questions 19–20.

Liz recorded the number of customers in her store each day for 20 days. The box-and-whisker graph below summarizes the information.

19. What was the median number of customers in the store?
7 customers

20. What was the greatest number of customers in the store on any of the 20 days?
10 customers

Use the table below to answer questions 21–23.

Senior Class Elections

Candidate	Girls' Votes	Boys' Votes
John	85	70
Thomas	40	85
Darlene	100	50
Vera	95	75
Total	320	280

21. Which candidate received the most votes?
Vera

22. How many of the boys' votes did John receive?
70 votes

23. If half of the boys who voted for Darlene had voted for John, who would have received the most votes?
John

24. A manager wants to draw a graph comparing the monthly sales of male and female sales staff. Ring the type of graph that she should draw.
(double-bar graph)
circle graph
box-and-whisker graph
stem-and-leaf plot

Stop!

Name

Posttest

CHAPTER 2
PAGE 1

Read each question. Find the answer.

Use the bar graph below to answer questions 1–2.

1990 Unemployment Rates in Four Cities

1. Which city had the highest rate of unemployment?
City B

2. The percent of unemployment in City A was how much less than the percent of unemployment in City B?
4 percent

Use the circle graph below to answer questions 3–4.

Students Taking Foreign Language

3. How many students are taking French?
25 students

4. How many students are taking Spanish or German?
67 students

Use the stem-and-leaf plot below to answer questions 5–6.

Number of Points Scored by a Basketball Team in 25 Games

Stem	Leaves
6	0 1 3 5 7
7	2 4 5 6 7 8 9 9
8	1 3 3 3 5 6 7
9	1 4 5 8 9

5. What was the greatest number of points scored by the team?
99 points

6. In how many of its games did the team score 95 or more points?
3 games

Go on to the next page.

Name

Posttest

CHAPTER 2
PAGE 2

Use the information and the double-line graph below to answer questions 7–10.

Each month, a real-estate agent predicts how many homes his office will sell that month. The double-line graph below shows the predicted and actual sales for a five-month period.

Predicted and Actual Home Sales: January–May 1990

7. For which month were actual sales greatest?
April

8. For which month were actual sales *most* behind predicted sales?
May

9. By how many homes did actual sales increase between March and April?
10 homes

10. During April, how many more homes were sold than predicted?
20 homes

Use the table below to answer questions 11–14.

Weight Loss in Pounds During First Week in Diet Class

Weight Loss	Frequency	Cumulative Frequency
0	2	2
1	4	6
2	6	12
3	8	20
4	12	32
5	5	37
6	3	40

11. Which weight loss was most frequent?
4 pounds

12. How many of the dieters lost 2 pounds?
6 dieters

13. How many of the dieters lost 4 or fewer pounds?
32 dieters

14. How many dieters lost no weight?
2 dieters

Go on to the next page.

Name

Posttest

CHAPTER 2
PAGE 3

Use the information below to answer questions 15–18.

These are the scores of 9 students on a math test.

15, 16, 16, 19, 21, 23, 24, 25, 30

15. What is the mode for the scores?
16

16. What is the range of the scores?
15

17. What is the mean of the scores?
21

18. What is the median score?
21

Use the information and the box-and-whisker graph below to answer questions 19–20.

Helen recorded the number of suits she sold each day for 30 days. The box-and-whisker graph below summarizes the sales.

19. What was the median number of suits sold?
14 suits

20. What was the greatest number of suits sold on any of the 30 days?
19 suits

Use the table below to answer questions 21–23.

Music Preference Among Taft High School Students

Type of Music	Boys	Girls
Classical	30	90
Country	20	20
Jazz	40	60
Rock	150	100
Total	240	270

21. Which music received more votes from boys than from girls?
rock

22. How many of the girls preferred classical music?
90 girls

23. What music was *least* preferred by both groups?
country

24. A writer wants to draw a graph showing the favorite sports of teenage girls. Ring the type of graph that she should draw.
(bar graph)
line graph
box-and-whisker graph
stem-and-leaf plot

Stop!

Name

Pretest

CHAPTER 3
PAGE 1

Read each question. Find the answer.

1. Ring the prime number.

(5) 10 15 35

2. Ring the composite number.

11 17 19 **(21)**

3. Ring the prime number.

26 **(29)** 35 51

4. How many factors does 37 have?

2 factors

5. What are the prime factors of 30?

2, 3, 5

6. Ring the prime factors of 52.

(2, 2, 13) 2, 26
4, 13 50, 2

7. Ring the prime factorization of 24.

($2^3 \times 3$) $2^2 \times 6$ $4 \times 2 \times 3$

8. Ring the prime factorization of 1,000.

$2^3 \times 25^2$ **($2^3 \times 5^3$)**
$2^2 \times 250$ 2×500

9. What is the greatest common factor of 24 and 36?

12

10. What is the greatest common factor of 32 and 72?

8

11. What is the least common multiple of 5 and 4?

20

12. What is the least common multiple of 6 and 9?

18

13. Ring the fraction that is expressed in simplest form.

$\frac{3}{15}$ $\frac{5}{15}$
$\frac{55}{100}$ **($\frac{99}{100}$)**

14. Ring the fraction that is *not* expressed in simplest form.

$\frac{1}{2}$ $\frac{2}{3}$ **($\frac{4}{8}$)** $\frac{11}{12}$

15. Ring the fraction that is greater than $1\frac{3}{4}$.

$\frac{27}{16}$ $\frac{32}{20}$
$\frac{173}{100}$ **($\frac{352}{200}$)**

16. Ring the group that is ordered from least to greatest.

$\frac{2}{5}, \frac{1}{2}, \frac{3}{10}$ **($\frac{1}{3}, \frac{1}{2}, \frac{3}{4}$)**
$\frac{3}{4}, \frac{3}{5}, \frac{3}{6}$ $\frac{2}{3}, \frac{4}{5}, \frac{7}{12}$

17. Write <, >, or =. $\frac{9}{27}$ **(=)** $\frac{1}{3}$

18. Write <, >, or =. $3\frac{1}{7}$ **(<)** $3\frac{9}{10}$

19. Write $\frac{75}{90}$ in simplest form.

$\frac{5}{6}$

Go on to the next page.

Name

Pretest

CHAPTER 3
PAGE 2

20. Ring the fraction that is equivalent to $\frac{2}{3}$.

$\frac{3}{4}$ $\frac{5}{7}$
$\frac{7}{9}$ **($\frac{6}{9}$)**

Use the information below to answer questions 21–24.

A taxi charges $1.00 for the first mile and $0.20 for each additional $\frac{1}{6}$ mile.

21. How much does it cost to ride $2\frac{1}{2}$ miles?

$2.80

22. How far can you travel for $5.00?

$4\frac{1}{3}$ miles

23. Jo rode 2 miles and Sam rode 3 miles. How much more did Sam pay?

$1.20

24. Rashawn and a friend take a taxi for $1\frac{1}{2}$ miles. The two riders split the fare evenly. How much does each pay?

$0.80

25. Ring the fraction that falls between $\frac{4}{10}$ and $\frac{5}{10}$.

($\frac{9}{20}$) $\frac{11}{20}$ $\frac{16}{40}$ $\frac{9}{10}$

26. Ring the fraction that falls between $\frac{3}{8}$ and $\frac{4}{8}$.

$\frac{6}{16}$ $\frac{8}{16}$ **($\frac{14}{32}$)** $\frac{17}{32}$

27. Ring the fraction that falls between $\frac{1}{5}$ and $\frac{2}{5}$.

($\frac{3}{10}$) $\frac{4}{10}$
$\frac{1}{20}$ $\frac{9}{20}$

28. Ring the fraction that falls between $\frac{1}{3}$ and $\frac{1}{2}$.

$\frac{3}{6}$ $\frac{2}{6}$
($\frac{5}{12}$) $\frac{7}{12}$

Use the information below to answer questions 29–32.

At Jones Pier the lake is 20 feet deep. Every $\frac{1}{4}$ mile from the pier the lake becomes 1 foot deeper until it reaches a maximum depth of 30 feet.

29. How deep is the water $1\frac{1}{2}$ miles from the pier?

26 feet

30. How many miles from the pier does the lake reach its maximum depth?

$2\frac{1}{2}$ miles

31. How much deeper is the water 2 miles from the pier than it is 1 mile out?

4 feet

32. How far from the pier does the water reach 22 feet in depth?

$\frac{1}{2}$ mile

Stop!

Name

Posttest

CHAPTER 3
PAGE 1

Read each question. Find the answer.

1. Ring the prime number.

(3) 6 10 12

2. Ring the composite number.

13 23 **(33)** 43

3. Ring the prime number.

42 **(59)** 65 81

4. How many factors does 41 have?

2 factors

5. What are the prime factors of 42?

2, 3, 7

6. Ring the prime factors of 68.

(2, 2, 17) 2, 34
4, 17 1, 68

7. Ring the prime factorization of 48.

($2^4 \times 3$) $2^3 \times 6$
$4 \times 4 \times 4$ 8×6

8. Ring the prime factorization of 100.

$2^2 \times 25^2$ **($2^2 \times 5^2$)**
$2^2 \times 25$ $2 \times 5 \times 10$

9. What is the greatest common factor of 32 and 40?

8

10. What is the greatest common factor of 16 and 24?

8

11. What is the least common multiple of 3 and 4?

12

12. What is the least common multiple of 9 and 12?

36

13. Ring the fraction that is expressed in simplest form.

$\frac{4}{18}$ $\frac{6}{20}$
$\frac{55}{200}$ **($\frac{97}{102}$)**

14. Ring the fraction that is *not* expressed in simplest form.

$\frac{1}{3}$ $\frac{2}{5}$ **($\frac{4}{10}$)** $\frac{5}{11}$

15. Ring the fraction that is greater than $1\frac{1}{4}$.

$\frac{24}{20}$ $\frac{32}{30}$
$\frac{123}{100}$ **($\frac{257}{200}$)**

16. Ring the group that is ordered from least to greatest.

$\frac{1}{5}, \frac{2}{3}, \frac{4}{10}$ **($\frac{1}{4}, \frac{1}{2}, \frac{3}{4}$)**
$\frac{3}{6}, \frac{3}{7}, \frac{3}{8}$ $\frac{3}{4}, \frac{7}{8}, \frac{6}{7}$

17. Write <, >, or =. $\frac{13}{18}$ **(>)** $\frac{3}{8}$

18. Write <, >, or =. $2\frac{7}{8}$ **(=)** $2\frac{21}{24}$

19. Write $\frac{7}{28}$ in simplest form.

$\frac{1}{4}$

Go on to the next page.

Name

Posttest

CHAPTER 3
PAGE 2

20. Ring the fraction that is equivalent to $\frac{3}{4}$.

$\frac{4}{5}$ $\frac{5}{7}$ $\frac{6}{9}$ **($\frac{6}{8}$)**

Use the information below to answer questions 21–24.

A parking lot charges $3.00 for the first hour and $1.00 for each additional $\frac{1}{2}$ hour.

21. How much does it cost to park for $2\frac{1}{2}$ hours?

$6.00

22. For how long can you park for $4.00?

$1\frac{1}{2}$ hours

23. Alan parks for 4 hours and Alexa parks for 6 hours. How much more does Alexa pay than Alan?

$4.00

24. Justin and Julie drive to work together. They split the cost of parking. If they park for $7\frac{1}{2}$ hours, how much will each pay?

$8.00

25. Ring the fraction that falls between $\frac{4}{12}$ and $\frac{5}{12}$.

($\frac{9}{24}$) $\frac{10}{24}$
$\frac{16}{48}$ $\frac{20}{48}$

26. Ring the fraction that falls between $\frac{3}{10}$ and $\frac{4}{10}$.

$\frac{6}{20}$ $\frac{8}{20}$ **($\frac{13}{40}$)** $\frac{16}{40}$

27. Ring the fraction that falls between $\frac{1}{8}$ and $\frac{2}{8}$.

($\frac{3}{16}$) $\frac{4}{16}$
$\frac{15}{32}$ $\frac{17}{32}$

28. Ring the fraction that falls between $\frac{1}{4}$ and $\frac{1}{3}$.

$\frac{3}{12}$ $\frac{4}{12}$
($\frac{7}{24}$) $\frac{6}{24}$

Use the information below to answer questions 29–32.

Bob and Peg began a running program. Bob ran 1 mile on day 1 and increased the distance by $\frac{1}{10}$ mile each day. Peg ran 2 miles on day 1 and increased the distance by $\frac{1}{20}$ mile each day.

29. How far did Bob run on day 5?

$1\frac{2}{5}$ miles

30. How far did Peg run on day 6?

$2\frac{1}{4}$ miles

31. How far did Bob run on day 11?

2 miles

32. How far did Peg run on day 15?

$2\frac{7}{10}$ miles

Stop!

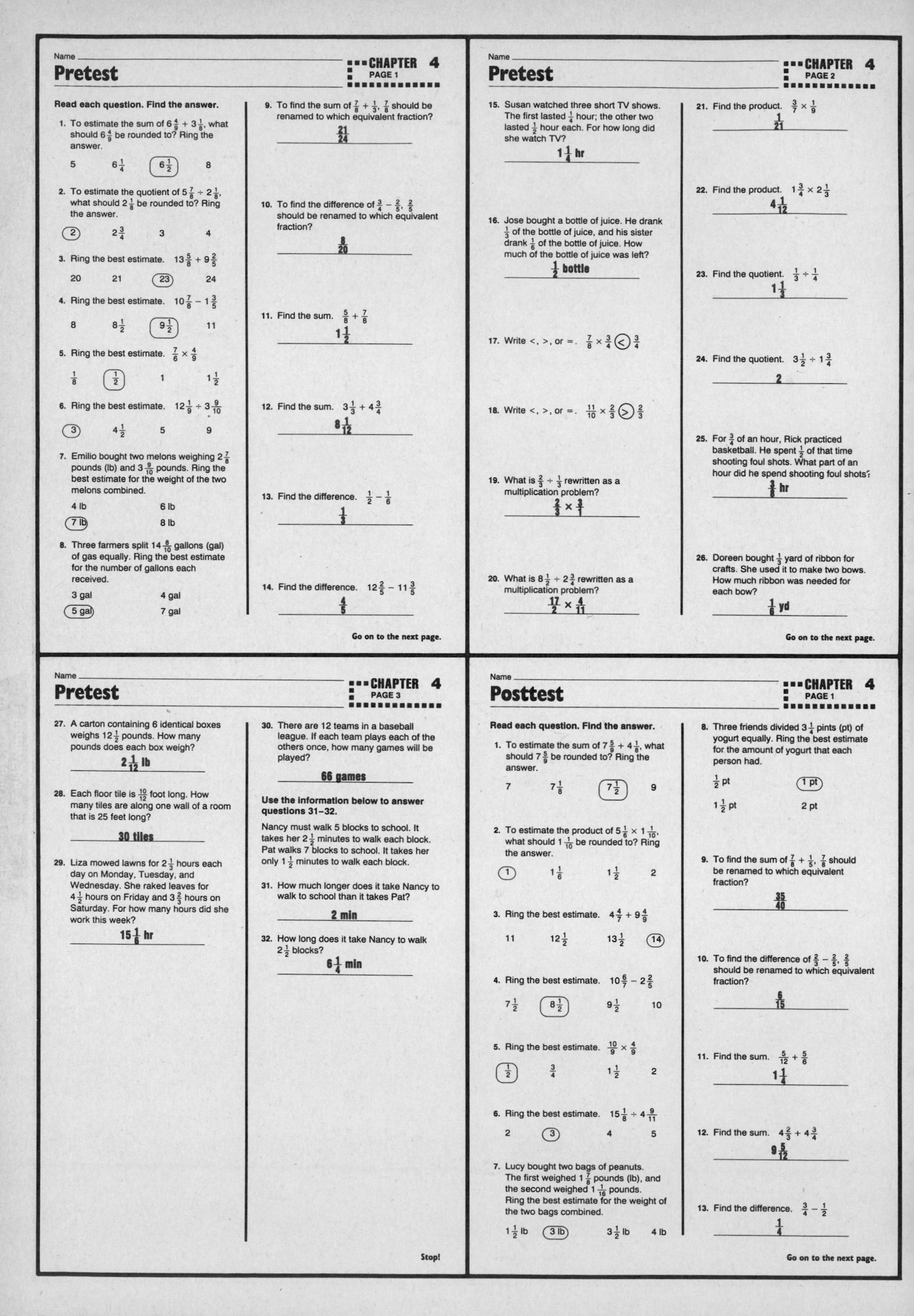

Name

Pretest

CHAPTER 4 PAGE 1

Read each question. Find the answer.

1. To estimate the sum of $6\frac{4}{9} + 3\frac{1}{8}$, what should $6\frac{4}{9}$ be rounded to? Ring the answer.

   5 $6\frac{1}{4}$ **$6\frac{1}{2}$** 8

2. To estimate the quotient of $5\frac{7}{8} \div 2\frac{1}{8}$, what should $2\frac{1}{8}$ be rounded to? Ring the answer.

   2 $2\frac{3}{4}$ 3 4

3. Ring the best estimate. $13\frac{5}{8} + 9\frac{2}{5}$

   20 21 **23** 24

4. Ring the best estimate. $10\frac{7}{8} - 1\frac{3}{5}$

   8 $8\frac{1}{2}$ **$9\frac{1}{2}$** 11

5. Ring the best estimate. $\frac{7}{6} \times \frac{4}{9}$

   $\frac{1}{8}$ **$\frac{1}{2}$** 1 $1\frac{1}{2}$

6. Ring the best estimate. $12\frac{1}{9} \div 3\frac{9}{10}$

   3 $4\frac{1}{2}$ 5 9

7. Emilio bought two melons weighing $2\frac{7}{8}$ pounds (lb) and $3\frac{9}{10}$ pounds. Ring the best estimate for the weight of the two melons combined.

   4 lb 6 lb **7 lb** 8 lb

8. Three farmers split $14\frac{8}{10}$ gallons (gal) of gas equally. Ring the best estimate for the number of gallons each received.

   3 gal 4 gal **5 gal** 7 gal

9. To find the sum of $\frac{7}{8} + \frac{1}{3}$, $\frac{7}{8}$ should be renamed to which equivalent fraction?

   $\frac{21}{24}$

10. To find the difference of $\frac{3}{4} - \frac{2}{5}$, $\frac{2}{5}$ should be renamed to which equivalent fraction?

    $\frac{8}{20}$

11. Find the sum. $\frac{5}{8} + \frac{7}{8}$

    $1\frac{1}{2}$

12. Find the sum. $3\frac{1}{3} + 4\frac{3}{4}$

    $8\frac{1}{12}$

13. Find the difference. $\frac{1}{2} - \frac{1}{6}$

    $\frac{1}{3}$

14. Find the difference. $12\frac{2}{5} - 11\frac{3}{5}$

    $\frac{4}{5}$

Go on to the next page.

Name

Pretest

CHAPTER 4 PAGE 2

15. Susan watched three short TV shows. The first lasted $\frac{1}{4}$ hour; the other two lasted $\frac{1}{2}$ hour each. For how long did she watch TV?

    $1\frac{1}{4}$ hr

16. Jose bought a bottle of juice. He drank $\frac{1}{3}$ of the bottle of juice, and his sister drank $\frac{1}{6}$ of the bottle of juice. How much of the bottle of juice was left?

    $\frac{1}{2}$ bottle

17. Write <, >, or =. $\frac{7}{8} \times \frac{3}{4}$ **<** $\frac{3}{4}$

18. Write <, >, or =. $\frac{11}{10} \times \frac{2}{3}$ **>** $\frac{2}{3}$

19. What is $\frac{2}{3} \div \frac{1}{3}$ rewritten as a multiplication problem?

    $\frac{2}{3} \times \frac{3}{1}$

20. What is $8\frac{1}{2} \div 2\frac{3}{4}$ rewritten as a multiplication problem?

    $\frac{17}{2} \times \frac{4}{11}$

21. Find the product. $\frac{3}{7} \times \frac{1}{9}$

    $\frac{1}{21}$

22. Find the product. $1\frac{3}{4} \times 2\frac{1}{3}$

    $4\frac{1}{12}$

23. Find the quotient. $\frac{1}{3} \div \frac{1}{4}$

    $1\frac{1}{3}$

24. Find the quotient. $3\frac{1}{2} \div 1\frac{3}{4}$

    2

25. For $\frac{3}{4}$ of an hour, Rick practiced basketball. He spent $\frac{1}{2}$ of that time shooting foul shots. What part of an hour did he spend shooting foul shots?

    $\frac{3}{8}$ hr

26. Doreen bought $\frac{1}{3}$ yard of ribbon for crafts. She used it to make two bows. How much ribbon was needed for each bow?

    $\frac{1}{6}$ yd

Go on to the next page.

Name

Pretest

CHAPTER 4 PAGE 3

27. A carton containing 6 identical boxes weighs $12\frac{1}{2}$ pounds. How many pounds does each box weigh?

    $2\frac{1}{12}$ lb

28. Each floor tile is $\frac{10}{12}$ foot long. How many tiles are along one wall of a room that is 25 feet long?

    30 tiles

29. Liza mowed lawns for $2\frac{1}{3}$ hours each day on Monday, Tuesday, and Wednesday. She raked leaves for $4\frac{1}{2}$ hours on Friday and $3\frac{2}{3}$ hours on Saturday. For how many hours did she work this week?

    $15\frac{1}{6}$ hr

30. There are 12 teams in a baseball league. If each team plays each of the others once, how many games will be played?

    66 games

Use the information below to answer questions 31–32.

Nancy must walk 5 blocks to school. It takes her $2\frac{1}{2}$ minutes to walk each block. Pat walks 7 blocks to school. It takes her only $1\frac{1}{2}$ minutes to walk each block.

31. How much longer does it take Nancy to walk to school than it takes Pat?

    2 min

32. How long does it take Nancy to walk $2\frac{1}{2}$ blocks?

    $6\frac{1}{4}$ min

Stop!

Name

Posttest

CHAPTER 4 PAGE 1

Read each question. Find the answer.

1. To estimate the sum of $7\frac{5}{9} + 4\frac{1}{8}$, what should $7\frac{5}{9}$ be rounded to? Ring the answer.

   7 $7\frac{1}{8}$ **$7\frac{1}{2}$** 9

2. To estimate the product of $5\frac{1}{6} \times 1\frac{1}{10}$, what should $1\frac{1}{10}$ be rounded to? Ring the answer.

   1 $1\frac{1}{6}$ $1\frac{1}{2}$ 2

3. Ring the best estimate. $4\frac{4}{7} + 9\frac{4}{9}$

   11 $12\frac{1}{2}$ $13\frac{1}{2}$ **14**

4. Ring the best estimate. $10\frac{6}{7} - 2\frac{2}{5}$

   $7\frac{1}{2}$ **$8\frac{1}{2}$** $9\frac{1}{2}$ 10

5. Ring the best estimate. $\frac{10}{9} \times \frac{4}{9}$

   $\frac{1}{2}$ $\frac{3}{4}$ $1\frac{1}{2}$ 2

6. Ring the best estimate. $15\frac{1}{8} \div 4\frac{9}{11}$

   2 **3** 4 5

7. Lucy bought two bags of peanuts. The first weighed $1\frac{7}{8}$ pounds (lb), and the second weighed $1\frac{1}{16}$ pounds. Ring the best estimate for the weight of the two bags combined.

   $1\frac{1}{2}$ lb **3 lb** $3\frac{1}{2}$ lb 4 lb

8. Three friends divided $3\frac{1}{4}$ pints (pt) of yogurt equally. Ring the best estimate for the amount of yogurt that each person had.

   $\frac{1}{2}$ pt **1 pt** $1\frac{1}{2}$ pt 2 pt

9. To find the sum of $\frac{7}{8} + \frac{1}{5}$, $\frac{7}{8}$ should be renamed to which equivalent fraction?

   $\frac{35}{40}$

10. To find the difference of $\frac{2}{3} - \frac{2}{5}$, $\frac{2}{5}$ should be renamed to which equivalent fraction?

    $\frac{6}{15}$

11. Find the sum. $\frac{5}{12} + \frac{5}{6}$

    $1\frac{1}{4}$

12. Find the sum. $4\frac{2}{3} + 4\frac{3}{4}$

    $9\frac{5}{12}$

13. Find the difference. $\frac{3}{4} - \frac{1}{2}$

    $\frac{1}{4}$

Go on to the next page.

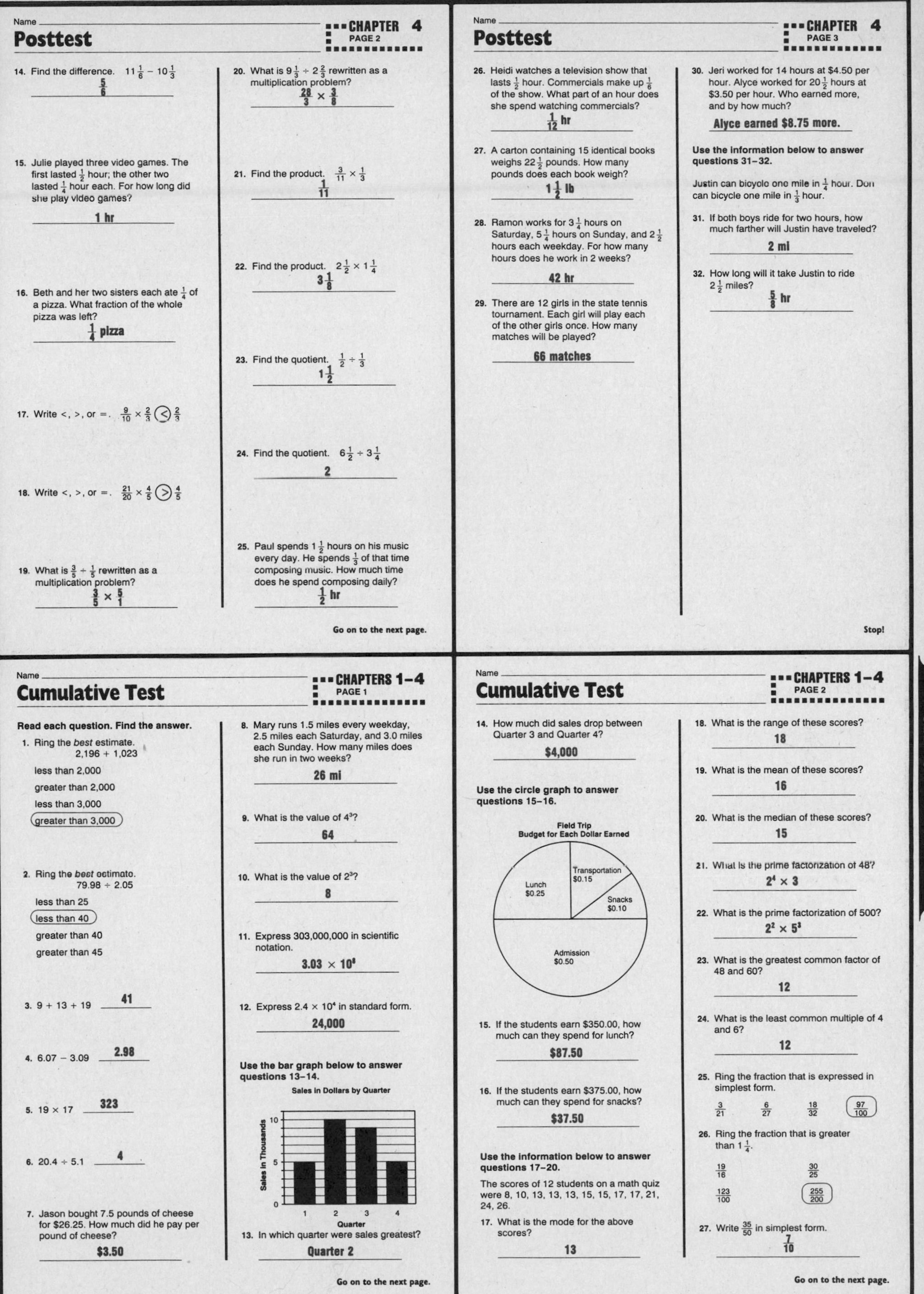

Name

Posttest

CHAPTER 4
PAGE 2

14. Find the difference. $11\frac{1}{6} - 10\frac{1}{3}$
$\frac{5}{6}$

15. Julie played three video games. The first lasted $\frac{1}{2}$ hour; the other two lasted $\frac{1}{4}$ hour each. For how long did she play video games?
1 hr

16. Beth and her two sisters each ate $\frac{1}{4}$ of a pizza. What fraction of the whole pizza was left?
$\frac{1}{4}$ pizza

17. Write <, >, or =. $\frac{9}{10} \times \frac{2}{3}$ (<) $\frac{2}{3}$

18. Write <, >, or =. $\frac{21}{20} \times \frac{4}{5}$ (>) $\frac{4}{5}$

19. What is $\frac{3}{5} \div \frac{1}{5}$ rewritten as a multiplication problem?
$\frac{3}{5} \times \frac{5}{1}$

20. What is $9\frac{1}{3} \div 2\frac{2}{3}$ rewritten as a multiplication problem?
$\frac{28}{3} \times \frac{3}{8}$

21. Find the product. $\frac{3}{11} \times \frac{1}{3}$
$\frac{1}{11}$

22. Find the product. $2\frac{1}{2} \times 1\frac{1}{4}$
$3\frac{1}{8}$

23. Find the quotient. $\frac{1}{2} \div \frac{1}{3}$
$1\frac{1}{2}$

24. Find the quotient. $6\frac{1}{2} \div 3\frac{1}{4}$
2

25. Paul spends $1\frac{1}{2}$ hours on his music every day. He spends $\frac{1}{3}$ of that time composing music. How much time does he spend composing daily?
$\frac{1}{2}$ hr

Go on to the next page.

Name

Posttest

CHAPTER 4
PAGE 3

26. Heidi watches a television show that lasts $\frac{1}{2}$ hour. Commercials make up $\frac{1}{6}$ of the show. What part of an hour does she spend watching commercials?
$\frac{1}{12}$ hr

27. A carton containing 15 identical books weighs $22\frac{1}{2}$ pounds. How many pounds does each book weigh?
$1\frac{1}{2}$ lb

28. Ramon works for $3\frac{1}{4}$ hours on Saturday, $5\frac{1}{4}$ hours on Sunday, and $2\frac{1}{2}$ hours each weekday. For how many hours does he work in 2 weeks?
42 hr

29. There are 12 girls in the state tennis tournament. Each girl will play each of the other girls once. How many matches will be played?
66 matches

30. Jeri worked for 14 hours at \$4.50 per hour. Alyce worked for $20\frac{1}{2}$ hours at \$3.50 per hour. Who earned more, and by how much?
Alyce earned \$8.75 more.

Use the information below to answer questions 31–32.

Justin can bicycle one mile in $\frac{1}{4}$ hour. Don can bicycle one mile in $\frac{1}{3}$ hour.

31. If both boys ride for two hours, how much farther will Justin have traveled?
2 mi

32. How long will it take Justin to ride $2\frac{1}{2}$ miles?
$\frac{5}{8}$ hr

Stop!

Name

Cumulative Test

CHAPTERS 1–4
PAGE 1

Read each question. Find the answer.

1. Ring the *best* estimate.
2,196 + 1,023
less than 2,000
greater than 2,000
less than 3,000
(greater than 3,000)

2. Ring the *best* estimate.
79.98 ÷ 2.05
less than 25
(less than 40)
greater than 40
greater than 45

3. 9 + 13 + 19 41

4. 6.07 − 3.09 2.98

5. 19 × 17 323

6. 20.4 ÷ 5.1 4

7. Jason bought 7.5 pounds of cheese for \$26.25. How much did he pay per pound of cheese?
\$3.50

8. Mary runs 1.5 miles every weekday, 2.5 miles each Saturday, and 3.0 miles each Sunday. How many miles does she run in two weeks?
26 mi

9. What is the value of 4^3?
64

10. What is the value of 2^3?
8

11. Express 303,000,000 in scientific notation.
3.03×10^8

12. Express 2.4×10^4 in standard form.
24,000

Use the bar graph below to answer questions 13–14.

13. In which quarter were sales greatest?
Quarter 2

Go on to the next page.

Name

Cumulative Test

CHAPTERS 1–4
PAGE 2

14. How much did sales drop between Quarter 3 and Quarter 4?
\$4,000

Use the circle graph to answer questions 15–16.

15. If the students earn \$350.00, how much can they spend for lunch?
\$87.50

16. If the students earn \$375.00, how much can they spend for snacks?
\$37.50

Use the information below to answer questions 17–20.

The scores of 12 students on a math quiz were 8, 10, 13, 13, 13, 15, 15, 17, 17, 21, 24, 26.

17. What is the mode for the above scores?
13

18. What is the range of these scores?
18

19. What is the mean of these scores?
16

20. What is the median of these scores?
15

21. What is the prime factorization of 48?
$2^4 \times 3$

22. What is the prime factorization of 500?
$2^2 \times 5^3$

23. What is the greatest common factor of 48 and 60?
12

24. What is the least common multiple of 4 and 6?
12

25. Ring the fraction that is expressed in simplest form.
$\frac{3}{21}$ $\frac{6}{27}$ $\frac{18}{32}$ ($\frac{97}{100}$)

26. Ring the fraction that is greater than $1\frac{1}{4}$.
$\frac{19}{16}$ $\frac{30}{25}$
$\frac{123}{100}$ ($\frac{255}{200}$)

27. Write $\frac{35}{50}$ in simplest form.
$\frac{7}{10}$

Go on to the next page.

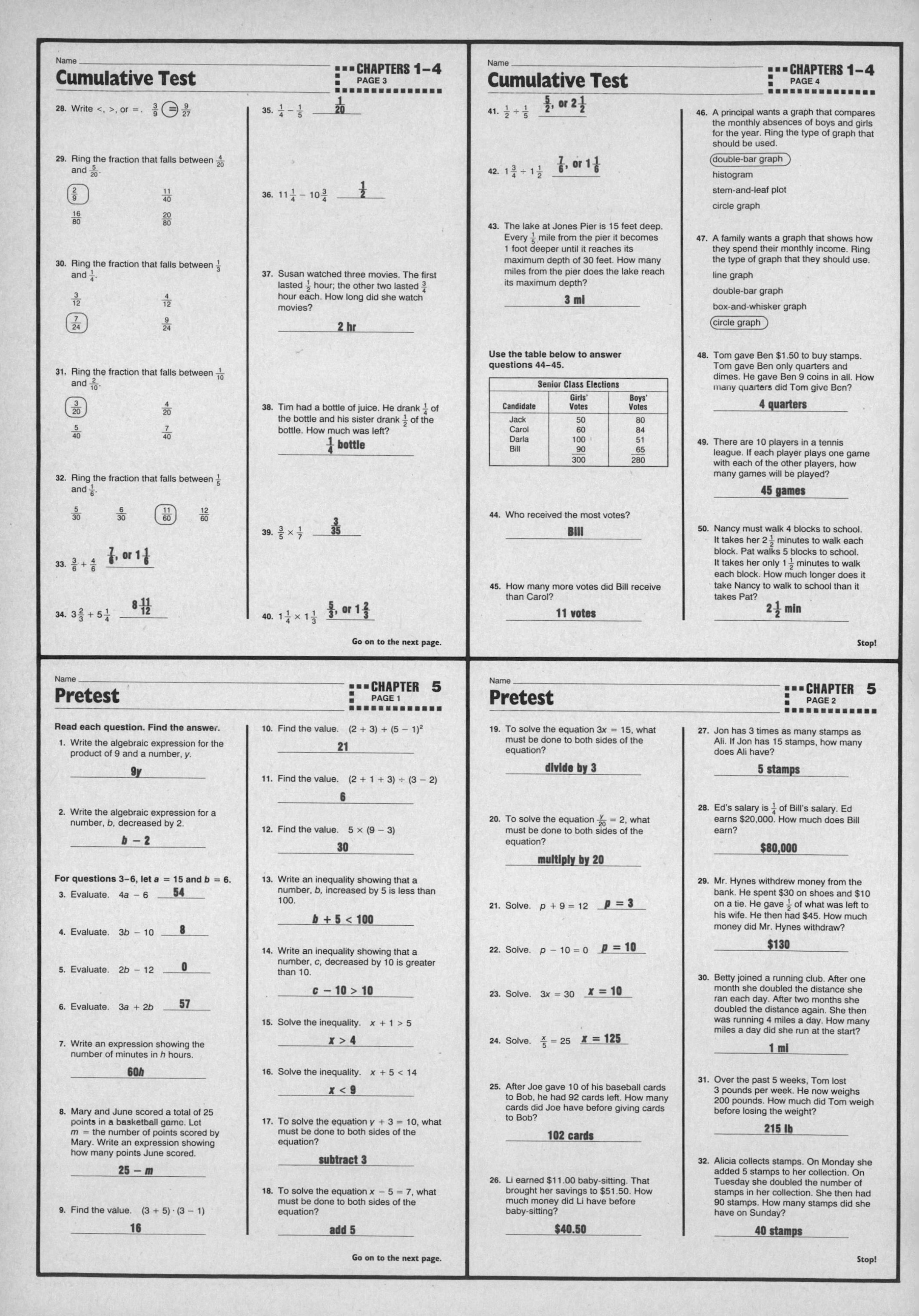

Name ______________________

Cumulative Test

CHAPTERS 1–4
PAGE 3

28. Write <, >, or =. $\frac{3}{9}$ (=) $\frac{9}{27}$

29. Ring the fraction that falls between $\frac{4}{20}$ and $\frac{5}{20}$.

(ringed: $\frac{2}{9}$) $\frac{11}{40}$ $\frac{16}{80}$ $\frac{20}{80}$

30. Ring the fraction that falls between $\frac{1}{3}$ and $\frac{1}{4}$.

$\frac{3}{12}$ $\frac{4}{12}$ (ringed: $\frac{7}{24}$) $\frac{9}{24}$

31. Ring the fraction that falls between $\frac{1}{10}$ and $\frac{2}{10}$.

(ringed: $\frac{3}{20}$) $\frac{4}{20}$ $\frac{5}{40}$ $\frac{7}{40}$

32. Ring the fraction that falls between $\frac{1}{5}$ and $\frac{1}{6}$.

$\frac{5}{30}$ $\frac{6}{30}$ (ringed: $\frac{11}{60}$) $\frac{12}{60}$

33. $\frac{3}{6} + \frac{4}{6}$ **$\frac{7}{6}$, or $1\frac{1}{6}$**

34. $3\frac{2}{3} + 5\frac{1}{4}$ **$8\frac{11}{12}$**

35. $\frac{1}{4} - \frac{1}{5}$ **$\frac{1}{20}$**

36. $11\frac{1}{4} - 10\frac{3}{4}$ **$\frac{1}{2}$**

37. Susan watched three movies. The first lasted $\frac{1}{2}$ hour; the other two lasted $\frac{3}{4}$ hour each. How long did she watch movies?

2 hr

38. Tim had a bottle of juice. He drank $\frac{1}{4}$ of the bottle and his sister drank $\frac{1}{2}$ of the bottle. How much was left?

$\frac{1}{4}$ bottle

39. $\frac{3}{5} \times \frac{1}{7}$ **$\frac{3}{35}$**

40. $1\frac{1}{4} \times 1\frac{1}{3}$ **$\frac{5}{3}$, or $1\frac{2}{3}$**

Go on to the next page.

Name ______________________

Cumulative Test

CHAPTERS 1–4
PAGE 4

41. $\frac{1}{2} \div \frac{1}{5}$ **$\frac{5}{2}$, or $2\frac{1}{2}$**

42. $1\frac{3}{4} \div 1\frac{1}{2}$ **$\frac{7}{6}$, or $1\frac{1}{6}$**

43. The lake at Jones Pier is 15 feet deep. Every $\frac{1}{5}$ mile from the pier it becomes 1 foot deeper until it reaches its maximum depth of 30 feet. How many miles from the pier does the lake reach its maximum depth?

3 mi

Use the table below to answer questions 44–45.

Senior Class Elections		
Candidate	Girls' Votes	Boys' Votes
Jack	50	80
Carol	60	84
Darla	100	51
Bill	90	65
	300	280

44. Who received the most votes?

Bill

45. How many more votes did Bill receive than Carol?

11 votes

46. A principal wants a graph that compares the monthly absences of boys and girls for the year. Ring the type of graph that should be used.

(ringed: double-bar graph)
histogram
stem-and-leaf plot
circle graph

47. A family wants a graph that shows how they spend their monthly income. Ring the type of graph that they should use.

line graph
double-bar graph
box-and-whisker graph
(ringed: circle graph)

48. Tom gave Ben $1.50 to buy stamps. Tom gave Ben only quarters and dimes. He gave Ben 9 coins in all. How many quarters did Tom give Ben?

4 quarters

49. There are 10 players in a tennis league. If each player plays one game with each of the other players, how many games will be played?

45 games

50. Nancy must walk 4 blocks to school. It takes her $2\frac{1}{2}$ minutes to walk each block. Pat walks 5 blocks to school. It takes her only $1\frac{1}{2}$ minutes to walk each block. How much longer does it take Nancy to walk to school than it takes Pat?

$2\frac{1}{2}$ min

Stop!

Name ______________________

Pretest

CHAPTER 5
PAGE 1

Read each question. Find the answer.

1. Write the algebraic expression for the product of 9 and a number, y.

$9y$

2. Write the algebraic expression for a number, b, decreased by 2.

$b - 2$

For questions 3–6, let $a = 15$ and $b = 6$.

3. Evaluate. $4a - 6$ **54**

4. Evaluate. $3b - 10$ **8**

5. Evaluate. $2b - 12$ **0**

6. Evaluate. $3a + 2b$ **57**

7. Write an expression showing the number of minutes in h hours.

$60h$

8. Mary and June scored a total of 25 points in a basketball game. Let m = the number of points scored by Mary. Write an expression showing how many points June scored.

$25 - m$

9. Find the value. $(3 + 5) \cdot (3 - 1)$

16

10. Find the value. $(2 + 3) + (5 - 1)^2$

21

11. Find the value. $(2 + 1 + 3) \div (3 - 2)$

6

12. Find the value. $5 \times (9 - 3)$

30

13. Write an inequality showing that a number, b, increased by 5 is less than 100.

$b + 5 < 100$

14. Write an inequality showing that a number, c, decreased by 10 is greater than 10.

$c - 10 > 10$

15. Solve the inequality. $x + 1 > 5$

$x > 4$

16. Solve the inequality. $x + 5 < 14$

$x < 9$

17. To solve the equation $y + 3 = 10$, what must be done to both sides of the equation?

subtract 3

18. To solve the equation $x - 5 = 7$, what must be done to both sides of the equation?

add 5

Go on to the next page.

Name ______________________

Pretest

CHAPTER 5
PAGE 2

19. To solve the equation $3x = 15$, what must be done to both sides of the equation?

divide by 3

20. To solve the equation $\frac{y}{20} = 2$, what must be done to both sides of the equation?

multiply by 20

21. Solve. $p + 9 = 12$ **$p = 3$**

22. Solve. $p - 10 = 0$ **$p = 10$**

23. Solve. $3x = 30$ **$x = 10$**

24. Solve. $\frac{x}{5} = 25$ **$x = 125$**

25. After Joe gave 10 of his baseball cards to Bob, he had 92 cards left. How many cards did Joe have before giving cards to Bob?

102 cards

26. Li earned $11.00 baby-sitting. That brought her savings to $51.50. How much money did Li have before baby-sitting?

$40.50

27. Jon has 3 times as many stamps as Ali. If Jon has 15 stamps, how many does Ali have?

5 stamps

28. Ed's salary is $\frac{1}{4}$ of Bill's salary. Ed earns $20,000. How much does Bill earn?

$80,000

29. Mr. Hynes withdrew money from the bank. He spent $30 on shoes and $10 on a tie. He gave $\frac{1}{2}$ of what was left to his wife. He then had $45. How much money did Mr. Hynes withdraw?

$130

30. Betty joined a running club. After one month she doubled the distance she ran each day. After two months she doubled the distance again. She then was running 4 miles a day. How many miles a day did she run at the start?

1 mi

31. Over the past 5 weeks, Tom lost 3 pounds per week. He now weighs 200 pounds. How much did Tom weigh before losing the weight?

215 lb

32. Alicia collects stamps. On Monday she added 5 stamps to her collection. On Tuesday she doubled the number of stamps in her collection. She then had 90 stamps. How many stamps did she have on Sunday?

40 stamps

Stop!

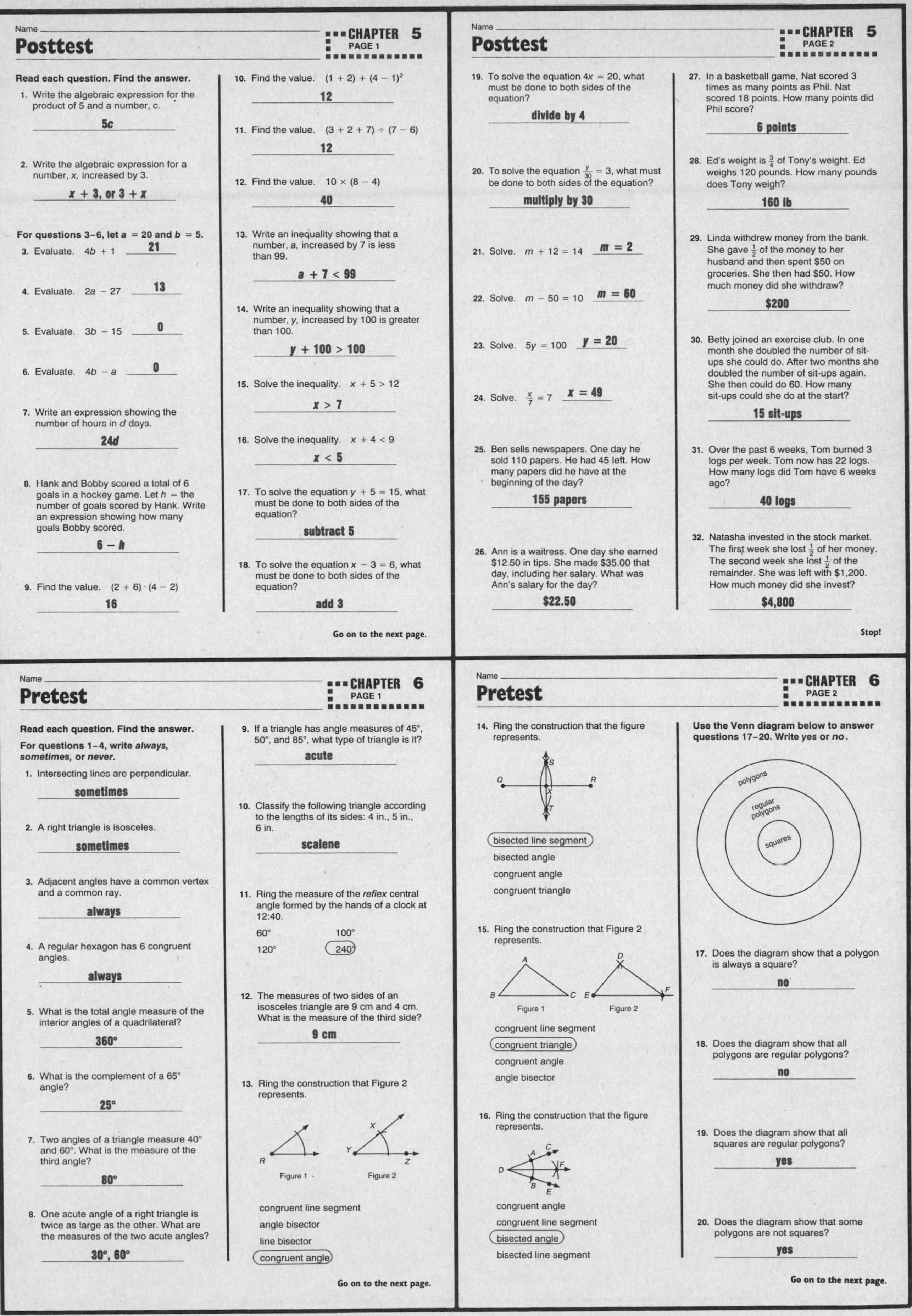

Name

Posttest

CHAPTER 5
PAGE 1

Read each question. Find the answer.

1. Write the algebraic expression for the product of 5 and a number, *c*.
5c

2. Write the algebraic expression for a number, *x*, increased by 3.
$x + 3$, or $3 + x$

For questions 3–6, let $a = 20$ and $b = 5$.

3. Evaluate. $4b + 1$ **21**

4. Evaluate. $2a - 27$ **13**

5. Evaluate. $3b - 15$ **0**

6. Evaluate. $4b - a$ **0**

7. Write an expression showing the number of hours in *d* days.
24d

8. Hank and Bobby scored a total of 6 goals in a hockey game. Let h = the number of goals scored by Hank. Write an expression showing how many goals Bobby scored.
$6 - h$

9. Find the value. $(2 + 6) \cdot (4 - 2)$
16

10. Find the value. $(1 + 2) + (4 - 1)^2$
12

11. Find the value. $(3 + 2 + 7) \div (7 - 6)$
12

12. Find the value. $10 \times (8 - 4)$
40

13. Write an inequality showing that a number, *a*, increased by 7 is less than 99.
$a + 7 < 99$

14. Write an inequality showing that a number, *y*, increased by 100 is greater than 100.
$y + 100 > 100$

15. Solve the inequality. $x + 5 > 12$
$x > 7$

16. Solve the inequality. $x + 4 < 9$
$x < 5$

17. To solve the equation $y + 5 = 15$, what must be done to both sides of the equation?
subtract 5

18. To solve the equation $x - 3 = 6$, what must be done to both sides of the equation?
add 3

Go on to the next page.

Name

Posttest

CHAPTER 5
PAGE 2

19. To solve the equation $4x = 20$, what must be done to both sides of the equation?
divide by 4

20. To solve the equation $\frac{y}{30} = 3$, what must be done to both sides of the equation?
multiply by 30

21. Solve. $m + 12 = 14$ **$m = 2$**

22. Solve. $m - 50 = 10$ **$m = 60$**

23. Solve. $5y = 100$ **$y = 20$**

24. Solve. $\frac{x}{7} = 7$ **$x = 49$**

25. Ben sells newspapers. One day he sold 110 papers. He had 45 left. How many papers did he have at the beginning of the day?
155 papers

26. Ann is a waitress. One day she earned \$12.50 in tips. She made \$35.00 that day, including her salary. What was Ann's salary for the day?
\$22.50

27. In a basketball game, Nat scored 3 times as many points as Phil. Nat scored 18 points. How many points did Phil score?
6 points

28. Ed's weight is $\frac{3}{4}$ of Tony's weight. Ed weighs 120 pounds. How many pounds does Tony weigh?
160 lb

29. Linda withdrew money from the bank. She gave $\frac{1}{2}$ of the money to her husband and then spent \$50 on groceries. She then had \$50. How much money did she withdraw?
\$200

30. Betty joined an exercise club. In one month she doubled the number of sit-ups she could do. After two months she doubled the number of sit-ups again. She then could do 60. How many sit-ups could she do at the start?
15 sit-ups

31. Over the past 6 weeks, Tom burned 3 logs per week. Tom now has 22 logs. How many logs did Tom have 6 weeks ago?
40 logs

32. Natasha invested in the stock market. The first week she lost $\frac{1}{2}$ of her money. The second week she lost $\frac{1}{2}$ of the remainder. She was left with \$1,200. How much money did she invest?
\$4,800

Stop!

Name

Pretest

CHAPTER 6
PAGE 1

Read each question. Find the answer.

For questions 1–4, write *always*, *sometimes*, or *never*.

1. Intersecting lines are perpendicular.
sometimes

2. A right triangle is isosceles.
sometimes

3. Adjacent angles have a common vertex and a common ray.
always

4. A regular hexagon has 6 congruent angles.
always

5. What is the total angle measure of the interior angles of a quadrilateral?
360°

6. What is the complement of a 65° angle?
25°

7. Two angles of a triangle measure 40° and 60°. What is the measure of the third angle?
80°

8. One acute angle of a right triangle is twice as large as the other. What are the measures of the two acute angles?
30°, 60°

9. If a triangle has angle measures of 45°, 50°, and 85°, what type of triangle is it?
acute

10. Classify the following triangle according to the lengths of its sides: 4 in., 5 in., 6 in.
scalene

11. Ring the measure of the *reflex* central angle formed by the hands of a clock at 12:40.
60° 100°
120° (240°)

12. The measures of two sides of an isosceles triangle are 9 cm and 4 cm. What is the measure of the third side?
9 cm

13. Ring the construction that Figure 2 represents.
R; Y, X, Z
Figure 1 Figure 2
congruent line segment
angle bisector
line bisector
(congruent angle)

Go on to the next page.

Name

Pretest

CHAPTER 6
PAGE 2

14. Ring the construction that the figure represents.
S, Q, X, R, T
(bisected line segment)
bisected angle
congruent angle
congruent triangle

15. Ring the construction that Figure 2 represents.
A, B, C; D, E, F
Figure 1 Figure 2
congruent line segment
(congruent triangle)
congruent angle
angle bisector

16. Ring the construction that the figure represents.
C, A, D, F, B, E
congruent angle
congruent line segment
(bisected angle)
bisected line segment

Use the Venn diagram below to answer questions 17–20. Write *yes* or *no*.

polygons
regular polygons
squares

17. Does the diagram show that a polygon is always a square?
no

18. Does the diagram show that all polygons are regular polygons?
no

19. Does the diagram show that all squares are regular polygons?
yes

20. Does the diagram show that some polygons are not squares?
yes

Go on to the next page.

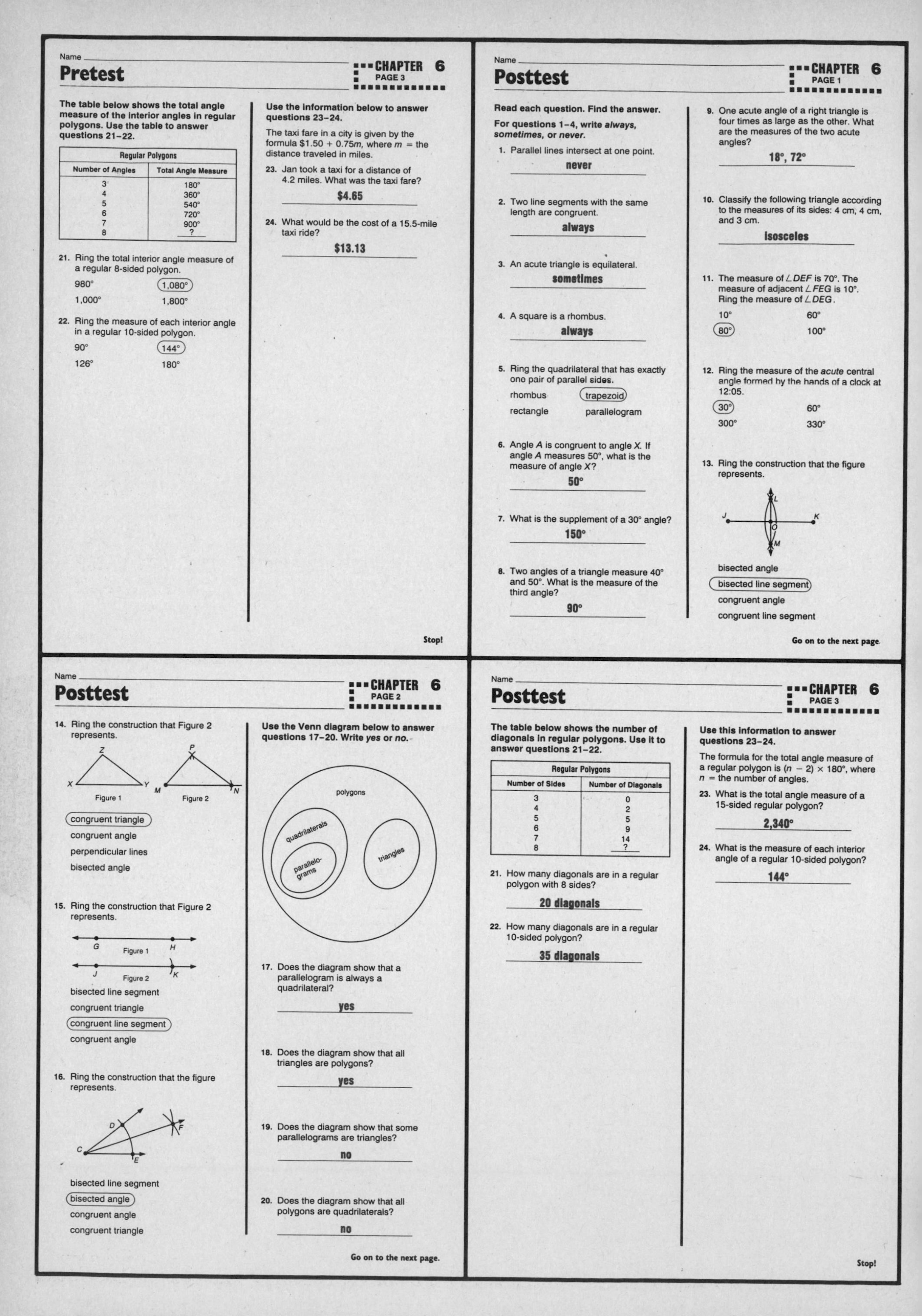

Name

Pretest

CHAPTER 6
PAGE 3

The table below shows the total angle measure of the interior angles in regular polygons. Use the table to answer questions 21–22.

Regular Polygons	
Number of Angles	Total Angle Measure
3	180°
4	360°
5	540°
6	720°
7	900°
8	?

21. Ring the total interior angle measure of a regular 8-sided polygon.

980° (1,080°)
1,000° 1,800°

22. Ring the measure of each interior angle in a regular 10-sided polygon.

90° (144°)
126° 180°

Use the information below to answer questions 23–24.

The taxi fare in a city is given by the formula $1.50 + 0.75*m*, where *m* = the distance traveled in miles.

23. Jan took a taxi for a distance of 4.2 miles. What was the taxi fare?

$4.65

24. What would be the cost of a 15.5-mile taxi ride?

$13.13

Stop!

Name

Posttest

CHAPTER 6
PAGE 1

Read each question. Find the answer.

For questions 1–4, write *always*, *sometimes*, or *never*.

1. Parallel lines intersect at one point.

never

2. Two line segments with the same length are congruent.

always

3. An acute triangle is equilateral.

sometimes

4. A square is a rhombus.

always

5. Ring the quadrilateral that has exactly one pair of parallel sides.

rhombus (trapezoid)
rectangle parallelogram

6. Angle *A* is congruent to angle *X*. If angle *A* measures 50°, what is the measure of angle *X*?

50°

7. What is the supplement of a 30° angle?

150°

8. Two angles of a triangle measure 40° and 50°. What is the measure of the third angle?

90°

9. One acute angle of a right triangle is four times as large as the other. What are the measures of the two acute angles?

18°, 72°

10. Classify the following triangle according to the measures of its sides: 4 cm, 4 cm, and 3 cm.

isosceles

11. The measure of $\angle DEF$ is 70°. The measure of adjacent $\angle FEG$ is 10°. Ring the measure of $\angle DEG$.

10° 60°
(80°) 100°

12. Ring the measure of the *acute* central angle formed by the hands of a clock at 12:05.

(30°) 60°
300° 330°

13. Ring the construction that the figure represents.

bisected angle
(bisected line segment)
congruent angle
congruent line segment

Go on to the next page.

Name

Posttest

CHAPTER 6
PAGE 2

14. Ring the construction that Figure 2 represents.

Figure 1 Figure 2

(congruent triangle)
congruent angle
perpendicular lines
bisected angle

15. Ring the construction that Figure 2 represents.

Figure 1
Figure 2

bisected line segment
congruent triangle
(congruent line segment)
congruent angle

16. Ring the construction that the figure represents.

bisected line segment
(bisected angle)
congruent angle
congruent triangle

Use the Venn diagram below to answer questions 17–20. Write *yes* or *no*.

17. Does the diagram show that a parallelogram is always a quadrilateral?

yes

18. Does the diagram show that all triangles are polygons?

yes

19. Does the diagram show that some parallelograms are triangles?

no

20. Does the diagram show that all polygons are quadrilaterals?

no

Go on to the next page.

Name

Posttest

CHAPTER 6
PAGE 3

The table below shows the number of diagonals in regular polygons. Use it to answer questions 21–22.

Regular Polygons	
Number of Sides	Number of Diagonals
3	0
4	2
5	5
6	9
7	14
8	?

21. How many diagonals are in a regular polygon with 8 sides?

20 diagonals

22. How many diagonals are in a regular 10-sided polygon?

35 diagonals

Use this information to answer questions 23–24.

The formula for the total angle measure of a regular polygon is $(n - 2) \times 180°$, where n = the number of angles.

23. What is the total angle measure of a 15-sided regular polygon?

2,340°

24. What is the measure of each interior angle of a regular 10-sided polygon?

144°

Stop!

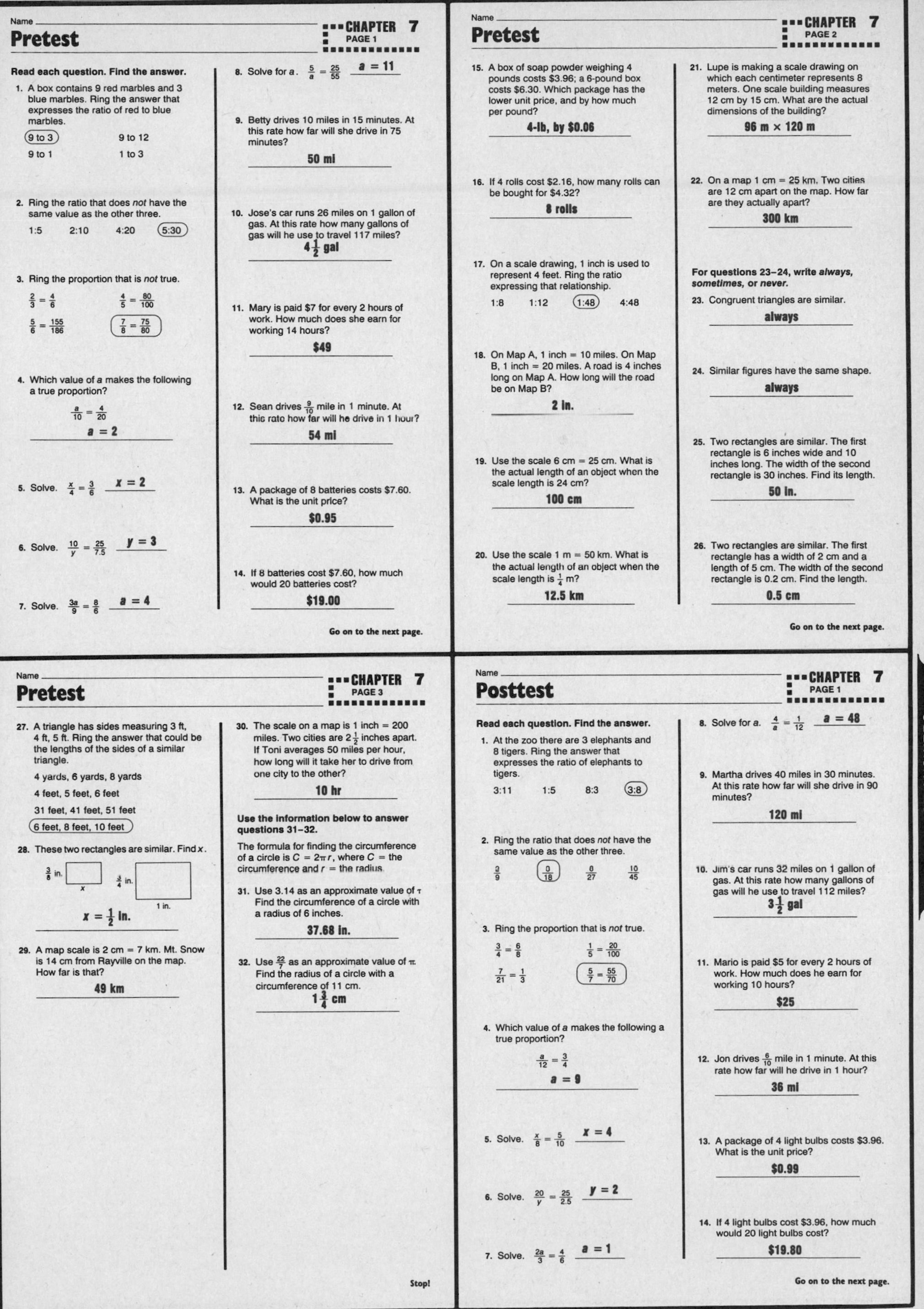

Name

Pretest

CHAPTER 7 PAGE 1

Read each question. Find the answer.

1. A box contains 9 red marbles and 3 blue marbles. Ring the answer that expresses the ratio of red to blue marbles.

   (9 to 3) 9 to 12 9 to 1 1 to 3

2. Ring the ratio that does *not* have the same value as the other three.

   1:5 2:10 4:20 (5:30)

3. Ring the proportion that is *not* true.

   $\frac{2}{3} = \frac{4}{6}$ $\frac{4}{5} = \frac{80}{100}$ $\frac{5}{6} = \frac{155}{186}$ ($\frac{7}{8} = \frac{75}{80}$)

4. Which value of *a* makes the following a true proportion?

   $\frac{a}{10} = \frac{4}{20}$

   $a = 2$

5. Solve. $\frac{x}{4} = \frac{3}{6}$ **$x = 2$**

6. Solve. $\frac{10}{y} = \frac{25}{7.5}$ **$y = 3$**

7. Solve. $\frac{3a}{9} = \frac{8}{6}$ **$a = 4$**

8. Solve for *a*. $\frac{5}{a} = \frac{25}{55}$ **$a = 11$**

9. Betty drives 10 miles in 15 minutes. At this rate how far will she drive in 75 minutes?

   50 mi

10. Jose's car runs 26 miles on 1 gallon of gas. At this rate how many gallons of gas will he use to travel 117 miles?

    $4\frac{1}{2}$ gal

11. Mary is paid $7 for every 2 hours of work. How much does she earn for working 14 hours?

    $49

12. Sean drives $\frac{9}{10}$ mile in 1 minute. At this rate how far will he drive in 1 hour?

    54 mi

13. A package of 8 batteries costs $7.60. What is the unit price?

    $0.95

14. If 8 batteries cost $7.60, how much would 20 batteries cost?

    $19.00

Go on to the next page.

Name

Pretest

CHAPTER 7 PAGE 2

15. A box of soap powder weighing 4 pounds costs $3.96; a 6-pound box costs $6.30. Which package has the lower unit price, and by how much per pound?

    4-lb, by $0.06

16. If 4 rolls cost $2.16, how many rolls can be bought for $4.32?

    8 rolls

17. On a scale drawing, 1 inch is used to represent 4 feet. Ring the ratio expressing that relationship.

    1:8 1:12 (1:48) 4:48

18. On Map A, 1 inch = 10 miles. On Map B, 1 inch = 20 miles. A road is 4 inches long on Map A. How long will the road be on Map B?

    2 in.

19. Use the scale 6 cm = 25 cm. What is the actual length of an object when the scale length is 24 cm?

    100 cm

20. Use the scale 1 m = 50 km. What is the actual length of an object when the scale length is $\frac{1}{4}$ m?

    12.5 km

21. Lupe is making a scale drawing on which each centimeter represents 8 meters. One scale building measures 12 cm by 15 cm. What are the actual dimensions of the building?

    96 m × 120 m

22. On a map 1 cm = 25 km. Two cities are 12 cm apart on the map. How far are they actually apart?

    300 km

For questions 23–24, write *always*, *sometimes*, or *never*.

23. Congruent triangles are similar.

    always

24. Similar figures have the same shape.

    always

25. Two rectangles are similar. The first rectangle is 6 inches wide and 10 inches long. The width of the second rectangle is 30 inches. Find its length.

    50 in.

26. Two rectangles are similar. The first rectangle has a width of 2 cm and a length of 5 cm. The width of the second rectangle is 0.2 cm. Find the length.

    0.5 cm

Go on to the next page.

Name

Pretest

CHAPTER 7 PAGE 3

27. A triangle has sides measuring 3 ft, 4 ft, 5 ft. Ring the answer that could be the lengths of the sides of a similar triangle.

    4 yards, 6 yards, 8 yards
    4 feet, 5 feet, 6 feet
    31 feet, 41 feet, 51 feet
    (6 feet, 8 feet, 10 feet)

28. These two rectangles are similar. Find *x*.

    $\frac{3}{8}$ in. x $\frac{3}{4}$ in. 1 in.

    $x = \frac{1}{2}$ in.

29. A map scale is 2 cm = 7 km. Mt. Snow is 14 cm from Rayville on the map. How far is that?

    49 km

30. The scale on a map is 1 inch = 200 miles. Two cities are $2\frac{1}{2}$ inches apart. If Toni averages 50 miles per hour, how long will it take her to drive from one city to the other?

    10 hr

Use the information below to answer questions 31–32.

The formula for finding the circumference of a circle is $C = 2\pi r$, where C = the circumference and r = the radius.

31. Use 3.14 as an approximate value of π. Find the circumference of a circle with a radius of 6 inches.

    37.68 in.

32. Use $\frac{22}{7}$ as an approximate value of π. Find the radius of a circle with a circumference of 11 cm.

    $1\frac{3}{4}$ cm

Stop!

Name

Posttest

CHAPTER 7 PAGE 1

Read each question. Find the answer.

1. At the zoo there are 3 elephants and 8 tigers. Ring the answer that expresses the ratio of elephants to tigers.

   3:11 1:5 8:3 (3:8)

2. Ring the ratio that does *not* have the same value as the other three.

   $\frac{2}{9}$ ($\frac{0}{18}$) $\frac{6}{27}$ $\frac{10}{45}$

3. Ring the proportion that is *not* true.

   $\frac{3}{4} = \frac{6}{8}$ $\frac{1}{5} = \frac{20}{100}$ $\frac{7}{21} = \frac{1}{3}$ ($\frac{5}{7} = \frac{55}{70}$)

4. Which value of *a* makes the following a true proportion?

   $\frac{a}{12} = \frac{3}{4}$

   $a = 9$

5. Solve. $\frac{x}{8} = \frac{5}{10}$ **$x = 4$**

6. Solve. $\frac{20}{y} = \frac{25}{2.5}$ **$y = 2$**

7. Solve. $\frac{2a}{3} = \frac{4}{6}$ **$a = 1$**

8. Solve for *a*. $\frac{4}{a} = \frac{1}{12}$ **$a = 48$**

9. Martha drives 40 miles in 30 minutes. At this rate how far will she drive in 90 minutes?

   120 mi

10. Jim's car runs 32 miles on 1 gallon of gas. At this rate how many gallons of gas will he use to travel 112 miles?

    $3\frac{1}{2}$ gal

11. Mario is paid $5 for every 2 hours of work. How much does he earn for working 10 hours?

    $25

12. Jon drives $\frac{6}{10}$ mile in 1 minute. At this rate how far will he drive in 1 hour?

    36 mi

13. A package of 4 light bulbs costs $3.96. What is the unit price?

    $0.99

14. If 4 light bulbs cost $3.96, how much would 20 light bulbs cost?

    $19.80

Go on to the next page.

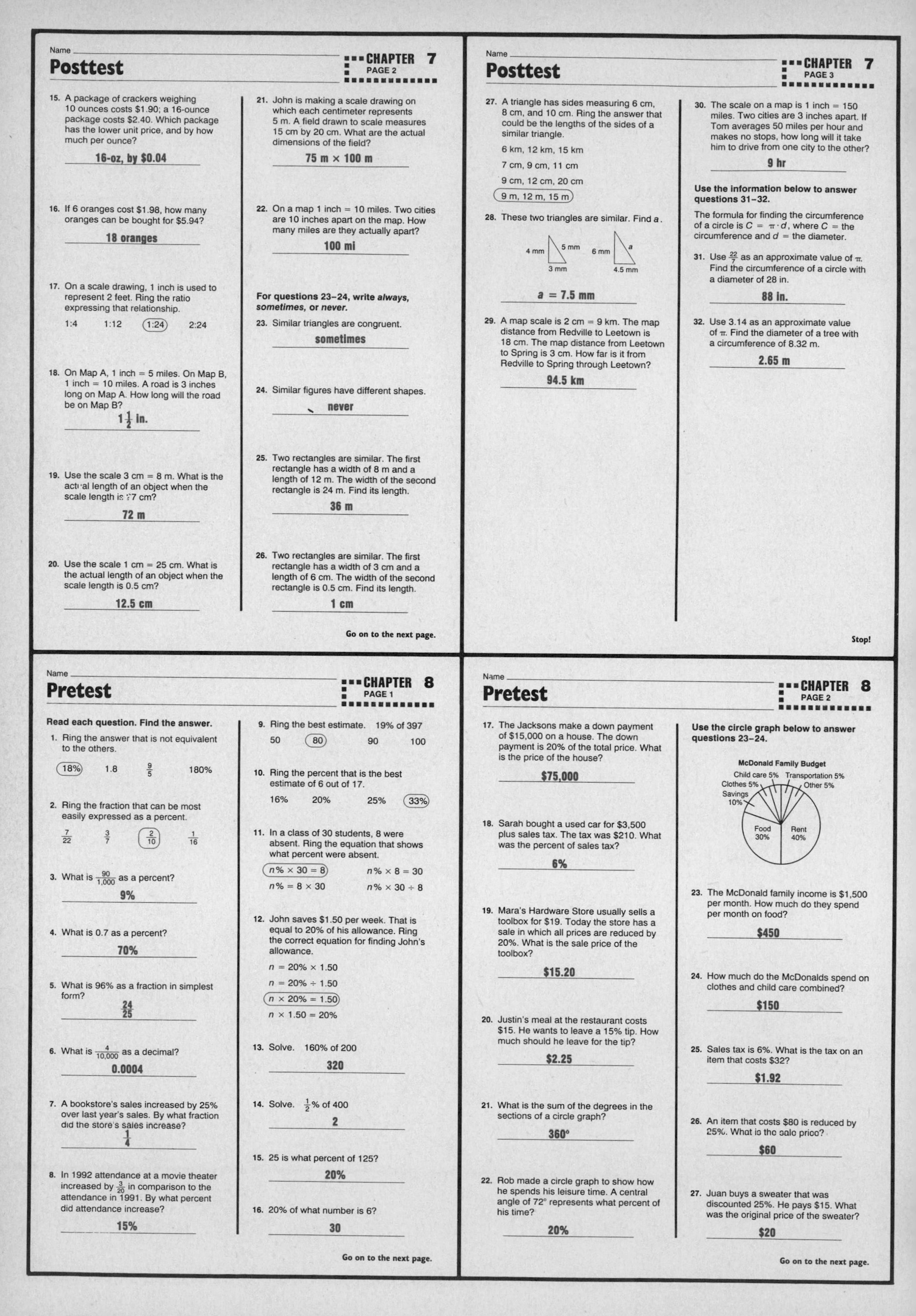

Name

Posttest

CHAPTER 7
PAGE 2

15. A package of crackers weighing 10 ounces costs $1.90; a 16-ounce package costs $2.40. Which package has the lower unit price, and by how much per ounce?

16-oz, by $0.04

16. If 6 oranges cost $1.98, how many oranges can be bought for $5.94?

18 oranges

17. On a scale drawing, 1 inch is used to represent 2 feet. Ring the ratio expressing that relationship.

1:4 1:12 (1:24) 2:24

18. On Map A, 1 inch = 5 miles. On Map B, 1 inch = 10 miles. A road is 3 inches long on Map A. How long will the road be on Map B?

$1\frac{1}{2}$ in.

19. Use the scale 3 cm = 8 m. What is the actual length of an object when the scale length is 27 cm?

72 m

20. Use the scale 1 cm = 25 cm. What is the actual length of an object when the scale length is 0.5 cm?

12.5 cm

21. John is making a scale drawing on which each centimeter represents 5 m. A field drawn to scale measures 15 cm by 20 cm. What are the actual dimensions of the field?

75 m × 100 m

22. On a map 1 inch = 10 miles. Two cities are 10 inches apart on the map. How many miles are they actually apart?

100 mi

For questions 23–24, write *always*, *sometimes*, or *never*.

23. Similar triangles are congruent.

sometimes

24. Similar figures have different shapes.

never

25. Two rectangles are similar. The first rectangle has a width of 8 m and a length of 12 m. The width of the second rectangle is 24 m. Find its length.

36 m

26. Two rectangles are similar. The first rectangle has a width of 3 cm and a length of 6 cm. The width of the second rectangle is 0.5 cm. Find its length.

1 cm

Go on to the next page.

Name

Posttest

CHAPTER 7
PAGE 3

27. A triangle has sides measuring 6 cm, 8 cm, and 10 cm. Ring the answer that could be the lengths of the sides of a similar triangle.

6 km, 12 km, 15 km
7 cm, 9 cm, 11 cm
9 cm, 12 cm, 20 cm
(9 m, 12 m, 15 m)

28. These two triangles are similar. Find *a*.

4 mm, 5 mm, 3 mm; 6 mm, *a*, 4.5 mm

a = 7.5 mm

29. A map scale is 2 cm = 9 km. The map distance from Redville to Leetown is 18 cm. The map distance from Leetown to Spring is 3 cm. How far is it from Redville to Spring through Leetown?

94.5 km

30. The scale on a map is 1 inch = 150 miles. Two cities are 3 inches apart. If Tom averages 50 miles per hour and makes no stops, how long will it take him to drive from one city to the other?

9 hr

Use the information below to answer questions 31–32.

The formula for finding the circumference of a circle is $C = \pi \cdot d$, where C = the circumference and d = the diameter.

31. Use $\frac{22}{7}$ as an approximate value of π. Find the circumference of a circle with a diameter of 28 in.

88 in.

32. Use 3.14 as an approximate value of π. Find the diameter of a tree with a circumference of 8.32 m.

2.65 m

Stop!

Name

Pretest

CHAPTER 8
PAGE 1

Read each question. Find the answer.

1. Ring the answer that is not equivalent to the others.

(18%) 1.8 $\frac{9}{5}$ 180%

2. Ring the fraction that can be most easily expressed as a percent.

$\frac{7}{22}$ $\frac{3}{7}$ ($\frac{2}{10}$) $\frac{1}{16}$

3. What is $\frac{90}{1,000}$ as a percent?

9%

4. What is 0.7 as a percent?

70%

5. What is 96% as a fraction in simplest form?

$\frac{24}{25}$

6. What is $\frac{4}{10,000}$ as a decimal?

0.0004

7. A bookstore's sales increased by 25% over last year's sales. By what fraction did the store's sales increase?

$\frac{1}{4}$

8. In 1992 attendance at a movie theater increased by $\frac{3}{20}$ in comparison to the attendance in 1991. By what percent did attendance increase?

15%

9. Ring the best estimate. 19% of 397

50 (80) 90 100

10. Ring the percent that is the best estimate of 6 out of 17.

16% 20% 25% (33%)

11. In a class of 30 students, 8 were absent. Ring the equation that shows what percent were absent.

(n% × 30 = 8) n% × 8 = 30
n% = 8 × 30 n% × 30 ÷ 8

12. John saves $1.50 per week. That is equal to 20% of his allowance. Ring the correct equation for finding John's allowance.

n = 20% × 1.50
n = 20% ÷ 1.50
(n × 20% = 1.50)
n × 1.50 = 20%

13. Solve. 160% of 200

320

14. Solve. $\frac{1}{2}$% of 400

2

15. 25 is what percent of 125?

20%

16. 20% of what number is 6?

30

Go on to the next page.

Name

Pretest

CHAPTER 8
PAGE 2

17. The Jacksons make a down payment of $15,000 on a house. The down payment is 20% of the total price. What is the price of the house?

$75,000

18. Sarah bought a used car for $3,500 plus sales tax. The tax was $210. What was the percent of sales tax?

6%

19. Mara's Hardware Store usually sells a toolbox for $19. Today the store has a sale in which all prices are reduced by 20%. What is the sale price of the toolbox?

$15.20

20. Justin's meal at the restaurant costs $15. He wants to leave a 15% tip. How much should he leave for the tip?

$2.25

21. What is the sum of the degrees in the sections of a circle graph?

360°

22. Rob made a circle graph to show how he spends his leisure time. A central angle of 72° represents what percent of his time?

20%

Use the circle graph below to answer questions 23–24.

McDonald Family Budget

Child care 5%, Transportation 5%, Clothes 5%, Other 5%, Savings 10%, Food 30%, Rent 40%

23. The McDonald family income is $1,500 per month. How much do they spend per month on food?

$450

24. How much do the McDonalds spend on clothes and child care combined?

$150

25. Sales tax is 6%. What is the tax on an item that costs $32?

$1.92

26. An item that costs $80 is reduced by 25%. What is the sale price?

$60

27. Juan buys a sweater that was discounted 25%. He pays $15. What was the original price of the sweater?

$20

Go on to the next page.

Name

Pretest

CHAPTER 8
PAGE 3

28. Alexa borrows $800 for 6 months. The interest rate is 11% per year. How much interest must Alexa pay?

$44

29. Two girls' ages add to 20. One girl is three times as old as the other. How old is the older girl?

15 years old

30. Two numbers add to 52. One of the numbers is 10 more than the other. What are the numbers?

21, 31

Use the circle graph below to answer questions 31–32.

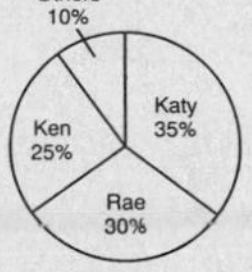

31. If 440 students voted, how many votes did Katy get?

154 votes

32. Only two students other than Katy, Rae, and Ken received any of the 440 votes. If Sy received 18 of those votes, how many did Julia receive?

26 votes

Stop!

Name

Posttest

CHAPTER 8
PAGE 1

Read each question. Find the answer.

1. Ring the answer that is not equivalent to the others.

0.9% $\frac{9}{1,000}$

0.009 (0.09%)

2. Ring the answer with the greatest value.

$\frac{4}{1,000}$ 0.3% (0.02) 0.06%

3. What is 0.17 as a percent?

17%

4. What is 0.15 as a percent?

15%

5. What is 85% as a fraction in simplest form?

$\frac{17}{20}$

6. What is 150% as a decimal?

1.50

7. A company's January sales dropped by $\frac{1}{5}$ compared to December sales. January sales were what percent of December sales?

80%

8. In 1992 a magazine's subscribers doubled in number. What was the percent increase in subscribers?

100%

9. Ring the best estimate. 30% of 189

35 (60) 90 700

10. Ring the percent that is the best estimate of 26 out of 36.

50% 60% (75%) 90%

11. Danielle had $18.50. She spent $15.00 on clothes. Ring the correct equation for finding what percent of her money Danielle spent.

($n\% = \frac{15}{18.5}$)

$n\% = \frac{18.5}{15}$

$n\% \times 15 = 18.5$

$n\% = 15 \times 18.5$

12. Phil invited 12 friends to a party. Only 50% of them could come. Ring the equation that shows how many came to the party.

$n = \frac{12}{50}$

$n \times 0.5 = 12$

$n = \frac{0.5}{12}$

($n = 0.5 \times 12$)

13. Solve. 125% of 300 **375**

14. Solve. 30% of 4 **1.2**

15. 2.5 is what percent of 50?

5%

16. 15% of what number is 60?

400

Go on to the next page.

Name

Posttest

CHAPTER 8
PAGE 2

17. Mr. Mosconi receives an 8% raise. The amount of the raise is $2,800. What was his salary before the raise?

$35,000

18. Sandy bought a ring on sale for $560. Its original price was $700. What was the percent of discount?

20%

19. Anna's Dress Shop is having a sale with all prices reduced by 20%. What is the sale price of a sweater that originally cost $35?

$28

20. George buys a concert ticket through the mail. He pays $25 plus a 5% service charge. How much does George pay for the ticket?

$26.25

21. A family makes a circle graph to show its budget. One section has a central angle of 54°. What percent of the budget is represented by the angle?

15%

22. How many degrees are in 75% of a circle graph?

270°

Use the circle graph below to answer questions 23–24.

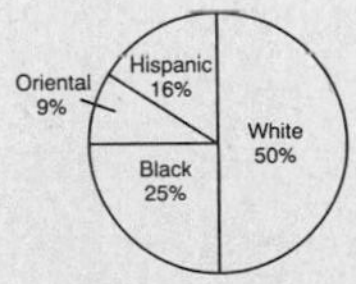

Ethnic Breakdown at Lincoln High

23. There are 1,050 black students at Lincoln High. What is the total number of white students at the school?

2,100 students

24. One half of the Hispanic students at Lincoln High take a biology class. If there are 1,200 students in the school, how many Hispanic students take a biology class?

96 students

25. The sales tax is 5%. What is the tax on an item that costs $24?

$1.20

26. An item that costs $90 is discounted by 60%. What is the sale price?

$36

27. Betty buys a bicycle that was discounted 20%. She pays $100. What was the original price of the bicycle?

$125

Go on to the next page.

Name

Posttest

CHAPTER 8
PAGE 3

28. Alex borrows $500 for 24 months. The interest rate is 9% per year. How much interest must Alex pay?

$90

29. The total number of points scored in a football game was 30. The winning team scored 4 times as many points as the losing team. How many points did the winning team score?

24 points

30. Henry and Anne went fishing. For every 2 fish that Henry caught, Anne caught 3. Together, they caught 15 fish. How many fish did Anne catch?

9 fish

Use the circle graph below to answer questions 31–32.

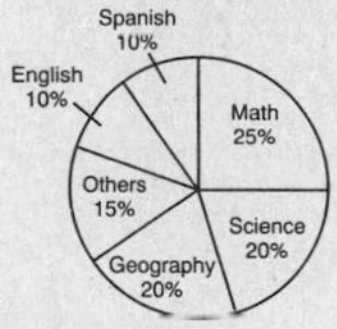

31. How many students chose science as their favorite subject?

72 students

32. If 20 students chose health as their favorite subject, what is the largest number of students who could have chosen French?

34 students

Stop!

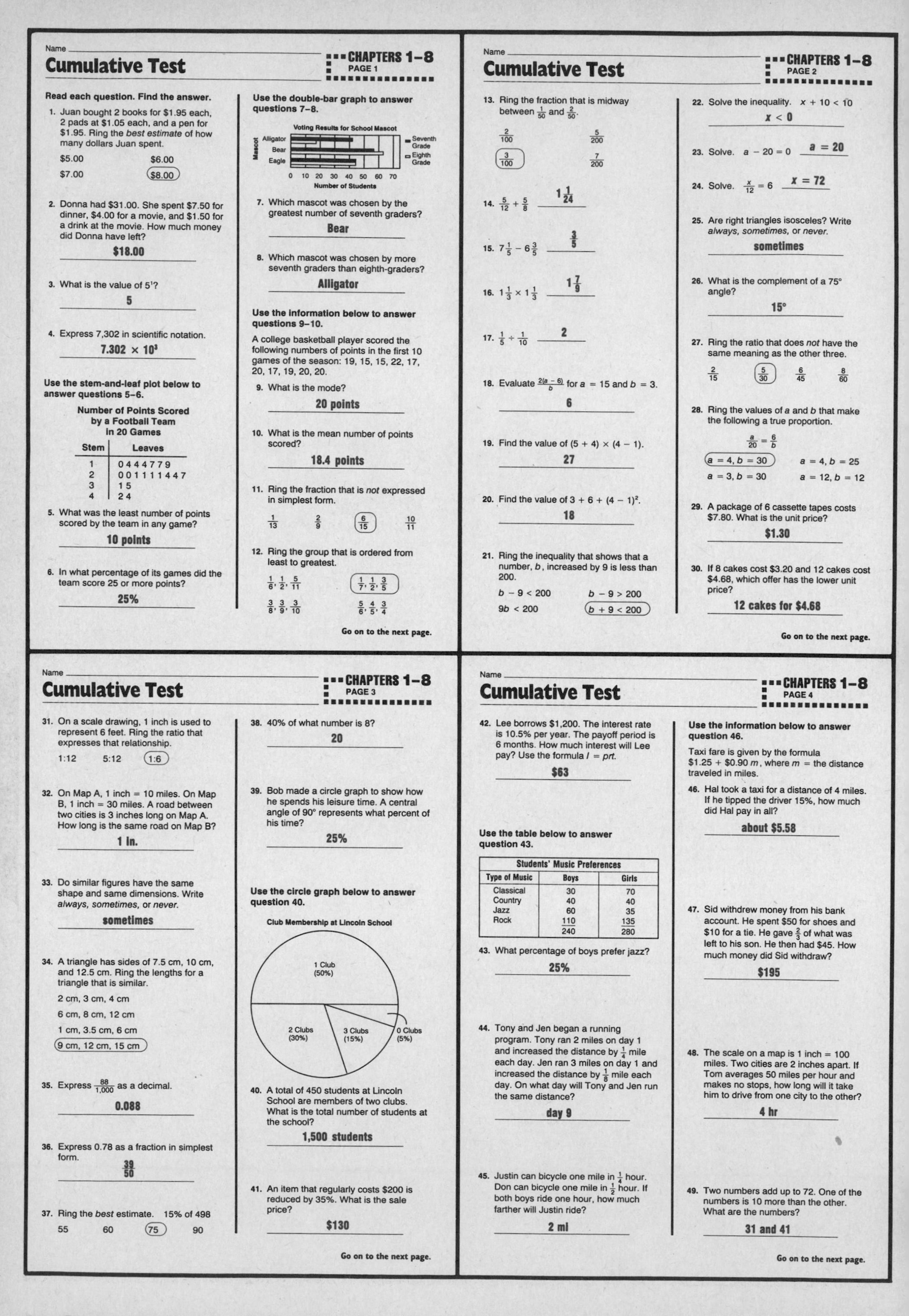

Name

Cumulative Test

CHAPTERS 1–8
PAGE 1

Read each question. Find the answer.

1. Juan bought 2 books for \$1.95 each, 2 pads at \$1.05 each, and a pen for \$1.95. Ring the *best estimate* of how many dollars Juan spent.

   \$5.00 \$6.00
   \$7.00 \$8.00

2. Donna had \$31.00. She spent \$7.50 for dinner, \$4.00 for a movie, and \$1.50 for a drink at the movie. How much money did Donna have left?

   \$18.00

3. What is the value of 5^1?

   5

4. Express 7,302 in scientific notation.

   7.302×10^3

Use the stem-and-leaf plot below to answer questions 5–6.

Number of Points Scored by a Football Team in 20 Games

Stem	Leaves
1	0 4 4 4 7 7 9
2	0 0 1 1 1 1 4 4 7
3	1 5
4	2 4

5. What was the least number of points scored by the team in any game?

   10 points

6. In what percentage of its games did the team score 25 or more points?

   25%

Use the double-bar graph to answer questions 7–8.

Voting Results for School Mascot
Mascot: Alligator, Bear, Eagle
Seventh Grade; Eighth Grade
0 10 20 30 40 50 60 70
Number of Students

7. Which mascot was chosen by the greatest number of seventh graders?

   Bear

8. Which mascot was chosen by more seventh graders than eighth-graders?

   Alligator

Use the information below to answer questions 9–10.

A college basketball player scored the following numbers of points in the first 10 games of the season: 19, 15, 15, 22, 17, 20, 17, 19, 20, 20.

9. What is the mode?

   20 points

10. What is the mean number of points scored?

    18.4 points

11. Ring the fraction that is *not* expressed in simplest form.

    $\frac{1}{13}$ $\frac{2}{9}$ $\frac{6}{15}$ $\frac{10}{11}$

12. Ring the group that is ordered from least to greatest.

    $\frac{1}{6}, \frac{1}{2}, \frac{5}{11}$ $\frac{1}{7}, \frac{1}{2}, \frac{3}{5}$
    $\frac{3}{8}, \frac{3}{9}, \frac{3}{10}$ $\frac{5}{6}, \frac{4}{5}, \frac{3}{4}$

Go on to the next page.

Name

Cumulative Test

CHAPTERS 1–8
PAGE 2

13. Ring the fraction that is midway between $\frac{1}{50}$ and $\frac{2}{50}$.

    $\frac{2}{100}$ $\frac{5}{200}$
    $\frac{3}{100}$ $\frac{7}{200}$

14. $\frac{5}{12} + \frac{5}{8}$ **$1\frac{1}{24}$**

15. $7\frac{1}{5} - 6\frac{3}{5}$ **$\frac{3}{5}$**

16. $1\frac{1}{3} \times 1\frac{1}{3}$ **$1\frac{7}{9}$**

17. $\frac{1}{5} \div \frac{1}{10}$ **2**

18. Evaluate $\frac{2(a-6)}{b}$ for $a = 15$ and $b = 3$.

    6

19. Find the value of $(5 + 4) \times (4 - 1)$.

    27

20. Find the value of $3 + 6 + (4 - 1)^2$.

    18

21. Ring the inequality that shows that a number, b, increased by 9 is less than 200.

    $b - 9 < 200$ $b - 9 > 200$
    $9b < 200$ $b + 9 < 200$

22. Solve the inequality. $x + 10 < 10$

    $x < 0$

23. Solve. $a - 20 = 0$ **$a = 20$**

24. Solve. $\frac{x}{12} = 6$ **$x = 72$**

25. Are right triangles isosceles? Write *always, sometimes,* or *never.*

    sometimes

26. What is the complement of a 75° angle?

    15°

27. Ring the ratio that does *not* have the same meaning as the other three.

    $\frac{2}{15}$ $\frac{5}{30}$ $\frac{6}{45}$ $\frac{8}{60}$

28. Ring the values of a and b that make the following a true proportion.

    $\frac{a}{20} = \frac{6}{b}$

    $a = 4, b = 30$ $a = 4, b = 25$
    $a = 3, b = 30$ $a = 12, b = 12$

29. A package of 6 cassette tapes costs \$7.80. What is the unit price?

    \$1.30

30. If 8 cakes cost \$3.20 and 12 cakes cost \$4.68, which offer has the lower unit price?

    12 cakes for \$4.68

Go on to the next page.

Name

Cumulative Test

CHAPTERS 1–8
PAGE 3

31. On a scale drawing, 1 inch is used to represent 6 feet. Ring the ratio that expresses that relationship.

    1:12 5:12 1:6

32. On Map A, 1 inch = 10 miles. On Map B, 1 inch = 30 miles. A road between two cities is 3 inches long on Map A. How long is the same road on Map B?

    1 in.

33. Do similar figures have the same shape and same dimensions. Write *always, sometimes,* or *never.*

    sometimes

34. A triangle has sides of 7.5 cm, 10 cm, and 12.5 cm. Ring the lengths for a triangle that is similar.

    2 cm, 3 cm, 4 cm
    6 cm, 8 cm, 12 cm
    1 cm, 3.5 cm, 6 cm
    9 cm, 12 cm, 15 cm

35. Express $\frac{88}{1,000}$ as a decimal.

    0.088

36. Express 0.78 as a fraction in simplest form.

    $\frac{39}{50}$

37. Ring the *best* estimate. 15% of 498

    55 60 75 90

38. 40% of what number is 8?

    20

39. Bob made a circle graph to show how he spends his leisure time. A central angle of 90° represents what percent of his time?

    25%

Use the circle graph below to answer question 40.

Club Membership at Lincoln School
1 Club (50%)
2 Clubs (30%)
3 Clubs (15%)
0 Clubs (5%)

40. A total of 450 students at Lincoln School are members of two clubs. What is the total number of students at the school?

    1,500 students

41. An item that regularly costs \$200 is reduced by 35%. What is the sale price?

    \$130

Go on to the next page.

Name

Cumulative Test

CHAPTERS 1–8
PAGE 4

42. Lee borrows \$1,200. The interest rate is 10.5% per year. The payoff period is 6 months. How much interest will Lee pay? Use the formula $I = prt$.

    \$63

Use the table below to answer question 43.

Students' Music Preferences		
Type of Music	Boys	Girls
Classical	30	70
Country	40	40
Jazz	60	35
Rock	110	135
	240	280

43. What percentage of boys prefer jazz?

    25%

44. Tony and Jen began a running program. Tony ran 2 miles on day 1 and increased the distance by $\frac{1}{4}$ mile each day. Jen ran 3 miles on day 1 and increased the distance by $\frac{1}{8}$ mile each day. On what day will Tony and Jen run the same distance?

    day 9

45. Justin can bicycle one mile in $\frac{1}{4}$ hour. Don can bicycle one mile in $\frac{1}{2}$ hour. If both boys ride one hour, how much farther will Justin ride?

    2 mi

Use the information below to answer question 46.

Taxi fare is given by the formula \$1.25 + \$0.90m, where m = the distance traveled in miles.

46. Hal took a taxi for a distance of 4 miles. If he tipped the driver 15%, how much did Hal pay in all?

    about \$5.58

47. Sid withdrew money from his bank account. He spent \$50 for shoes and \$10 for a tie. He gave $\frac{2}{3}$ of what was left to his son. He then had \$45. How much money did Sid withdraw?

    \$195

48. The scale on a map is 1 inch = 100 miles. Two cities are 2 inches apart. If Tom averages 50 miles per hour and makes no stops, how long will it take him to drive from one city to the other?

    4 hr

49. Two numbers add up to 72. One of the numbers is 10 more than the other. What are the numbers?

    31 and 41

Go on to the next page.

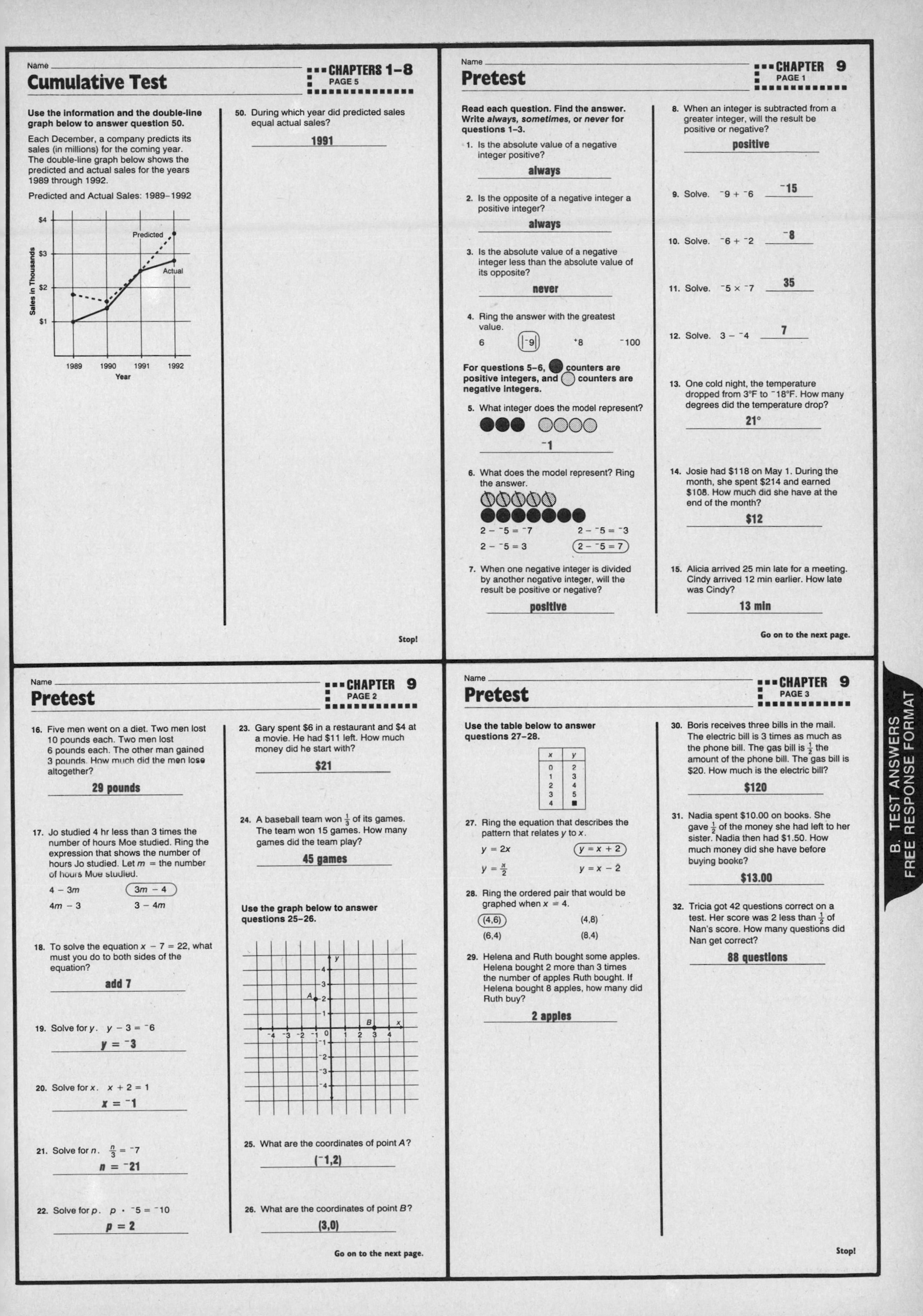

Name

Cumulative Test

CHAPTERS 1–8
PAGE 5

Use the information and the double-line graph below to answer question 50.

Each December, a company predicts its sales (in millions) for the coming year. The double-line graph below shows the predicted and actual sales for the years 1989 through 1992.

Predicted and Actual Sales: 1989–1992

50. During which year did predicted sales equal actual sales?

1991

Stop!

Name

Pretest

CHAPTER 9
PAGE 1

Read each question. Find the answer. Write *always, sometimes,* or *never* for questions 1–3.

1. Is the absolute value of a negative integer positive?

always

2. Is the opposite of a negative integer a positive integer?

always

3. Is the absolute value of a negative integer less than the absolute value of its opposite?

never

4. Ring the answer with the greatest value.

6   (|⁻9|)   ⁺8   ⁻100

For questions 5–6, ● counters are positive integers, and ○ counters are negative integers.

5. What integer does the model represent?

●●● ○○○○

⁻1

6. What does the model represent? Ring the answer.

$2 - {}^-5 = {}^-7$ $2 - {}^-5 = {}^-3$

$2 - {}^-5 = 3$ ($2 - {}^-5 = 7$)

7. When one negative integer is divided by another negative integer, will the result be positive or negative?

positive

8. When an integer is subtracted from a greater integer, will the result be positive or negative?

positive

9. Solve. $^-9 + {}^-6$ **⁻15**

10. Solve. $^-6 + {}^-2$ **⁻8**

11. Solve. $^-5 \times {}^-7$ **35**

12. Solve. $3 - {}^-4$ **7**

13. One cold night, the temperature dropped from 3°F to ⁻18°F. How many degrees did the temperature drop?

21°

14. Josie had $118 on May 1. During the month, she spent $214 and earned $108. How much did she have at the end of the month?

$12

15. Alicia arrived 25 min late for a meeting. Cindy arrived 12 min earlier. How late was Cindy?

13 min

Go on to the next page.

Name

Pretest

CHAPTER 9
PAGE 2

16. Five men went on a diet. Two men lost 10 pounds each. Two men lost 6 pounds each. The other man gained 3 pounds. How much did the men lose altogether?

29 pounds

17. Jo studied 4 hr less than 3 times the number of hours Moe studied. Ring the expression that shows the number of hours Jo studied. Let m = the number of hours Moe studied.

$4 - 3m$ ($3m - 4$)

$4m - 3$ $3 - 4m$

18. To solve the equation $x - 7 = 22$, what must you do to both sides of the equation?

add 7

19. Solve for y. $y - 3 = {}^-6$

$y = {}^-3$

20. Solve for x. $x + 2 = 1$

$x = {}^-1$

21. Solve for n. $\frac{n}{3} = {}^-7$

$n = {}^-21$

22. Solve for p. $p \cdot {}^-5 = {}^-10$

$p = 2$

23. Gary spent $6 in a restaurant and $4 at a movie. He had $11 left. How much money did he start with?

$21

24. A baseball team won $\frac{1}{3}$ of its games. The team won 15 games. How many games did the team play?

45 games

Use the graph below to answer questions 25–26.

25. What are the coordinates of point A?

(⁻1,2)

26. What are the coordinates of point B?

(3,0)

Go on to the next page.

Name

Pretest

CHAPTER 9
PAGE 3

Use the table below to answer questions 27–28.

x	y
0	2
1	3
2	4
3	5
4	■

27. Ring the equation that describes the pattern that relates y to x.

$y = 2x$ ($y = x + 2$)

$y = \frac{x}{2}$ $y = x - 2$

28. Ring the ordered pair that would be graphed when $x = 4$.

((4,6))   (4,8)

(6,4)   (8,4)

29. Helena and Ruth bought some apples. Helena bought 2 more than 3 times the number of apples Ruth bought. If Helena bought 8 apples, how many did Ruth buy?

2 apples

30. Boris receives three bills in the mail. The electric bill is 3 times as much as the phone bill. The gas bill is $\frac{1}{2}$ the amount of the phone bill. The gas bill is $20. How much is the electric bill?

$120

31. Nadia spent $10.00 on books. She gave $\frac{1}{2}$ of the money she had left to her sister. Nadia then had $1.50. How much money did she have before buying books?

$13.00

32. Tricia got 42 questions correct on a test. Her score was 2 less than $\frac{1}{2}$ of Nan's score. How many questions did Nan get correct?

88 questions

Stop!

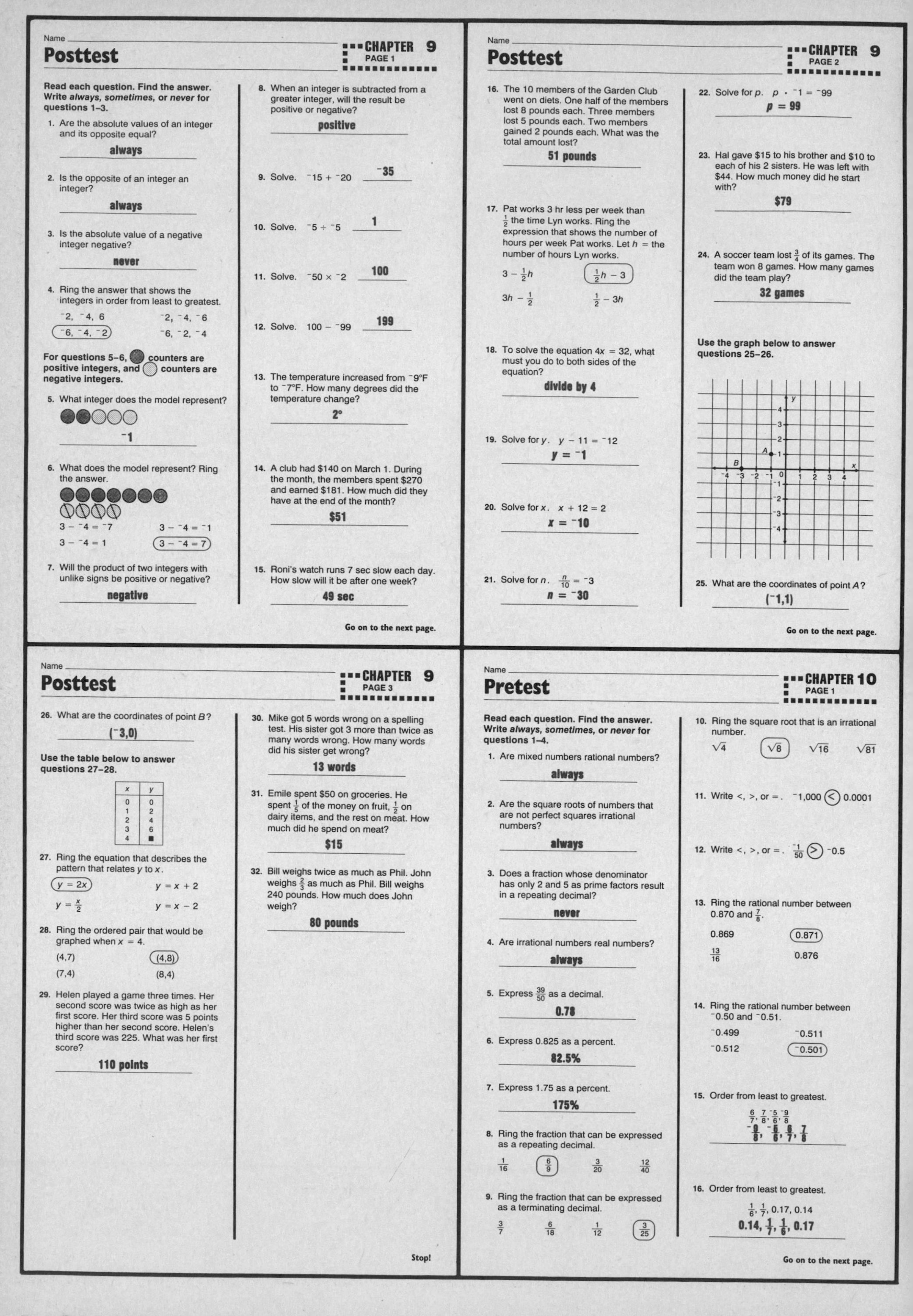

Name

Posttest

CHAPTER 9 PAGE 1

Read each question. Find the answer. Write *always*, *sometimes*, or *never* for questions 1–3.

1. Are the absolute values of an integer and its opposite equal?

always

2. Is the opposite of an integer an integer?

always

3. Is the absolute value of a negative integer negative?

never

4. Ring the answer that shows the integers in order from least to greatest.

$^-2, ^-4, 6$ $^-2, ^-4, ^-6$

($^-6, ^-4, ^-2$) $^-6, ^-2, ^-4$

For questions 5–6, ● counters are positive integers, and ○ counters are negative integers.

5. What integer does the model represent?

$^-1$

6. What does the model represent? Ring the answer.

$3 - ^-4 = ^-7$ $3 - ^-4 = ^-1$

$3 - ^-4 = 1$ ($3 - ^-4 = 7$)

7. Will the product of two integers with unlike signs be positive or negative?

negative

8. When an integer is subtracted from a greater integer, will the result be positive or negative?

positive

9. Solve. $^-15 + ^-20$ **$^-35$**

10. Solve. $^-5 \div ^-5$ **1**

11. Solve. $^-50 \times ^-2$ **100**

12. Solve. $100 - ^-99$ **199**

13. The temperature increased from $^-9$°F to $^-7$°F. How many degrees did the temperature change?

2°

14. A club had $140 on March 1. During the month, the members spent $270 and earned $181. How much did they have at the end of the month?

$51

15. Roni's watch runs 7 sec slow each day. How slow will it be after one week?

49 sec

Go on to the next page.

Name

Posttest

CHAPTER 9 PAGE 2

16. The 10 members of the Garden Club went on diets. One half of the members lost 8 pounds each. Three members lost 5 pounds each. Two members gained 2 pounds each. What was the total amount lost?

51 pounds

17. Pat works 3 hr less per week than $\frac{1}{2}$ the time Lyn works. Ring the expression that shows the number of hours per week Pat works. Let h = the number of hours Lyn works.

$3 - \frac{1}{2}h$ ($\frac{1}{2}h - 3$)

$3h - \frac{1}{2}$ $\frac{1}{2} - 3h$

18. To solve the equation $4x = 32$, what must you do to both sides of the equation?

divide by 4

19. Solve for y. $y - 11 = ^-12$

$y = ^-1$

20. Solve for x. $x + 12 = 2$

$x = ^-10$

21. Solve for n. $\frac{n}{10} = ^-3$

$n = ^-30$

22. Solve for p. $p \cdot ^-1 = ^-99$

$p = 99$

23. Hal gave $15 to his brother and $10 to each of his 2 sisters. He was left with $44. How much money did he start with?

$79

24. A soccer team lost $\frac{3}{4}$ of its games. The team won 8 games. How many games did the team play?

32 games

Use the graph below to answer questions 25–26.

25. What are the coordinates of point A?

($^-1$,1)

Go on to the next page.

Name

Posttest

CHAPTER 9 PAGE 3

26. What are the coordinates of point B?

($^-3$,0)

Use the table below to answer questions 27–28.

x	y
0	0
1	2
2	4
3	6
4	■

27. Ring the equation that describes the pattern that relates y to x.

($y = 2x$) $y = x + 2$

$y = \frac{x}{2}$ $y = x - 2$

28. Ring the ordered pair that would be graphed when $x = 4$.

(4,7)   ((4,8))

(7,4)   (8,4)

29. Helen played a game three times. Her second score was twice as high as her first score. Her third score was 5 points higher than her second score. Helen's third score was 225. What was her first score?

110 points

30. Mike got 5 words wrong on a spelling test. His sister got 3 more than twice as many words wrong. How many words did his sister get wrong?

13 words

31. Emile spent $50 on groceries. He spent $\frac{1}{5}$ of the money on fruit, $\frac{1}{2}$ on dairy items, and the rest on meat. How much did he spend on meat?

$15

32. Bill weighs twice as much as Phil. John weighs $\frac{2}{3}$ as much as Phil. Bill weighs 240 pounds. How much does John weigh?

80 pounds

Stop!

Name

Pretest

CHAPTER 10 PAGE 1

Read each question. Find the answer. Write *always*, *sometimes*, or *never* for questions 1–4.

1. Are mixed numbers rational numbers?

always

2. Are the square roots of numbers that are not perfect squares irrational numbers?

always

3. Does a fraction whose denominator has only 2 and 5 as prime factors result in a repeating decimal?

never

4. Are irrational numbers real numbers?

always

5. Express $\frac{39}{50}$ as a decimal.

0.78

6. Express 0.825 as a percent.

82.5%

7. Express 1.75 as a percent.

175%

8. Ring the fraction that can be expressed as a repeating decimal.

$\frac{1}{16}$ ($\frac{6}{9}$) $\frac{3}{20}$ $\frac{12}{40}$

9. Ring the fraction that can be expressed as a terminating decimal.

$\frac{3}{7}$ $\frac{6}{18}$ $\frac{1}{12}$ ($\frac{3}{25}$)

10. Ring the square root that is an irrational number.

$\sqrt{4}$ ($\sqrt{8}$) $\sqrt{16}$ $\sqrt{81}$

11. Write <, >, or =. $^-1{,}000$ (<) 0.0001

12. Write <, >, or =. $\frac{^-1}{50}$ (>) $^-0.5$

13. Ring the rational number between 0.870 and $\frac{7}{8}$.

0.869   (0.871)

$\frac{13}{16}$ 0.876

14. Ring the rational number between $^-0.50$ and $^-0.51$.

$^-0.499$ $^-0.511$

$^-0.512$ ($^-0.501$)

15. Order from least to greatest.

$\frac{6}{7}, \frac{7}{8}, \frac{^-5}{6}, \frac{^-9}{8}$

$\frac{^-9}{8}, \frac{^-5}{6}, \frac{6}{7}, \frac{7}{8}$

16. Order from least to greatest.

$\frac{1}{6}, \frac{1}{7}, 0.17, 0.14$

$0.14, \frac{1}{7}, \frac{1}{6}, 0.17$

Go on to the next page.

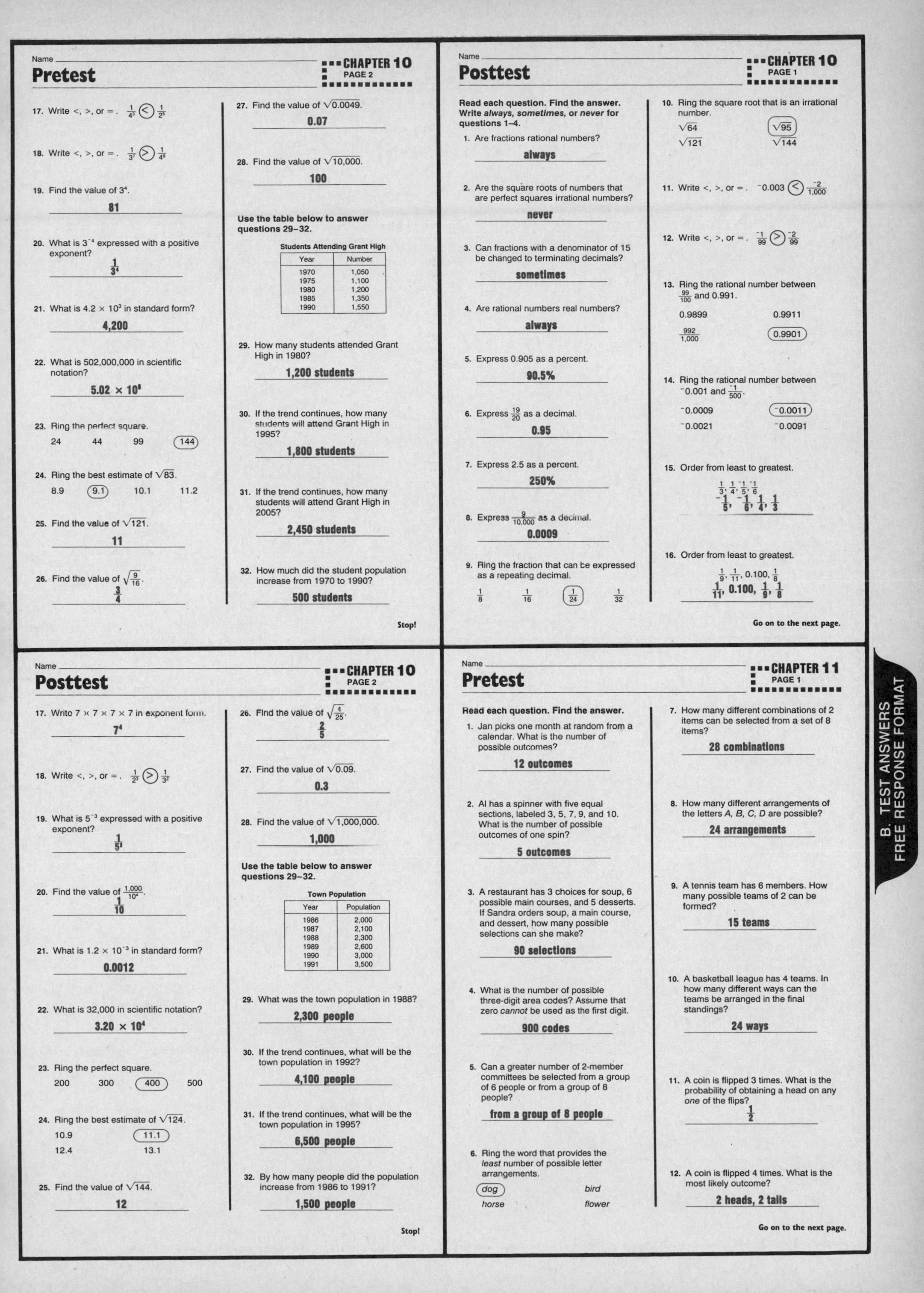

Name

Pretest

CHAPTER 10
PAGE 2

17. Write <, >, or =. $\frac{1}{4^3}$ (<) $\frac{1}{2^5}$

18. Write <, >, or =. $\frac{1}{3^7}$ (>) $\frac{1}{4^8}$

19. Find the value of 3^4.
81

20. What is 3^{-4} expressed with a positive exponent?
$\frac{1}{3^4}$

21. What is 4.2×10^3 in standard form?
4,200

22. What is 502,000,000 in scientific notation?
5.02×10^8

23. Ring the perfect square.
24 44 99 (144)

24. Ring the best estimate of $\sqrt{83}$.
8.9 (9.1) 10.1 11.2

25. Find the value of $\sqrt{121}$.
11

26. Find the value of $\sqrt{\frac{9}{16}}$.
$\frac{3}{4}$

27. Find the value of $\sqrt{0.0049}$.
0.07

28. Find the value of $\sqrt{10{,}000}$.
100

Use the table below to answer questions 29–32.

Students Attending Grant High

Year	Number
1970	1,050
1975	1,100
1980	1,200
1985	1,350
1990	1,550

29. How many students attended Grant High in 1980?
1,200 students

30. If the trend continues, how many students will attend Grant High in 1995?
1,800 students

31. If the trend continues, how many students will attend Grant High in 2005?
2,450 students

32. How much did the student population increase from 1970 to 1990?
500 students

Stop!

Name

Posttest

CHAPTER 10
PAGE 1

Read each question. Find the answer. Write *always*, *sometimes*, or *never* for questions 1–4.

1. Are fractions rational numbers?
always

2. Are the square roots of numbers that are perfect squares irrational numbers?
never

3. Can fractions with a denominator of 15 be changed to terminating decimals?
sometimes

4. Are rational numbers real numbers?
always

5. Express 0.905 as a percent.
90.5%

6. Express $\frac{19}{20}$ as a decimal.
0.95

7. Express 2.5 as a percent.
250%

8. Express $\frac{9}{10{,}000}$ as a decimal.
0.0009

9. Ring the fraction that can be expressed as a repeating decimal.
$\frac{1}{8}$ $\frac{1}{16}$ ($\frac{1}{24}$) $\frac{1}{32}$

10. Ring the square root that is an irrational number.
$\sqrt{64}$ ($\sqrt{95}$)
$\sqrt{121}$ $\sqrt{144}$

11. Write <, >, or =. $^-0.003$ (<) $\frac{^-2}{1{,}000}$

12. Write <, >, or =. $\frac{^-1}{99}$ (>) $\frac{^-2}{99}$

13. Ring the rational number between $\frac{99}{100}$ and 0.991.
0.9899 0.9911
$\frac{992}{1{,}000}$ (0.9901)

14. Ring the rational number between $^-0.001$ and $\frac{^-1}{500}$.
$^-0.0009$ ($^-0.0011$)
$^-0.0021$ $^-0.0091$

15. Order from least to greatest.
$\frac{1}{3}, \frac{1}{4}, \frac{^-1}{5}, \frac{^-1}{6}$
$-\frac{1}{5}, -\frac{1}{6}, \frac{1}{4}, \frac{1}{3}$

16. Order from least to greatest.
$\frac{1}{9}, \frac{1}{11}, 0.100, \frac{1}{8}$
$\frac{1}{11}$, 0.100, $\frac{1}{9}, \frac{1}{8}$

Go on to the next page.

Name

Posttest

CHAPTER 10
PAGE 2

17. Write $7 \times 7 \times 7 \times 7$ in exponent form.
7^4

18. Write <, >, or =. $\frac{1}{2^3}$ (>) $\frac{1}{3^2}$

19. What is 5^{-3} expressed with a positive exponent?
$\frac{1}{5^3}$

20. Find the value of $\frac{1{,}000}{10^4}$.
$\frac{1}{10}$

21. What is 1.2×10^{-3} in standard form?
0.0012

22. What is 32,000 in scientific notation?
3.20×10^4

23. Ring the perfect square.
200 300 (400) 500

24. Ring the best estimate of $\sqrt{124}$.
10.9 (11.1)
12.4 13.1

25. Find the value of $\sqrt{144}$.
12

26. Find the value of $\sqrt{\frac{4}{25}}$.
$\frac{2}{5}$

27. Find the value of $\sqrt{0.09}$.
0.3

28. Find the value of $\sqrt{1{,}000{,}000}$.
1,000

Use the table below to answer questions 29–32.

Town Population

Year	Population
1986	2,000
1987	2,100
1988	2,300
1989	2,600
1990	3,000
1991	3,500

29. What was the town population in 1988?
2,300 people

30. If the trend continues, what will be the town population in 1992?
4,100 people

31. If the trend continues, what will be the town population in 1995?
6,500 people

32. By how many people did the population increase from 1986 to 1991?
1,500 people

Stop!

Name

Pretest

CHAPTER 11
PAGE 1

Read each question. Find the answer.

1. Jan picks one month at random from a calendar. What is the number of possible outcomes?
12 outcomes

2. Al has a spinner with five equal sections, labeled 3, 5, 7, 9, and 10. What is the number of possible outcomes of one spin?
5 outcomes

3. A restaurant has 3 choices for soup, 6 possible main courses, and 5 desserts. If Sandra orders soup, a main course, and dessert, how many possible selections can she make?
90 selections

4. What is the number of possible three-digit area codes? Assume that zero *cannot* be used as the first digit.
900 codes

5. Can a greater number of 2-member committees be selected from a group of 6 people or from a group of 8 people?
from a group of 8 people

6. Ring the word that provides the *least* number of possible letter arrangements.
(*dog*) *bird*
horse *flower*

7. How many different combinations of 2 items can be selected from a set of 8 items?
28 combinations

8. How many different arrangements of the letters *A, B, C, D* are possible?
24 arrangements

9. A tennis team has 6 members. How many possible teams of 2 can be formed?
15 teams

10. A basketball league has 4 teams. In how many different ways can the teams be arranged in the final standings?
24 ways

11. A coin is flipped 3 times. What is the probability of obtaining a head on any *one* of the flips?
$\frac{1}{2}$

12. A coin is flipped 4 times. What is the most likely outcome?
2 heads, 2 tails

Go on to the next page.

Name

Pretest

CHAPTER 11 PAGE 2

13. Hector randomly picks an integer between 0 and 9. Ring the answer that is the most likely outcome.

The number is 5.

The number is 9.

The number is odd.

(The number is not 1.)

14. Ring the answer that has the highest probability on one roll of a number cube.

a number greater than 4

a 4

(a number less than 4)

a 1 or a 2

15. A coin is flipped 4 times. What is the probability of obtaining 4 heads?

$\frac{1}{16}$

Use the information below to answer questions 16–18.

A spinner has 10 equal sections, numbered from 1 to 10.

16. What is the probability of spinning an odd number?

$\frac{5}{10}$, or $\frac{1}{2}$

17. What is the probability of spinning a number greater than 6?

$\frac{4}{10}$, or $\frac{2}{5}$

18. What is the best prediction of how many times the number 10 will be obtained in 900 spins?

90 times

Use the information below to answer questions 19–20.

A large jar is filled with red beans and blue beans. It holds 100 beans in all. Bea shakes the jar and removes a handful. Her handful contains 12 red beans and 18 blue beans.

19. What ratio should Bea use to predict the number of blue beans in the jar?

$\frac{18}{30}$, or $\frac{3}{5}$

20. What is the best prediction of how many red beans are in the jar?

40 beans

21. How should you compute the probability of two independent events both happening?

multiply the two probabilities

22. Ring the answer that does *not* describe independent events.

flipping a coin 2 times

rolling a number cube 2 times

spinning a spinner 2 times

(drawing 2 balls from a box without replacing the first ball)

Go on to the next page.

Name

Pretest

CHAPTER 11 PAGE 3

To answer questions 23–24, assume that a jar contains 3 red balls and 3 black balls. The first ball drawn is *not* replaced in the jar.

23. Helen takes 2 balls from the jar. What is the probability that both balls are black?

$\frac{1}{5}$

24. Mike takes 2 balls from the jar. What is the probability that they are the same color?

$\frac{2}{5}$

25. Juan flips a coin and rolls a number cube. What is the probability that the coin lands on heads and the number cube shows a 1?

$\frac{1}{12}$

26. Olga flips a coin 2 times. What is the probability that both flips come up heads?

$\frac{1}{4}$

27. Two baseball teams play 2 games. Since Team A is the better team, its chance of winning each game is $\frac{2}{3}$. What is the probability that Team A will win both games?

$\frac{4}{9}$

28. A jar contains 5 red and 5 white marbles. Jo takes 1 marble from the jar, does not replace it, and then removes another marble. What is the probability that both marbles are white?

$\frac{4}{18}$, or $\frac{2}{9}$

29. Sam, Tina, and Ray each have a collection. Ray and the stamp collector are best friends. The coin collector lives next to Tina, who enjoys trading bells from her collection. Who has the coin collection?

Ray

Use the information below to answer questions 30–32.

Joe and Edna must take music, health, and art courses during a two-year period. Each student is randomly assigned one of the courses for the current semester. Assume that the probability of assignment to each course is $\frac{1}{3}$.

30. What is the probability that both students will be assigned the art course?

$\frac{1}{9}$

31. What is the probability that both students will be assigned any of the three courses together?

$\frac{3}{9}$, or $\frac{1}{3}$

32. What is the probability that neither student will be assigned the music course?

$\frac{4}{9}$

Stop!

Name

Posttest

CHAPTER 11 PAGE 1

Read each question. Find the answer.

1. Lisa picks one day of the week from the calendar. What is the number of possible outcomes?

7 outcomes

2. John has a spinner with 8 equal sections, labeled 1, 2, 3, 4, 5, 6, 7, and 8. What is the number of possible outcomes?

8 outcomes

3. Len is at a fruit stand. He has 6 kinds of fruit and 5 kinds of vegetables to choose from. If he buys 1 fruit and 1 vegetable, how many outcomes are possible?

30 outcomes

4. What is the number of possible four-digit numbers? Do *not* use zero in the first place.

9,000 numbers

5. Can a greater number of 2-member teams be selected from a group of 4 girls or from a group of 5 girls?

from a group of 5 girls

6. Ring the word that provides the *greatest* number of possible letter arrangements.

hat *noon*

seven (*jacket*)

7. How many different combinations of 2 items can be selected from a set of 6 items?

15 combinations

8. How many different arrangements of the letters *A, B, C, D, E* are possible?

120 arrangements

9. A baseball team has 12 members. The team needs 2 co-captains. How many different pairs of co-captains can be selected from the 12 members?

66 pairs

10. A basketball league has 7 teams. In how many different ways can the teams be arranged in the final standings?

5,040 ways

Go on to the next page.

Name

Posttest

CHAPTER 11 PAGE 2

11. What is the probability of an event that is impossible?

0

12. A coin is flipped 2 times. What is the most likely outcome?

1 head, 1 tail

13. Lupe randomly picks one letter of the alphabet. Ring the answer that is the most likely outcome.

The letter is *A*.

The letter is *Z*.

The letter is a vowel.

(The letter is a consonant.)

14. Ring the answer that has the highest probability on one roll of a number cube.

an odd number

an even number

(a number less than 5)

a number greater than 5

15. A coin is flipped 3 times. What is the probability of obtaining 3 heads?

$\frac{1}{8}$

Use the information below to answer questions 16–18.

A spinner has 10 equal sections, numbered from 1 to 10.

16. What is the probability of spinning 1, 4, or 9?

$\frac{3}{10}$

17. What is the probability of spinning a number greater than 2?

$\frac{8}{10}$, or $\frac{4}{5}$

18. What is the best prediction of how many times the number 2 will be obtained in 1,000 spins?

100 times

Use the information below to answer questions 19–20.

A large jar is filled with black candies and white candies. It holds 200 candies in all. Hal shakes the jar and removes a handful. His handful contains 10 white candies and 15 black candies.

19. What ratio should Hal use to predict the number of white candies in the jar?

$\frac{10}{25}$, or $\frac{2}{5}$

20. What is the best prediction of the total number of black candies in the jar?

120 candies

21. When the outcome of an event is *not* affected by the outcome of an earlier event, what are the events called?

independent

Go on to the next page.

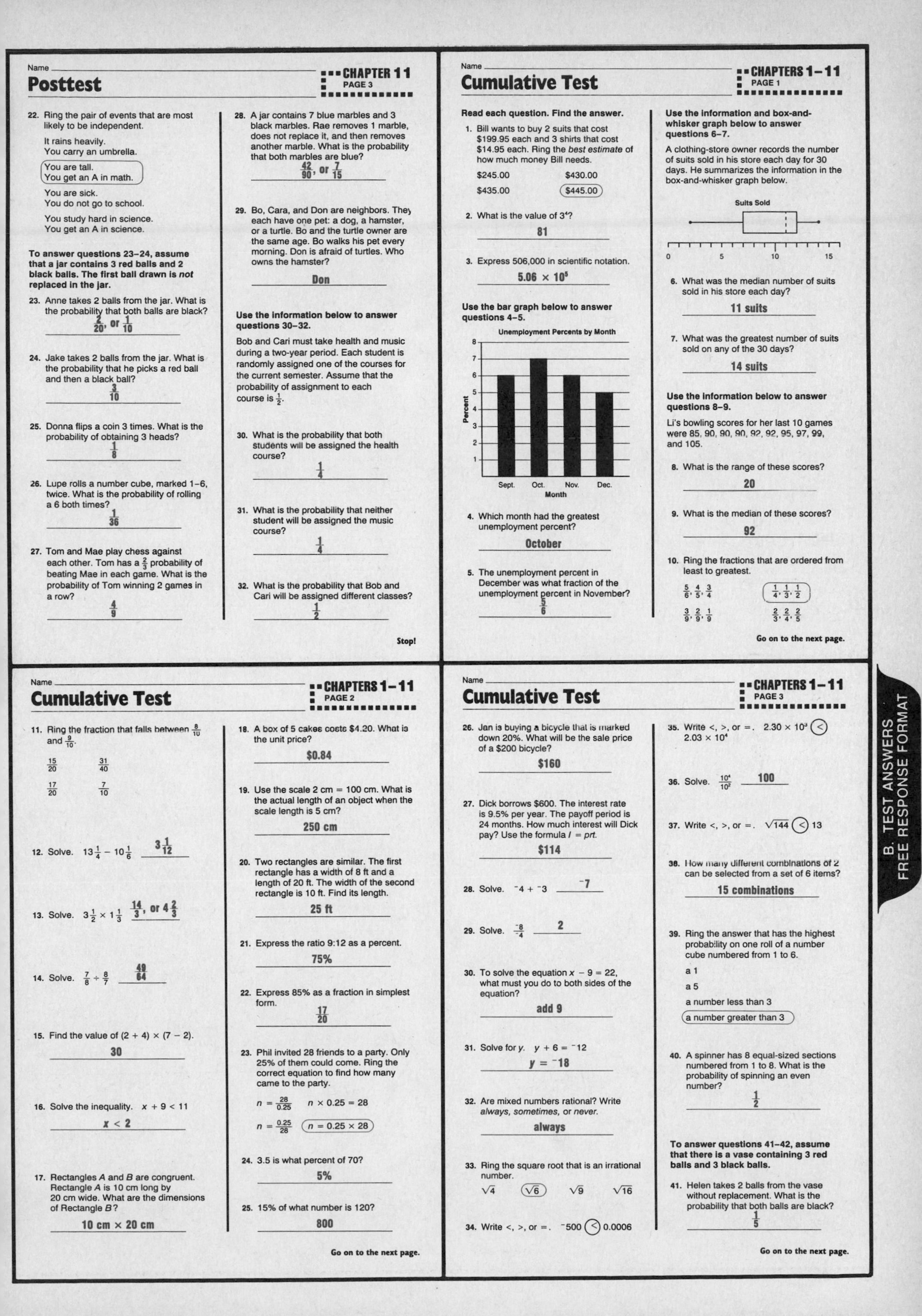

Name

Posttest

CHAPTER 11
PAGE 3

22. Ring the pair of events that are most likely to be independent.

It rains heavily.
You carry an umbrella.

**You are tall.
You get an A in math.**

You are sick.
You do not go to school.

You study hard in science.
You get an A in science.

To answer questions 23–24, assume that a jar contains 3 red balls and 2 black balls. The first ball drawn is *not* replaced in the jar.

23. Anne takes 2 balls from the jar. What is the probability that both balls are black?
$\frac{2}{20}$, or $\frac{1}{10}$

24. Jake takes 2 balls from the jar. What is the probability that he picks a red ball and then a black ball?
$\frac{3}{10}$

25. Donna flips a coin 3 times. What is the probability of obtaining 3 heads?
$\frac{1}{8}$

26. Lupe rolls a number cube, marked 1–6, twice. What is the probability of rolling a 6 both times?
$\frac{1}{36}$

27. Tom and Mae play chess against each other. Tom has a $\frac{2}{3}$ probability of beating Mae in each game. What is the probability of Tom winning 2 games in a row?
$\frac{4}{9}$

28. A jar contains 7 blue marbles and 3 black marbles. Rae removes 1 marble, does not replace it, and then removes another marble. What is the probability that both marbles are blue?
$\frac{42}{90}$, or $\frac{7}{15}$

29. Bo, Cara, and Don are neighbors. They each have one pet: a dog, a hamster, or a turtle. Bo and the turtle owner are the same age. Bo walks his pet every morning. Don is afraid of turtles. Who owns the hamster?
Don

Use the information below to answer questions 30–32.

Bob and Cari must take health and music during a two-year period. Each student is randomly assigned one of the courses for the current semester. Assume that the probability of assignment to each course is $\frac{1}{2}$.

30. What is the probability that both students will be assigned the health course?
$\frac{1}{4}$

31. What is the probability that neither student will be assigned the music course?
$\frac{1}{4}$

32. What is the probability that Bob and Cari will be assigned different classes?
$\frac{1}{2}$

Stop!

Name

Cumulative Test

CHAPTERS 1–11
PAGE 1

Read each question. Find the answer.

1. Bill wants to buy 2 suits that cost \$199.95 each and 3 shirts that cost \$14.95 each. Ring the *best estimate* of how much money Bill needs.

\$245.00 \$430.00
\$435.00 **\$445.00**

2. What is the value of 3^4?
81

3. Express 506,000 in scientific notation.
5.06×10^5

Use the bar graph below to answer questions 4–5.

Unemployment Percents by Month

Month	Percent
Sept.	6
Oct.	7
Nov.	6
Dec.	5

4. Which month had the greatest unemployment percent?
October

5. The unemployment percent in December was what fraction of the unemployment percent in November?
$\frac{5}{6}$

Use the information and box-and-whisker graph below to answer questions 6–7.

A clothing-store owner records the number of suits sold in his store each day for 30 days. He summarizes the information in the box-and-whisker graph below.

Suits Sold
0 5 10 15

6. What was the median number of suits sold in his store each day?
11 suits

7. What was the greatest number of suits sold on any of the 30 days?
14 suits

Use the information below to answer questions 8–9.

Li's bowling scores for her last 10 games were 85, 90, 90, 90, 92, 92, 95, 97, 99, and 105.

8. What is the range of these scores?
20

9. What is the median of these scores?
92

10. Ring the fractions that are ordered from least to greatest.

$\frac{5}{6}, \frac{5}{5}, \frac{3}{4}$ **$\frac{1}{4}, \frac{1}{3}, \frac{1}{2}$**
$\frac{3}{9}, \frac{2}{9}, \frac{1}{9}$ $\frac{2}{3}, \frac{2}{4}, \frac{2}{5}$

Go on to the next page.

Name

Cumulative Test

CHAPTERS 1–11
PAGE 2

11. Ring the fraction that falls between $\frac{8}{10}$ and $\frac{9}{10}$.

$\frac{15}{20}$ $\frac{31}{40}$
$\frac{17}{20}$ $\frac{7}{10}$

12. Solve. $13\frac{1}{4} - 10\frac{1}{6}$ **$3\frac{1}{12}$**

13. Solve. $3\frac{1}{2} \times 1\frac{1}{3}$ **$\frac{14}{3}$, or $4\frac{2}{3}$**

14. Solve. $\frac{7}{8} \div \frac{8}{7}$ **$\frac{49}{64}$**

15. Find the value of $(2 + 4) \times (7 - 2)$.
30

16. Solve the inequality. $x + 9 < 11$
$x < 2$

17. Rectangles *A* and *B* are congruent. Rectangle *A* is 10 cm long by 20 cm wide. What are the dimensions of Rectangle *B*?
10 cm × 20 cm

18. A box of 5 cakes costs \$4.20. What is the unit price?
\$0.84

19. Use the scale 2 cm = 100 cm. What is the actual length of an object when the scale length is 5 cm?
250 cm

20. Two rectangles are similar. The first rectangle has a width of 8 ft and a length of 20 ft. The width of the second rectangle is 10 ft. Find its length.
25 ft

21. Express the ratio 9:12 as a percent.
75%

22. Express 85% as a fraction in simplest form.
$\frac{17}{20}$

23. Phil invited 28 friends to a party. Only 25% of them could come. Ring the correct equation to find how many came to the party.

$n = \frac{28}{0.25}$ $n \times 0.25 = 28$
$n = \frac{0.25}{28}$ **$n = 0.25 \times 28$**

24. 3.5 is what percent of 70?
5%

25. 15% of what number is 120?
800

Go on to the next page.

Name

Cumulative Test

CHAPTERS 1–11
PAGE 3

26. Jan is buying a bicycle that is marked down 20%. What will be the sale price of a \$200 bicycle?
\$160

27. Dick borrows \$600. The interest rate is 9.5% per year. The payoff period is 24 months. How much interest will Dick pay? Use the formula $I = prt$.
\$114

28. Solve. $^-4 + {^-3}$ **$^-7$**

29. Solve. $\frac{^-8}{^-4}$ **2**

30. To solve the equation $x - 9 = 22$, what must you do to both sides of the equation?
add 9

31. Solve for *y*. $y + 6 = {^-12}$
$y = {^-18}$

32. Are mixed numbers rational? Write *always, sometimes,* or *never.*
always

33. Ring the square root that is an irrational number.

$\sqrt{4}$ **$\sqrt{6}$** $\sqrt{9}$ $\sqrt{16}$

34. Write <, >, or =. $^-500$ **<** 0.0006

35. Write <, >, or =. 2.30×10^3 **<** 2.03×10^4

36. Solve. $\frac{10^4}{10^2}$ **100**

37. Write <, >, or =. $\sqrt{144}$ **<** 13

38. How many different combinations of 2 can be selected from a set of 6 items?
15 combinations

39. Ring the answer that has the highest probability on one roll of a number cube numbered from 1 to 6.

a 1
a 5
a number less than 3
a number greater than 3

40. A spinner has 8 equal-sized sections numbered from 1 to 8. What is the probability of spinning an even number?
$\frac{1}{2}$

To answer questions 41–42, assume that there is a vase containing 3 red balls and 3 black balls.

41. Helen takes 2 balls from the vase without replacement. What is the probability that both balls are black?
$\frac{1}{5}$

Go on to the next page.

B. TEST ANSWERS
FREE RESPONSE FORMAT

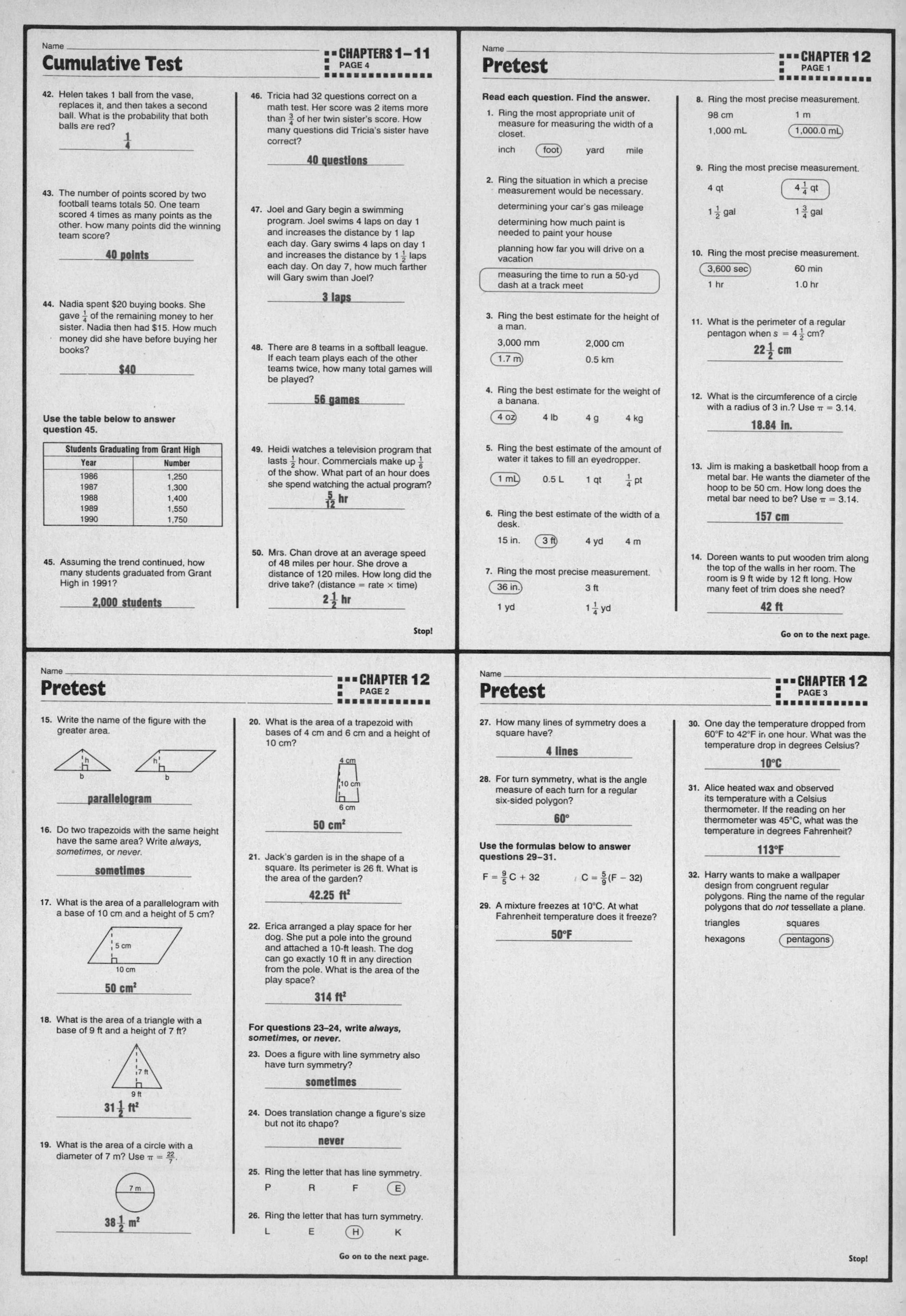

Name

Cumulative Test

CHAPTERS 1–11
PAGE 4

42. Helen takes 1 ball from the vase, replaces it, and then takes a second ball. What is the probability that both balls are red?

$\frac{1}{4}$

43. The number of points scored by two football teams totals 50. One team scored 4 times as many points as the other. How many points did the winning team score?

40 points

44. Nadia spent $20 buying books. She gave $\frac{1}{4}$ of the remaining money to her sister. Nadia then had $15. How much money did she have before buying her books?

$40

Use the table below to answer question 45.

Students Graduating from Grant High	
Year	**Number**
1986	1,250
1987	1,300
1988	1,400
1989	1,550
1990	1,750

45. Assuming the trend continued, how many students graduated from Grant High in 1991?

2,000 students

46. Tricia had 32 questions correct on a math test. Her score was 2 items more than $\frac{3}{4}$ of her twin sister's score. How many questions did Tricia's sister have correct?

40 questions

47. Joel and Gary begin a swimming program. Joel swims 4 laps on day 1 and increases the distance by 1 lap each day. Gary swims 4 laps on day 1 and increases the distance by $1\frac{1}{2}$ laps each day. On day 7, how much farther will Gary swim than Joel?

3 laps

48. There are 8 teams in a softball league. If each team plays each of the other teams twice, how many total games will be played?

56 games

49. Heidi watches a television program that lasts $\frac{1}{2}$ hour. Commercials make up $\frac{1}{6}$ of the show. What part of an hour does she spend watching the actual program?

$\frac{5}{12}$ hr

50. Mrs. Chan drove at an average speed of 48 miles per hour. She drove a distance of 120 miles. How long did the drive take? (distance = rate × time)

$2\frac{1}{2}$ hr

Stop!

Name

Pretest

CHAPTER 12
PAGE 1

Read each question. Find the answer.

1. Ring the most appropriate unit of measure for measuring the width of a closet.

inch (foot) yard mile

2. Ring the situation in which a precise measurement would be necessary.

determining your car's gas mileage
determining how much paint is needed to paint your house
planning how far you will drive on a vacation
(measuring the time to run a 50-yd dash at a track meet)

3. Ring the best estimate for the height of a man.

3,000 mm 2,000 cm (1.7 m) 0.5 km

4. Ring the best estimate for the weight of a banana.

(4 oz) 4 lb 4 g 4 kg

5. Ring the best estimate of the amount of water it takes to fill an eyedropper.

(1 mL) 0.5 L 1 qt $\frac{1}{4}$ pt

6. Ring the best estimate of the width of a desk.

15 in. (3 ft) 4 yd 4 m

7. Ring the most precise measurement.

(36 in.) 3 ft 1 yd $1\frac{1}{4}$ yd

8. Ring the most precise measurement.

98 cm 1 m 1,000 mL (1,000.0 mL)

9. Ring the most precise measurement.

4 qt ($4\frac{1}{4}$ qt) $1\frac{1}{2}$ gal $1\frac{3}{4}$ gal

10. Ring the most precise measurement.

(3,600 sec) 60 min 1 hr 1.0 hr

11. What is the perimeter of a regular pentagon when $s = 4\frac{1}{2}$ cm?

$22\frac{1}{2}$ cm

12. What is the circumference of a circle with a radius of 3 in.? Use $\pi = 3.14$.

18.84 in.

13. Jim is making a basketball hoop from a metal bar. He wants the diameter of the hoop to be 50 cm. How long does the metal bar need to be? Use $\pi = 3.14$.

157 cm

14. Doreen wants to put wooden trim along the top of the walls in her room. The room is 9 ft wide by 12 ft long. How many feet of trim does she need?

42 ft

Go on to the next page.

Name

Pretest

CHAPTER 12
PAGE 2

15. Write the name of the figure with the greater area.

h b h b

parallelogram

16. Do two trapezoids with the same height have the same area? Write *always*, *sometimes*, or *never*.

sometimes

17. What is the area of a parallelogram with a base of 10 cm and a height of 5 cm?

5 cm 10 cm

50 cm²

18. What is the area of a triangle with a base of 9 ft and a height of 7 ft?

7 ft 9 ft

$31\frac{1}{2}$ ft²

19. What is the area of a circle with a diameter of 7 m? Use $\pi = \frac{22}{7}$.

7 m

$38\frac{1}{2}$ m²

20. What is the area of a trapezoid with bases of 4 cm and 6 cm and a height of 10 cm?

4 cm 10 cm 6 cm

50 cm²

21. Jack's garden is in the shape of a square. Its perimeter is 26 ft. What is the area of the garden?

42.25 ft²

22. Erica arranged a play space for her dog. She put a pole into the ground and attached a 10-ft leash. The dog can go exactly 10 ft in any direction from the pole. What is the area of the play space?

314 ft²

For questions 23–24, write *always*, *sometimes*, or *never*.

23. Does a figure with line symmetry also have turn symmetry?

sometimes

24. Does translation change a figure's size but not its shape?

never

25. Ring the letter that has line symmetry.

P R F (E)

26. Ring the letter that has turn symmetry.

L E (H) K

Go on to the next page.

Name

Pretest

CHAPTER 12
PAGE 3

27. How many lines of symmetry does a square have?

4 lines

28. For turn symmetry, what is the angle measure of each turn for a regular six-sided polygon?

60°

Use the formulas below to answer questions 29–31.

$F = \frac{9}{5}C + 32$ $C = \frac{5}{9}(F - 32)$

29. A mixture freezes at 10°C. At what Fahrenheit temperature does it freeze?

50°F

30. One day the temperature dropped from 60°F to 42°F in one hour. What was the temperature drop in degrees Celsius?

10°C

31. Alice heated wax and observed its temperature with a Celsius thermometer. If the reading on her thermometer was 45°C, what was the temperature in degrees Fahrenheit?

113°F

32. Harry wants to make a wallpaper design from congruent regular polygons. Ring the name of the regular polygons that do *not* tessellate a plane.

triangles squares hexagons (pentagons)

Stop!

Name

Posttest

CHAPTER 12
PAGE 1

Read each question. Find the answer.

1. Ring the most appropriate unit of measure for measuring the distance between two cities.
   millimeter centimeter
   meter (kilometer)

2. Ring the situation in which an estimation would be most appropriate.
   pouring a dose of medicine
   measuring your ring size for a jeweler
   (measuring the temperature of meat you are cooking)
   measuring a table leg that you must replace

3. Ring the best estimate for the height of a room in a home.
   1,000 mm 5,000 cm
   (3.5 m) 0.1 km

4. Ring the best estimate for the weight of an egg.
   (2 oz) 1 lb 2 g 1 kg

5. Ring the best estimate of the amount of water it takes to fill a cup.
   10 mL 1 L (8 oz) 2 pt

6. Ring the best estimate of the width of a refrigerator.
   15 in. (3 ft) 4 yd 5 m

7. Ring the most precise measurement.
   35 in. ($36\frac{1}{2}$ in.)
   3 ft $3\frac{1}{2}$ ft

8. Ring the most precise measurement.
   98 cm (98.1 cm)
   1 m 1.13 m

9. Ring the most precise measurement.
   (128 oz) 8 pt
   2 qt $\frac{1}{2}$ gal

10. Ring the most precise measurement.
    2 sec 1.1 sec
    $1\frac{1}{2}$ sec (1.001 sec)

11. What is the perimeter of a regular hexagon when $s = 2.5$ cm?
    15 cm

12. What is the circumference of a circle with a radius of 2 ft? Use $\pi = 3.14$.
    12.56 ft

13. Harold wants to make a hoop with a circumference of 16 ft. What will be the diameter? Use $\pi = 3.14$.
    5.10 ft

14. The perimeter of a rectangle is 10 cm. The shorter sides are each 2 cm. What is the length of each of the longer sides?
    3 cm

Go on to the next page.

Name

Posttest

CHAPTER 12
PAGE 2

15. Which triangle has the greater area? Ring the answer.
    the equilateral triangle
    the right triangle
    (The areas are equal.)

16. Do two circles with the same circumference have the same area? Write *always, sometimes,* or *never.*
    always

17. What is the area of a parallelogram with a base of 20 cm and a height of 15 cm?
    15 cm; 20 cm
    300 cm²

18. What is the area of a triangle with a base of 3 ft and a height of 4 ft?
    4 ft; 3 ft
    6 ft²

19. What is the area of a circle with a diameter of 10 m? Use $\pi = 3.14$.
    10 m
    78.5 m²

20. What is the area of a trapezoid with bases of 10 cm and 20 cm and a height of 10 cm?
    10 cm; 10 cm; 20 cm
    150 cm²

21. Jack's garden is in the shape of a square. Its area is 36 ft². What is the perimeter of the garden?
    24 ft

22. Hilda's garden is circular. The area is 3.14 ft². What is the diameter?
    2 ft

For questions 23–24, write *always, sometimes,* or *never.*

23. Does a circle have more than 4 lines of symmetry?
    always

24. Does reflection change both the size and the shape of a figure?
    never

25. Ring the letter that has line symmetry.
    J Q G (H)

26. Ring the letter that has turn symmetry.
    A P W (X)

27. How many lines of symmetry does an isosceles triangle have?
    1 line

Go on to the next page.

Name

Posttest

CHAPTER 12
PAGE 3

28. For turn symmetry, what is the angle measure of each turn for a square?
    90°

Use the formulas below to answer questions 29–31.

$F = \frac{9}{5}C + 32$ $C = \frac{5}{9}(F - 32)$

29. Water boils at 100°C. At what Fahrenheit temperature does it boil?
    212°F

30. One day the temperature increased from 70°F to 79°F in one hour. What was the temperature increase in degrees Celsius?
    5°C

31. Berto chilled water and observed its temperature with a Fahrenheit thermometer. If the reading on his thermometer was 50°F, what was the temperature in degrees Celsius?
    10°C

32. John wants to make a quilt design from congruent regular polygons. He can use any regular polygon for which the measures of the angles meeting at a vertex have a sum of how many degrees?
    360°

Stop!

Name

Pretest

CHAPTER 13
PAGE 1

Read the question. Find the answer.

1. Paco is drawing a pattern for a solid figure. The base of the figure is a polygon. Each side of the base is the base of an isosceles triangle. Ring the figure that Paco is drawing.
   cone cylinder
   (pyramid) triangular prism

2. Helen built a solid figure with 2 flat surfaces and 1 curved surface. Ring the figure that she built.
   (cylinder) sphere
   cone polyhedron

3. Ring the figure that has no flat surfaces.
   cone cylinder
   pyramid (sphere)

4. What do prisms and pyramids have in common? Ring the answer.
   Their lateral faces are all triangles.
   (They are polyhedrons.)
   They have congruent, parallel bases.
   None of their surfaces are polygons.

5. How many faces does a triangular prism have?
   5 faces

6. A polyhedron has 6 faces and 8 vertices. How many edges does it have?
   12 edges

7. How many faces does a rectangular pyramid have?
   5 faces

8. A solid figure has 6 faces: 5 triangles and 1 pentagon. Ring the figure.
   (pentagonal pyramid)
   triangular prism
   pentagonal prism
   not here

9. What is the surface area of this cylinder? Use $\pi = 3.14$.
   3 cm; 10 cm
   244.92 cm²

10. The dimensions of a rectangular prism are 5 ft × 4 ft × 4 ft. What is its surface area?
    112 ft²

11. A cube has a surface area of 600 cm². What is the length of each face?
    10 cm

12. What is the surface area of this square pyramid?
    12 in.; 5 in.; 5 in.
    145 in.²

Go on to the next page.

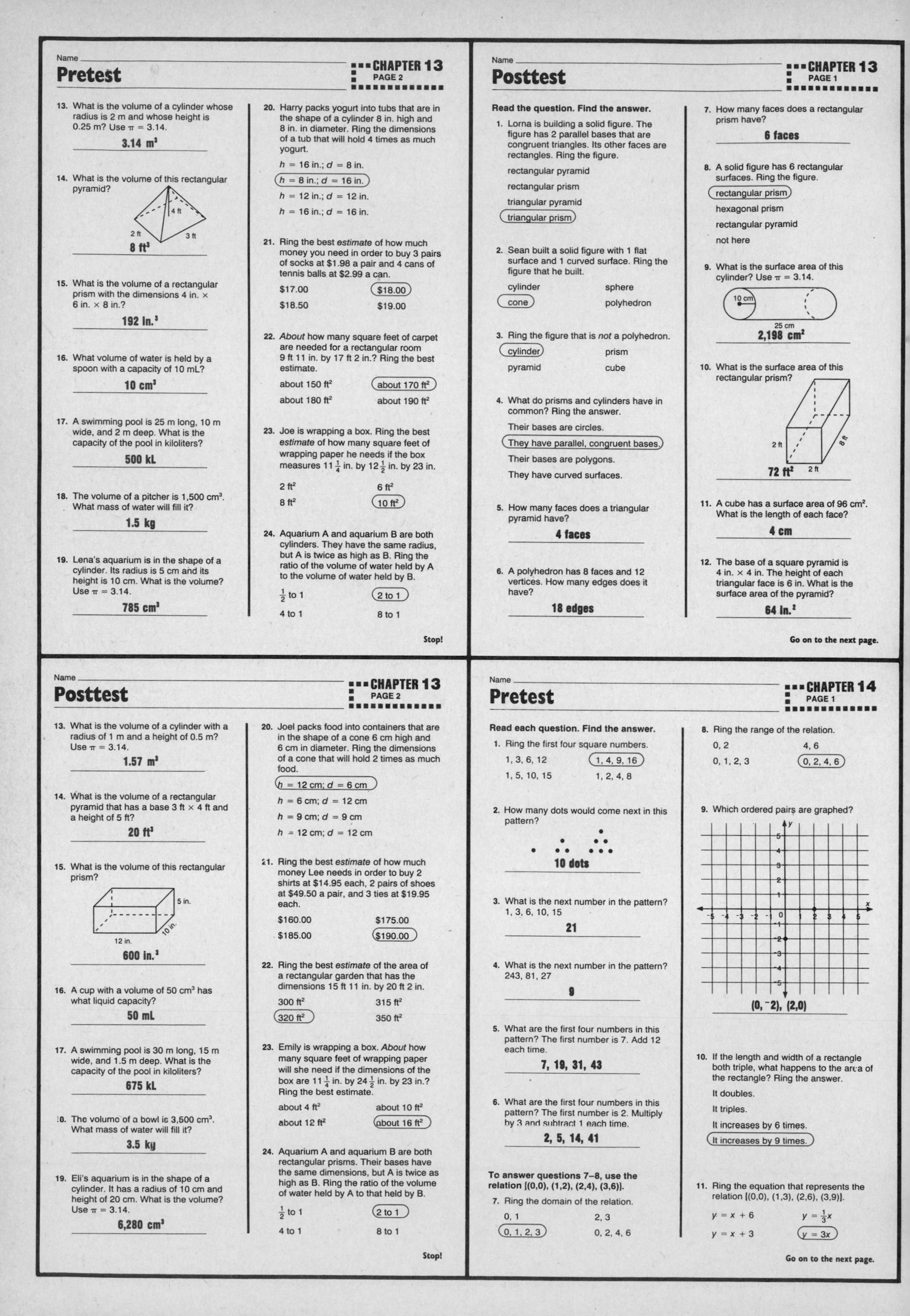

Name

Pretest

CHAPTER 13 PAGE 2

13. What is the volume of a cylinder whose radius is 2 m and whose height is 0.25 m? Use $\pi = 3.14$.

3.14 m³

14. What is the volume of this rectangular pyramid?

8 ft³

15. What is the volume of a rectangular prism with the dimensions 4 in. × 6 in. × 8 in.?

192 in.³

16. What volume of water is held by a spoon with a capacity of 10 mL?

10 cm³

17. A swimming pool is 25 m long, 10 m wide, and 2 m deep. What is the capacity of the pool in kiloliters?

500 kL

18. The volume of a pitcher is 1,500 cm³. What mass of water will fill it?

1.5 kg

19. Lena's aquarium is in the shape of a cylinder. Its radius is 5 cm and its height is 10 cm. What is the volume? Use $\pi = 3.14$.

785 cm³

20. Harry packs yogurt into tubs that are in the shape of a cylinder 8 in. high and 8 in. in diameter. Ring the dimensions of a tub that will hold 4 times as much yogurt.

$h = 16$ in.; $d = 8$ in.
$h = 8$ in.; $d = 16$ in.
$h = 12$ in.; $d = 12$ in.
$h = 16$ in.; $d = 16$ in.

21. Ring the best *estimate* of how much money you need in order to buy 3 pairs of socks at \$1.98 a pair and 4 cans of tennis balls at \$2.99 a can.

\$17.00 **\$18.00**
\$18.50 \$19.00

22. *About* how many square feet of carpet are needed for a rectangular room 9 ft 11 in. by 17 ft 2 in.? Ring the best estimate.

about 150 ft² **about 170 ft²**
about 180 ft² about 190 ft²

23. Joe is wrapping a box. Ring the best *estimate* of how many square feet of wrapping paper he needs if the box measures $11\frac{1}{4}$ in. by $12\frac{1}{2}$ in. by 23 in.

2 ft² 6 ft²
8 ft² **10 ft²**

24. Aquarium A and aquarium B are both cylinders. They have the same radius, but A is twice as high as B. Ring the ratio of the volume of water held by A to the volume of water held by B.

$\frac{1}{2}$ to 1 **2 to 1**
4 to 1 8 to 1

Stop!

Name

Posttest

CHAPTER 13 PAGE 1

Read the question. Find the answer.

1. Lorna is building a solid figure. The figure has 2 parallel bases that are congruent triangles. Its other faces are rectangles. Ring the figure.

rectangular pyramid
rectangular prism
triangular pyramid
triangular prism

2. Sean built a solid figure with 1 flat surface and 1 curved surface. Ring the figure that he built.

cylinder sphere
cone polyhedron

3. Ring the figure that is *not* a polyhedron.

cylinder prism
pyramid cube

4. What do prisms and cylinders have in common? Ring the answer.

Their bases are circles.
They have parallel, congruent bases.
Their bases are polygons.
They have curved surfaces.

5. How many faces does a triangular pyramid have?

4 faces

6. A polyhedron has 8 faces and 12 vertices. How many edges does it have?

18 edges

7. How many faces does a rectangular prism have?

6 faces

8. A solid figure has 6 rectangular surfaces. Ring the figure.

rectangular prism
hexagonal prism
rectangular pyramid
not here

9. What is the surface area of this cylinder? Use $\pi = 3.14$.

2,198 cm²

10. What is the surface area of this rectangular prism?

72 ft²

11. A cube has a surface area of 96 cm². What is the length of each face?

4 cm

12. The base of a square pyramid is 4 in. × 4 in. The height of each triangular face is 6 in. What is the surface area of the pyramid?

64 in.²

Go on to the next page.

Name

Posttest

CHAPTER 13 PAGE 2

13. What is the volume of a cylinder with a radius of 1 m and a height of 0.5 m? Use $\pi = 3.14$.

1.57 m³

14. What is the volume of a rectangular pyramid that has a base 3 ft × 4 ft and a height of 5 ft?

20 ft³

15. What is the volume of this rectangular prism?

600 in.³

16. A cup with a volume of 50 cm³ has what liquid capacity?

50 mL

17. A swimming pool is 30 m long, 15 m wide, and 1.5 m deep. What is the capacity of the pool in kiloliters?

675 kL

18. The volume of a bowl is 3,500 cm³. What mass of water will fill it?

3.5 kg

19. Eli's aquarium is in the shape of a cylinder. It has a radius of 10 cm and height of 20 cm. What is the volume? Use $\pi = 3.14$.

6,280 cm³

20. Joel packs food into containers that are in the shape of a cone 6 cm high and 6 cm in diameter. Ring the dimensions of a cone that will hold 2 times as much food.

$h = 12$ cm; $d = 6$ cm
$h = 6$ cm; $d = 12$ cm
$h = 9$ cm; $d = 9$ cm
$h = 12$ cm; $d = 12$ cm

21. Ring the best *estimate* of how much money Lee needs in order to buy 2 shirts at \$14.95 each, 2 pairs of shoes at \$49.50 a pair, and 3 ties at \$19.95 each.

\$160.00 \$175.00
\$185.00 **\$190.00**

22. Ring the best *estimate* of the area of a rectangular garden that has the dimensions 15 ft 11 in. by 20 ft 2 in.

300 ft² 315 ft²
320 ft² 350 ft²

23. Emily is wrapping a box. *About* how many square feet of wrapping paper will she need if the dimensions of the box are $11\frac{1}{4}$ in. by $24\frac{1}{2}$ in. by 23 in.? Ring the best estimate.

about 4 ft² about 10 ft²
about 12 ft² **about 16 ft²**

24. Aquarium A and aquarium B are both rectangular prisms. Their bases have the same dimensions, but A is twice as high as B. Ring the ratio of the volume of water held by A to that held by B.

$\frac{1}{2}$ to 1 **2 to 1**
4 to 1 8 to 1

Stop!

Name

Pretest

CHAPTER 14 PAGE 1

Read each question. Find the answer.

1. Ring the first four square numbers.

1, 3, 6, 12 **1, 4, 9, 16**
1, 5, 10, 15 1, 2, 4, 8

2. How many dots would come next in this pattern?

10 dots

3. What is the next number in the pattern? 1, 3, 6, 10, 15

21

4. What is the next number in the pattern? 243, 81, 27

9

5. What are the first four numbers in this pattern? The first number is 7. Add 12 each time.

7, 19, 31, 43

6. What are the first four numbers in this pattern? The first number is 2. Multiply by 3 and subtract 1 each time.

2, 5, 14, 41

To answer questions 7–8, use the relation {(0,0), (1,2), (2,4), (3,6)}.

7. Ring the domain of the relation.

0, 1 2, 3
0, 1, 2, 3 0, 2, 4, 6

8. Ring the range of the relation.

0, 2 4, 6
0, 1, 2, 3 **0, 2, 4, 6**

9. Which ordered pairs are graphed?

(0, ⁻2), (2,0)

10. If the length and width of a rectangle both triple, what happens to the area of the rectangle? Ring the answer.

It doubles.
It triples.
It increases by 6 times.
It increases by 9 times.

11. Ring the equation that represents the relation {(0,0), (1,3), (2,6), (3,9)}.

$y = x + 6$ $y = \frac{1}{3}x$
$y = x + 3$ **$y = 3x$**

Go on to the next page.

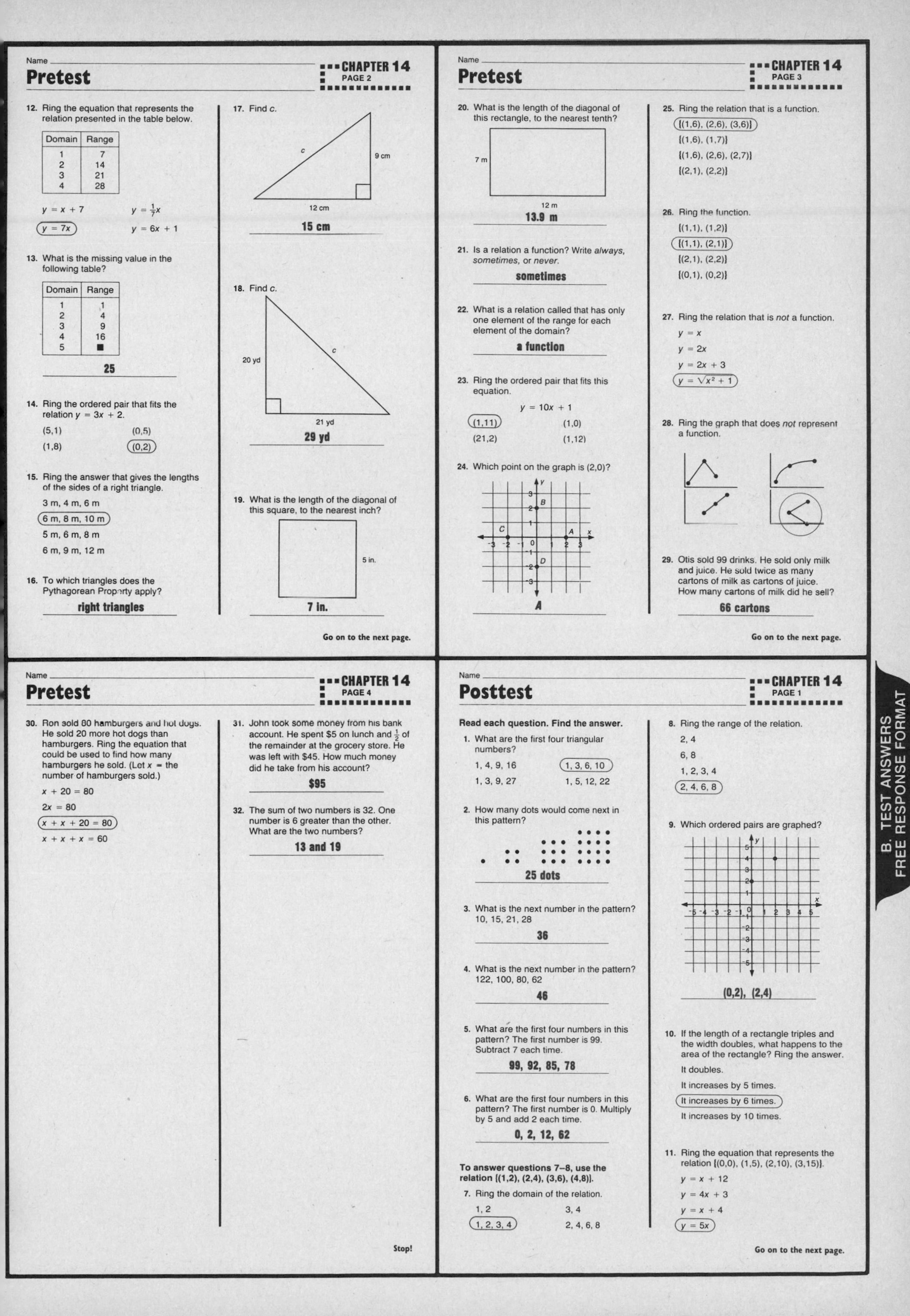

Name

Pretest

CHAPTER 14
PAGE 2

12. Ring the equation that represents the relation presented in the table below.

Domain	Range
1	7
2	14
3	21
4	28

$y = x + 7$ $\quad y = \frac{1}{7}x$

$y = 7x$ $\quad y = 6x + 1$

13. What is the missing value in the following table?

Domain	Range
1	1
2	4
3	9
4	16
5	■

25

14. Ring the ordered pair that fits the relation $y = 3x + 2$.

(5,1) (0,5)

(1,8) (0,2)

15. Ring the answer that gives the lengths of the sides of a right triangle.

3 m, 4 m, 6 m

6 m, 8 m, 10 m

5 m, 6 m, 8 m

6 m, 9 m, 12 m

16. To which triangles does the Pythagorean Property apply?

right triangles

17. Find c.

c, 9 cm, 12 cm

15 cm

18. Find c.

20 yd, c, 21 yd

29 yd

19. What is the length of the diagonal of this square, to the nearest inch?

5 in.

7 in.

Go on to the next page.

Name

Pretest

CHAPTER 14
PAGE 3

20. What is the length of the diagonal of this rectangle, to the nearest tenth?

7 m, 12 m

13.9 m

21. Is a relation a function? Write *always*, *sometimes*, or *never*.

sometimes

22. What is a relation called that has only one element of the range for each element of the domain?

a function

23. Ring the ordered pair that fits this equation.

$y = 10x + 1$

(1,11) (1,0)

(21,2) (1,12)

24. Which point on the graph is (2,0)?

A

25. Ring the relation that is a function.

{(1,6), (2,6), (3,6)}

{(1,6), (1,7)}

{(1,6), (2,6), (2,7)}

{(2,1), (2,2)}

26. Ring the function.

{(1,1), (1,2)}

{(1,1), (2,1)}

{(2,1), (2,2)}

{(0,1), (0,2)}

27. Ring the relation that is *not* a function.

$y = x$

$y = 2x$

$y = 2x + 3$

$y = \sqrt{x^2 + 1}$

28. Ring the graph that does *not* represent a function.

29. Otis sold 99 drinks. He sold only milk and juice. He sold twice as many cartons of milk as cartons of juice. How many cartons of milk did he sell?

66 cartons

Go on to the next page.

Name

Pretest

CHAPTER 14
PAGE 4

30. Ron sold 80 hamburgers and hot dogs. He sold 20 more hot dogs than hamburgers. Ring the equation that could be used to find how many hamburgers he sold. (Let x = the number of hamburgers sold.)

$x + 20 = 80$

$2x = 80$

$x + x + 20 = 80$

$x + x + x = 60$

31. John took some money from his bank account. He spent $5 on lunch and $\frac{1}{2}$ of the remainder at the grocery store. He was left with $45. How much money did he take from his account?

$95

32. The sum of two numbers is 32. One number is 6 greater than the other. What are the two numbers?

13 and 19

Stop!

Name

Posttest

CHAPTER 14
PAGE 1

Read each question. Find the answer.

1. What are the first four triangular numbers?

1, 4, 9, 16 1, 3, 6, 10

1, 3, 9, 27 1, 5, 12, 22

2. How many dots would come next in this pattern?

25 dots

3. What is the next number in the pattern? 10, 15, 21, 28

36

4. What is the next number in the pattern? 122, 100, 80, 62

46

5. What are the first four numbers in this pattern? The first number is 99. Subtract 7 each time.

99, 92, 85, 78

6. What are the first four numbers in this pattern? The first number is 0. Multiply by 5 and add 2 each time.

0, 2, 12, 62

To answer questions 7–8, use the relation {(1,2), (2,4), (3,6), (4,8)}.

7. Ring the domain of the relation.

1, 2 3, 4

1, 2, 3, 4 2, 4, 6, 8

8. Ring the range of the relation.

2, 4

6, 8

1, 2, 3, 4

2, 4, 6, 8

9. Which ordered pairs are graphed?

(0,2), (2,4)

10. If the length of a rectangle triples and the width doubles, what happens to the area of the rectangle? Ring the answer.

It doubles.

It increases by 5 times.

It increases by 6 times.

It increases by 10 times.

11. Ring the equation that represents the relation {(0,0), (1,5), (2,10), (3,15)}.

$y = x + 12$

$y = 4x + 3$

$y = x + 4$

$y = 5x$

Go on to the next page.

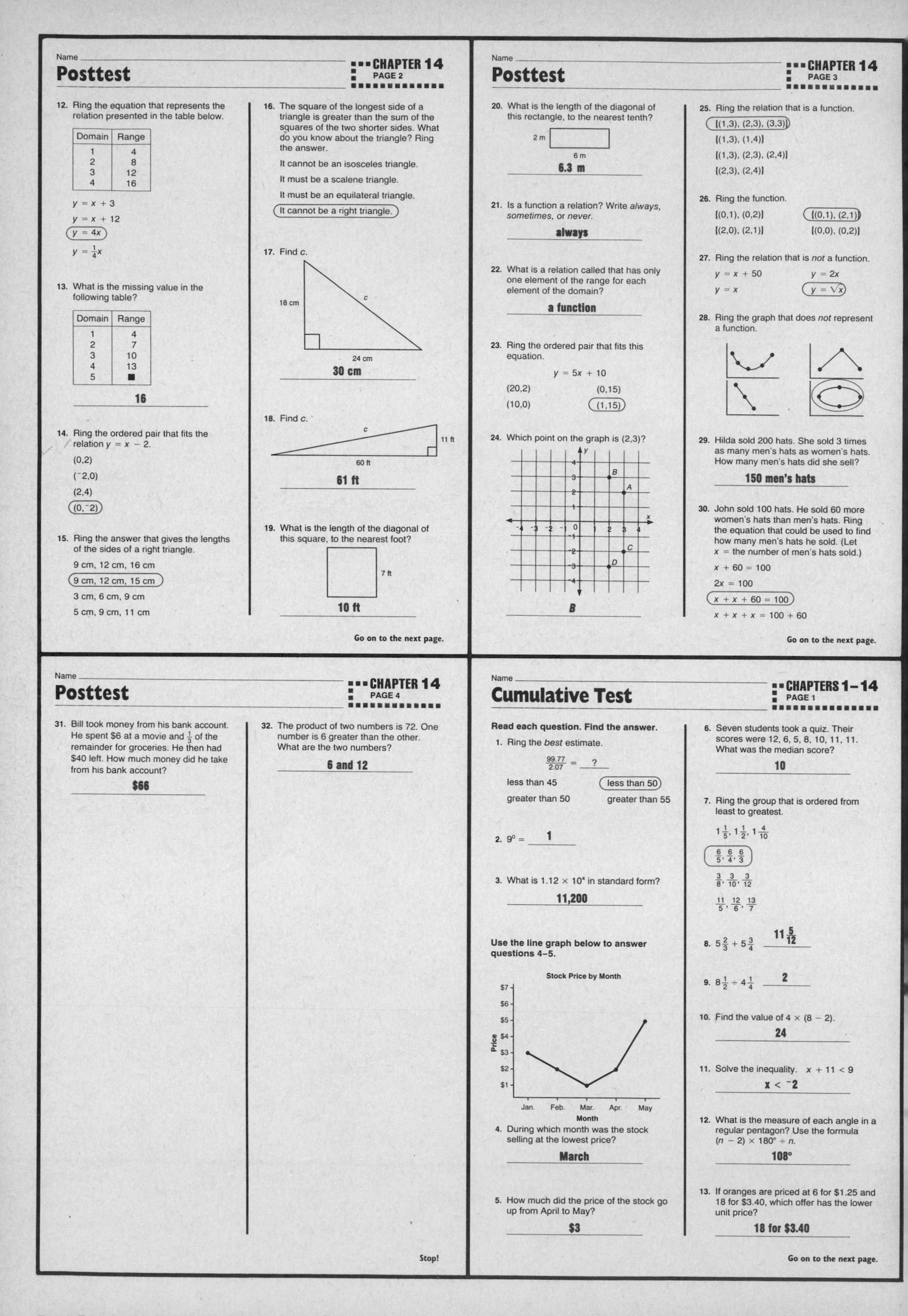

Name

Posttest

CHAPTER 14
PAGE 2

12. Ring the equation that represents the relation presented in the table below.

Domain	Range
1	4
2	8
3	12
4	16

$y = x + 3$

$y = x + 12$

($y = 4x$) — ringed

$y = \frac{1}{4}x$

13. What is the missing value in the following table?

Domain	Range
1	4
2	7
3	10
4	13
5	■

16

14. Ring the ordered pair that fits the relation $y = x - 2$.

(0,2)

($^-2$,0)

(2,4)

((0,$^-2$)) — ringed

15. Ring the answer that gives the lengths of the sides of a right triangle.

9 cm, 12 cm, 16 cm

(9 cm, 12 cm, 15 cm) — ringed

3 cm, 6 cm, 9 cm

5 cm, 9 cm, 11 cm

16. The square of the longest side of a triangle is greater than the sum of the squares of the two shorter sides. What do you know about the triangle? Ring the answer.

It cannot be an isosceles triangle.

It must be a scalene triangle.

It must be an equilateral triangle.

(It cannot be a right triangle.) — ringed

17. Find *c*.

18 cm, *c*, 24 cm

30 cm

18. Find *c*.

c, 11 ft, 60 ft

61 ft

19. What is the length of the diagonal of this square, to the nearest foot?

7 ft

10 ft

Go on to the next page.

Name

Posttest

CHAPTER 14
PAGE 3

20. What is the length of the diagonal of this rectangle, to the nearest tenth?

2 m, 6 m

6.3 m

21. Is a function a relation? Write *always*, *sometimes*, or *never*.

always

22. What is a relation called that has only one element of the range for each element of the domain?

a function

23. Ring the ordered pair that fits this equation.

$y = 5x + 10$

(20,2) (0,15)

(10,0) ((1,15)) — ringed

24. Which point on the graph is (2,3)?

B

25. Ring the relation that is a function.

({(1,3), (2,3), (3,3)}) — ringed

{(1,3), (1,4)}

{(1,3), (2,3), (2,4)}

{(2,3), (2,4)}

26. Ring the function.

{(0,1), (0,2)} ({(0,1), (2,1)}) — ringed

{(2,0), (2,1)} {(0,0), (0,2)}

27. Ring the relation that is *not* a function.

$y = x + 50$ $y = 2x$

$y = x$ ($y = \sqrt{x}$) — ringed

28. Ring the graph that does *not* represent a function.

29. Hilda sold 200 hats. She sold 3 times as many men's hats as women's hats. How many men's hats did she sell?

150 men's hats

30. John sold 100 hats. He sold 60 more women's hats than men's hats. Ring the equation that could be used to find how many men's hats he sold. (Let x = the number of men's hats sold.)

$x + 60 = 100$

$2x = 100$

($x + x + 60 = 100$) — ringed

$x + x + x = 100 + 60$

Go on to the next page.

Name

Posttest

CHAPTER 14
PAGE 4

31. Bill took money from his bank account. He spent \$6 at a movie and $\frac{1}{3}$ of the remainder for groceries. He then had \$40 left. How much money did he take from his bank account?

\$66

32. The product of two numbers is 72. One number is 6 greater than the other. What are the two numbers?

6 and 12

Stop!

Name

Cumulative Test

CHAPTERS 1–14
PAGE 1

Read each question. Find the answer.

1. Ring the *best* estimate.

$\frac{99.77}{2.07} =$?

less than 45 (less than 50) — ringed

greater than 50 greater than 55

2. $9^0 =$ **1**

3. What is 1.12×10^4 in standard form?

11,200

Use the line graph below to answer questions 4–5.

4. During which month was the stock selling at the lowest price?

March

5. How much did the price of the stock go up from April to May?

\$3

6. Seven students took a quiz. Their scores were 12, 6, 5, 8, 10, 11, 11. What was the median score?

10

7. Ring the group that is ordered from least to greatest.

$1\frac{1}{5}, 1\frac{1}{2}, 1\frac{4}{10}$

($\frac{6}{5}, \frac{6}{4}, \frac{6}{3}$) — ringed

$\frac{3}{8}, \frac{3}{10}, \frac{3}{12}$

$\frac{11}{5}, \frac{12}{6}, \frac{13}{7}$

8. $5\frac{2}{3} + 5\frac{3}{4}$ **$11\frac{5}{12}$**

9. $8\frac{1}{2} \div 4\frac{1}{4}$ **2**

10. Find the value of $4 \times (8 - 2)$.

24

11. Solve the inequality. $x + 11 < 9$

$x < {}^-2$

12. What is the measure of each angle in a regular pentagon? Use the formula $(n - 2) \times 180° \div n$.

108°

13. If oranges are priced at 6 for \$1.25 and 18 for \$3.40, which offer has the lower unit price?

18 for \$3.40

Go on to the next page.

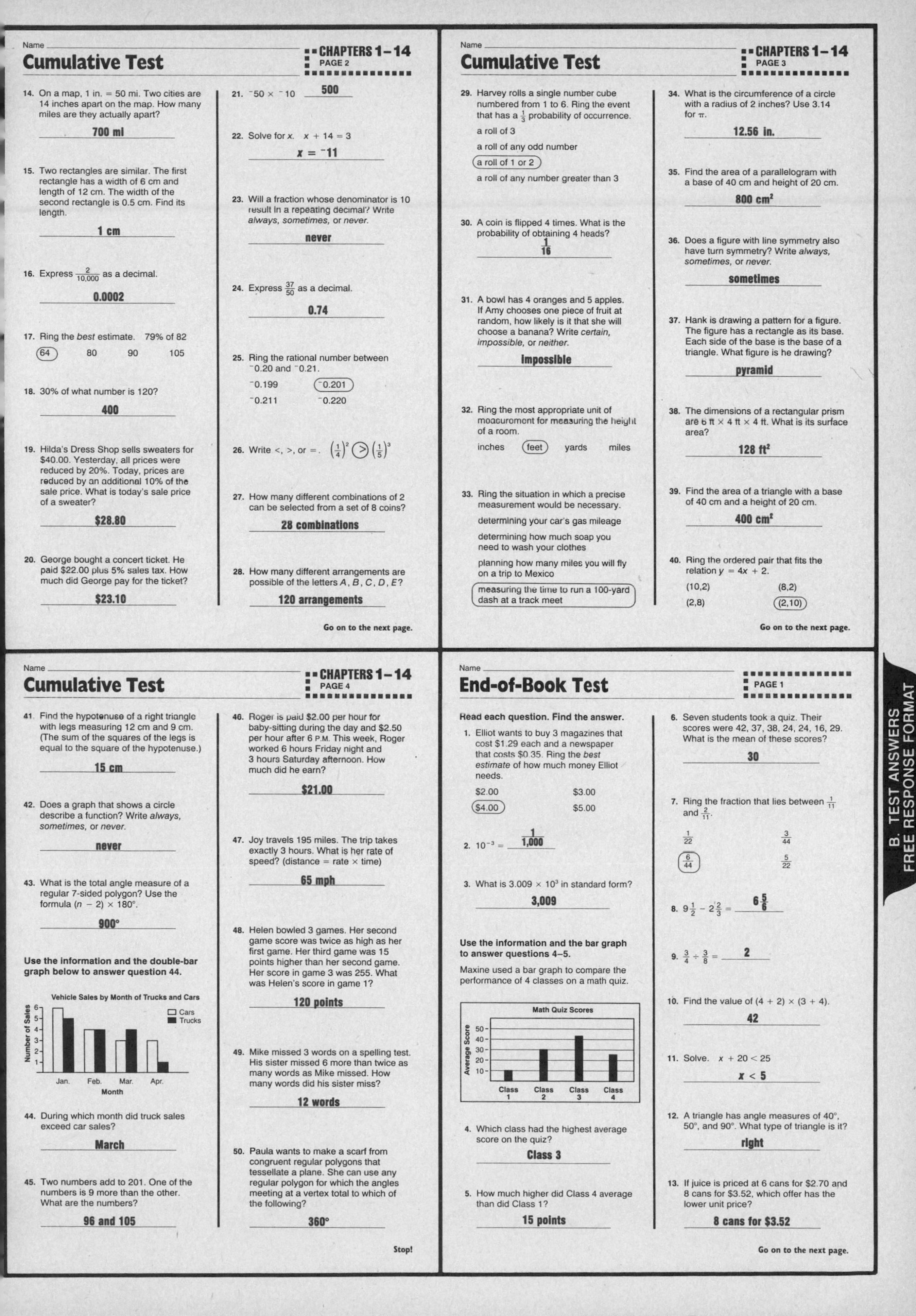

Name

Cumulative Test

CHAPTERS 1–14
PAGE 2

14. On a map, 1 in. = 50 mi. Two cities are 14 inches apart on the map. How many miles are they actually apart?

700 mi

15. Two rectangles are similar. The first rectangle has a width of 6 cm and length of 12 cm. The width of the second rectangle is 0.5 cm. Find its length.

1 cm

16. Express $\frac{2}{10,000}$ as a decimal.

0.0002

17. Ring the *best* estimate. 79% of 82

(64) 80 90 105

18. 30% of what number is 120?

400

19. Hilda's Dress Shop sells sweaters for $40.00. Yesterday, all prices were reduced by 20%. Today, prices are reduced by an additional 10% of the sale price. What is today's sale price of a sweater?

$28.80

20. George bought a concert ticket. He paid $22.00 plus 5% sales tax. How much did George pay for the ticket?

$23.10

21. $^-50 \times {}^-10$ **500**

22. Solve for *x*. $x + 14 = 3$

$x = {}^-11$

23. Will a fraction whose denominator is 10 result in a repeating decimal? Write *always, sometimes,* or *never.*

never

24. Express $\frac{37}{50}$ as a decimal.

0.74

25. Ring the rational number between $^-0.20$ and $^-0.21$.

$^-0.199$ ($^-0.201$)

$^-0.211$ $^-0.220$

26. Write <, >, or =. $\left(\frac{1}{4}\right)^2$ (>) $\left(\frac{1}{5}\right)^3$

27. How many different combinations of 2 can be selected from a set of 8 coins?

28 combinations

28. How many different arrangements are possible of the letters *A, B, C, D, E*?

120 arrangements

Go on to the next page.

Name

Cumulative Test

CHAPTERS 1–14
PAGE 3

29. Harvey rolls a single number cube numbered from 1 to 6. Ring the event that has a $\frac{1}{3}$ probability of occurrence.

a roll of 3

a roll of any odd number

(a roll of 1 or 2)

a roll of any number greater than 3

30. A coin is flipped 4 times. What is the probability of obtaining 4 heads?

$\frac{1}{16}$

31. A bowl has 4 oranges and 5 apples. If Amy chooses one piece of fruit at random, how likely is it that she will choose a banana? Write *certain, impossible,* or *neither.*

impossible

32. Ring the most appropriate unit of measurement for measuring the height of a room.

inches (feet) yards miles

33. Ring the situation in which a precise measurement would be necessary.

determining your car's gas mileage

determining how much soap you need to wash your clothes

planning how many miles you will fly on a trip to Mexico

(measuring the time to run a 100-yard dash at a track meet)

34. What is the circumference of a circle with a radius of 2 inches? Use 3.14 for π.

12.56 in.

35. Find the area of a parallelogram with a base of 40 cm and height of 20 cm.

800 cm²

36. Does a figure with line symmetry also have turn symmetry? Write *always, sometimes,* or *never.*

sometimes

37. Hank is drawing a pattern for a figure. The figure has a rectangle as its base. Each side of the base is the base of a triangle. What figure is he drawing?

pyramid

38. The dimensions of a rectangular prism are 6 ft × 4 ft × 4 ft. What is its surface area?

128 ft²

39. Find the area of a triangle with a base of 40 cm and a height of 20 cm.

400 cm²

40. Ring the ordered pair that fits the relation $y = 4x + 2$.

(10,2) (8,2)

(2,8) ((2,10))

Go on to the next page.

Name

Cumulative Test

CHAPTERS 1–14
PAGE 4

41. Find the hypotenuse of a right triangle with legs measuring 12 cm and 9 cm. (The sum of the squares of the legs is equal to the square of the hypotenuse.)

15 cm

42. Does a graph that shows a circle describe a function? Write *always, sometimes,* or *never.*

never

43. What is the total angle measure of a regular 7-sided polygon? Use the formula $(n - 2) \times 180°$.

900°

Use the information and the double-bar graph below to answer question 44.

44. During which month did truck sales exceed car sales?

March

45. Two numbers add to 201. One of the numbers is 9 more than the other. What are the numbers?

96 and 105

46. Roger is paid $2.00 per hour for baby-sitting during the day and $2.50 per hour after 6 P.M. This week, Roger worked 6 hours Friday night and 3 hours Saturday afternoon. How much did he earn?

$21.00

47. Joy travels 195 miles. The trip takes exactly 3 hours. What is her rate of speed? (distance = rate × time)

65 mph

48. Helen bowled 3 games. Her second game score was twice as high as her first game. Her third game was 15 points higher than her second game. Her score in game 3 was 255. What was Helen's score in game 1?

120 points

49. Mike missed 3 words on a spelling test. His sister missed 6 more than twice as many words as Mike missed. How many words did his sister miss?

12 words

50. Paula wants to make a scarf from congruent regular polygons that tessellate a plane. She can use any regular polygon for which the angles meeting at a vertex total to which of the following?

360°

Stop!

Name

End-of-Book Test

PAGE 1

Read each question. Find the answer.

1. Elliot wants to buy 3 magazines that cost $1.29 each and a newspaper that costs $0.35. Ring the *best estimate* of how much money Elliot needs.

$2.00 $3.00

($4.00) $5.00

2. $10^{-3} = \frac{1}{1,000}$

3. What is 3.009×10^3 in standard form?

3,009

Use the information and the bar graph to answer questions 4–5.

Maxine used a bar graph to compare the performance of 4 classes on a math quiz.

4. Which class had the highest average score on the quiz?

Class 3

5. How much higher did Class 4 average than did Class 1?

15 points

6. Seven students took a quiz. Their scores were 42, 37, 38, 24, 24, 16, 29. What is the mean of these scores?

30

7. Ring the fraction that lies between $\frac{1}{11}$ and $\frac{2}{11}$.

$\frac{1}{22}$ $\frac{3}{44}$

($\frac{6}{44}$) $\frac{5}{22}$

8. $9\frac{1}{2} - 2\frac{2}{3} = 6\frac{5}{6}$

9. $\frac{3}{4} \div \frac{3}{8} = 2$

10. Find the value of $(4 + 2) \times (3 + 4)$.

42

11. Solve. $x + 20 < 25$

$x < 5$

12. A triangle has angle measures of 40°, 50°, and 90°. What type of triangle is it?

right

13. If juice is priced at 6 cans for $2.70 and 8 cans for $3.52, which offer has the lower unit price?

8 cans for $3.52

Go on to the next page.

B. TEST ANSWERS FREE RESPONSE FORMAT

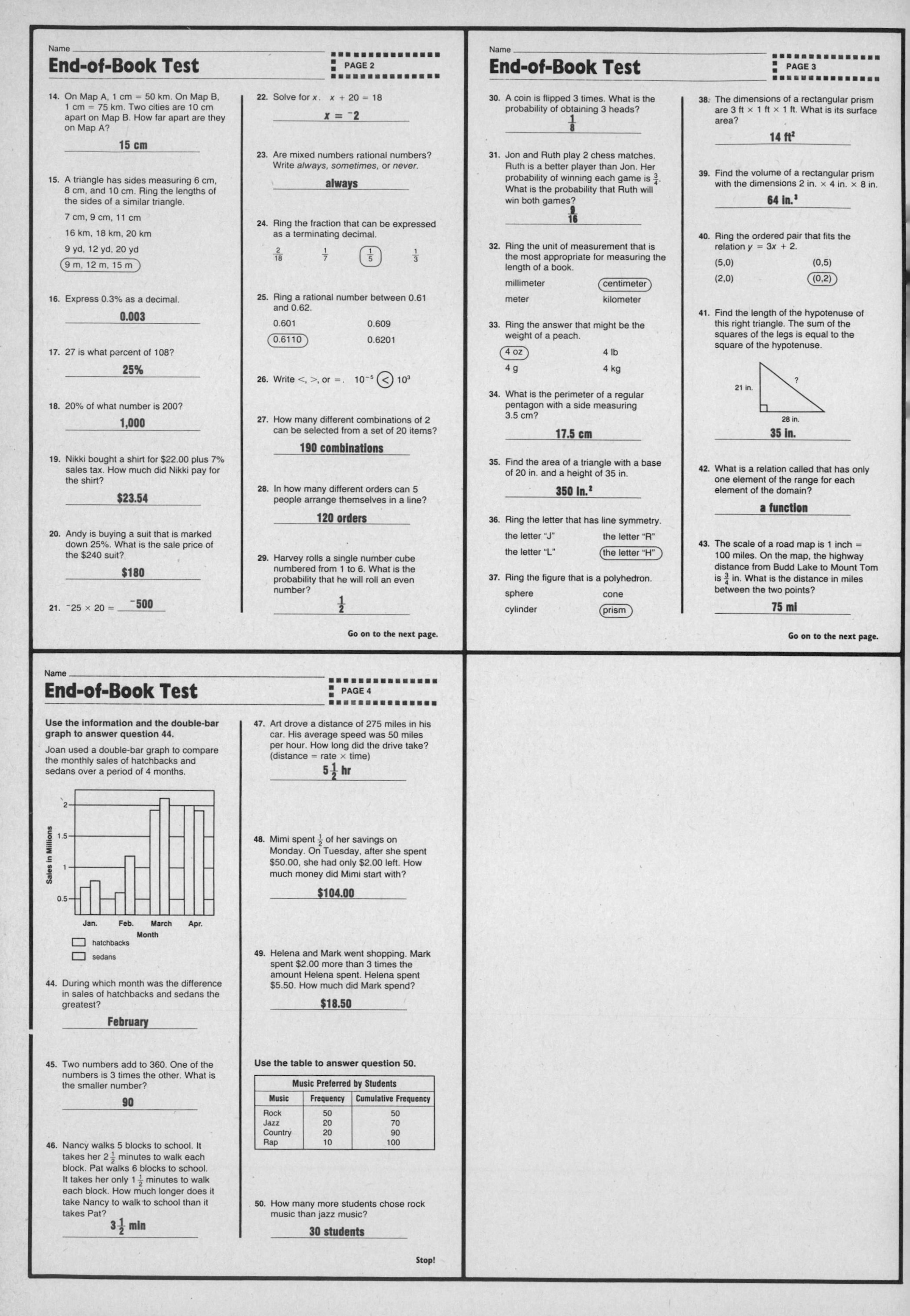

Name ____________________

End-of-Book Test

PAGE 2

14. On Map A, 1 cm = 50 km. On Map B, 1 cm = 75 km. Two cities are 10 cm apart on Map B. How far apart are they on Map A?

15 cm

15. A triangle has sides measuring 6 cm, 8 cm, and 10 cm. Ring the lengths of the sides of a similar triangle.

7 cm, 9 cm, 11 cm
16 km, 18 km, 20 km
9 yd, 12 yd, 20 yd
9 m, 12 m, 15 m

16. Express 0.3% as a decimal.

0.003

17. 27 is what percent of 108?

25%

18. 20% of what number is 200?

1,000

19. Nikki bought a shirt for $22.00 plus 7% sales tax. How much did Nikki pay for the shirt?

$23.54

20. Andy is buying a suit that is marked down 25%. What is the sale price of the $240 suit?

$180

21. $^-25 \times 20 =$ **$^-500$**

22. Solve for x. $x + 20 = 18$

$x = {}^-2$

23. Are mixed numbers rational numbers? Write *always, sometimes,* or *never.*

always

24. Ring the fraction that can be expressed as a terminating decimal.

$\frac{2}{18}$ $\frac{1}{7}$ **$\frac{1}{5}$** $\frac{1}{3}$

25. Ring a rational number between 0.61 and 0.62.

0.601 0.609
0.6110 0.6201

26. Write <, >, or =. 10^{-5} **<** 10^3

27. How many different combinations of 2 can be selected from a set of 20 items?

190 combinations

28. In how many different orders can 5 people arrange themselves in a line?

120 orders

29. Harvey rolls a single number cube numbered from 1 to 6. What is the probability that he will roll an even number?

$\frac{1}{2}$

Go on to the next page.

Name ____________________

End-of-Book Test

PAGE 3

30. A coin is flipped 3 times. What is the probability of obtaining 3 heads?

$\frac{1}{8}$

31. Jon and Ruth play 2 chess matches. Ruth is a better player than Jon. Her probability of winning each game is $\frac{3}{4}$. What is the probability that Ruth will win both games?

$\frac{9}{16}$

32. Ring the unit of measurement that is the most appropriate for measuring the length of a book.

millimeter **centimeter**
meter kilometer

33. Ring the answer that might be the weight of a peach.

4 oz 4 lb
4 g 4 kg

34. What is the perimeter of a regular pentagon with a side measuring 3.5 cm?

17.5 cm

35. Find the area of a triangle with a base of 20 in. and a height of 35 in.

350 in.²

36. Ring the letter that has line symmetry.

the letter "J" the letter "R"
the letter "L" **the letter "H"**

37. Ring the figure that is a polyhedron.

sphere cone
cylinder **prism**

38. The dimensions of a rectangular prism are 3 ft × 1 ft × 1 ft. What is its surface area?

14 ft²

39. Find the volume of a rectangular prism with the dimensions 2 in. × 4 in. × 8 in.

64 in.³

40. Ring the ordered pair that fits the relation $y = 3x + 2$.

(5,0) (0,5)
(2,0) **(0,2)**

41. Find the length of the hypotenuse of this right triangle. The sum of the squares of the legs is equal to the square of the hypotenuse.

21 in.
?
28 in.

35 in.

42. What is a relation called that has only one element of the range for each element of the domain?

a function

43. The scale of a road map is 1 inch = 100 miles. On the map, the highway distance from Budd Lake to Mount Tom is $\frac{3}{4}$ in. What is the distance in miles between the two points?

75 mi

Go on to the next page.

Name ____________________

End-of-Book Test

PAGE 4

Use the information and the double-bar graph to answer question 44.

Joan used a double-bar graph to compare the monthly sales of hatchbacks and sedans over a period of 4 months.

Sales in Millions
2
1.5
1
0.5
Jan.
Feb.
March
Apr.
Month
hatchbacks
sedans

44. During which month was the difference in sales of hatchbacks and sedans the greatest?

February

45. Two numbers add to 360. One of the numbers is 3 times the other. What is the smaller number?

90

46. Nancy walks 5 blocks to school. It takes her $2\frac{1}{2}$ minutes to walk each block. Pat walks 6 blocks to school. It takes her only $1\frac{1}{2}$ minutes to walk each block. How much longer does it take Nancy to walk to school than it takes Pat?

$3\frac{1}{2}$ min

47. Art drove a distance of 275 miles in his car. His average speed was 50 miles per hour. How long did the drive take? (distance = rate × time)

$5\frac{1}{2}$ hr

48. Mimi spent $\frac{1}{2}$ of her savings on Monday. On Tuesday, after she spent $50.00, she had only $2.00 left. How much money did Mimi start with?

$104.00

49. Helena and Mark went shopping. Mark spent $2.00 more than 3 times the amount Helena spent. Helena spent $5.50. How much did Mark spend?

$18.50

Use the table to answer question 50.

Music Preferred by Students		
Music	Frequency	Cumulative Frequency
Rock	50	50
Jazz	20	70
Country	20	90
Rap	10	100

50. How many more students chose rock music than jazz music?

30 students

Stop!

III

Management Forms

Answer Sheet

This Copying Master is an individual recording sheet for up to 50 items on the standardized-format test.

Grading Made Easy

This percent converter can be used for all quizzes and tests. The percents given are based on all problems having equal value. Percents are rounded to the nearest whole percent giving the benefit of 0.5 percent.

Individual Record Form

One Copying Master for each chapter is provided. Criterion scores for each tested objective are given for the Pretest and Posttest. The student's total scores are recorded at the top of the page. The scores for each objective can also be recorded. You can use the Review Options in the Pupil's Edition, Teacher's Edition, and Workbooks that are listed on the form to assign additional review for the student unable to pass the test.

Formal Assessment Class Record Form

The scores for all tests can be recorded for your class on these record forms. The Criterion Score for each test is given.

Cumulative Record Form

Individual student progress can be recorded on the Copying Masters that are provided. Copies of each grade's tested objectives and criteria are provided. Using these forms, you can find the objectives that will be covered in the next or previous grade.

Name ____________________ Date ____________________

Test Answer Sheet

MATHEMATICS PLUS

Test Title ______________________________

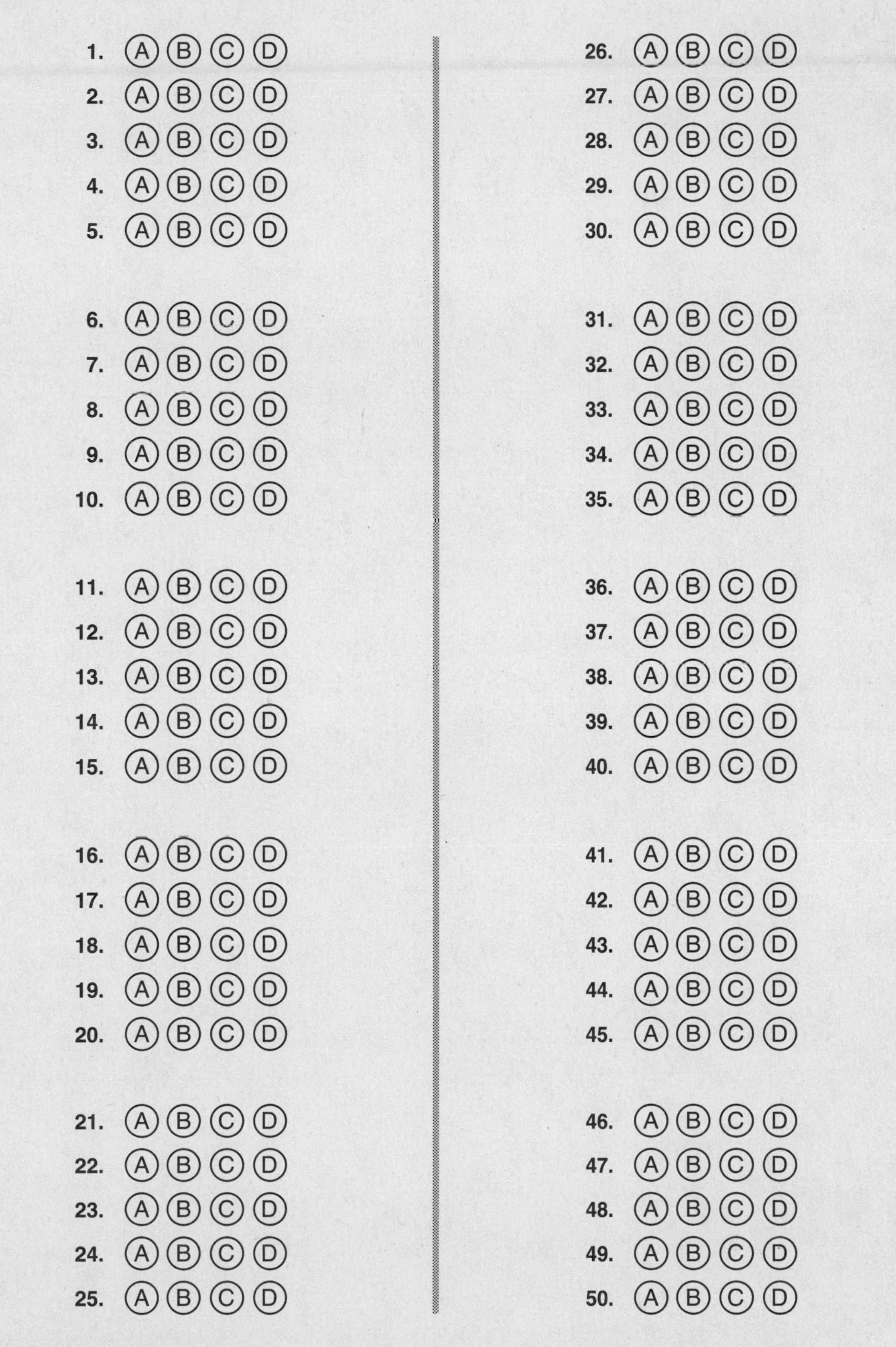

1. (A) (B) (C) (D)
2. (A) (B) (C) (D)
3. (A) (B) (C) (D)
4. (A) (B) (C) (D)
5. (A) (B) (C) (D)
6. (A) (B) (C) (D)
7. (A) (B) (C) (D)
8. (A) (B) (C) (D)
9. (A) (B) (C) (D)
10. (A) (B) (C) (D)
11. (A) (B) (C) (D)
12. (A) (B) (C) (D)
13. (A) (B) (C) (D)
14. (A) (B) (C) (D)
15. (A) (B) (C) (D)
16. (A) (B) (C) (D)
17. (A) (B) (C) (D)
18. (A) (B) (C) (D)
19. (A) (B) (C) (D)
20. (A) (B) (C) (D)
21. (A) (B) (C) (D)
22. (A) (B) (C) (D)
23. (A) (B) (C) (D)
24. (A) (B) (C) (D)
25. (A) (B) (C) (D)
26. (A) (B) (C) (D)
27. (A) (B) (C) (D)
28. (A) (B) (C) (D)
29. (A) (B) (C) (D)
30. (A) (B) (C) (D)
31. (A) (B) (C) (D)
32. (A) (B) (C) (D)
33. (A) (B) (C) (D)
34. (A) (B) (C) (D)
35. (A) (B) (C) (D)
36. (A) (B) (C) (D)
37. (A) (B) (C) (D)
38. (A) (B) (C) (D)
39. (A) (B) (C) (D)
40. (A) (B) (C) (D)
41. (A) (B) (C) (D)
42. (A) (B) (C) (D)
43. (A) (B) (C) (D)
44. (A) (B) (C) (D)
45. (A) (B) (C) (D)
46. (A) (B) (C) (D)
47. (A) (B) (C) (D)
48. (A) (B) (C) (D)
49. (A) (B) (C) (D)
50. (A) (B) (C) (D)

Total Number of Test Items

Number of Test Items Wrong	4	5	6	7	8	9	10	11	12	13	14	15	16	17	18	19	20	21	22	23	24	25	26	27	28	29	30	32	50
1	75	80	83	86	88	89	90	91	92	92	93	93	94	94	94	95	95	95	95	96	96	96	96	96	96	97	97	97	98
2	50	60	67	71	75	78	80	82	83	85	86	87	88	88	89	89	90	90	91	91	92	92	92	93	93	93	93	94	96
3	25	40	50	57	63	67	70	73	75	77	79	80	81	82	83	84	85	86	86	87	88	88	88	89	89	90	90	91	94
4	0	20	33	43	50	56	60	64	67	69	71	73	75	76	78	79	80	81	82	83	83	84	85	85	86	86	87	88	92
5		0	17	29	38	44	50	55	58	62	64	67	69	71	72	74	75	76	77	78	79	80	81	81	82	83	83	84	90
6			0	14	25	33	40	45	50	54	57	60	63	65	67	68	70	71	73	74	75	76	77	78	79	79	80	81	88
7				0	13	22	30	36	42	46	50	53	56	59	61	63	65	67	68	70	71	72	73	74	75	76	77	78	86
8					0	11	20	27	33	38	43	47	50	53	56	58	60	62	64	65	67	68	69	70	71	72	73	75	84
9						0	10	18	25	31	36	40	44	47	50	53	55	57	59	61	63	64	65	67	68	69	70	72	82
10							0	9	17	23	29	33	38	41	44	47	50	52	55	57	58	60	62	63	64	66	67	69	80
11								0	8	15	21	27	31	35	39	42	45	48	50	52	54	56	58	59	61	62	63	66	78
12									0	8	14	20	25	29	33	37	40	43	45	48	50	52	54	56	57	59	60	63	76
13										0	7	13	19	24	28	32	35	38	41	43	46	48	50	52	54	55	57	59	74
14											0	7	13	18	22	26	30	33	36	39	42	44	46	48	50	52	53	56	72
15												0	6	12	17	21	25	29	32	35	38	40	42	44	46	48	50	53	70
16													0	6	11	16	20	24	27	30	33	36	38	41	43	45	47	50	68
17														0	6	11	15	19	23	26	29	32	35	37	39	41	43	47	66
18															0	5	10	14	18	22	25	28	31	33	36	38	40	44	64
19																0	5	10	14	17	21	24	27	30	32	34	37	41	62
20																	0	5	9	13	17	20	23	26	29	31	33	38	60
21																		0	5	9	13	16	19	22	25	28	30	34	58
22																			0	4	8	12	15	19	21	24	27	31	56
23																				0	4	8	12	15	18	21	23	28	54
24																					0	4	8	11	14	17	20	25	52
25																						0	4	7	11	14	17	22	50
26																							0	4	7	10	13	19	48
27																								0	4	7	10	16	46
28																									0	3	7	13	44
29																										0	3	9	42
30																											0	6	40
31																												3	38
32																												0	36

Individual Record Form

MATHEMATICS PLUS

Grade 7 • Chapter 1

	Date	Score
Inventory Test		
Pretest		
Posttest		

Student Name: ______________________________

TESTED OBJECTIVES		CHAPTER TESTS						REVIEW OPTIONS					
Obj. No.	Tested Objective	Test Items: Concept	Test Items: Skill	Test Items: Problem Solving	Criterion Scores	Pre-test Score	Post-test Score	Lesson Page Numbers	Pupil's Edition	Teacher's Edition	Workbooks: P	Workbooks: R	Workbooks: E
1–A	To estimate sums, differences, products, and quotients of whole numbers and decimals	1–4	5–8	9–10	7/10			4–5	Br, pp. H2–H3	ATS, p. C1	1.2	1.2	1.2
								6–7	MP, p. H34	ATS, p. C1	1.3	1.3	1.3
								8–9	MP, p. H35	ATS, p. C2	1.4	1.4	1.4
								10–11	MP, p. H35	MIN, p. 10	1.5	1.5	1.5
								18–19	MP, p. H36	MIN, p. 18	1.8	1.8	1.8
1–B	To use addition, subtraction, multiplication, and division to solve whole-number and decimal problems	11–14	15–18	19–20	7/10			12–13	MP, p. H35	ATS, p. C2	1.6	1.6	1.6
								20–21	MP, p. H36	PS Link, p. 14A	1.9	1.9	1.9
								22–23	MP, p. H36	MIN, p. 22	1.10	1.10	1.10
								24–25	MP, p. H36	Man. Link, p. 2B	1.11	1.11	1.11
1–C	To identify and evaluate numbers in standard form and exponent form		21–24		3/4			28–29	MP, p. H37	MIN, p. 28	1.13	1.13	1.13
1–D	To express and evaluate numbers in scientific notation		25–28		3/4			30–31	Br, pp. H4–H5	MIN, p. 30	1.14	1.14	1.14
1–E	To solve problems by using a diagram and by guessing and checking			29–32	3/4			14–15	MP, p. H35	MIN, p. 14	1.7	1.7	1.7
								26–27	MP, p. H37	MIN, p. 26	1.12	1.12	1.12

KEY: MP—More Practice Br—Bridge Lesson PS Link—Problem-Solving Link MIN—Meeting Individual Needs (First Activity)
ATS—Alternative Teaching Strategies Man. Link—Manipulative Link
Workbooks: P—Practice R—Reteaching E—Enrichment

Individual Record Form

MATHEMATICS PLUS

Grade 7 • Chapter 2

	Date	Score
Pretest		
Posttest		

Student Name: ______________________________

TESTED OBJECTIVES		CHAPTER TESTS						REVIEW OPTIONS					
Obj. No.	Tested Objective	Test Items			Criterion Scores	Pre-test Score	Post-test Score	Lesson Page Numbers	Pupil's Edition	Teacher's Edition	Workbooks		
		Concept	Skill	Problem Solving							P	R	E
2–A	To read and interpret bar graphs, line graphs, circle graphs, histograms, and stem-and-leaf plots	1, 3, 5	2, 4, 6	7–10	7/10			44–45	MP, p. H38	ATS, p. C3	2.3	2.3	2.3
								46–47	MP, p. H39	ATS, p. C3	2.4	2.4	2.4
								52–53	Br, pp. H6–H7	MIN, p. 52	2.6	2.6	2.6
								54–55	MP, p. H40	Man. Link, p. 40A	2.7	2.7	2.7
								56–57	MP, p. H40	ATS, p. C4	2.8	2.8	2.8
2–B	To use frequency distribution tables; to use unbiased surveys	11–12	13–14		3/4			40–41	MP, p. H38	MIN, p. 40	2.1	2.1	2.1
								42–43	MP, p. H38	Man. Link, p. 40B	2.2	2.2	2.2
2–C	To analyze sets of data by using range, quartiles, mean, median, and mode; to read and interpret box-and-whisker graphs	15–16	17–18	19–20	4/6			58–59	MP, p. H40	MIN, p. 58	2.9	2.9	2.9
								60–61	MP, p. H41	ATS, p. C4	2.10	2.10	2.10
								62–63	MP, p. H41	MIN, p. 62	2.11	2.11	2.11
2–D	To solve problems by using tables and graphs and by selecting a graph for a set of data			21–24	3/4			48–49	MP, p. H39	MIN, p. 48	2.5	2.5	2.5
								64–65	MP, p. H41	MIN, p. 64	2.12	2.12	2.12

KEY: MP—More Practice Br—Bridge Lesson PS Link—Problem-Solving Link MIN—Meeting Individual Needs (First Activity)
ATS—Alternative Teaching Strategies Man. Link—Manipulative Link
Workbooks: P—Practice R—Reteaching E—Enrichment

Individual Record Form

Grade 7 • Chapter 3

	Date	Score
Pretest		
Posttest		

Student Name: ______________________

TESTED OBJECTIVES		CHAPTER TESTS						REVIEW OPTIONS					
Obj. No.	Tested Objective	Test Items: Concept	Test Items: Skill	Test Items: Problem Solving	Criterion Scores	Pre-test Score	Post-test Score	Lesson Page Numbers	Pupil's Edition	Teacher's Edition	Workbooks: P	Workbooks: R	Workbooks: E
3–A	To find the prime factorization of a composite number	1–4	5–8		$5/8$			76–77	MP, p. H42	ATS, p. C5	3.1	3.1	3.1
								78–79	MP, p. H42	ATS, p. C5	3.2	3.2	3.2
								80–81	MP, p. H42	MIN, p. 80	3.3	3.3	3.3
3–B	To find the greatest common factor and the least common multiple of two or more numbers		9–12		$3/4$			82–83	MP, p. H43	Man. Link, p. 76B	3.4	3.4	3.4
								90–91	Br, pp. H8–H9	ATS, p. C6	3.7	3.7	3.7
3–C	To identify equivalent fractions and fractions in simplest form; to compare and order two or more fractions and mixed numbers; to identify one or two fractions between any two given fractions	13–16	17–20	21–24	$8/12$			84–85	MP, p. H43	ATS, p. C6	3.5	3.5	3.5
								90–91	Br, pp. H8–H9	ATS, p. C6	3.7	3.7	3.7
								92–93	MP, p. H45	MIN, p. 92	3.8	3.8	3.8
3–D	To identify one or two fractions between any two given fractions	25–26	27–28		$3/4$			94–95	MP, p. H45	MIN, p. 94	3.9	3.9	3.9
3–E	To solve problems by finding a pattern and by using a diagram			29–32	$3/4$			86–87	MP, p. H44	MIN, p. 86	3.6	3.6	3.6
								96–97	MP, p. H45	MIN, p. 96	3.10	3.10	3.10

KEY: MP—More Practice Br—Bridge Lesson PS Link—Problem-Solving Link MIN—Meeting Individual Needs (First Activity)
ATS—Alternative Teaching Strategies Man. Link—Manipulative Link
Workbooks: P—Practice R—Reteaching E—Enrichment

Individual Record Form

MATHEMATICS PLUS

Grade 7 • Chapter 4

Student Name: ____________________

	Date	Score
Pretest		
Posttest		
Cumulative Test		

TESTED OBJECTIVES		CHAPTER TESTS						REVIEW OPTIONS					
Obj. No.	Tested Objective	Test Items: Concept	Test Items: Skill	Test Items: Problem Solving	Criterion Scores	Pre-test Score	Post-test Score	Lesson Page Numbers	Pupil's Edition	Teacher's Edition	Workbooks: P	Workbooks: R	Workbooks: E
4–A	To estimate sums, differences, products, and quotients of fractions	1–2	3–6	7–8	5/8			106–107	MP, p. H46	ATS, p. C7	4.1	4.1	4.1
								116–117	MP, p. H47	MIN, p. 116	4.6	4.6	4.6
								124–125	MP, p. H48	MIN, p. 124	4.9	4.9	4.9
4–B	To add and subtract fractions and mixed numbers with and without regrouping	9–10	11–14	15–16	5/8			108–109	Br, pp. H10–H11	ATS, p. C7	4.2	4.2	4.2
								110–111	MP, p. H46	MIN, p. 110	4.3	4.3	4.3
								112–113	MP, p. H47	ATS, p. C8	4.4	4.4	4.4
4–C	To multiply and divide fractions and mixed numbers	17–20	21–24	25–28	8/12			120–121	MP, p. H48	Man. Link, p. 106A	4.7	4.7	4.7
								122–123	MP, p. H48	ATS, p. C8	4.8	4.8	4.8
								126–127	MP, p. H48	MIN, p. 126	4.10	4.10	4.10
								128–129	MP, p. H49	PS Link, p. 114B	4.11	4.11	4.11
								130–131	MP, p. H49	MIN, p. 130	4.12	4.12	4.12
4–D	To solve multistep problems and to solve problems by solving a simpler problem			29–32	3/4			114–115	MP, p. H47	MIN, p. 114	4.5	4.5	4.5
								132–133	MP, p. H49	MIN, p. 132	4.13	4.13	4.13

KEY: MP—More Practice Br—Bridge Lesson PS Link—Problem-Solving Link MIN—Meeting Individual Needs (First Activity)
ATS—Alternative Teaching Strategies Man. Link—Manipulative Link
Workbooks: P—Practice R—Reteaching E—Enrichment

Individual Record Form

Grade 7 • Chapter 5

	Date	Score
Pretest		
Posttest		

Student Name: ____________________

TESTED OBJECTIVES		CHAPTER TESTS						REVIEW OPTIONS					
Obj. No.	Tested Objective	Test Items			Criterion Scores	Pre-test Score	Post-test Score	Lesson Page Numbers	Pupil's Edition	Teacher's Edition	Workbooks		
		Concept	Skill	Problem Solving							P	R	E
5–A	To identify and evaluate expressions	1–2	3–6	7–8	5/8			142–143	MP, p. H50	MIN, p. 142	5.1	5.1	5.1
								146–147	MP, p. H50	ATS, p. C9	5.3	5.3	5.3
								158–159	MP, p. H52	MIN, p. 158	5.8	5.8	5.8
5–B	To solve problems, using the order of operations		9–12		3/4			144–145	MP, p. H50	MIN, p. 144	5.2	5.2	5.2
5–C	To solve a one-step inequality involving addition	13–14	15–16		3/4			142–143	MP, p. H50	MIN, p. 142	5.1	5.1	5.1
								166–167	MP, p. H53	ATS, p. C10	5.12	5.12	5.12
5–D	To solve one-step equations	17–20	21–24	25–28	8/12			142–143	MP, p. H50	MIN, p. 142	5.1	5.1	5.1
								150–151	MP, p. H51	MIN, p. 150	5.5	5.5	5.5
								152–153	MP, p. H51	Man. Link, p. 142A	5.6	5.6	5.6
								154–155	MP, p. H52	ATS, p. C9	5.7	5.7	5.7
								160–161	MP, p. H52	MIN, p. 160	5.9	5.9	5.9
								164–165	MP, p. H53	ATS, p. C10	5.11	5.11	5.11
5–E	To solve problems by using a formula and by working backward			29–32	3/4			162–163	MP, p. H53	MIN, p. 162	5.10	5.10	5.10
								168–169	MP, p. H53	MIN, p. 168	5.13	5.13	5.13

KEY: MP—More Practice Br—Bridge Lesson PS Link—Problem-Solving Link MIN—Meeting Individual Needs (First Activity)
ATS—Alternative Teaching Strategies Man. Link—Manipulative Link
Workbooks: P—Practice R—Reteaching E—Enrichment

Individual Record Form

MATHEMATICS PLUS

Grade 7 • Chapter 6

	Date	Score
Pretest		
Posttest		

Student Name: ____________________

TESTED OBJECTIVES		CHAPTER TESTS						REVIEW OPTIONS					
Obj. No.	Tested Objective	Test Items: Concept	Skill	Problem Solving	Criterion Scores	Pre-test Score	Post-test Score	Lesson Page Numbers	Pupil's Edition	Teacher's Edition	Workbooks: P	R	E
6–A	To identify, measure, and classify lines, angles, triangles, polygons, and quadrilaterals; to identify congruent figures and geometric relationships	1–4	5–12		8/12			178–179	MP, p. H54	MIN, p. 178	6.1	6.1	6.1
								180–181	MP, p. H54	MIN, p. 180	6.2	6.2	6.2
								182–185	MP, p. H54	ATS, p. C11	6.3	6.3	6.3
								186–187	Br, pp. H12–H13	ATS, p. C11	6.4	6.4	6.4
								188–190	Br, pp. H14–H15	ATS, p. C12	6.5	6.5	6.5
								192–193	MP, p. H55	ATS, p. C12	6.6	6.6	6.6
								198–199	MP, p. H56	MIN, p. 198	6.8	6.8	6.8
								200–201	MP, p. H56	MIN, p. 200	6.9	6.9	6.9
								204–205	MP, p. H57	MIN, p. 204	6.11	6.11	6.11
6–B	To identify constructions of congruent angles, congruent triangles, line bisectors, anda congruent line segments		13–16		3/4			182–185	MP, p. H54	ATS, p. C11	6.3	6.3	6.3
								188–190	Br, pp. H14–H15	ATS, p. C12	6.5	6.5	6.5
								200–201	MP, p. H56	MIN, p. 200	6.9	6.9	6.9
6–C	To solve problems by using logical reasoning and Venn diagrams	17–18		19–20	3/4			202–203	MP, p. H57	MIN, p. 202	6.10	6.10	6.10
6–D	To solve problems by finding a pattern and by using a formula			21–24	3/4			194–195	MP, p. H56	MIN, p. 194	6.7	6.7	6.7
								206–207	MP, p. H57	MIN, p. 206	6.12	6.12	6.12

KEY: MP—More Practice Br—Bridge Lesson PS Link—Problem-Solving Link MIN—Meeting Individual Needs (First Activity)
ATS—Alternative Teaching Strategies Man. Link—Manipulative Link
Workbooks: P—Practice R—Reteaching E—Enrichment

Individual Record Form

Grade 7 • Chapter 7

	Date	Score
Pretest		
Posttest		

Student Name: ______________________

TESTED OBJECTIVES		CHAPTER TESTS						REVIEW OPTIONS					
Obj. No.	Tested Objective	Test Items			Criterion Scores	Pre-test Score	Post-test Score	Lesson Page Numbers	Pupil's Edition	Teacher's Edition	Workbooks		
		Concept	Skill	Problem Solving							P	R	E
7–A	To identify ratios and rates; to complete equal ratios; to solve proportions	1–4	5–8	9–12	8/12			216–217	Br, pp. H16–H17	ATS, p. C13	7.1	7.1	7.1
								218–219	MP, p. H58	MIN, p. 218	7.2	7.2	7.2
								220–221	MP, p. H58	Man. Link, p. 216A	7.3	7.3	7.3
								222–223	MP, p. H59	ATS, p. C13	7.4	7.4	7.4
7–B	To use ratios and proportions to compute unit prices and to solve problems			13–16	3/4			226–227	MP, p. H59	MIN, p. 226	7.6	7.6	7.6
7–C	To use scale drawings to solve problems	17–18	19–20	21–22	4/6			234–235	MP, p. H60	ATS, p. C14	7.9	7.9	7.9
7–D	To identify corresponding parts of similar figures; to find the ratio of corresponding sides; to use proportions to find the missing measures of similar figures	23–24	25–28		4/6			230–231	MP, p. H60	ATS, p. C14	7.7	7.7	7.7
								232–233	MP, p. H60	MIN, p. 232	7.8	7.8	7.8
7–E	To solve problems by using a map and by using a formula			29–32	3/4			224–225	MP, p. H59	MIN, p. 224	7.5	7.5	7.5
								236–237	MP, p. H61	MIN, p. 236	7.10	7.10	7.10

KEY: MP—More Practice Br—Bridge Lesson PS Link—Problem-Solving Link MIN—Meeting Individual Needs (First Activity)
ATS—Alternative Teaching Strategies Man. Link—Manipulative Link
Workbooks: P—Practice R—Reteaching E—Enrichment

Individual Record Form

MATHEMATICS PLUS

Grade 7 • Chapter 8

	Date	Score
Pretest		
Posttest		
Cumulative Test		

Student Name: ____________________

TESTED OBJECTIVES		CHAPTER TESTS						REVIEW OPTIONS					
		Test Items									Workbooks		
Obj. No.	Tested Objective	Concept	Skill	Problem Solving	Criterion Scores	Pre-test Score	Post-test Score	Lesson Page Numbers	Pupil's Edition	Teacher's Edition	P	R	E
8–A	To identify fraction, decimal, ratio, and percent equivalencies	1–2	3–6	7–8	5/8			248–249	Br, pp. H18–H19	MIN, p. 248	8.1	8.1	8.1
								250–251	MP, p. H62	ATS, p. C15	8.2	8.2	8.2
								252–253	MP, p. H62	Man. Link, p. 248A	8.3	8.3	8.3
								254–255	MP, p. H62	PS Link, p. 258A	8.4	8.4	8.4
8–B	To estimate and find the percent of a number and the percent one number is of another; to find a number when a percent of it is known, by using estimation, proportions, and equations	9–12	13–16	17–20	8/12			256–257	MP, p. H63	ATS, p. C15	8.5	8.5	8.5
								268–269	MP, p. H64	MIN, p. 268	8.10	8.10	8.10
								270–271	MP, p. H64	ATS, p. C16	8.11	8.11	8.11
								272–273	MP, p. H65	ATS, p. C16	8.12	8.12	8.12
								274–275	MP, p. H65	MIN, p. 274	8.13	8.13	8.13
8–C	To analyze a circle graph	21–22	23–24		3/4			260–261	MP, p. H63	MIN, p. 260	8.7	8.7	8.7
8–D	To solve sales-tax, discount, and simple-interest problems		25–26	27–28	3/4			264–265	MP, p. H63	MIN, p. 264	8.8	8.8	8.8
								266–267	MP, p. H64	MIN, p. 266	8.9	8.9	8.9
8–E	To solve problems by using a graph and by guessing and checking			29–32	3/4			258–259	MP, p. H63	MIN, p. 258	8.6	8.6	8.6
								276–277	MP, p. H65	MIN, p. 276	8.14	8.14	8.14

KEY: MP—More Practice Br—Bridge Lesson PS Link—Problem-Solving Link MIN—Meeting Individual Needs (First Activity)
ATS—Alternative Teaching Strategies Man. Link—Manipulative Link
Workbooks: P—Practice R—Reteaching E—Enrichment

Individual Record Form

MATHEMATICS PLUS

Grade 7 • Chapter 9

	Date	Score
Pretest		
Posttest		

Student Name: ______________________________

TESTED OBJECTIVES		CHAPTER TESTS						REVIEW OPTIONS					
		Test Items									Workbooks		
Obj. No.	Tested Objective	Concept	Skill	Problem Solving	Criterion Scores	Pre-test Score	Post-test Score	Lesson Page Numbers	Pupil's Edition	Teacher's Edition	P	R	E
9–A	To define or identify integers; to determine the absolute value of an integer	1–4			3/4			286–287	Br, pp. H20–H21	ATS, p. C17	9.1	9.1	9.1
9–B	To add, subtract, multiply, and divide integers	5–8	9–12	13–16	8/12			288–289	MP, p. H66	Man. Link, p. 286A	9.2	9.2	9.2
								290–291	MP, p. H66	ATS, p. C17	9.3	9.3	9.3
								292–293	MP, p. H66	ATS, p. C18	9.4	9.4	9.4
								294–295	MP, p. H67	MIN, p. 294	9.5	9.5	9.5
								296–297	MP, p. H67	MIN, p. 296	9.6	9.6	9.6
								300–301	MP, p. H68	MIN, p. 300	9.8	9.8	9.8
								304–305	MP, p. H68	MIN, p. 304	9.9	9.9	9.9
9–C	To write and solve algebraic expressions and equations using integers	17–18	19–22	23–24	5/8			306–307	MP, p. H68	Man. Link, p. 286B	9.10	9.10	9.10
								308–309	MP, p. H68	PS Link, p. 298A	9.11	9.11	9.11
9–D	To graph ordered pairs on a coordinate plane		25–28		3/4			310–311	MP, p. H69	ATS, p. C18	9.12	9.12	9.12
								312–313	MP, p. H69	MIN, p. 312	9.13	9.13	9.13
9–E	To solve problems by working backward and by writing an equation			29–32	3/4			298–299	MP, p. H67	MIN, p. 298	9.7	9.7	9.7
								314–315	MP, p. H69	PS Link, p. 298B	9.14	9.14	9.14

KEY: MP—More Practice Br—Bridge Lesson PS Link—Problem-Solving Link MIN—Meeting Individual Needs (First Activity)
ATS—Alternative Teaching Strategies Man. Link—Manipulative Link
Workbooks: P—Practice R—Reteaching E—Enrichment

Individual Record Form

MATHEMATICS PLUS

Grade 7 • Chapter 10

Student Name: ______________________

	Date	Score
Pretest		
Posttest		

TESTED OBJECTIVES		CHAPTER TESTS						REVIEW OPTIONS					
Obj. No.	Tested Objective	Test Items			Criterion Scores	Pre-test Score	Post-test Score	Lesson Page Numbers	Pupil's Edition	Teacher's Edition	Workbooks		
		Concept	Skill	Problem Solving							P	R	E
10–A	To identify rational, irrational, and real numbers; to identify terminating and repeating decimals; to change decimals to percents, and fractions to decimals	1–4	5–10		7/10			324–325	MP, p. H70	MIN, p. 324	10.1	10.1	10.1
								326–327	MP, p. H70	MIN, p. 326	10.2	10.2	10.2
								328–329	MP, p. H70	ATS, p. C19	10.3	10.3	10.3
								332–333	MP, p. H71	MIN, p. 332	10.5	10.5	10.5
								348–349	MP, p. H73	MIN, p. 348	10.11	10.11	10.11
10–B	To compare and order rational numbers; to apply the density property	11–12	13–16		4/6			334–335	MP, p. H71	ATS, p. C19	10.6	10.6	10.6
								338–339	MP, p. H72	PS Link, p. 330A	10.7	10.7	10.7
10–C	To write numbers, using integers as exponents and using scientific notation	17–18	19–22		4/6			340–341	MP, p. H72	ATS, p. C20	10.8	10.8	10.8
								342–343	MP, p. H72	ATS, p. C20	10.9	10.9	10.9
10–D	To identify perfect squares; to find the square roots of positive rational numbers	23–24	25–28		4/6			344–347	MP, p. H73	PS Link, p. 330B	10.10	10.10	10.10
10–E	To solve problems by using a table			29–32	3/4			350–351	MP, p. H73	MIN, p. 350	10.12	10.12	10.12

KEY: MP—More Practice Br—Bridge Lesson PS Link—Problem-Solving Link MIN—Meeting Individual Needs (First Activity)
ATS—Alternative Teaching Strategies Man. Link—Manipulative Link
Workbooks: P—Practice R—Reteaching E—Enrichment

Individual Record Form

Grade 7 • Chapter 11

	Date	Score
Pretest		
Posttest		
Cumulative Test		

Student Name: ______________________

TESTED OBJECTIVES		CHAPTER TESTS						REVIEW OPTIONS					
Obj. No.	Tested Objective	Test Items: Concept	Test Items: Skill	Test Items: Problem Solving	Criterion Scores	Pre-test Score	Post-test Score	Lesson Page Numbers	Pupil's Edition	Teacher's Edition	Workbooks: P	Workbooks: R	Workbooks: E
11–A	To identify a sample space and use a tree diagram to find outcomes	1–2	3–4		3/4			360–361	MP, p. H74	ATS, p. C21	11.1	11.1	11.1
11–B	To identify the number of combinations and permutations	5–6	7–8	9–10	4/6			362–363	MP, p. H74	MIN, p. 362	11.2	11.2	11.2
								364–365	MP, p. H74	ATS, p. C21	11.3	11.3	11.3
11–C	To identify the mathematical probability that an event will occur; to use mathematical probability to predict expected outcomes	11–14	15–18	19–20	7/10			366–367	MP, p. H75	MIN, p. 366	11.4	11.4	11.4
								368–369	MP, p. H75	Man. Link, p. 360A	11.5	11.5	11.5
								374–375	MP, p. H76	MIN, p. 374	11.7	11.7	11.7
								380–381	MP, p. H77	MIN, p. 380	11.10	11.10	11.10
11–D	To find the probability of independent and dependent events	21–22	23–26	27–28	5/8			376–377	MP, p. H76	ATS, p. C22	11.8	11.8	11.8
								378–379	MP, pp. H76–H77	ATS, p. C22	11.9	11.9	11.9
11–E	To solve problems by making an organized list			29–32	3/4			370–371	MP, p. H75	MIN, p. 370	11.6	11.6	11.6

KEY: MP—More Practice Br—Bridge Lesson PS Link—Problem-Solving Link MIN—Meeting Individual Needs (First Activity)
ATS—Alternative Teaching Strategies Man. Link—Manipulative Link
Workbooks: P—Practice R—Reteaching E—Enrichment

Individual Record Form

MATHEMATICS PLUS

Grade 7 • Chapter 12

Student Name: ______________________

	Date	Score
Pretest		
Posttest		

TESTED OBJECTIVES		CHAPTER TESTS						REVIEW OPTIONS					
		Test Items									Workbooks		
Obj. No.	Tested Objective	Concept	Skill	Problem Solving	Criterion Scores	Pre-test Score	Post-test Score	Lesson Page Numbers	Pupil's Edition	Teacher's Edition	P	R	E
12–A	To choose an appropriate metric or customary unit of measure; to choose between an estimate or an exact measurement for a given situation	1–2	3–6		4/6			392–393	Br, pp. H22–H23	MIN, p. 392	12.1	12.1	12.1
								394–395	Br, pp. H24–H25	MIN, p. 394	12.2	12.2	12.2
12–B	To determine the more precise unit of measure		7–10		3/4			396–397	MP, p. H78	MIN, p. 396	12.3	12.3	12.3
12–C	To find the perimeter of a polygon; to find the circumference of a circle		11–12	13–14	3/4			398–399	MP, p. H79	Man. Link, p. 392A	12.4	12.4	12.4
								400–401	MP, p. H79	ATS, p. C23	12.5	12.5	12.5
12–D	To find the area of parallelograms, triangles, circles, and trapezoids	15–16	17–20	21–22	5/8			406–407	MP, p. H80	ATS, p. C23	12.7	12.7	12.7
								408–409	MP, p. H80	ATS, p. C24	12.8	12.8	12.8
								410–411	MP, p. H80	ATS, p. C24	12.9	12.9	12.9
12–E	To identify figures with line symmetry and turn symmetry; to identify translations, reflections, and rotation images	23–24	25–28		4/6			414–415	MP, p. H81	MIN, p. 414	12.11	12.11	12.11
								416–419	MP, p. H81	MIN, p. 418	12.12	12.12	12.12
								420–421	MP, p. H81	MIN, p. 420	12.13	12.13	12.13
12–F	To solve problems by using a formula and by using a model			29–32	3/4			402–403	MP, p. H79	MIN, p. 402	12.6	12.6	12.6
								412–413	MP, p. H80	MIN, p. 412	12.10	12.10	12.10

KEY: MP—More Practice Br—Bridge Lesson PS Link—Problem-Solving Link MIN—Meeting Individual Needs (First Activity)
ATS—Alternative Teaching Strategies Man. Link—Manipulative Link
Workbooks: P—Practice R—Reteaching E—Enrichment

Individual Record Form

MATHEMATICS PLUS

Grade 7 • Chapter 13

	Date	Score
Pretest		
Posttest		

Student Name: ______________________

TESTED OBJECTIVES		CHAPTER TESTS						REVIEW OPTIONS					
		Test Items									Workbooks		
Obj. No.	Tested Objective	Concept	Skill	Problem Solving	Criterion Scores	Pre-test Score	Post-test Score	Lesson Page Numbers	Pupil's Edition	Teacher's Edition	P	R	E
13–A	To identify solid figures	1–4	5–8		5/8			430–433	MP, p. H82	MIN, p. 432	13.1	13.1	13.1
								434–435	MP, p. H82	MIN, p. 434	13.2	13.2	13.2
13–B	To find the surface area of a solid figure		9–12		3/4			436–437	MP, p. H82	ATS, p. C25	13.3	13.3	13.3
								438–439	MP, p. H83	MIN, p. 438	13.4	13.4	13.4
13–C	To find the volume, capacity, and mass of a solid figure		13–16	17–20	5/8			444–445	MP, p. H84	ATS, p. C26	13.6	13.6	13.6
								446–447	MP, p. H84	PS Link, p. 440A	13.7	13.7	13.7
								448–449	MP, p. H85	ATS, p. C26	13.8	13.8	13.8
13–D	To solve problems by using estimation and by using a model			21–24	3/4			440–441	MP, p. H83	ATS, p. C25	13.5	13.5	13.5
								450–451	MP, p. H85	MIN, p. 450	13.9	13.9	13.9

KEY: MP—More Practice Br—Bridge Lesson PS Link—Problem-Solving Link MIN—Meeting Individual Needs (First Activity)
ATS—Alternative Teaching Strategies Man. Link—Manipulative Link
Workbooks: P—Practice R—Reteaching E—Enrichment

Individual Record Form

MATHEMATICS PLUS

Grade 7 • Chapter 14

Student Name: ______________________

	Date	Score
Pretest		
Posttest		
Cumulative Test		
End-of-Book Test		

TESTED OBJECTIVES		CHAPTER TESTS						REVIEW OPTIONS					
		Test Items									Workbooks		
Obj. No.	Tested Objective	Concept	Skill	Problem Solving	Criterion Scores	Pre-test Score	Post-test Score	Lesson Page Numbers	Pupil's Edition	Teacher's Edition	P	R	E
14–A	To identify patterns to solve problems	1–2	3–6		4/6			460–461	MP, p. H86	MIN, p. 460	14.1	14.1	14.1
								464–465	MP, p. H86	Man. Link, p. 460A	14.3	14.3	14.3
14–B	To identify relations; to use relations to solve problems	7–10	11–14		5/8			468–471	MP, p. H87	MIN, p. 470	14.5	14.5	14.5
								472–473	MP, p. H87	MIN, p. 472	14.6	14.6	14.6
14–C	To identify and use the Pythagorean Property to solve problems	15–16	17–20		4/6			480–481	MP, p. H88	ATS, p. C27	14.7	14.7	14.7
								482–483	MP, p. H88	Man. Link, p. 460B	14.8	14.8	14.8
14–D	To identify whether or not a relation is a function	21–24	25–28		5/8			476–477	MP, p. H88	ATS, p. C28	14.9	14.9	14.9
								478–479	MP, p. H89	PS Link, p. 466A	14.10	14.10	14.10
14–E	To solve problems by solving a simpler problem and to choose a strategy to solve a problem			29–32	3/4			466–467	MP, p. H86	ATS, p. C27	14.4	14.4	14.4
								484–485	MP, p. H89	ATS, p. C28	14.11	14.11	14.11

KEY: MP—More Practice Br—Bridge Lesson PS Link—Problem-Solving Link MIN—Meeting Individual Needs (First Activity)
ATS—Alternative Teaching Strategies Man. Link—Manipulative Link
Workbooks: P—Practice R—Reteaching E—Enrichment

Formal Assessment

MATHEMATICS PLUS
Grade 7

Class Record Form

School / Teacher	Inventory	CHAPTER 1		CHAPTER 2		CHAPTER 3		CHAPTER 4		Cumulative	CHAPTER 5		CHAPTER 6		CHAPTER 7	
		Pretest	Posttest	Pretest	Posttest	Pretest	Posttest	Pretest	Posttest		Pretest	Posttest	Pretest	Posttest	Pretest	Posttest
Criterion Score	35/50	24/32	24/32	18/24	18/24	24/32	24/32	24/32	24/32	35/50	24/32	24/32	18/24	18/24	24/32	24/32
NAMES Date																

continued

Formal Assessment

MATHEMATICS PLUS
Grade 7

Class Record Form (continued)

School	CHAPTER 8		Cumulative	CHAPTER 9		CHAPTER 10		CHAPTER 11		Cumulative	CHAPTER 12		CHAPTER 13		CHAPTER 14		Cumulative	End-of-Book
Teacher	Pretest	Posttest		Pretest	Posttest	Pretest	Posttest	Pretest	Posttest		Pretest	Posttest	Pretest	Posttest	Pretest	Posttest		
Criterion Score	24/32	24/32	35/50	24/32	24/32	24/32	24/32	24/32	24/32	35/50	24/32	24/32	18/24	18/24	24/32	24/32	35/50	35/50
NAMES Date																		

Cumulative Record Form

MATHEMATICS PLUS
Grade 1

Student ______________________ Teacher ______________________

INVENTORY TEST

Criteria 35/50 Test Score ________

CHAPTER 1

Test Scores:	Criteria 15/20	Pretest	Posttest	Needs More Work	Accom-plished
☐☐ 1–A	To sort by two attributes and to extend a simple pattern				
☐☐ 1–B	To identify groups of 0 to 10 objects				
☐☐ 1–C	To compare and order numbers through 10				
☐☐ 1–D	To solve problems by drawing a picture and reading a pictograph				

CHAPTER 2

Test Scores:	Criteria 18/24	Pretest	Posttest	Needs More Work	Accom-plished
☐☐ 2–A	To identify an addition sentence represented by a model				
☐☐ 2–B	To add basic facts with sums to 6 in horizontal and vertical format				
☐☐ 2–C	To identify combinations of addends with sums to 6				
☐☐ 2–D	To solve problems by using a picture and acting out problems				

CHAPTER 3

Test Scores:	Criteria 18/24	Pretest	Posttest	Needs More Work	Accom-plished
☐☐ 3–A	To identify a subtraction sentence represented by a model				
☐☐ 3–B	To subtract basic facts to 6 in horizontal and vertical format				
☐☐ 3–C	To identify families of facts				
☐☐ 3–D	To solve problems by choosing the operation or choosing an appropriate question				

CUMULATIVE TEST CHAPTERS 1–3

Criteria 35/50 Test Score ________

CHAPTER 4

Test Scores:	Criteria 18/24	Pretest	Posttest	Needs More Work	Accom-plished
☐☐ 4–A	To add basic facts to 10				
☐☐ 4–B	To add three addends with sums to 10				
☐☐ 4–C	To solve problems by using pictures and identifying irrelevant information				

CHAPTER 5

Test Scores:	Criteria 18/24	Pretest	Posttest	Needs More Work	Accom-plished
☐☐ 5–A	To subtract basic facts through 10				
☐☐ 5–B	To use inverse operations and identify families of facts to recall sums and differences through 10				
☐☐ 5–C	To solve problems by guessing and checking and using models				

continued

Cumulative Record Form

MATHEMATICS PLUS
Grade 1

Student ______________________ Teacher ______________________

CHAPTER 6

Test Scores:	Criteria 18/24	Pretest	Posttest	Needs More Work	Accom-plished
☐☐ 6–A	To identify plane and solid figures				
☐☐ 6–B	To identify whether an object is inside, outside, or on a plane figure; to identify open and closed figures				
☐☐ 6–C	To count the number of sides and corners of plane figures				
☐☐ 6–D	To identify congruent figures and to identify lines of symmetry				
☐☐ 6–E	To identify and extend patterns				
☐☐ 6–F	To solve problems by using graphs and identifying a pattern				

CUMULATIVE TEST CHAPTERS 1–6

Criteria 35/50 Test Score ________

CHAPTER 7

Test Scores:	Criteria 18/24	Pretest	Posttest	Needs More Work	Accom-plished
☐☐ 7–A	To identify numbers to 99				
☐☐ 7–B	To compare and order numbers to 100				
☐☐ 7–C	To identify ordinal numbers first through tenth				
☐☐ 7–D	To skip count by twos, fives, and tens				
☐☐ 7–E	To solve problems by using estimation and by using patterns				

CHAPTER 8

Test Scores:	Criteria 24/32	Pretest	Posttest	Needs More Work	Accom-plished
☐☐ 8–A	To recall basic facts with sums to 12				
☐☐ 8–B	To recall basic facts with differences from 12				
☐☐ 8–C	To add and subtract pennies; to count pennies to find the total amount and count change				
☐☐ 8–D	To use inverse operations and to identify families of facts				
☐☐ 8–E	To solve problems by using bar graphs and choosing the operation				

CHAPTER 9

Test Scores:	Criteria 24/32	Pretest	Posttest	Needs More Work	Accom-plished
☐☐ 9–A	To estimate and measure length using nonstandard units, inches, and centimeters				
☐☐ 9–B	To estimate weight, capacity, and temperature				
☐☐ 9–C	To identify equal parts, halves, thirds, and fourths of a region or group of objects				
☐☐ 9–D	To solve problems by using tables and visualizing results				

CUMULATIVE TEST CHAPTERS 1–9

Criteria 35/50 Test Score ________

continued

Cumulative Record Form

MATHEMATICS PLUS
Grade 1

Student ______________________ Teacher ______________________

CHAPTER 10

Test Scores:	Criteria 24/32	Pretest	Posttest	Needs More Work	Accom-plished
☐☐ 10–A	To sequence events; to estimate which event takes more or less time				
☐☐ 10–B	To tell time to the hour and half hour				
☐☐ 10–C	To read a calendar				
☐☐ 10–D	To find the value of a group of coins: pennies, nickels, dimes, and quarters				
☐☐ 10–E	To determine equivalent groups of coins				
☐☐ 10–F	To solve problems by using a picture and by using a model				

CHAPTER 11

Test Scores:	Criteria 24/32	Pretest	Posttest	Needs More Work	Accom-plished
☐☐ 11–A	To add and subtract basic facts with sums to 12				
☐☐ 11–B	To identify a two-digit number represented by groups of tens and ones				
☐☐ 11–C	To add two-digit numbers without regrouping				
☐☐ 11–D	To subtract two-digit numbers without regrouping				
☐☐ 11–E	To solve problems by using estimation and by choosing a sensible answer				

CHAPTER 12

Test Scores:	Criteria 24/32	Pretest	Posttest	Needs More Work	Accom-plished
☐☐ 12–A	To add basic facts with sums to 18				
☐☐ 12–B	To subtract basic facts with sums to 18				
☐☐ 12–C	To identify fact families with sums to 18				
☐☐ 12–D	To solve problems by using a model and by choosing the strategy				

CUMULATIVE TEST CHAPTERS 1–12

Criteria 35/50 Test Score ________

END-OF-BOOK TEST

Criteria 35/50 Test Score ________

Cumulative Record Form

MATHEMATICS PLUS
Grade 2

Student ______________________ Teacher ______________________

INVENTORY TEST

Criteria 35/50 Test Score ________

CHAPTER 1

Test Scores:	Criteria 18/24	Pretest	Posttest	Needs More Work	Accom-plished
☐☐ 1–A	To add basic facts with sums to 10				
☐☐ 1–B	To subtract basic facts with minuends to 10				
☐☐ 1–C	To use inverse operations and to identify families of facts to 10				
☐☐ 1–D	To solve problems by using a model or writing a number sentence				

CHAPTER 2

Test Scores:	Criteria 18/24	Pretest	Posttest	Needs More Work	Accom-plished
☐☐ 2–A	To add basic facts with sums to 18				
☐☐ 2–B	To subtract basic facts with minuends to 18				
☐☐ 2–C	To use inverse operations and to identify families of facts to 18				
☐☐ 2–D	To solve problems by choosing the operation and identifying irrelevant information				

CHAPTER 3

Test Scores:	Criteria 24/32	Pretest	Posttest	Needs More Work	Accom-plished
☐☐ 3–A	To identify groups of ten; to identify tens and ones to 99				
☐☐ 3–B	To compare and order numbers to 100				
☐☐ 3–C	To identify ordinal numbers first through twentieth				
☐☐ 3–D	To count by twos, threes, fives, and tens				
☐☐ 3–E	To identify odd and even numbers				
☐☐ 3–F	To solve problems by finding a pattern and using a table				

CUMULATIVE TEST CHAPTERS 1–3

Criteria 35/50 Test Score ________

CHAPTER 4

Test Scores:	Criteria 18/24	Pretest	Posttest	Needs More Work	Accom-plished
☐☐ 4–A	To count by tens, fives, and ones				
☐☐ 4–B	To identify amounts of money made up of pennies, nickels, dimes, quarters, and half-dollars				
☐☐ 4–C	To identify the same amount of money using different combinations of coins; to determine the change left when purchasing an item				
☐☐ 4–D	To solve problems by using a model and making a decision				

continued

Cumulative Record Form

MATHEMATICS PLUS
Grade 2

Student ______________________ Teacher ______________________

CHAPTER 5

Test Scores:	Criteria 18/24	Pretest	Posttest	Needs More Work	Accom-plished
☐☐ 5–A	To tell time to the hour, half-hour, and minute				
☐☐ 5–B	To estimate the amount of time needed to complete an activity				
☐☐ 5–C	To read a calendar				
☐☐ 5–D	To solve problems by using data and by using a picture				

CHAPTER 6

Test Scores:	Criteria 18/24	Pretest	Posttest	Needs More Work	Accom-plished
☐☐ 6–A	To add two-digit numbers with and without regrouping				
☐☐ 6–B	To estimate sums to 99				
☐☐ 6–C	To add money amounts to 99¢				
☐☐ 6–D	To solve problems by using data and to identify whether a problem has enough information to solve				

CUMULATIVE TEST CHAPTERS 1–6

Criteria 35/50 Test Score ________

CHAPTER 7

Test Scores:	Criteria 18/24	Pretest	Posttest	Needs More Work	Accom-plished
☐☐ 7–A	To subtract two-digit numbers with and without regrouping				
☐☐ 7–B	To use addition to check subtraction				
☐☐ 7–C	To add and subtract money amounts to 99¢; to add and subtract two-digit numbers				
☐☐ 7–D	To solve problems by making and using graphs, by choosing the operation, and by using data				

CHAPTER 8

Test Scores:	Criteria 24/32	Pretest	Posttest	Needs More Work	Accom-plished
☐☐ 8–A	To identify plane and solid figures				
☐☐ 8–B	To identify and extend patterns				
☐☐ 8–C	To identify congruent and symmetric figures				
☐☐ 8–D	To identify equal parts; to identify fractions that represent part of a whole and part of a group; to compare fractions				
☐☐ 8–E	To solve problems by using a table and by using a picture				

CHAPTER 9

Test Scores:	Criteria 24/32	Pretest	Posttest	Needs More Work	Accom-plished
☐☐ 9–A	To estimate and measure length to the nearest inch, foot, and centimeter; to measure perimeter in centimeters				
☐☐ 9–B	To measure weight with nonstandard units, pounds, and kilograms				
☐☐ 9–C	To measure capacity with quarts, pints, and cups; to estimate and measure the volume of a cube				
☐☐ 9–D	To use a thermometer to measure temperature				
☐☐ 9–E	To solve problems by choosing a reasonable answer and by guessing and checking the area of a rectangle				

CUMULATIVE TEST CHAPTERS 1–9

Criteria 35/50 Test Score ________

continued

Cumulative Record Form

MATHEMATICS PLUS
Grade 2

Student ______________________ Teacher ______________________

CHAPTER 10

Test Scores:	Criteria 24/32	Pretest	Posttest	Needs More Work	Accom-plished
☐☐ 10–A	To identify numbers to 1,000; to identify place value				
☐☐ 10–B	To compare and order numbers to 1,000				
☐☐ 10–C	To count by ones, fives, tens, and hundreds				
☐☐ 10–D	To identify combinations of coins that equal $1.00; to identify money amounts expressed as dollars and cents				
☐☐ 10–E	To solve problems by using a bar graph and by using a table to choose an appropriate question				

CHAPTER 11

Test Scores:	Criteria 18/24	Pretest	Posttest	Needs More Work	Accom-plished
☐☐ 11–A	To add three-digit numbers with and without regrouping				
☐☐ 11–B	To subtract three-digit numbers with and without regrouping				
☐☐ 11–C	To estimate sums and differences by rounding				
☐☐ 11–D	To solve problems by using a graph and by identifying reasonable results				

CHAPTER 12

Test Scores:	Criteria 24/32	Pretest	Posttest	Needs More Work	Accom-plished
☐☐ 12–A	To find products using objects, pictures, and repeated addition				
☐☐ 12–B	To multiply basic facts with factors of 0–5				
☐☐ 12–C	To find quotients using objects, pictures, and repeated subtraction				
☐☐ 12–D	To solve problems by using a graph and by using a drawing				

CUMULATIVE TEST CHAPTERS 1–12

Criteria 35/50 ______ Test Score ______

END-OF-BOOK TEST

Criteria 35/50 ______ Test Score ______

Cumulative Record Form

MATHEMATICS PLUS
Grade 3

Student ______________________ Teacher ______________________

INVENTORY TEST

Criteria 35/50 Test Score ________

CHAPTER 1

Test Scores:	Criteria 18/24	Pretest	Posttest	Needs More Work	Accom-plished
☐☐ 1–A	To add and subtract basic facts with sums to 18 and differences from 18				
☐☐ 1–B	To use inverse operations to identify fact families and to find missing addends				
☐☐ 1–C	To choose the correct operation to solve the problem				
☐☐ 1–D	To solve a problem by acting it out or guessing and checking				

CHAPTER 2

Test Scores:	Criteria 24/32	Pretest	Posttest	Needs More Work	Accom-plished
☐☐ 2–A	To identify odd and even numbers				
☐☐ 2–B	To read and identify numbers through hundred thousands; to identify place value				
☐☐ 2–C	To compare and order numbers through hundreds				
☐☐ 2–D	To round numbers to the nearest ten or nearest hundred				
☐☐ 2–E	To identify ordinal numbers first through fiftieth				
☐☐ 2–F	To solve problems by using a table				

CHAPTER 3

Test Scores:	Criteria 18/24	Pretest	Posttest	Needs More Work	Accom-plished
☐☐ 3–A	To estimate sums and differences by rounding				
☐☐ 3–B	To add and subtract one- and two-digit numbers with and without regrouping				
☐☐ 3–C	To find the sum of more than two addends				
☐☐ 3–D	To solve problems with too much or too little information and to solve problems by organizing data				

CUMULATIVE TEST CHAPTERS 1–3

Criteria 35/50 Test Score ________

CHAPTER 4

Test Scores:	Criteria 18/24	Pretest	Posttest	Needs More Work	Accom-plished
☐☐ 4–A	To estimate sums and differences by rounding				
☐☐ 4–B	To add three- and four-digit numbers and money amounts with regrouping				
☐☐ 4–C	To subtract three- and four-digit numbers and money amounts with regrouping				
☐☐ 4–D	To solve problems by choosing the operation and by choosing a number sentence				

continued

Cumulative Record Form

MATHEMATICS PLUS
Grade 3

Student ______________________ Teacher ______________________

CHAPTER 5

Test Scores:	Criteria 24/32	Pretest	Posttest	Needs More Work	Accom-plished
☐☐ 5–A	To use a calendar				
☐☐ 5–B	To tell time to the hour, half hour, quarter hour, and minute				
☐☐ 5–C	To use a schedule				
☐☐ 5–D	To count money amounts to $9.99; to compare amounts of money; to solve problems using money				
☐☐ 5–E	To solve problems by working backward and by acting them out				

CHAPTER 6

Test Scores:	Criteria 15/20	Pretest	Posttest	Needs More Work	Accom-plished
☐☐ 6–A	To relate multiplication to skip counting, repeated addition, and a number line				
☐☐ 6–B	To multiply basic facts when one of the factors is 0, 1, 2, 3, 4, or 5				
☐☐ 6–C	To solve problems by finding a pattern or by using a model				

CUMULATIVE TEST CHAPTERS 1–6

Criteria 35/50 Test Score ________

CHAPTER 7

Test Scores:	Criteria 18/24	Pretest	Posttest	Needs More Work	Accom-plished
☐☐ 7–A	To multiply basic facts when one of the factors is 6, 7, 8, or 9				
☐☐ 7–B	To multiply 3 factors				
☐☐ 7–C	To solve problems with too much or too little information and to solve problems by making choices				

CHAPTER 8

Test Scores:	Criteria 18/24	Pretest	Posttest	Needs More Work	Accom-plished
☐☐ 8–A	To demonstrate an understanding of division using models and repeated subtraction				
☐☐ 8–B	To use inverse operations to identify fact families and to find factors				
☐☐ 8–C	To divide basic facts with 0, 1, 2, 3, 4, and 5 as factors				
☐☐ 8–D	To solve multistep problems and to solve problems by writing a number sentence				

CHAPTER 9

Test Scores:	Criteria 18/24	Pretest	Posttest	Needs More Work	Accom-plished
☐☐ 9–A	To divide basic facts				
☐☐ 9–B	To divide a two-digit number by a one-digit number to get a quotient with a remainder				
☐☐ 9–C	To solve problems by choosing the operation and by choosing the method of computation				

CUMULATIVE TEST CHAPTERS 1–9

Criteria 35/50 Test Score ________

continued

Cumulative Record Form

MATHEMATICS PLUS
Grade 3

Student ______________________ Teacher ______________________

CHAPTER 10

Test Scores:	Criteria 24/32	Pretest	Posttest	Needs More Work	Accom-plished
☐☐ 10–A	To identify solid and plane figures and compare their attributes				
☐☐ 10–B	To identify lines, line segments, angles, and right angles				
☐☐ 10–C	To identify congruent figures and symmetrical figures				
☐☐ 10–D	To find the perimeter, area, and volume of figures using nonstandard units				
☐☐ 10–E	To use ordered pairs to locate points on a grid				
☐☐ 10–F	To solve problems by finding a pattern and by using a pictograph				

CHAPTER 11

Test Scores:	Criteria 24/32	Pretest	Posttest	Needs More Work	Accom-plished
☐☐ 11–A	To estimate and measure length using customary and metric units				
☐☐ 11–B	To estimate and measure capacity using customary and metric units				
☐☐ 11–C	To estimate and measure weight using customary and metric units				
☐☐ 11–D	To estimate and measure temperature using custom-ary and metric units				
☐☐ 11–E	To solve problems by using a picture and by using a bar graph				

CHAPTER 12

Test Scores:	Criteria 24/32	Pretest	Posttest	Needs More Work	Accom-plished
☐☐ 12–A	To identify and find part of a whole and part of a group				
☐☐ 12–B	To identify equivalent fractions				
☐☐ 12–C	To compare fractions				
☐☐ 12–D	To identify mixed numbers				
☐☐ 12–E	To read and write decimals to hundredths				
☐☐ 12–F	To add and subtract decimals				
☐☐ 12–G	To solve multistep problems; to solve problems by using a picture				

CHAPTER 13

Test Scores:	Criteria 24/32	Pretest	Posttest	Needs More Work	Accom-plished
☐☐ 13–A	To multiply and divide by tens and hundreds				
☐☐ 13–B	To multiply two- and three-digit numbers				
☐☐ 13–C	To estimate products				
☐☐ 13–D	To divide with one- and two-digit numbers to find quotients with and without remainders				
☐☐ 13–E	To solve problems by using estimation and by choosing a strategy				

CUMULATIVE TEST CHAPTERS 1–13

Criteria 35/50 Test Score ________

END-OF-BOOK TEST

Criteria 35/50 Test Score ________

Cumulative Record Form

MATHEMATICS PLUS
Grade 4

Student ______________________ Teacher ______________________

INVENTORY TEST

Criteria 35/50 Test Score ______

CHAPTER 1

Test Scores:	Criteria 18/24	Pretest	Posttest	Needs More Work	Accom-plished
☐☐ 1–A	To read and identify numbers in expanded and standard form through hundred millions; to identify place value				
☐☐ 1–B	To compare and order numbers through millions				
☐☐ 1–C	To estimate numbers by rounding to the nearest 10, 100, or 1,000				
☐☐ 1–D	To identify ordinal numbers through hundredth				
☐☐ 1–E	To solve problems by using a table and finding a pattern				

CHAPTER 2

Test Scores:	Criteria 18/24	Pretest	Posttest	Needs More Work	Accom-plished
☐☐ 2–A	To add or subtract basic facts				
☐☐ 2–B	To estimate sums or differences for reasonableness, using front-end digits and rounding				
☐☐ 2–C	To add or subtract whole numbers and money amounts with and without regrouping				
☐☐ 2–D	To solve problems by using a picture or using a table to analyze data				

CHAPTER 3

Test Scores:	Criteria 24/32	Pretest	Posttest	Needs More Work	Accom-plished
☐☐ 3–A	To multiply basic facts				
☐☐ 3–B	To divide basic facts				
☐☐ 3–C	To use inverse operations and to identify fact families				
☐☐ 3–D	To solve problems by writing a number sentence and identifying needed or extraneous information				

CUMULATIVE TEST CHAPTERS 1–3

Criteria 35/50 Test Score ______

CHAPTER 4

Test Scores:	Criteria 24/32	Pretest	Posttest	Needs More Work	Accom-plished
☐☐ 4–A	To tell time				
☐☐ 4–B	To compute elapsed time using an analog clock, a digital clock, and a calendar				
☐☐ 4–C	To use and interpret tally tables, frequency tables, pictographs, bar graphs, and line graphs				
☐☐ 4–D	To use ordered pairs to find points on a coordinate grid				
☐☐ 4–E	To solve problems by using a table or schedule and analyzing data to make decisions				

continued

Cumulative Record Form

MATHEMATICS PLUS
Grade 4

Student ____________________ Teacher ____________________

CHAPTER 5

Test Scores:	Criteria 18/24	Pretest	Posttest	Needs More Work	Accomplished
☐☐ 5–A	To multiply by multiples of 0, 100, and 1,000				
☐☐ 5–B	To estimate products				
☐☐ 5–C	To multiply two-, three-, and four-digit numbers by one-digit numbers				
☐☐ 5–D	To solve problems by working backward and choosing a method of computation				

CHAPTER 6

Test Scores:	Criteria 24/32	Pretest	Posttest	Needs More Work	Accomplished
☐☐ 6–A	To use mental math to multiply by multiples of 10				
☐☐ 6–B	To estimate products				
☐☐ 6–C	To multiply numbers through thousands, or money amounts by a two-digit number				
☐☐ 6–D	To solve problems by using a graph and by guessing and checking				

CUMULATIVE TEST CHAPTERS 1–6

Criteria 35/50 Test Score ________

CHAPTER 7

Test Scores:	Criteria 18/24	Pretest	Posttest	Needs More Work	Accomplished
☐☐ 7–A	To choose and use appropriate metric units of length, capacity, and mass				
☐☐ 7–B	To choose and use appropriate customary units of length, capacity, and weight				
☐☐ 7–C	To find the perimeter				
☐☐ 7–D	To convert from smaller to larger customary units and vice versa				
☐☐ 7–E	To solve multistep problems and to solve problems by using a picture				

CHAPTER 8

Test Scores:	Criteria 24/32	Pretest	Posttest	Needs More Work	Accomplished
☐☐ 8–A	To estimate quotients with one-digit divisors				
☐☐ 8–B	To identify multiplication and division as inverse operations				
☐☐ 8–C	To divide one-, two-, and three-digit numbers and money amounts by a one-digit number				
☐☐ 8–D	To find the median, range, and average of a set of data				
☐☐ 8–E	To choose appropriate strategies to solve problems and to choose the method of computation				

CHAPTER 9

Test Scores:	Criteria 24/32	Pretest	Posttest	Needs More Work	Accomplished
☐☐ 9–A	To identify and distinguish between plane and solid figures and their properties				
☐☐ 9–B	To measure area in square units and multiply to find area; to measure volume in cubic units and multiply to find volume				
☐☐ 9–C	To identify lines, line segments, rays, perpendicular and parallel lines, angles, circles, and parts of a circle				
☐☐ 9–D	To identify congruent and similar figures; to identify lines of symmetry; to identify a slide, flip, and turn of a figure				
☐☐ 9–E	To solve multistep problems and to solve problems by using a model				

CUMULATIVE TEST CHAPTERS 1–9

Criteria 35/50 Test Score ________

continued

Cumulative Record Form

MATHEMATICS PLUS
Grade 4

Student ______________________ Teacher ______________________

CHAPTER 10

Test Scores:	Criteria 24/32	Pretest	Posttest	Needs More Work	Accom-plished
☐☐ 10–A	To identify a fractional part of a group; to identify the fractional part of a number				
☐☐ 10–B	To identify equivalent fractions; to simplify fractions				
☐☐ 10–C	To compare fractions with like and unlike denominators				
☐☐ 10–D	To identify fractions as mixed numbers				
☐☐ 10–E	To solve problems by acting them out and by choosing appropriate strategies				

CHAPTER 11

Test Scores:	Criteria 24/32	Pretest	Posttest	Needs More Work	Accom-plished
☐☐ 11–A	To estimate the fractional part of a number and sums and differences of fractions				
☐☐ 11–B	To add and subtract fractions with like and unlike denominators; to add and subtract mixed numbers with like denominators				
☐☐ 11–C	To identify the length of an object to a fractional part of an inch				
☐☐ 11–D	To identify possible combinations and arrangements using tree diagrams; to find the probability of an event				
☐☐ 11–E	To solve problems by using a model and by making an organized list				

CHAPTER 12

Test Scores:	Criteria 24/32	Pretest	Posttest	Needs More Work	Accom-plished
☐☐ 12–A	To read and identify decimals with tenths and hundredths; to relate fractions and decimals				
☐☐ 12–B	To identify equivalent decimals; to compare and order decimals				
☐☐ 12–C	To estimate decimals by rounding to the nearest whole number; to estimate decimal sums and differences				
☐☐ 12–D	To add and subtract decimals				
☐☐ 12–E	To solve problems by working backward and by using estimation				

CHAPTER 13

Test Scores:	Criteria 18/24	Pretest	Posttest	Needs More Work	Accom-plished
☐☐ 13–A	To estimate the quotient of a number when dividing by a two-digit divisor				
☐☐ 13–B	To divide by multiples of 10 with and without remainders				
☐☐ 13–C	To divide with two-digit divisors with and without remainders				
☐☐ 13–D	To solve problems by finding the hidden question and by interpreting the remainder				

CUMULATIVE TEST CHAPTERS 1–13

Criteria 35/50 Test Score ________

END-OF-BOOK TEST

Criteria 35/50 Test Score ________

Cumulative Record Form

MATHEMATICS PLUS
Grade 5

Student ______________________ Teacher ______________________

INVENTORY TEST
Criteria 35/50 Test Score ________

CHAPTER 1

Test Scores:	Criteria 18/24	Pretest	Posttest	Needs More Work	Accom-plished
☐☐ 1–A	To read and identify numbers in expanded and standard form from hundred millions to thousandths; to identify place value				
☐☐ 1–B	To compare and order whole numbers and decimals				
☐☐ 1–C	To estimate whole numbers and decimals by rounding				
☐☐ 1–D	To solve problems by using a table or a drawing				

CHAPTER 2

Test Scores:	Criteria 24/32	Pretest	Posttest	Needs More Work	Accom-plished
☐☐ 2–A	To use addition and subtraction as inverse operations				
☐☐ 2–B	To estimate whole-number and decimal sums and differences				
☐☐ 2–C	To add and subtract whole numbers				
☐☐ 2–D	To add and subtract decimals to thousandths				
☐☐ 2–E	To solve problems by identifying and solving hidden questions and choosing a number sentence				

CHAPTER 3

Test Scores:	Criteria 18/24	Pretest	Posttest	Needs More Work	Accom-plished
☐☐ 3–A	To estimate products				
☐☐ 3–B	To multiply with multiples of 10				
☐☐ 3–C	To multiply whole numbers				
☐☐ 3–D	To solve multistep problems and to solve problems by finding a pattern				

CUMULATIVE TEST CHAPTERS 1–3
Criteria 35/50 Test Score ________

CHAPTER 4

Test Scores:	Criteria 24/32	Pretest	Posttest	Needs More Work	Accom-plished
☐☐ 4–A	To identify and classify lines, rays, angles, polygons, triangles, quadrilaterals, and circles				
☐☐ 4–B	To identify congruent, similar, and symmetrical figures				
☐☐ 4–C	To identify slides, flips, and turns				
☐☐ 4–D	To identify solid figures and their attributes				
☐☐ 4–E	To solve problems by guessing and checking and by solving a simpler problem				

continued

Cumulative Record Form

MATHEMATICS PLUS
Grade 5

Student ______________________ Teacher ______________________

CHAPTER 5

Test Scores:	Criteria 24/32	Pretest	Posttest	Needs More Work	Accomplished
☐☐ 5–A	To determine divisibility by 2, 3, 5, and 10				
☐☐ 5–B	To estimate quotients with one-digit divisors				
☐☐ 5–C	To divide whole numbers by a one-digit divisor with and without a remainder				
☐☐ 5–D	To find the average of a set of data				
☐☐ 5–E	To solve problems by choosing the operation and working backward				

CHAPTER 6

Test Scores:	Criteria 18/24	Pretest	Posttest	Needs More Work	Accomplished
☐☐ 6–A	To estimate quotients with a whole-number divisor				
☐☐ 6–B	To divide whole numbers by two-digit divisors				
☐☐ 6–C	To solve problems by using estimation and by choosing a strategy to solve a problem				

CUMULATIVE TEST CHAPTERS 1–6

Criteria 35/50 Test Score ________

CHAPTER 7

Test Scores:	Criteria 18/24	Pretest	Posttest	Needs More Work	Accomplished
☐☐ 7–A	To find the mean, median, mode, and range of a set of data				
☐☐ 7–B	To read and interpret bar graphs, pictographs, line graphs, and circle graphs				
☐☐ 7–C	To graph ordered pairs				
☐☐ 7–D	To solve problems by using a table and using logical reasoning				

CHAPTER 8

Test Scores:	Criteria 24/32	Pretest	Posttest	Needs More Work	Accomplished
☐☐ 8–A	To estimate decimal products				
☐☐ 8–B	To multiply decimals by whole numbers and decimals by decimals				
☐☐ 8–C	To divide decimals by whole numbers				
☐☐ 8–D	To solve problems by choosing the method of computation and identifying relevant and irrelevant information				

CHAPTER 9

Test Scores:	Criteria 24/32	Pretest	Posttest	Needs More Work	Accomplished
☐☐ 9–A	To identify fractions; to identify equivalent fractions; to express fractions in simplest form				
☐☐ 9–B	To identify numbers as prime or composite; to find common factors and multiples of whole numbers				
☐☐ 9–C	To rename mixed numbers and improper fractions as mixed numbers or whole numbers				
☐☐ 9–D	To compare and order fractions				
☐☐ 9–E	To solve problems by using a picture and by acting it out				

CUMULATIVE TEST CHAPTERS 1–9

Criteria 35/50 Test Score ________

continued

Cumulative Record Form

MATHEMATICS PLUS
Grade 5

Student ______________________ Teacher ______________________

CHAPTER 10

Test Scores:	Criteria 24/32	Pretest	Posttest	Needs More Work	Accom-plished
☐☐ 10–A	To estimate the sums and differences of fractions				
☐☐ 10–B	To add and subtract like and unlike fractions and mixed numbers				
☐☐ 10–C	To multiply fractions and whole numbers; fractions and fractions; fractions and mixed numbers				
☐☐ 10–D	To divide fractions				
☐☐ 10–E	To solve problems by using tables and diagrams				

CHAPTER 11

Test Scores:	Criteria 24/32	Pretest	Posttest	Needs More Work	Accom-plished
☐☐ 11–A	To choose appropriate metric and customary units of length, capacity, and mass				
☐☐ 11–B	To compute using metric and customary measures and intervals of time				
☐☐ 11–C	To identify the length of an object to the nearest $\frac{1}{16}$ of an inch and to the nearest 1 millimeter				
☐☐ 11–D	To solve problems by evaluating answers for reasonableness and by using a schedule				

CHAPTER 12

Test Scores:	Criteria 24/32	Pretest	Posttest	Needs More Work	Accom-plished
☐☐ 12–A	To find the perimeter of a polygon and the circumference of a circle				
☐☐ 12–B	To find the area of parallelograms, triangles, and complex and curved figures				
☐☐ 12–C	To estimate and find the volume of a rectangular prism				
☐☐ 12–D	To solve problems by using a model and using a formula				

CHAPTER 13

Test Scores:	Criteria 18/24	Pretest	Posttest	Needs More Work	Accom-plished
☐☐ 13–A	To identify ratios and equivalent ratios; to use scale drawings to find actual distances				
☐☐ 13–B	To express a ratio as a percent, and a percent as a fraction and as a decimal				
☐☐ 13–C	To determine possible outcomes and to find the probability of an event				
☐☐ 13–D	To solve problems by making an organized list and by conducting a simulation				

CUMULATIVE TEST CHAPTERS 1–13

Criteria 35/50 Test Score ________

END-OF-BOOK TEST

Criteria 35/50 Test Score ________

Cumulative Record Form

MATHEMATICS PLUS
Grade 6

Student ____________________ Teacher ____________________

INVENTORY TEST
Criteria 35/50 Test Score ______

CHAPTER 1

Test Scores:	Criteria 24/32	Pretest	Posttest	Needs More Work	Accomplished
☐☐ 1–A	To read and identify numbers in standard and expanded form from hundred billions to ten-thousandths; to identify place value				
☐☐ 1–B	To compare and order whole numbers and decimals				
☐☐ 1–C	To round whole numbers through billions; to round money amounts				
☐☐ 1–D	To solve problems using the order of operations				
☐☐ 1–E	To identify powers, exponents, squares, and square roots				
☐☐ 1–F	To solve problems by finding a pattern or by using a table				

CHAPTER 2

Test Scores:	Criteria 15/20	Pretest	Posttest	Needs More Work	Accomplished
☐☐ 2–A	To estimate sums and differences of whole numbers and decimals				
☐☐ 2–B	To find sums and differences of whole numbers and decimals				
☐☐ 2–C	To identify and evaluate algebraic expressions				
☐☐ 2–D	To solve problems by guessing and checking and by choosing the method of computation				

CHAPTER 3

Test Scores:	Criteria 15/20	Pretest	Posttest	Needs More Work	Accomplished
☐☐ 3–A	To estimate whole-number and/or decimal products				
☐☐ 3–B	To multiply a whole number by a one-, two-, or three-digit number				
☐☐ 3–C	To multiply a decimal by a decimal or by a whole number				
☐☐ 3–D	To solve multistep problems and to solve problems by making decisions				

CHAPTER 4

Test Scores:	Criteria 24/32	Pretest	Posttest	Needs More Work	Accomplished
☐☐ 4–A	To determine divisibility by 2, 3, 5, 9, or 10				
☐☐ 4–B	To estimate whole-number quotients; to round decimal quotients				
☐☐ 4–C	To divide a whole number by a whole number				
☐☐ 4–D	To divide a decimal by a whole number and a decimal by a decimal				
☐☐ 4–E	To identify and evaluate algebraic expressions				
☐☐ 4–F	To solve problems by choosing an equation and by working backward				

CUMULATIVE TEST CHAPTERS 1–4
Criteria 35/50 Test Score ______

continued

Cumulative Record Form

MATHEMATICS PLUS
Grade 6

Student ______________________ Teacher ______________________

CHAPTER 5

Test Scores:	Criteria 24/32	Pretest	Posttest	Needs More Work	Accom-plished
☐☐ 5–A	To collect and organize data				
☐☐ 5–B	To analyze data using bar graphs, histograms, line graphs, circle graphs, and stem-and-leaf plots				
☐☐ 5–C	To find the range, mean, median, and mode for a collection of data				
☐☐ 5–D	To use the counting principle and tree diagrams to identify the number of possible outcomes				
☐☐ 5–E	To determine the probability of events				
☐☐ 5–F	To solve problems by finding a pattern and choosing an appropriate graph				

CHAPTER 6

Test Scores:	Criteria 24/32	Pretest	Posttest	Needs More Work	Accom-plished
☐☐ 6–A	To identify factors, prime and composite numbers, prime factors, greatest common factor, and least common multiple				
☐☐ 6–B	To find equivalent fractions; to identify least common denominator; to identify a fraction in simplest form				
☐☐ 6–C	To compare and order fractions				
☐☐ 6–D	To identify a mixed number as a fraction and vice versa				
☐☐ 6–E	To solve problems by acting them out and by choosing an appropriate strategy				

CHAPTER 7

Test Scores:	Criteria 18/24	Pretest	Posttest	Needs More Work	Accom-plished
☐☐ 7–A	To estimate sums and differences of fractions and mixed numbers				
☐☐ 7–B	To add and subtract like and unlike fractions				
☐☐ 7–C	To add and subtract mixed numbers with like and unlike denominators, with and without renaming				
☐☐ 7–D	To solve problems by using a diagram and by choosing an appropriate strategy				

CHAPTER 8

Test Scores:	Criteria 24/32	Pretest	Posttest	Needs More Work	Accom-plished
☐☐ 8–A	To multiply fractions, mixed numbers, and whole numbers; to simplify fractions before multiplying				
☐☐ 8–B	To estimate the product of fractions by fractions and fractions by whole numbers				
☐☐ 8–C	To divide fractions, whole numbers, and mixed numbers by fractions or mixed numbers				
☐☐ 8–D	To change a fraction to a decimal and vice versa				
☐☐ 8–E	To solve problems by writing an equation and solving a simpler problem				

CUMULATIVE TEST CHAPTERS 1–8

Criteria 35/50 Test Score ________

continued

Cumulative Record Form

MATHEMATICS PLUS
Grade 6

Student ______________________ Teacher ______________________

CHAPTER 9

Test Scores:	Criteria 24⁄32 ____ Pretest ____ Posttest ____	Needs More Work	Accom-plished
☐☐ 9–A	To choose and use appropriate metric units of length, capacity, and mass		
☐☐ 9–B	To determine the more precise measure		
☐☐ 9–C	To compute using customary and metric measure and intervals of time		
☐☐ 9–D	To choose and use appropriate customary units of length, capacity, and mass		
☐☐ 9–E	To solve problems by using estimation and using a schedule		

CHAPTER 10

Test Scores:	Criteria 24⁄32 ____ Pretest ____ Posttest ____	Needs More Work	Accom-plished
☐☐ 10–A	To identify a ratio comparing two numbers; to identify equivalent ratios and unit rates		
☐☐ 10–B	To write and solve proportions		
☐☐ 10–C	To find the decimal or fraction equal to a percent or vice versa		
☐☐ 10–D	To estimate or find the percent of a number		
☐☐ 10–E	To solve problems by using a scale drawing		

CHAPTER 11

Test Scores:	Criteria 24⁄32 ____ Pretest ____ Posttest ____	Needs More Work	Accom-plished
☐☐ 11–A	To identify and classify lines, angles, polygons, and solid figures		
☐☐ 11–B	To identify constructions of congruent line segments, line bisectors, and congruent angles		
☐☐ 11–C	To identify similar, congruent, and symmetrical figures		
☐☐ 11–D	To identify translations of figures		
☐☐ 11–E	To solve problems by using a circle graph and by finding a pattern		

CUMULATIVE TEST CHAPTERS 1–11

Criteria 35⁄50 Test Score ________

continued

Cumulative Record Form

MATHEMATICS PLUS
Grade 6

Student ______________________ Teacher ______________________

CHAPTER 12

Test Scores:	Criteria 24/32	Pretest	Posttest	Needs More Work	Accom-plished
☐☐ 12–A	To find perimeter of regular and irregular polygons				
☐☐ 12–B	To find the area of rectangles, parallelograms, triangles, and irregular figures; to relate perimeter and area				
☐☐ 12–C	To identify parts of a circle; to find circumference and area of a circle				
☐☐ 12–D	To find the surface area of a rectangular prism; to find the volume of a rectangular prism				
☐☐ 12–E	To solve problems by using a formula and using a model				

CHAPTER 13

Test Scores:	Criteria 24/32	Pretest	Posttest	Needs More Work	Accom-plished
☐☐ 13–A	To identify integers; to compare and order integers				
☐☐ 13–B	To add and subtract integers				
☐☐ 13–C	To identify and graph ordered pairs on a coordinate plane				
☐☐ 13–D	To choose strategies to solve problems				

CHAPTER 14

Test Scores:	Criteria 18/24	Pretest	Posttest	Needs More Work	Accom-plished
☐☐ 14–A	To identify and evaluate algebraic expressions				
☐☐ 14–B	To identify rational numbers; to solve equations involving rational numbers				
☐☐ 14–C	To solve inequalities				
☐☐ 14–D	To use relations to extend patterns; to graph relations				
☐☐ 14–E	To solve problems by writing an equation and by using logical reasoning				

CUMULATIVE TEST CHAPTERS 1–14

Criteria 35/50 Test Score ________

END-OF-BOOK TEST

Criteria 35/50 Test Score ________

Cumulative Record Form

MATHEMATICS PLUS
Grade 7

Student ______________________ Teacher ______________________

INVENTORY TEST

Criteria $^{35}/_{50}$ Test Score ________

CHAPTER 1

Test Scores:	Criteria $^{24}/_{32}$	Pretest	Posttest	Needs More Work	Accom-plished
☐☐ 1–A	To estimate sums, differences, products, and quotients of whole numbers and decimals				
☐☐ 1–B	To use addition, subtraction, multiplication, and division to solve whole-number and decimal problems				
☐☐ 1–C	To identify and evaluate numbers in standard form and exponent form				
☐☐ 1–D	To express and evaluate numbers in scientific notation				
☐☐ 1–E	To solve problems by using a picture and by guessing and checking				

CHAPTER 2

Test Scores:	Criteria $^{18}/_{24}$	Pretest	Posttest	Needs More Work	Accom-plished
☐☐ 2–A	To read and interpret bar graphs, line graphs, circle graphs, histograms, and stem-and-leaf plots				
☐☐ 2–B	To use freguency distribution tables; to use unbiased surveys				
☐☐ 2–C	To analyze sets of data using range, quartiles, mean, median, and mode; to read and interpret box-and-whisker graphs				
☐☐ 2–D	To solve problems by using tables and graphs and by selecting different ways to graph a set of data				

CHAPTER 3

Test Scores:	Criteria $^{24}/_{32}$	Pretest	Posttest	Needs More Work	Accom-plished
☐☐ 3–A	To find the prime factorization of a composite number				
☐☐ 3–B	To find the greatest common factor and the least common multiple of two or more numbers				
☐☐ 3–C	To identify equivalent fractions; fractions in simplest form; to compare and order two or more fractions and mixed numbers; to identify one or two fractions between any two given fractions				
☐☐ 3–D	To identify one or two fractions between any two given fractions				
☐☐ 3–E	To solve problems by finding a pattern and by using a picture				

CHAPTER 4

Test Scores:	Criteria $^{24}/_{32}$	Pretest	Posttest	Needs More Work	Accom-plished
☐☐ 4–A	To estimate sums, differences, products, and quotients of fractions				
☐☐ 4–B	To add and subtract fractions and mixed numbers with and without regrouping				
☐☐ 4–C	To multiply and divide fractions and mixed numbers				
☐☐ 4–D	To solve multistep problems and to solve problems by solving a simpler problem				

CUMULATIVE TEST CHAPTERS 1–4

Criteria $^{35}/_{50}$ Test Score ________

continued

Cumulative Record Form

MATHEMATICS PLUS
Grade 7

Student ______________________ Teacher ______________________

CHAPTER 5

Test Scores:	Criteria 24/32	Pretest	Posttest	Needs More Work	Accomplished
☐☐ 5–A	To identify and evaluate expressions				
☐☐ 5–B	To solve problems using the order of operations				
☐☐ 5–C	To solve a one-step inequality involving addition				
☐☐ 5–D	To solve one-step equations				
☐☐ 5–E	To solve problems by using a formula and by working backward				

CHAPTER 6

Test Scores:	Criteria 18/24	Pretest	Posttest	Needs More Work	Accomplished
☐☐ 6–A	To identify, measure, and classify lines, angles, triangles, polygons, and quadrilaterals; to identify congruent figures and geometric relationships				
☐☐ 6–B	To identify constructions of congruent angles and triangles, and line bisectors				
☐☐ 6–C	To solve problems using logical reasoning and Venn diagrams				
☐☐ 6–D	To solve problems by finding a pattern and by using a formula				

CHAPTER 7

Test Scores:	Criteria 24/32	Pretest	Posttest	Needs More Work	Accomplished
☐☐ 7–A	To identify ratios and rates; to complete equal ratios and solve proportions				
☐☐ 7–B	To use ratios and proportions to compute unit prices and to solve problems				
☐☐ 7–C	To use scale drawings to solve problems				
☐☐ 7–D	To identify corresponding parts of similar figures; to find the ratio of corresponding sides; to use proportions to find the missing measures of similar figures				
☐☐ 7–E	To solve problems by using a map and by using a formula				

CHAPTER 8

Test Scores:	Criteria 24/32	Pretest	Posttest	Needs More Work	Accomplished
☐☐ 8–A	To identify fraction, decimal, ratio, and percent equivalencies				
☐☐ 8–B	To estimate and find the percent of a number, the percent one number is of another; to find a number when a percent of it is known by using estimation, proportions, and equations				
☐☐ 8–C	To analyze a circle graph				
☐☐ 8–D	To solve sales tax, discount, and simple-interest problems				
☐☐ 8–E	To solve problems by using a graph and by guessing and checking				

CUMULATIVE TEST CHAPTERS 1–8

Criteria 35/50 Test Score ________

continued

Cumulative Record Form

MATHEMATICS PLUS
Grade 7

Student ______________________ Teacher ______________________

CHAPTER 9

Test Scores:	Criteria 24/32	Pretest	Posttest	Needs More Work	Accomplished
☐☐ 9–A	To define or identify integers; to determine the absolute value of an integer				
☐☐ 9–B	To add, subtract, multiply, and divide integers				
☐☐ 9–C	To write and solve algebraic expressions and equations using integers				
☐☐ 9–D	To graph ordered pairs on a coordinate plane				
☐☐ 9–E	To solve problems by working backward and writing an equation				

CHAPTER 10

Test Scores:	Criteria 24/32	Pretest	Posttest	Needs More Work	Accomplished
☐☐ 10–A	To identify rational, irrational, and real numbers; to identify terminating and repeating decimals; to change decimals to percents, and fractions to decimals				
☐☐ 10–B	To compare and order rational numbers; to apply the Density Property				
☐☐ 10–C	To write numbers using integers as exponents and using scientific notation				
☐☐ 10–D	To identify perfect squares and to find the square roots of positive rational numbers				
☐☐ 10–E	To solve problems by acting them out and by using a table				

CHAPTER 11

Test Scores:	Criteria 24/32	Pretest	Posttest	Needs More Work	Accomplished
☐☐ 11–A	To identify sample spaces and use tree diagrams to find outcomes				
☐☐ 11–B	To identify the number of combinations and permutations				
☐☐ 11–C	To identify the mathematical probability that an event will occur; to use mathematical probability to predict expected outcomes				
☐☐ 11–D	To find the probability of independent and dependent events				
☐☐ 11–E	To solve problems by making an organized list				

CUMULATIVE TEST CHAPTERS 1–11

Criteria 35/50 ______ Test Score ______

CHAPTER 12

Test Scores:	Criteria 24/32	Pretest	Posttest	Needs More Work	Accomplished
☐☐ 12–A	To choose an appropriate metric or customary unit of measure; to choose between an estimate or an exact measurement for a given situation				
☐☐ 12–B	To determine the more precise unit of measure				
☐☐ 12–C	To find the perimeter of a polygon; to find the circumference of a circle				
☐☐ 12–D	To find the area of parallelograms, triangles, circles, and trapezoids				
☐☐ 12–E	To identify figures with line symmetry and turn symmetry; to identify translations, reflections, and rotation images				
☐☐ 12–F	To solve problems by using a formula and using a model				

continued

Cumulative Record Form

MATHEMATICS PLUS
Grade 7

Student ______________________ Teacher ______________________

CHAPTER 13

Test Scores:	Criteria 18/24	Pretest	Posttest	Needs More Work	Accom-plished
☐☐ 13–A	To identify solid figures				
☐☐ 13–B	To find the surface area of a solid figure				
☐☐ 13–C	To find the volume, capacity, and mass of a solid figure				
☐☐ 13–D	To solve problems by using estimation and using a model				

CHAPTER 14

Test Scores:	Criteria 24/32	Pretest	Posttest	Needs More Work	Accom-plished
☐☐ 14–A	To identify patterns to solve problems				
☐☐ 14–B	To identify relations and use relations to solve problems				
☐☐ 14–C	To identify and use the Pythagorean Property to solve problems				
☐☐ 14–D	To identify whether or not a relation is a function				
☐☐ 14–E	To solve problems by solving a simpler problem and to choose a strategy to solve a problem				

CUMULATIVE TEST CHAPTERS 1–14

Criteria 35/50 Test Score ________

END-OF-BOOK TEST

Criteria 35/50 Test Score ________

Cumulative Record Form

MATHEMATICS PLUS
Grade 8

Student ______________________ Teacher ______________________

INVENTORY TEST

Criteria 35/50 Test Score ________

CHAPTER 1

Test Scores:	Criteria 18/24	Pretest	Posttest	Needs More Work	Accom-plished
☐☐ 1–A	To interpret rounded numbers and to identify overestimates and underestimates				
☐☐ 1–B	To use addition, subtraction, multiplication, and division with whole numbers and decimals to solve problems				
☐☐ 1–C	To express and evaluate numbers in exponent form				
☐☐ 1–D	To evaluate numerical expressions using the rules for the order of operations				
☐☐ 1–E	To solve multistep problems and to solve problems by finding a pattern				

CHAPTER 2

Test Scores:	Criteria 18/24	Pretest	Posttest	Needs More Work	Accom-plished
☐☐ 2–A	To identify algebraic expressions for word expressions; to evaluate algebraic expressions				
☐☐ 2–B	To solve one- and two-step equations involving whole numbers and decimals; to identify an equation used to solve a problem				
☐☐ 2–C	To solve one- and two-step inequalities using whole number replacements; to identify an inequality used to solve a problem				
☐☐ 2–D	To solve problems by guessing and checking and by choosing an equation				

CHAPTER 3

Test Scores:	Criteria 24/32	Pretest	Posttest	Needs More Work	Accom-plished
☐☐ 3–A	To identify prime and composite numbers and to write the prime factoriza-ton of a number				
☐☐ 3–B	To use the GCF and the LCM to solve problems				
☐☐ 3–C	To write equivalent frac-tions; to express fractions in simplest form, fractions as whole numbers or mixed numbers and vice versa; to compare and order fractions and mixed numbers				
☐☐ 3–D	To estimate sums and differences of fractions and mixed numbers				
☐☐ 3–E	To add and subtract frac-tions and mixed numbers				
☐☐ 3–F	To solve problems by making choices and by using pictures				

CHAPTER 4

Test Scores:	Criteria 18/24	Pretest	Posttest	Needs More Work	Accom-plished
☐☐ 4–A	To estimate products and quotients of fractions and mixed numbers				
☐☐ 4–B	To multiply and divide fractions or mixed numbers				
☐☐ 4–C	To solve one- and two-step equations using fractions				
☐☐ 4–D	To identify a decimal equivalent to a fraction and vice versa				
☐☐ 4–E	To solve problems by solving a simpler problem and by choosing a strat-egy to solve a problem				

CUMULATIVE TEST CHAPTERS 1–4

Criteria 35/50 Test Score ________

continued

Cumulative Record Form

MATHEMATICS PLUS
Grade 8

Student ______________________ Teacher ______________________

CHAPTER 5

Test Scores:	Criteria 24/32	Pretest	Posttest	Needs More Work	Accom-plished
□□ 5–A	To identify points, lines, line segments, rays, angles, intersecting lines, and angle relationships between types of angle pairs				
□□ 5–B	To identify angles and angle relationships between types of angle pairs				
□□ 5–C	To identify constructions of congruent segments and angles, parallel and perpendicular lines, and bisections of segments and angles				
□□ 5–D	To identify the relationships of angles formed by parallel lines and transversals				
□□ 5–E	To identify properties of polygons: quadrilaterals and triangles				
□□ 5–F	To identify congruent figures and their corresponding parts				
□□ 5–G	To solve problems by finding a pattern and by making choices				

CHAPTER 6

Test Scores:	Criteria 24/32	Pretest	Posttest	Needs More Work	Accom-plished
□□ 6–A	To read, write, and simplify ratios and rates				
□□ 6–B	To identify and write proportions to solve problems; to use a scale drawing to find actual measurement or scale measurement				
□□ 6–C	To identify an equivalent ratio, decimal, or percent when one form is given				
□□ 6–D	To use estimation in solving percent problems; to solve percent problems				
□□ 6–E	To solve simple-interest problems				
□□ 6–F	To solve problems by using a map and by using estimation				

CHAPTER 7

Test Scores:	Criteria 24/32	Pretest	Posttest	Needs More Work	Accom-plished
□□ 7–A	To compare and order integers				
□□ 7–B	To add, subtract, multiply, and divide integers				
□□ 7–C	To write numbers using negative exponents, and to evaluate powers with negative exponents				
□□ 7–D	To multiply and divide powers				
□□ 7–E	To use scientific notation to name a number and vice versa				
□□ 7–F	To solve problems by using a table and by writing an equation				

continued

Cumulative Record Form

MATHEMATICS PLUS
Grade 8

Student ______________________ Teacher ______________________

CHAPTER 8

Test Scores:	Criteria 24/32	Pretest	Posttest	Needs More Work	Accom-plished
☐☐ 8–A	To identify rational and irrational numbers; to compare and order rational numbers				
☐☐ 8–B	To find the square and square root of a number				
☐☐ 8–C	To add, subtract, multiply, and divide rational numbers				
☐☐ 8–D	To solve one- and two-step equations involving rational numbers				
☐☐ 8–E	To solve and graph inequalities using real numbers				
☐☐ 8–F	To solve problems by using a table and using a formula				

CUMULATIVE TEST CHAPTERS 1–8

Criteria 35/50 Test Score ________

CHAPTER 9

Test Scores:	Criteria 24/32	Pretest	Posttest	Needs More Work	Accom-plished
☐☐ 9–A	To locate points on a coordinate plane; to graph points, linear equations, and inequalities on a coordinate plane				
☐☐ 9–B	To identify and graph relations and functions				
☐☐ 9–C	To find the slope of a line				
☐☐ 9–D	To solve a system of equations by graphing				
☐☐ 9–E	To identify transformations on a coordinate plane				
☐☐ 9–F	To solve problems by guessing and checking and by using a graph to estimate				

CHAPTER 10

Test Scores:	Criteria 24/32	Pretest	Posttest	Needs More Work	Accom-plished
☐☐ 10–A	To analyze sets of data using range, quartiles, extremes, mean, median, and mode; to read and interpret box-and-whisker graphs				
☐☐ 10–B	To read and interpret frequency tables, histograms, line graphs, bar graphs, circle graphs, and stem-and-leaf plots				
☐☐ 10–C	To use extrapolation and interpolation to make estimates				
☐☐ 10–D	To identify the type of correlation between two variables using a scattergram; to read and interpret scattergrams				
☐☐ 10–E	To distinguish between sufficient and insufficient data to solve problems and to choose an appropriate graph				

CHAPTER 11

Test Scores:	Criteria 24/32	Pretest	Posttest	Needs More Work	Accom-plished
☐☐ 11–A	To use factorials to find the number of choices, permutations, and combinations				
☐☐ 11–B	To find simple probabilities; to find probabilities of independent and dependent events				
☐☐ 11–C	To use random numbers to stimulate probability experiments and to make predictions using experimental probability				
☐☐ 11–D	To use Venn diagrams to show relationships among groups				
☐☐ 11–E	To solve problems by using a diagram and to solve problems by acting them out				

continued

Cumulative Record Form

MATHEMATICS PLUS
Grade 8

Student ______________________ Teacher ______________________

CUMULATIVE TEST CHAPTERS 1–11

Criteria 35/50 Test Score ________

CHAPTER 12

Test Scores:	Criteria 24/32	Pretest	Posttest	Needs More Work	Accom-plished
□□ 12–A	To choose an appropriate metric or customary unit of measure; to estimate measures using non-standard and standard units of measure				
□□ 12–B	To identify the precision, the greatest possible error, and the significant digits of a measurement				
□□ 12–C	To find the perimeter of polygons; to find the circumference of a circle				
□□ 12–D	To find the area of rect-angles, parallelograms, triangles, trapezoids, and circles				
□□ 12–E	To solve problems by making a model and by using a formula				

CHAPTER 13

Test Scores:	Criteria 24/32	Pretest	Posttest	Needs More Work	Accom-plished
□□ 13–A	To identify and describe solid figures				
□□ 13–B	To find the surface area of prisms, pyramids, cylinders, and cones				
□□ 13–C	To find the volume of prisms, pyramids, cylinders, and cones				
□□ 13–D	To identify relationships between metric units of volume, capacity, and mass				
□□ 13–E	To solve problems by using a table and by using a formula				

CHAPTER 14

Test Scores:	Criteria 24/32	Pretest	Posttest	Needs More Work	Accom-plished
□□ 14–A	To identify right triangles and to use properties of 45–45 right triangles and 30–60 right triangles to find unknown lengths				
□□ 14–B	To use proportions to find unknown lengths in similar figures				
□□ 14–C	To identify and use tan-gent, sine, and cosine ratios to find unknown lengths in right triangles				
□□ 14–D	To solve problems by making choices and by using a picture				

CUMULATIVE TEST CHAPTERS 1–14

Criteria 35/50 Test Score ________

END-OF-BOOK TEST

Criteria 35/50 Test Score